Scientific Research on Ancient Asian Metallurgy

Proceedings of the Fifth Forbes Symposium at the Freer Gallery of Art

Scientific Research on Ancient Asian Metallurgy

Proceedings of the Fifth Forbes Symposium at the Freer Gallery of Art

Edited by Paul Jett, Blythe McCarthy, and Janet G. Douglas

Archetype Publications

in association with the Freer Gallery of Art, Smithsonian Institution

First published 2012 by Archetype Publications Ltd.

Archetype Publications Ltd.
6 Fitzroy Square
London W1T 5HJ
www.archetype.co.uk

Smithsonian
Freer Gallery of Art

ISBN: 978-1-904982-72-2

British Library Cataloguing in Publication Data
A catalogue record for this book is available from the British Library.

Edited by Theresa Dowell Blackinton
Images edited by Jenna Vaccaro and Stacy Bowe
Cover design by Reid Hoffman
Cover illustration: X-radiographic image of the lid of a *ding*, China, late fifth–early sixth century BCE. F1947.20.

Typeset by PDQ Digital Media Solutions Ltd, Bungay
Printed and bound in Italy by Printer Trento srl

Contents

Foreword

These proceedings record the papers presented at a symposium held at the Freer Gallery of Art, Washington, D.C., in the fall of 2010. The meeting was the fifth in a series of symposia on scientific research in the field of Asian art that began in 2001, named in memory of Edward Waldo Forbes. Mr. Forbes, in his role as director of the Fogg Art Museum at Harvard University, was an early supporter of the technical study of the materials and methods that go into the making of art. With this meeting, we return to one of the museum's earliest technical research topics—the study of ancient Chinese bronzes and other copper-alloy metalwork from Asia. The technical studies volume of *The Freer Chinese Bronzes,* a catalogue published in 1969, was the result of more than ten years of work by museum staff and Rutherford John Gettens, head of what was then called the Technical Laboratory and is now known as the Department of Conservation and Scientific Research. The research that went into the making of the technical volume was only the beginning of many studies of copper-alloy objects from Asia. Gettens's successor, W. Thomas Chase, expanded Gettens's work on Chinese bronzes as well as his work in many other research areas. Another important aspect of the technical volume of the Chinese bronze catalogue was that, with its combination of art-historical and technical discussions, it came to serve as a model for later catalogues on the Islamic and pre-Islamic metalwork in the Freer collection and for the catalogues of other museums as well.

It was this type of research that led Jack Thacher, former director of Dumbarton Oaks Museum and Gardens (and former assistant director at the Fogg Museum under Mr. Forbes), to make his bequest to the Freer Gallery for the support of scientific research, a bequest that funds the Forbes symposia and the publication of the proceedings. The occasion of the Forbes symposium and the publication of the proceedings provide an ideal moment in which to recall Mr. Thacher's generosity. His bequest is fundamental to supporting scientific research at the Freer and Sackler galleries and has been amplified by the generosity of many other organizations, particularly that of the Andrew W. Mellon Foundation, which recently provided an endowment to support the position of a senior scientist. This support allows us to build upon the efforts to use scientific research for the study of works of art that began at the Freer Gallery sixty years ago and continue today with a diversity of projects, collaborations, and, we hope, contributions to the understanding of the artistic heritage of Asia.

Julian Raby
Dame Jillian Sackler Director of the
Arthur M. Sackler Gallery and the Freer Gallery of Art

Acknowledgments

As anyone who has organized a symposium knows, it is a tremendous amount of work best shared among a number of people. We are fortunate at the Freer and Sackler galleries to have many colleagues who are willing to pitch in and make meetings like the Forbes symposia successes. The staff of the Department of Conservation and Scientific Research contributed greatly to the preparations for this meeting and to its execution. In particular, departmental assistant Stacy Bowe saw to many of the arrangements for the speakers as well as countless other details. In order to keep the symposium running smoothly, we also had the help of many volunteers, including Sandra Rowland, Susan Lubick, Mariam Gheissari, Bonny Cochran, Yumi Shintani, Tomoko Kotajima, and Elizabeth Beesley.

The end of the symposium means the beginning of the publication process for the proceedings, and again, many people contribute to this effort. Co-editors Blythe McCarthy and Janet Douglas saw to the timely preparation of many of the papers, and we were helped in this by the editorial work of Theresa Dowell Blackinton. Jenna Vaccaro of the Department of Scholarly Programs and Publication devoted many days to checking the references and helping prepare the images and tables. Two volunteers also generously devoted time to preparing the manuscript. Catherine Magee entered corrections to the papers and also helped check references, while Isabel Chen assisted with Chinese translations. Additionally, Ms. Bowe acted as production manager for the task of getting the proceedings to press. Reid Hoffman of the Design Department, for these proceedings and a number of the preceding ones, applied his artistic talents to designing the cover. I would also like to thank Julian Raby, Dame Jillian Sackler Director of the Arthur M. Sackler Gallery and the Freer Gallery of Art, for his continuing support, and the authors and Archetype Publications for making the publication possible. Finally, I would like to add that if I have inadvertently left out mentioning someone who contributed their time to this effort, please accept my apology and my thanks.

Paul Jett
Emeritus Head, Conservation and Scientific Research
Freer Gallery of Art and Arthur M. Sackler Gallery

List of Contributors

Robert BAGLEY
Princeton University
Princeton, New Jersey, USA

Anna BENNETT
Analyseart
Brussels, Belgium

Francesca G. BEWER
Straus Center for Conservation and Technical Studies
Harvard Art Museums
Cambridge, Massachusetts, USA

David BOURGARIT
Center for Research and Restoration of
the French Museums (C2RMF)
Paris, France

W. T. CHASE
Chase Art Services
Woodstock, Connecticut, USA

CHEN Kunlong
University of Science and Technology
Beijing, China

Antoine COURCIER
National Center for Scientific Research (CNRS)
Paris, France

Paul T. CRADDOCK
The British Museum
London, UK

Katherine EREMIN
Straus Center for Conservation and Technical Studies
Harvard Art Museums
Cambridge, Massachusetts, USA

Lesley D. FRAME
Department of Materials Science and Engineering
University of Arizona
Tuscon, Arizona, USA

Ian GLOVER
Institute of Archaeology
University College
London, UK

Farhad GULIYEV
Institute of Archaeology and Ethnology
Baku, Azerbaijan

Duncan HOOK
The British Museum
London, UK

Paul JETT
Freer Gallery of Art and the Arthur M. Sackler Gallery
Smithsonian Institution
Washington, D.C., USA

JIANG Zhilong
Yunnan Provincial Institute of Culture
Relics and Archaeology
Kunming, China

JIN Zhengyao
Archaeometry Laboratory
University of Science and Technology
Hefei, China

Susanna KIRK
Conservation and Analytical Research
National Museums Scotland
Edinburgh, UK

Susan LA NIECE
The British Museum
London, UK

LI Gong
Archaeometry Laboratory
University of Science and Technology
Hefei, China

Bertille LYONNET
National Center for Scientific Research (CNRS)
Paris, France

MEI Jianjun
Institute of Historical Metallurgy and Materials
University of Science and Technology
Beijing, China

Benoît MILLE
Center for Research and Restoration of
the French Museums (C2RMF)
Paris, France

Vincent C. PIGOTT
Institute of Archaeology
University College
London, UK

Thomas Oliver PRYCE
Research Laboratory for Archaeology
and the History of Art (RLAHA)
University of Oxford
Oxford, UK

Megan RICHTERS
Centre for Archaeological and Forensic Analysis
Cranfield University
Shrivenham, UK

David A. SCOTT
Getty Program in Archaeological and
Ethnographic Conservation
University of California at Los Angeles
Los Angeles, California, USA

SHEN Lu
Institute of Historical Metallurgy and Materials
University of Science and Technology
Beijing, China

Andrew SHORTLAND
Centre for Archaeological and Forensic Analysis
Cranfield University
Shrivenham, UK

Donna STRAHAN
The Metropolitan Museum of Art
New York, New York, USA

TIAN Jianhua
Archaeometry Laboratory
University of Science and Technology
Hefei, China

TRINH Sinh
Institute of Archaeology
Hanoi, Vietnam

Brice VINCENT
University of Paris III–Sorbonne Nouvelle
Paris, France

WANG Haigang
Archaeometry Laboratory
University of Science and Technology
Hefei, China

WANG Hui
Gansu Provincial Institute of Cultural
Relics and Archaeology
Lanzhou, China

WANG Quanyu
The British Museum
London, UK

Rachel WARD
Department of Oriental Antiquities
The British Museum
London, UK

XU Jianwei
Institute of Historical Metallurgy and Materials
University of Science and Technology
Beijing, China

YAN Lifeng
Hefei National Laboratory for Physical Sciences at the
Microscale University of Science and Technology
Hefei, China

Introduction

When the planning began for the fifth Forbes symposium on scientific research in the field of Asian art, it was 2009 and the fortieth anniversary of the publication of the technical volume of the Freer Chinese bronze catalogue. Additionally, the symposium was expected to coincide with the exhibit *Gods of Angkor: Bronzes from the National Museum of Cambodia* at the Sackler Gallery and the opening of a newly designed gallery for ancient Chinese bronzes in the Freer. With this confluence of events, it seemed ordained that the theme of the meeting should be research on ancient Asian metallurgy.

While the subject is broad in scope and covers many periods and geographical areas, one idea that underlies a number of the papers and ties them together is that of the spread of metallurgical knowledge and practices across Asia. The debate over indigenous versus imported technological development is a continuing discussion that has gone on for decades; aspects of this debate ebb and flow as our knowledge grows with new archaeological discovery. For example, there is recent growing evidence that suggests bronze metallurgy was introduced to China from the West and, that once established there, it spread south into Southeast Asia (see the papers by Craddock and Pigott). Some of these theories are not entirely new but were once rejected (or at least discouraged) and have now made a comeback.

The study of copper-alloy artifacts in general has a relatively long history, with the analysis of ancient metals dating back to the end of the eighteenth century. In the late nineteenth and early twentieth century, the study of bronzes was undertaken by a variety of scientists both in Europe and the United States as part of the study of ancient cultures and the metallurgical arts and also as part of early conservation efforts. Beginning in the late 1920s, one of those scientists was Rutherford John Gettens, who studied bronzes, both Chinese and others, at Harvard University's Fogg Art Museum. Gettens, of course, continued his work on bronzes after he came to the Freer Gallery of Art, and his studies and those of his colleagues on Chinese bronzes resulted in the Freer Chinese bronze catalogue, a publication that seemed—then and now—revolutionary. The technical study of Chinese bronzes continued at the Freer with the work of W. T. Chase, who began as an intern working with Mr. Gettens on the bronze volume and then succeeded him as head of the laboratory; perhaps most important of Mr. Chase's work on the subject was his study of, and collaborative efforts with others on, the use of lead isotope ratio analyses. Of course, this was just one area of bronze technology and conservation that Mr. Chase studied. We hope this volume will be looked upon as a tribute to Messrs. Gettens and Chase and their valuable contributions to the Freer and Sackler galleries, the understanding of ancient Chinese bronzes, and the field of technical studies of art.

These proceedings present the papers from the meeting grouped under four themes: reminiscences on Mr. Gettens and the Freer Chinese bronzes technical volume, and recent studies of copper-alloy production and products from China, Southeast Asia, and West Asia. Some of the papers provide overviews of the various topics, while others are more focused. Any short symposium can only give a small sampling of the work going on in a particular field, but we hope this sampling is representative of the strong research, both performed in the past and currently being conducted, on the metalwork of ancient Asia.

Paul Jett

Reflections on R. J. Gettens and the Technical Study of Chinese Bronzes

A Chemist Under a Spell: Rutherford John Gettens's Early Encounters with Chinese Bronzes

Francesca G. Bewer

ABSTRACT Rutherford John Gettens, the scientist who helped establish the Freer Gallery of Art's Laboratory for Technical Studies in Art and Archaeology, became well known for his study of Chinese bronzes and especially his seminal publication, *Freer Chinese Bronzes*, volume 2, *Technical Studies* (Gettens 1969). This paper will lay out the work that preceded Gettens's study at the Freer: a number of earlier encounters with Chinese bronze artifacts during the 1930s and 1940s while he was the museum chemist at Harvard University's Fogg Museum.

The Very Beginnings

"It was not hard for a chemist to fall under the spell of early Oriental metallurgy," recalled Rutherford John Gettens, head of the Freer Gallery of Art Technical Laboratory, in the preface to the technical volume on the Freer's ritual Chinese bronze vessels. He credited his being "imbued with a special interest in antiquities of metal" to Langdon Warner, the Fogg Museum's curator of Asian art and a key figure in that field. Gettens had joined the Fogg, Harvard University's small teaching museum, as staff chemist of the country's first scientific research and conservation laboratory in 1928 (Gettens 1969) (fig. 1). Indeed, Chinese bronzes were to preoccupy him from early on in his career.

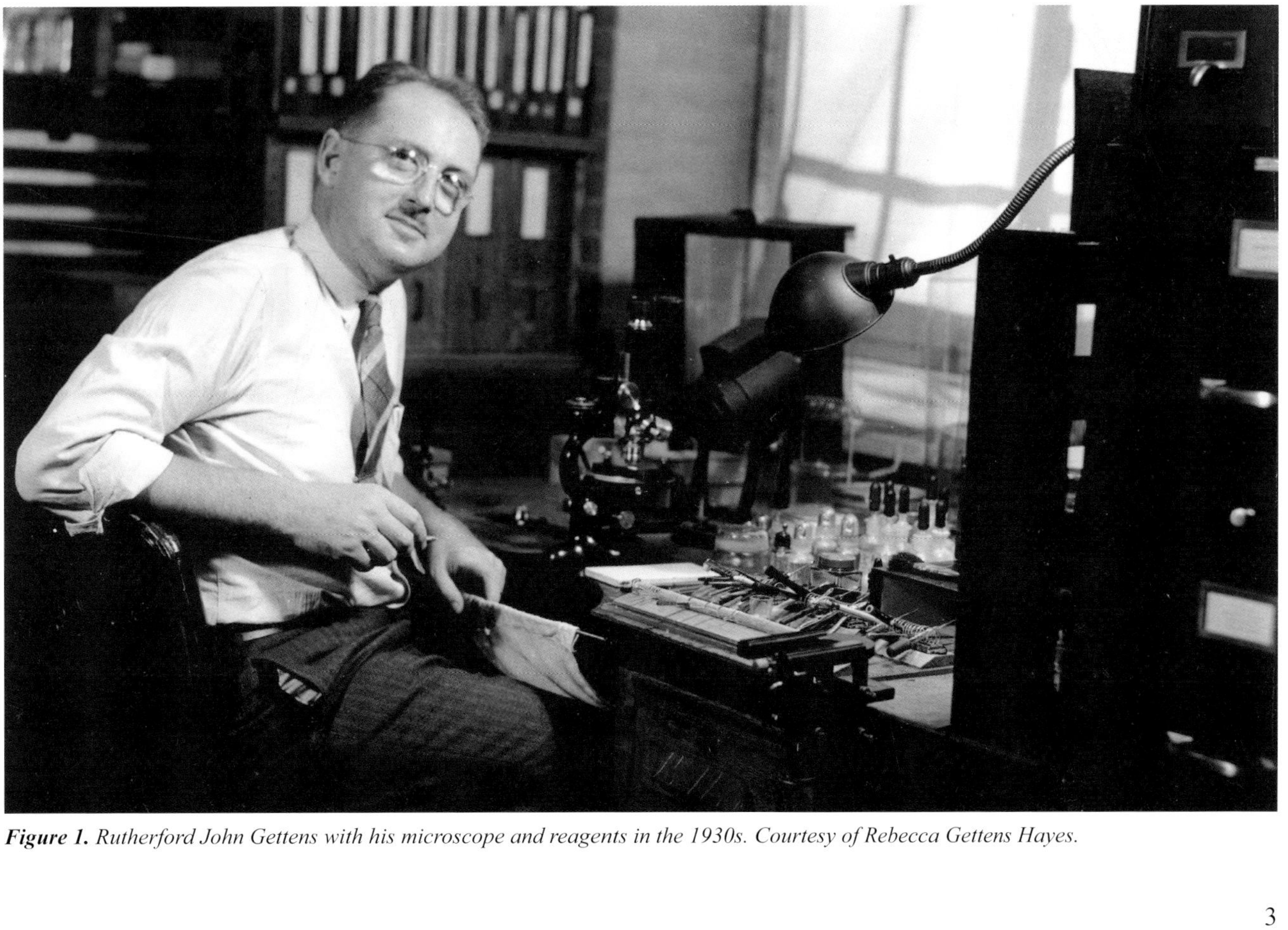

Figure 1. *Rutherford John Gettens with his microscope and reagents in the 1930s. Courtesy of Rebecca Gettens Hayes.*

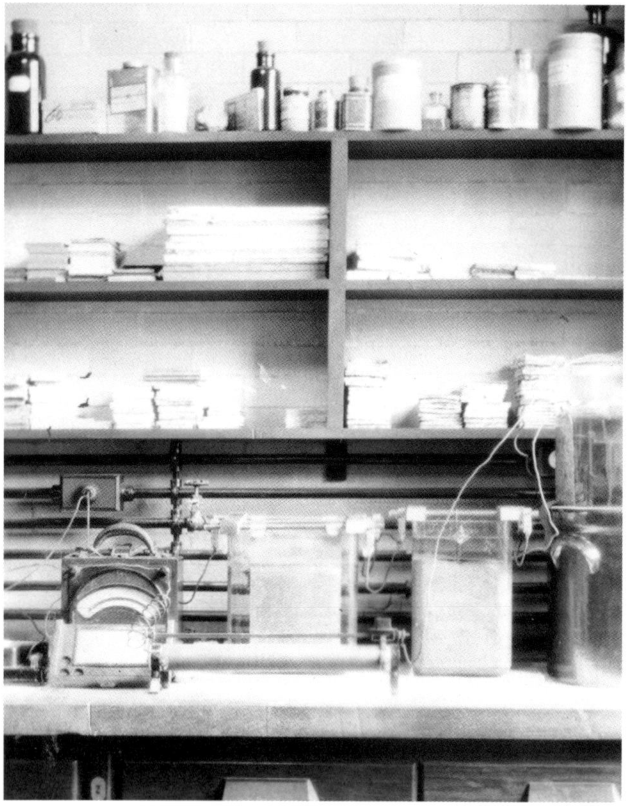

Figure 2. *Setup for the electrolytic treatment of archaeological bronzes in Gettens's laboratory in 1930. Digital Imaging and Visual Resources Department, Harvard Art Museums, © 2010 President and Fellows of Harvard College.*

One of Gettens's early tasks at the Fogg was to study "bronze disease" and its treatment on archaeological artifacts. By the late 1920s, it was understood that this condition was due to an electrochemical reaction, and in consequence, electrochemical and electrolytic treatments were found to be among the more expedient ways to stabilize or remove both the culprit chlorides and unwanted mineral encrustations (fig. 2). Gettens first experimented with known treatment methods on bronzes from Harvard's Peabody Museum of Archaeology and Ethnology and with objects from the Semitic Museum that had come from Nuzi, Iraq. In the course of this early work, he discovered the locus of active corrosion within the layered structure of the metal alterations. Publication of his findings helped to establish the Fogg's reputation for scientific research related to the preservation of art works and identified the Fogg as a reputable place for treatment (Gettens 1933; Bewer 2010; Eremin, in this vol.).

Among the works that Gettens was subsequently called upon to treat for outside collections was a group of Chinese bronzes owned by the William Rockhill Nelson Trust of Kansas City, which included two Han dynasty (206 BCE–220 CE) horses somewhat disfigured by corrosion. Gettens's experience with these two works was fraught with unexpected mishaps, which he noted in a remarkably candid treatment report for that era. Gettens initially attempted to break down the surface accretions of horse 1 by the electrolytic process (fig. 3). This met with disastrous results. The horse was not solid, as originally suspected, but hollow and heavily restored. It had been broken in many pieces

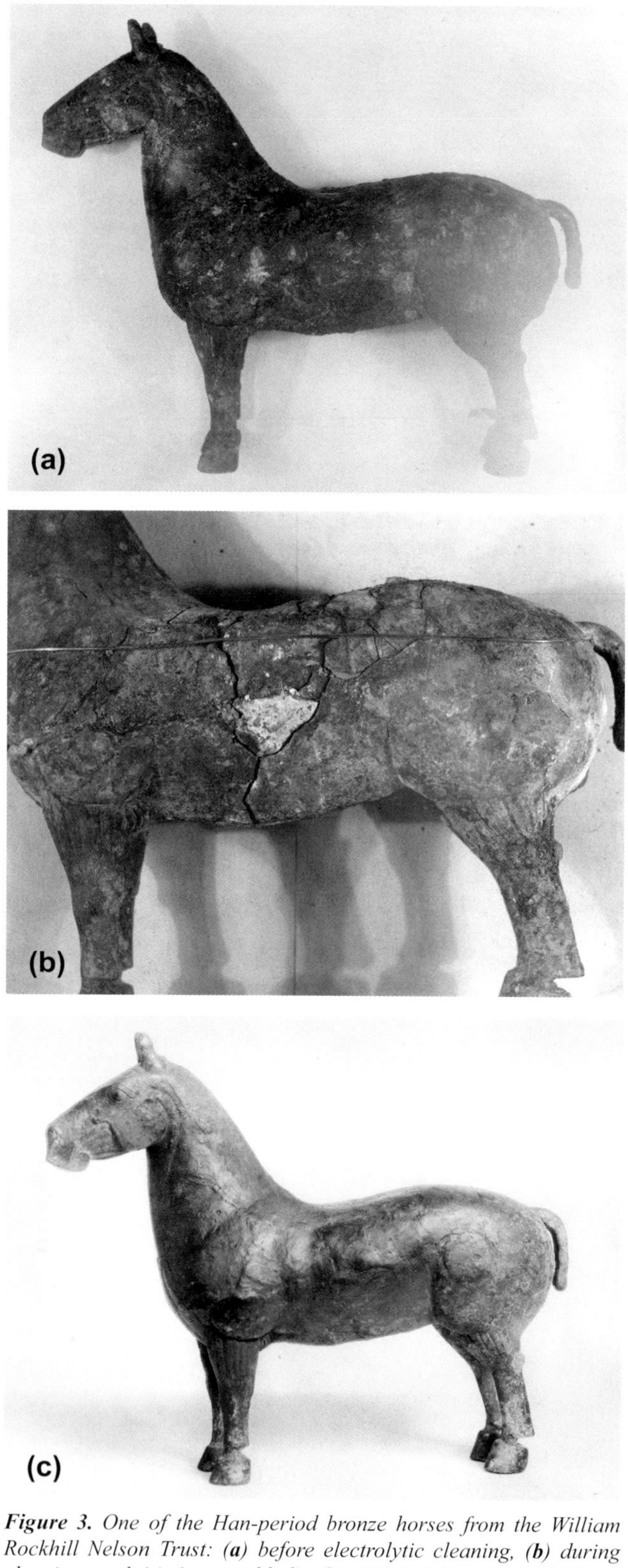

Figure 3. *One of the Han-period bronze horses from the William Rockhill Nelson Trust:* ***(a)*** *before electrolytic cleaning,* ***(b)*** *during cleaning, and* ***(c)*** *(reassembled) after. Digital Imaging and Visual Resources Department, Harvard Art Museums, © 2010 President and Fellows of Harvard College.*

and filled with plaster reinforced with brass strips, and the repairs had been covered with colored pigments and varnish to mimic corrosion. The electrolytic process had caused the plaster to swell, resulting in the disintegration of the restoration and causing the horse to fall into pieces. The fate of the companion piece was equally unfortunate: its

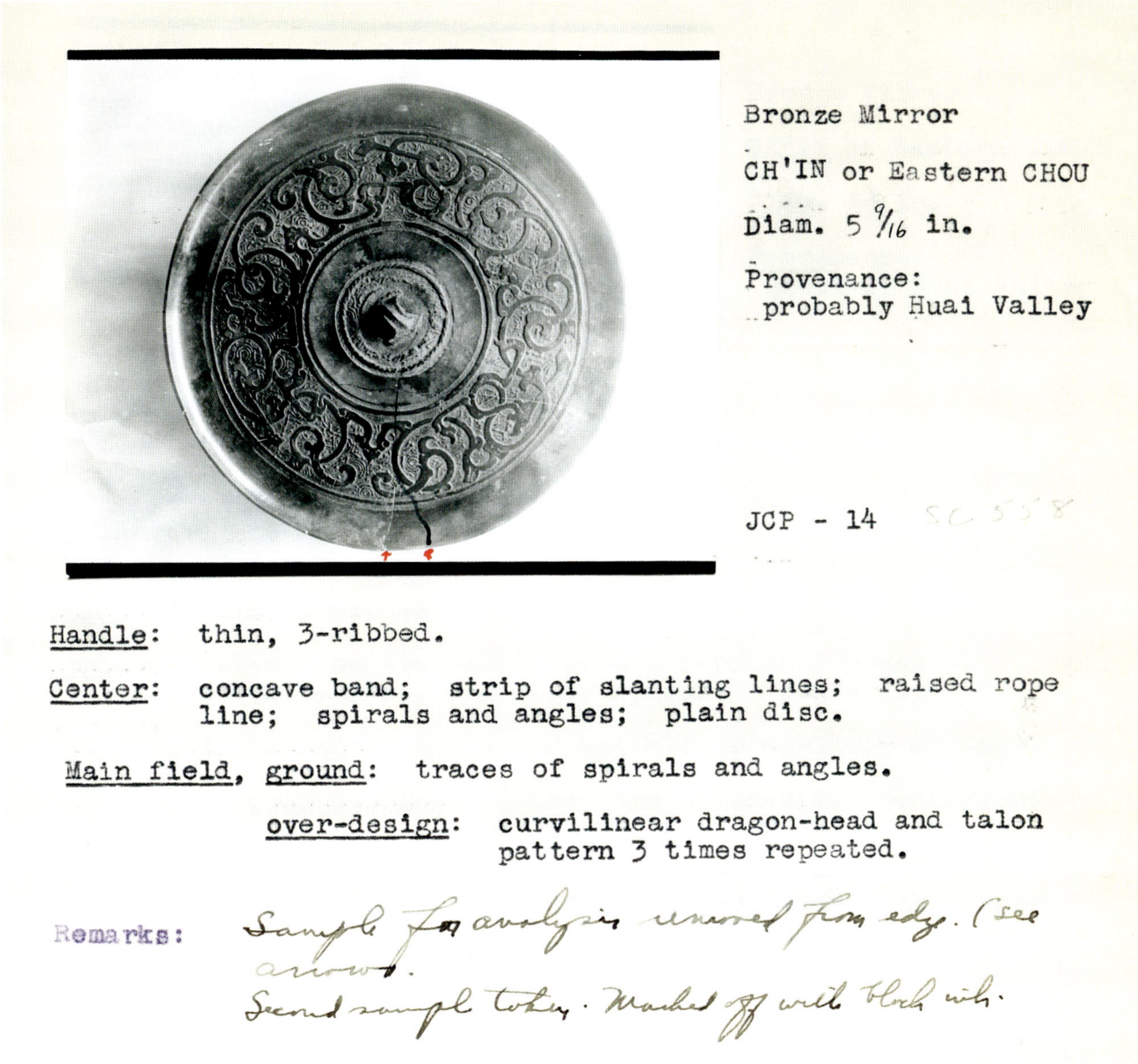
Bronze Mirror

CH'IN or Eastern CHOU

Diam. 5 9/16 in.

Provenance:
probably Huai Valley

JCP - 14

Handle: thin, 3-ribbed.

Center: concave band; strip of slanting lines; raised rope line; spirals and angles; plain disc.

Main field, ground: traces of spirals and angles.

over-design: curvilinear dragon-head and talon pattern 3 times repeated.

Remarks: Sample for analysis removed from edge. (See arrow). Second sample taken. Marked off with black ink.

Figure 4. *Detail of a page from the James Marshall and Carol Plumer catalogue of bronze mirrors showing how Gettens marked the areas he sampled. Freer Gallery of Art, Department of Conservation and Scientific Research.*

head was severed cleanly from the body at the neck when the horse was accidentally tipped over on the work bench during cleaning. Gettens found a silver lining in the experience, as it revealed the nature of the core, the thinness of the bronze, and the near-complete oxidized condition of the metal. These may constitute his earliest recorded observations relating to the casting technique of a Chinese bronze (Gettens 1934a).

Gettens's technical investigations of Chinese bronzes followed from his interest in corrosion mechanisms. He next focused predominantly on the black patina in a handful of Chinese bronze mirrors belonging to the Fogg. Hitherto, published technical information on Chinese mirrors had consisted primarily of results obtained by wet chemical analysis of their alloy composition and some speculation on their manufacture and patina (Chikashige 1920; Karlbeck 1926; Umehara 1927; Yetts 1930, 42; 1931; Anonymous 1930; Collins 1931; Komatsu and Yamanouchi 1933). The extent of the technical studies and the value of their results were, however, somewhat limited. Given the rarity of scientific excavations of archaeological remains in China to date, most of the objects that had been analyzed came from the market and therefore lacked a documented burial context. Furthermore, most of the analyzed works were "whole," as it was rare for fragments to be considered of commercial interest and integral objects were not (and still are not) willingly offered up for destructive analysis.

Nevertheless, the findings seemed to generally agree that most mirrors were made of a special type of high-tin bronze—known as "speculum metal"—in which the tin/copper ratio resulted in a eutectic alloy that had good reflecting power. The 6–8 wt.% lead content typically found in the alloy was no doubt added to help increase fluidity, reduce brittleness, and facilitate polishing. It was also speculated

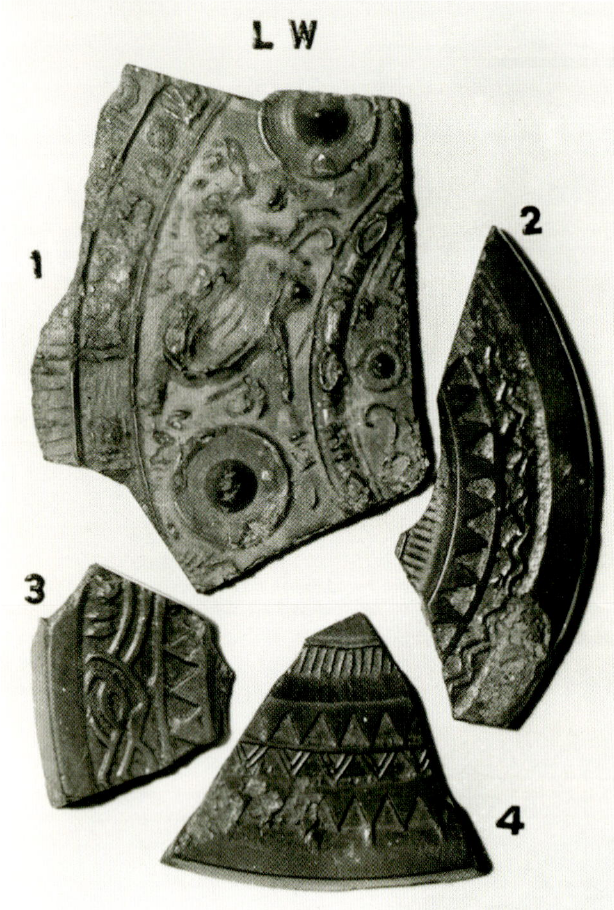

Figure 5. *Fragments of Chinese bronze mirrors that Langdon Warner provided for analysis alongside the Plumer bronzes. Freer Gallery of Art, Department of Conservation and Scientific Research.*

that the environmental conditions of burial were more of a determining factor in the color and amount of patina than the composition of metal. But many questions remained to be answered regarding metal composition, production, surface treatment, physical condition, and corrosion mechanisms. Gettens's publication contains the first images of metallographic sections of such works and affords preliminary observations on the unusual corrosion found on the mirrors. The article also clearly invited collectors to come forward with more material on which he might carry out in-depth metallographic and metallurgical research (Gettens 1934b).

The Plumer Mirrors

The answer to this wish came from James Marshall Plumer, a Harvard graduate who had studied with Langdon Warner, served the Chinese government in Maritime Customs since 1923, and maintained close ties with his alma mater—he was appointed assistant and tutor in Chinese language at Harvard for a year (1929–30) and acted as secretary of the Harvard-Yenching Institute (Stern and Usilton 1963). Plumer began to collect bronze mirrors in 1931 and ratcheted up his acquisition of such artifacts upon learning about Gettens's research interests from Warner—news that pleased Gettens immensely (Plumer and Plumer 1934; Warner 193[?]). Plumer and his wife meticulously catalogued around four dozen mirrors[1] and sent a copy of the collated manuscript along with the bronzes to the Fogg Museum "for the express purpose of being experimented upon [...and] with permission to sample the metal of each piece, to treat with chemical, etc, etc.[…]. Indeed some… might well be dropped in the melting pot entirely" (Plumer and Plumer 1934). This is the kind of directive of which most museum scientists only dream (fig. 4)!

Gettens launched into research on this new material, gathering information and seeking assistance from institutions and experts far and near, including the Field Museum of Natural History, Chicago, where analyses had been done on Chinese bronzes (Laufer 1934), and Anton Alexander Benedetti-Pichler, the father of microchemical analysis in America (Benedetti-Pichler 1935). But most important was the network of scientists in the Boston area that Gettens collaborated with on the study of the Plumer mirrors. Primary among these was Victor G. Mooradian of the Division of Industrial Cooperation at the Massachusetts Institute of Technology, who helped Gettens undertake a complete metallographic and microscopic investigation of the Plumer collection as well as a few mirror fragments that Langdon Warner donated for the project (Gettens 1935) (fig. 5). For alloy compositions, Gettens relied on the spectrographic analyses by colleagues at Boston University, and for x-ray diffraction on corrosion products, he turned to Harvard University's Department of Mineralogy (Vanderbelt 1936).

The aim of the investigation was to observe any similarity in structural condition, gather clues to the conditions or methods of casting, polishing, and working, and learn more about the mechanism of corrosion on the mirrors. Spectrographic analysis revealed that the mirrors were all high-tin bronzes with an average of 25 wt.% tin, 4 wt.% lead, and 70 wt.% copper, and relatively small amounts of impurities. It was surprising to find that the Tang (681–907 CE) mirrors were similar in composition to those of the Han period. The mirrors' metal structures were found to be quite similar, and Mooradian and Gettens attributed what

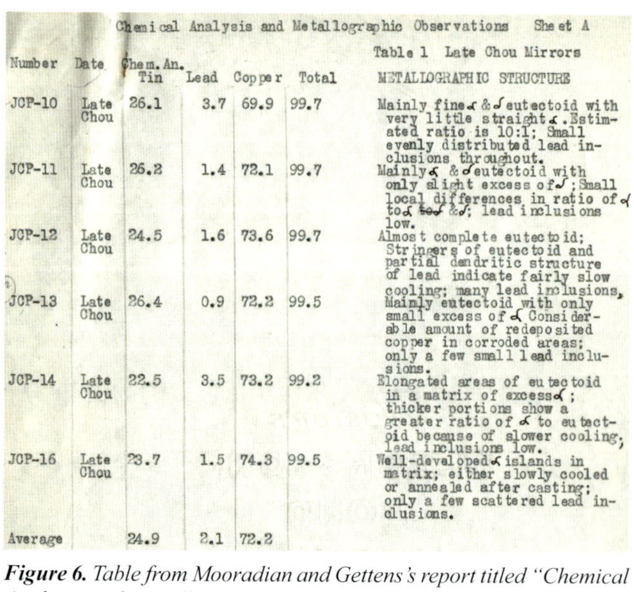

Chemical Analysis and Metallographic Observations Sheet A

Table 1 Late Chou Mirrors

Number	Date	Chem. An. Tin	Lead	Copper	Total	METALLOGRAPHIC STRUCTURE
JCP-10	Late Chou	26.1	3.7	69.9	99.7	Mainly fine α & δ eutectoid with very little straight α. Estimated ratio is 10:1; Small evenly distributed lead inclusions throughout.
JCP-11	Late Chou	26.2	1.4	72.1	99.7	Mainly α & δ eutectoid with only slight excess of δ; Small local differences in ratio of α to α ~~to δ~~ & δ; lead inclusions low.
JCP-12	Late Chou	24.5	1.6	73.6	99.7	Almost complete eutectoid; Stringers of eutectoid and partial dendritic structure of lead indicate fairly slow cooling; many lead inclusions,
JCP-13	Late Chou	26.4	0.9	72.2	99.5	Mainly eutectoid with only small excess of α Considerable amount of redeposited copper in corroded areas; only a few small lead inclusions.
JCP-14	Late Chou	22.5	3.5	73.2	99.2	Elongated areas of eutectoid in a matrix of excess α; thicker portions show a greater ratio of α to eutectoid because of slower cooling; lead inclusions low.
JCP-16	Late Chou	23.7	1.5	74.3	99.5	Well-developed α islands in matrix; either slowly cooled or annealed after casting; only a few scattered lead inclusions.
Average		24.9	2.1	72.2		

Figure 6. *Table from Mooradian and Gettens's report titled "Chemical Analysis and Metallographic Observations." Freer Gallery of Art, Department of Conservation and Scientific Research.*

differences there were to variations in casting technique or cooling rates, which they reconstructed from the metal phases (fig. 6). There was no evidence in the microstructure of any of the mirrors of extensive cold-working (Mooradian and Gettens n.d.).

The two scientists collated the disparate data on technique, alloy, and condition into one table organized by chronological dating of the bronzes based on stylistic features. Judging from the similarity of the alloy through the centuries (except for a lower average of lead in the late Zhou period [ca. 1050–ca. 221 BCE] than in the Han period), they concluded that the composition must have been considered optimal for over a thousand years! They interpreted the lack of impurities and relatively uniform manipulation of the metal as clear evidence of the great skill and metallurgical know-how of early Chinese metalworkers.

In the course of their examinations, Gettens and Mooradian encountered a variety of structures in the corrosion of the mirrors, including a unique blistering and scaling that they speculated might be due to the difference in metal structure between the outer edge and the interior of the specimen. They also found a few unusual corrosion products: among others, the much coveted, enamel-like, light green patina often referred to as "water patina," which they were the first to identify as anhydrous tin oxide (by microchemical testing and x-ray diffraction). Though they realized that it was too soon to draw any definitive conclusions given the lack of comparative data from other types of objects and information on metal sources of ancient China, Gettens and Mooradian's novel, multifaceted approach allowed them to correlate previously unconnected data. Their findings were presented at a number of venues, such as the annual meeting of the American Institute of Mining, Metallurgical and Petroleum Engineers (Smith 1939) and the Princeton Chemical Club (Gettens 1940a). In the process of soliciting feedback for their paper, Mooradian and Gettens made contact with Cyril Stanley Smith, an industrial research metallurgist and historian of technology who had also developed a great love for Chinese bronzes (Smith 1939). Despite repeated efforts (Kennedy 1938; Gettens 1939; Rhines 1939), Gettens and Mooradian did not succeed in publishing their findings until after the war, and then not in their entirety (Gettens 1948). Therefore, others working on Chinese bronzes at the time, like Harold Plenderleith at the British Museum, had little idea that the mystery of the jade-like patina was being tackled: the topic is almost glossed over in the British scientist's paper on the patina and encrustations of Chinese bronzes (Plenderleith 1939).

Plumer and Gettens also attempted a joint publication (Gettens 1940b; Plumer and Gettens 1941). Plumer, now at the University of Michigan as lecturer in Far Eastern art, was pursuing other avenues of investigation with these works: he corresponded with the Swedish expert of Chinese bronzes, Orvar Karlbeck (Karlbeck 1941), and even sent a mirror fragment to the Federal Bureau of Investigation to ascertain whether the tool marks in the decorative areas were made before or after casting (fig. 7). J. Edgar Hoover, the director of the bureau, signed the ensuing report, which presented evidence of cold-working on the sample piece (Plumer 1940; Hoover 1940). Plumer eventually published an art-historical article on the mirrors in which he integrated some of the scientific findings of Gettens and his colleagues (Plumer 1944).

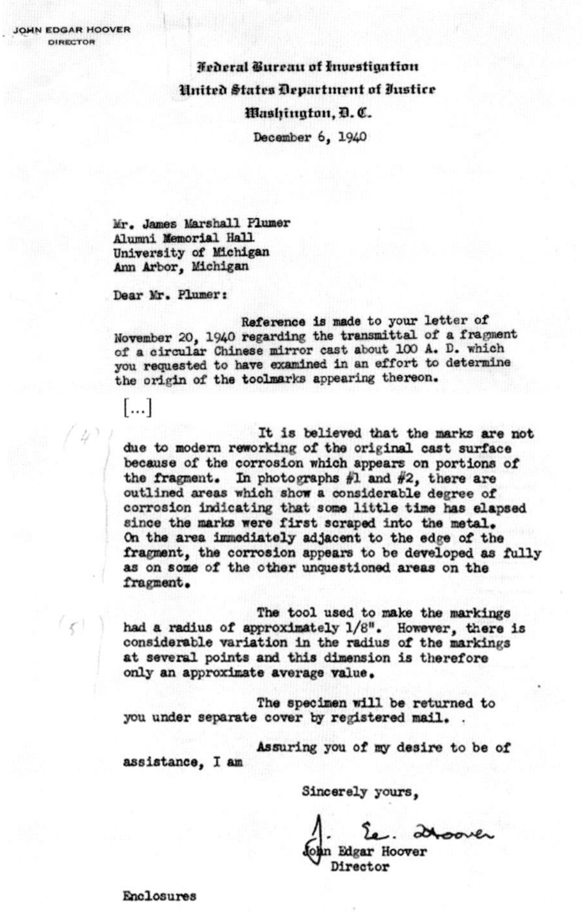
JOHN EDGAR HOOVER
DIRECTOR

Federal Bureau of Investigation
United States Department of Justice
Washington, D. C.
December 6, 1940

Mr. James Marshall Plumer
Alumni Memorial Hall
University of Michigan
Ann Arbor, Michigan

Dear Mr. Plumer:

Reference is made to your letter of November 20, 1940 regarding the transmittal of a fragment of a circular Chinese mirror cast about 100 A. D. which you requested to have examined in an effort to determine the origin of the toolmarks appearing thereon.

[...]

(4) It is believed that the marks are not due to modern reworking of the original cast surface because of the corrosion which appears on portions of the fragment. In photographs #1 and #2, there are outlined areas which show a considerable degree of corrosion indicating that some little time has elapsed since the marks were first scraped into the metal. On the area immediately adjacent to the edge of the fragment, the corrosion appears to be developed as fully as on some of the other unquestioned areas on the fragment.

(5) The tool used to make the markings had a radius of approximately 1/8". However, there is considerable variation in the radius of the markings at several points and this dimension is therefore only an approximate average value.

The specimen will be returned to you under separate cover by registered mail.

Assuring you of my desire to be of assistance, I am

Sincerely yours,

J. E. Hoover
John Edgar Hoover
Director

Enclosures

***Figure** 7. FBI document. Freer Gallery of Art, Department of Conservation and Scientific Research.*

During the late 1930s and early 1940s, while working on the Plumer material, Gettens was mostly busy on other projects, serving as a consultant to the Federal Art Project Laboratory of the Federal Works Projects Administration in Boston (Bewer 2010, 193–98), assistant to the directors of the Fogg Museum, and acting head of the Fogg's Department of Conservation. He did examine and treat a handful of Chinese bronzes from outside collections at the Fogg, though without much attention to technique.[2] Then, in 1944–45, perhaps as a result of his earlier contact with Cyril S. Smith, who had been placed in charge of metallurgy at the United States Engineers' Manhattan District Project, Gettens was recruited as a microchemist to work on the chemistry of fissionable materials at Los Alamos (Bewer 2010, 223; Rhodes 1986, 347). This wartime activity may account for Gettens not undertaking any significant research on the large and important collection of Chinese bronzes that came to the Fogg in 1943 as part of the Grenville L. Winthrop bequest (Fogg Art Museum 1943a, 1943b; Bewer 2010, 225) (fig. 8).

It was only well after the war that Gettens turned his attention once again to the technical study of Chinese bronzes. In 1948, he presented a paper on the corrosion

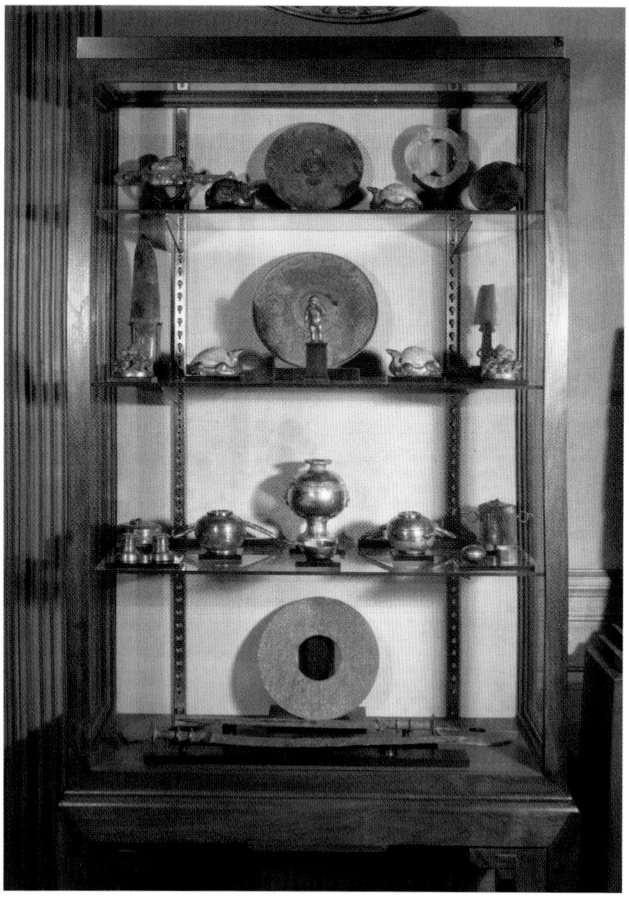

Figure 8. One of the many display cases containing Chinese bronzes in Grenville L. Winthrop's house in New York, early 1940s. Digital Imaging and Visual Resources Department, Harvard Art Museums, © 2010 President and Fellows of Harvard College.

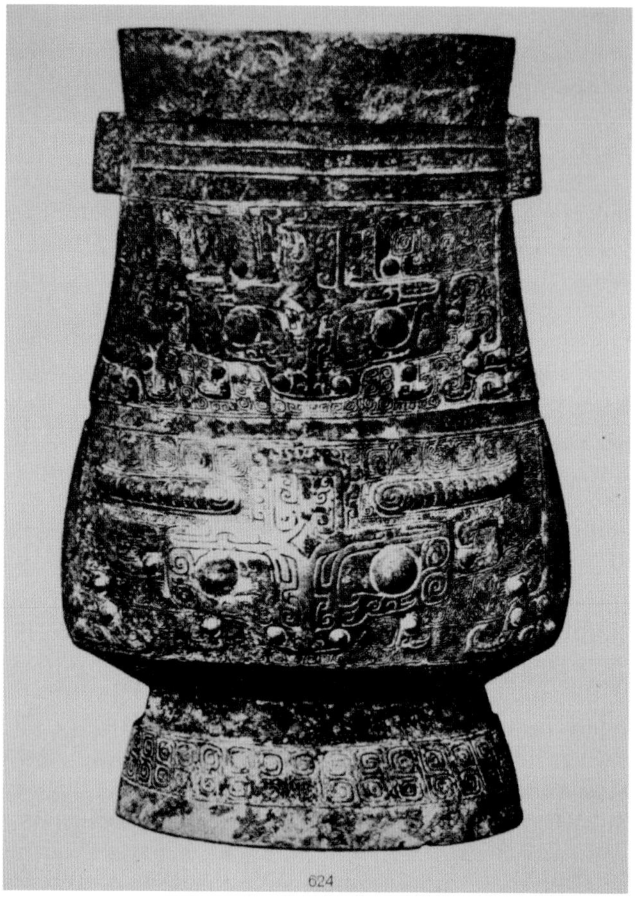

Figure 9. "Broad hu*" depicted in Karlgren 1937.*

products of high-tin bronzes that he had identified during his work on the Plumer mirrors (Gettens 1948). This was at a meeting of the Art Technical Section of the American Association of Museums (AAM), a specialty group spearheaded by the Fogg in the early 1930s that became a key national venue for the discussion of conservation and technical issues between conservators, restorers, scientists, art historians, museum directors, and curators (Bewer 2010, 171–72). His article on tin oxide corrosion products was published not long thereafter. In it, he laid out the various elements of their complex structure and pointed to the oxides of the tin-rich eutectoid phases as being responsible for the desirable jade-like "water patina," all the while leaving many questions open about the corrosion mechanisms (Gettens 1949).

The Kelley *Hu*

The last significant Chinese bronze research project that Gettens undertook at the Fogg was on a ceremonial "Broad *hu*"-type vessel from the Zhou period that he received in fragments in the late 1940s with the instruction that he was free to do as he pleased with them. These were loaned to Gettens by Charles Fabens Kelley, assistant director and curator of Oriental art at the Art Institute of Chicago. The vessel had been purchased as a whole object in China during the war but had been broken during shipment to this country. Nothing was known of the exact origin of the piece or the circumstances of its burial. Examination showed it was a restoration made up of old fragments and new parts that had been fashioned from pieces of copper sheet joined with soft solder to form a complete shape similar to a vessel depicted in an article by Karlgren (1937, 45 and plate 18) (fig. 9). The joins and reconstructed parts had been cleverly concealed with plaster and a fake patina. The solder had adhered poorly to the old corroded metal, and the object had fallen apart in transit. In its broken condition there was no difficulty in telling the old parts from the new. The original archaeological pieces were of chief interest because of their value for analytical purposes, and in fact, several fragments were also sent to the Boston Museum of Fine Arts around that time for analysis (Gettens 1946; Kelley 1946, 1947; Young 1947).

The main focus of the research was on the mechanism that led to the formation of the smooth, hard, gray-green patina, though Gettens's technical analysis also allowed him to determine that the base and rim of the object were from the same original piece (Kelley 1948). As with the Plumer research, Gettens gathered numerous collaborators around him to help with the analyses of the original parts of the Kelley vessel. Consulting metallurgist Daniel Cushing executed metallography on specimens prepared at the Department of Geology at Harvard University

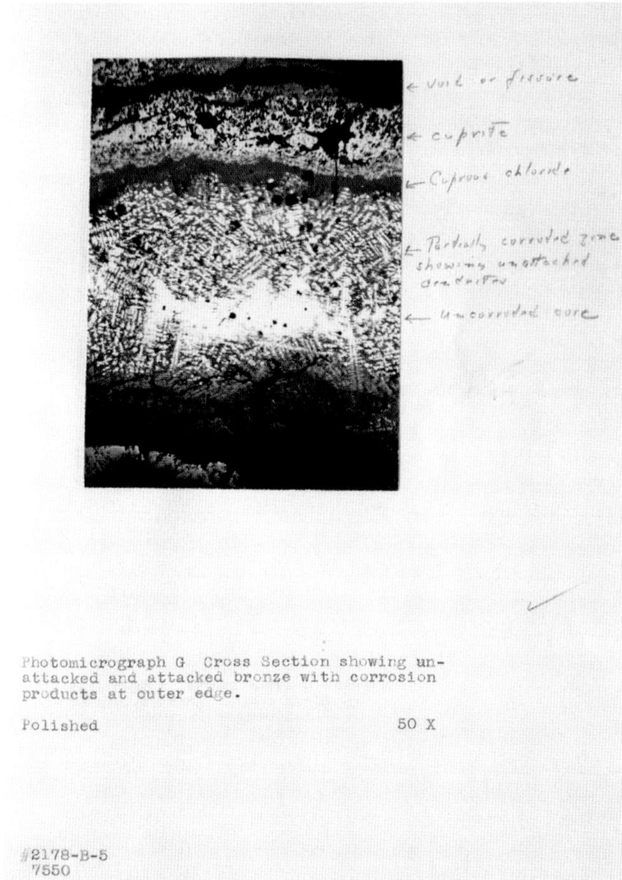

Figure 10. *Detail of a page from Daniel Cushing's metallography report with Gettens's handwritten annotations. Freer Gallery of Art, Department of Conservation and Scientific Research.*

(Cushing 1947) (fig. 10). Scientists from Harvard's Department of Mineralogy performed x-ray diffraction analyses, and others from the Rhode Island State College undertook spectrographic analysis. Their combined efforts resulted in a more thorough understanding of the complex structure and mechanisms of corrosion in a high-tin bronze with 4–5 wt.% lead (though not so high as to be a speculum metal).

Examining the polished metallographic samples, Gettens made a distinction between the partially mineralized intermediary zone he observed around the uncorroded metal core and the completely mineralized zone on the outer surface. In the first, the dendrites of the copper-rich alpha phase remained unscathed; the tin-rich eutectic phase had been converted to a black corrosion product containing the tin oxide cassiterite, which is usually white. He speculated that they might have been stained black by cupric oxide or hydroxide. From there on outward, only mineral corrosion products were found. At the interface between the intermediary and outer zones, cuprous chloride, which is responsible for "bronze disease," was found. Intermittent areas of redeposited copper in the next layer were evidence of the electrochemical nature of the corrosion process. A thick layer of cuprous oxide (cuprite) with streaky deposits of insoluble tin oxide in it formed the next layer, which was finally covered with the predominant exterior corrosion product—malachite (basic cupric carbonate) mixed with some basic cupric chloride (paratacamite). The role of lead in the corrosion process remained to be studied.

Building on the work of colleagues Colin Fink, professor of electrochemistry at Columbia University, and Earl R. Caley, analytical chemist with an interest in archaeology (at Princeton then at Ohio State), Gettens ultimately visualized the corrosion process as an invasion of the solid metal crystal structure by an ion-bearing solution and the forming and solidification of colloidal gels. Gettens presented these findings to the Art Technical Section of the AAM and published them a few years later (Gettens 1947, 1951).[3]

In 1950, due to budget cuts imposed by the university, Gettens's contract at the Fogg Museum was not renewed (Bewer 2010, 247). Jack Thacher, director of Dumbarton Oaks, Harvard's scholarly research institute and museum in Washington, D.C., spearheaded a campaign to bring Gettens to Washington (Thacher 1951; Stout 1951). Thacher was a Harvard graduate with close connections to the Fogg, and he had sent objects from Dumbarton Oaks to the Fogg's conservation department over the years, including a number of Chinese bronzes that Gettens had examined and treated (Gettens 1937 and 1941–42). To the great disappointment of the National Bureau of Standards, Gettens turned down their offer of a job (Wichers 1951; Saylor 1951) in order to join the Freer Gallery of Art, Smithsonian Institution, Washington, D.C., where he was appointed associate in technical research, effective April 1, 1951 (the position became permanent that fall). At the Freer, he set up the Laboratory for Technical Studies in Art and Archaeology, which opened the following year, and he continued to pursue his research there until the end of his professional life. His knowledge and interest in Asian bronzes bore a lasting fruit in the publication of the seminal technical volume on Chinese bronze ritual vessels in the Freer's collection, whose anniversary of publication we are celebrating with this symposium.

Acknowledgments

A very special thanks, first and foremost, to Tom Chase for sharing his invaluable insights and recollections on the technical study of Chinese bronzes and for being a wonderful "Cicerone" among the wealth of archival materials preserved in the Department of Conservation and Scientific Research (DCSR) of the Freer Gallery of Art and Arthur M. Sackler Gallery, Smithsonian Institution. I am also most grateful to colleagues and friends at the DCSR, in particular Paul Jett, Blythe McCarthy, and Stacy Bowes; and at the Straus Center for Conservation and Technical Studies, Harvard Art Museums, especially Ariel O'Connor. I am thankful to Elisabeth West FitzHugh for historical information and to Rebecca Gettens Hayes for some wonderful photographs. Many thanks to James Carder, archivist and house collection manager at the Dumbarton Oaks Research Library and Collection. And last but not least, I am indebted to Rutherford John Gettens, Edward Waldo Forbes, and their colleagues, who kept such good notes for posterity and whose work has inspired many and paved the way for the kind of technical research that we continue to pursue and are celebrating with this volume.

Notes

1. By 1939 there were more than eighty-eight mirrors in the Plumer collection. About a quarter of these were analyzed. A selection of Plumer mirrors eventually came to the Freer Gallery of Art after the collector's death and forms part of the study collection to this day.
2. These bronzes included works from the collections of Dumbarton Oaks, C. T. Loo, and General Sir Neil Malcolm. See Gettens's correspondence and treatment notes in the Harvard Art Museums, Straus Center for Conservation Object Treatment files.
3. The Kelley bronze research continued at the Freer and was revisited recently in a collaboration between scientists at the Art Institute of Chicago, the Freer Gallery of Art and Arthur M. Sackler Gallery, Department of Conservation and Scientific Research, and W. T. Chase (Young et al. 2010).

References

Anonymous. 1930. Analyzen chinesischer bronzen. *Ostasiatische Zeitschrift* 6:284.

Benedetti-Pichler, A. A. 1935. Letter to R. J. Gettens, January 24. Freer Gallery of Art and Arthur M. Sackler Gallery, Department of Conservation and Scientific Research, bronze archives, Chinese bronze mirrors, Mirrors—Correspondence.

Bewer, F. 2010. *A laboratory for art: Harvard's Fogg Museum and the emergence of conservation in America, 1900–1950.* Cambridge, MA, and New Haven, CT: Harvard Art Museums and Yale University Press.

Chikashige, M. 1920. Ancient Oriental chemistry and its allied arts. *Man* 20:161–68.

Collins, W. F. 1931. The corrosion of early Chinese bronzes. *The Journal of the Institute of Metals* 45(1):23–47.

Cushing, D. 1947. Photomicrographs of [an] ancient Chinese bronze, for [the] Fogg Art Museum. Freer Gallery of Art and Arthur M. Sackler Gallery, Department of Conservation and Scientific Research, Treatment files, SC-B-18 "Kelley bronze" notebook.

Fogg Art Museum. 1943a. *Chinese sculpture, bronzes, jades, paintings and drawings, Egyptian and Persian sculpture, pre-Columbian art: Selected from the collection of Grenville Lindall Winthrop*. Cambridge, MA: Fogg Art Museum.

———. 1943b. A special number devoted to the Grenville Lindall Winthrop bequest. *Bulletin of the Fogg Museum of Art* 10:2.

Gettens, R. J. 1933. Mineralization, electrolytic treatment and radiographic examination of copper and bronze objects from Nuzi. *Technical Studies in the Field of the Fine Arts* 1(3):118–42.

———. 1934a. Addendum to a report on the examination and cleaning of two bronze horses, owned by the William Rockhill Nelson Trust, Kansas City, Missouri, December 15. Harvard Art Museums, Straus Center for Conservation, Bronze—Wm. R. Nelson Gallery of Art.

———. 1934b. Some observations concerning the lustrous surface on certain ancient Eastern bronze mirrors. *Technical Studies in the Field of Fine Arts* 3(1):29–37.

———. 1935. Notes on Langdon Warner fragments (LW1-4) interspersed among Plumer mirror lab notes. Freer Gallery of Art and Arthur M. Sackler Gallery, Department of Conservation and Scientific Research, bronze archives, green loose leaf binder notebook marked "Plumer Mirrors General Notes (etc)."

———. 1937 and 1941–42. Harvard Art Museums, Straus Center for Conservation, Object Treatment Files, Dumbarton Oaks, Oriental Chinese bronze objects—Mrs. R. W. Bliss, (loans 4311abc); Dumbarton Oaks Collection, notes on the examination of metal objects; and Harvard Art Museums, Straus Center for Conservation, Bronze—Bliss—Restoration, Photographs.

———. 1939. Letter to J. M. Plumer, June 23. Freer Gallery of Art and Arthur M. Sackler Gallery, Department of Conservation and Scientific Research, bronze archives, box marked "save for Marinka—Bronze mirrors," notebook titled "Notes on a metallographic and microscopic study of ancient Chinese bronze mirrors (belonging to James Plumer, Ann Arbor)."

———. 1940a. Outline for "A chemical and metallographic study of ancient Chinese bronze mirrors" lecture given at the Princeton Chemical Club February 7. Harvard Art Museums, Straus Center for Conservation, Gettens papers, lecture notes.

———. 1940b. Letters to J. M. Plumer, January–February. Freer Gallery of Art and Arthur M. Sackler Gallery, Department of Conservation and Scientific Research, bronze archives, unmarked gray box, mirrors miscellaneous, mirrors important "chop suey" folder.

———. 1940c. Report on metallographic analysis, March. Harvard Art Museums, Straus Center for Conservation, Object Treatment Files, C. T. Loo, gilded bronze Chinese Buddha (5126).

———. 1946. Letter to C. F. Kelley, November 13. Freer Gallery of Art and Arthur M. Sackler Gallery, Department of Conservation and Scientific Research, bronze archives, SC-B-18 "Kelley Bronze" notebook.

———. 1947. "An interesting fraudulent early Chinese bronze," read at the Art Technical Section of the AMA, Quebec, Canada, May 31. Freer Gallery of Art and Arthur M. Sackler Gallery, Department of Conservation and Scientific Research, bronze archives, green loose leaf binder labeled SC. B. 18.

———. 1948. Lecture notes for "Corrosion products on certain high-tin bronzes," paper read at the Art Technical Session, American Association of Museums annual meeting, Fogg Museum, Cambridge, MA. Harvard Art Museums, Straus Center for Conservation, Gettens papers, lecture notes.

———. 1949. Tin-oxide patina of ancient high-tin bronze. *Fogg Museum of Art Bulletin* 2:16–26.

———. 1951. Corrosion products of an ancient Chinese bronze. *Journal of Chemical Education* 28(2):67–71.

———. 1969. *The Freer Chinese bronzes,* Vol. 2, *Technical studies*. Oriental Study Series 7. Washington, DC: Freer Gallery of Art, Smithsonian Institution.

Hoover, J. E. 1940. Letter to J. M. Plumer, December 6. Freer Gallery of Art and Arthur M. Sackler Gallery, Department of Conservation and Scientific Research, bronze archives, unmarked gray box, mirrors miscellaneous, mirrors important "chop suey."

Karlbeck, O. 1926. Notes on some early Chinese bronze mirrors. *The China Journal of Science and Arts* 4:3–9.

———. 1941. Letter to J. M. Plumer, August 27. Freer Gallery of Art and Arthur M. Sackler Gallery, Department of Conservation and Scientific Research, bronze archives, unmarked gray box, mirrors miscellaneous, mirrors important "chop suey" folder, envelope "Mirrors—Important Notes."

Karlgren, B. 1937. New studies on Chinese bronzes. *Bulletin of the Museum of Far Eastern Antiquities* 9:1–118.

Kelley, C. F. 1946. Letter to R. J. Gettens, November 15. Freer Gallery of Art and Arthur M. Sackler Gallery, Department of Conservation and Scientific Research, bronze archives, SC-B-18 "Kelley Bronze" notebook.

———. 1947. Letter to R. J. Gettens, April 14. Freer Gallery of Art and Arthur M. Sackler Gallery, Department of Conservation and Scientific Research, bronze archives, SC-B-18 "Kelley Bronze" notebook.

———. 1948. Letter to R. J. Gettens, April 20. Freer Gallery of Art and Arthur M. Sackler Gallery, Department of Conservation and Scientific Research, bronze archives, SC-B-18 "Kelley Bronze" notebook.

Kennedy, E. J., Jr. 1938. Letter to V. G. Mooradian, June 26. Freer Gallery of Art and Arthur M. Sackler Gallery, Department of Conservation and Scientific Research, bronze archives, box marked "save for Marinka—Bronze mirrors," notebook titled "Notes on a metallographic and microscopic study of ancient Chinese bronze mirrors (belonging to James Plumer, Ann Arbor)."

Komatsu, S., and Y. Yamanouchi. 1933. A chemical study of ancient Chinese bronzes. *Journal of Oriental Studies* 3:295–304.

Laufer, B. 1934. Letter to R. J. Gettens, May 24. Freer Gallery of Art and Arthur M. Sackler Gallery, Department of Conservation and Scientific Research, bronze archives, Chinese bronze mirrors, Mirrors—Correspondence.

Mooradian, V. G., and R. J. Gettens. n.d. A metallographic and microscopic study of ancient Chinese bronze mirrors. Freer Gallery of Art and Arthur M. Sackler Gallery, Department of Conservation and Scientific Research, bronze archives (green binder).

Plenderleith, H. 1939. Technical notes on Chinese bronzes with special reference to patina and incrustations. *Transactions of the Oriental Ceramic Society 1938–39* XVI:33–55.

Plumer, C., and J. M. Plumer. 1934. Catalogue of 42 early Chinese bronze mirrors. Research material collected in China. Freer Gallery of Art and Arthur M. Sackler Gallery, Department of Conservation and Scientific Research, bronze archives, JMP Mirrors Photographs (metal box).

Plumer, J. M. 1940. Letter to J. E. Hoover, November 20. Freer Gallery of Art and Arthur M. Sackler Gallery, Department of Conservation and Scientific Research, bronze archives, unmarked gray box, mirrors miscellaneous, mirrors important "chop suey" folder, manila envelope.

———. 1944. The Chinese bronze mirror: Two instruments in one. *The Art Quarterly* 7(2):90–108.

Plumer, J. M., and R. J. Gettens. 1941. Outline and incomplete manuscript of "Chinese bronze mirrors: A study in culture and craftsmanship together with chemical and microscopic analyses." Freer Gallery of Art and Arthur M. Sackler Gallery, Department of Conservation and Scientific Research, bronze archives, unmarked gray box, mirrors miscellaneous, mirrors important "chop suey" folder.

Rhines, F. N. 1939. Letter to G. Mooradian, April 26. Freer Gallery of Art and Arthur M. Sackler Gallery, Department of Conservation and Scientific Research, bronze archives, box marked "save for Marinka—Bronze mirrors," notebook titled "Notes on a metallographic and microscopic study of ancient Chinese bronze mirrors (belonging to James Plumer, Ann Arbor)."

Rhodes, R. 1986. *The making of the atomic bomb*. New York: Simon and Schuster.

Saylor, C. P. 1951. Letter to R. J. Gettens, March 29. Freer Gallery of Art and Arthur M. Sackler Gallery, Department of Conservation and Scientific Research, R. J. Gettens correspondence files, National Bureau of Standards.

Smith, C. S. 1939. Letter to V. Mooradian, March 20. Freer Gallery of Art and Arthur M. Sackler Gallery, Department of Conservation and Scientific Research, Bronze archive, box marked "save for Marinka—Bronze mirrors," notebook titled "Notes on a metallographic and microscopic study of ancient Chinese bronze mirrors (belonging to James Plumer, Ann Arbor)."

Stern, H. P., and B. M. Usilton. 1963. James Marshall Plumer. *Ars Orientalis* 5:329–36.

Stout, G. L. 1951. Letters to J. Thacher, February 26 and March 12. Dumbarton Oaks Research Library and Collections, Washington, DC, Archives, Administration files, John. S. Thacher correspondence, 1951.

Thacher, J. 1951. Letters to G. L. Stout, February 21 and March 6. Dumbarton Oaks Research Library and Collections, Washington, DC, Archives, Administration files, John. S. Thacher correspondence, 1951.

Umehara, S. 1927. L'analyse chimique des bronzes anciens de la Chine. *Artibus Asiae* 2(4):247–64.

Vanderbelt, J. M. 1936. Chemical analysis of a suite of old Chinese bronze mirrors. PhD diss., Boston University.

Warner, L. 193[?]. Letter to J. M. Plumer, June 22. Freer Gallery of Art and Arthur M. Sackler Gallery, Department of Conservation and Scientific Research, bronze archives, box marked "save for Marinka—Bronze mirrors," notebook titled "Notes on a metallographic and microscopic study of ancient Chinese bronze mirrors (belonging to James Plumer, Ann Arbor)."

Wichers, E. 1951. Letter to R. J. Gettens, April 2. Freer Gallery of Art and Arthur M. Sackler Gallery, Department of Conservation and Scientific Research, R. J. Gettens correspondence files, National Bureau of Standards.

Yetts, W. P. 1930. *The George Eumorfopoulos Collection catalogue of Chinese and Corean bronzes, sculpture, jades, jewellery and miscellaneous objects*. Vol. 2. London: E. Benn.

———. 1931. Problems of Chinese bronzes. *Royal Central Asian Society Journal* 18(3):399–402.

Young, M. L., F. Casadio, J. Marvin, W. T. Chase, and D. C. Dunand. 2010. An ancient Chinese bronze fragment re-examined after 50 years: Contributions from modern and traditional techniques. *Archaeometry* 52(6):1015–43.

Young, W. J. 1947. Letter to R. J. Gettens, May 15. Freer Gallery of Art and Arthur M. Sackler Gallery, Department of Conservation and Scientific Research, bronze archives, SC-B-18 "Kelley Bronze" notebook.

The Making of Volume 2 of the Freer Chinese Bronze Book

W. T. Chase

ABSTRACT This paper focuses on the genesis of the *Freer Chinese Bronzes* volumes, particularly volume 2, *Technical Studies*, which took over forty man-years to write. The laboratory of the Freer Gallery of Art and its equipment are described, chemical analysis procedures are mentioned, and some of the details of photomicrography and photomacrography are also covered.

Introduction

In this article I will reveal some of the history of the *Technical Studies* volume (Gettens 1969) of the *Freer Chinese Bronzes* and give a little of the flavor of what it was like to produce it and to work with John Gettens. I'm going to refer to this volume as BBII or the bronze book, but of course it is volume 2 of a catalogue. To emphasize this, figure 1 shows the first volume (the actual catalogue [Pope et al. 1967]) with BBII, both in their dust jackets.

The history of BBII was summed up beautifully by John Pope in his foreword to the volume (Gettens 1969, xiii).

> This volume of Technical Studies supplements the brief paragraphs headed Technical Observations that appear under each bronze in Volume 1, The Catalogue. Together they embody the results of the sustained and intensive research program carried on by Rutherford John Gettens, Head Curator of the Freer Technical Laboratory, and his staff over the last ten years and more. Acknowledgment of the help he has received from members of the Freer staff and from outside collaborators is made in the author's preface. It is appropriate here to record the special contribution of W. T. Chase who first took part in the investigations as an intern graduate student from the New York University Conservation Center of the Institute of Fine Arts, later joined the regular staff in the final year of the study, and who assumed direction of the Laboratory on Mr. Gettens' retirement this year. In addition to his contributions in the scientific field, he has been most helpful in the editorial work, and during the author's illness, he displayed efficiency and confidence in collaborating with the Editor and preparing the manuscript for the press.
>
> To the best of our knowledge, this is the first time a volume has been devoted to the detailed scientific analysis of a major collection of Chinese ceremonial bronzes. We publish it in the hope that it may contribute to the advancement of investigation into the methods and materials of ancient craftsman.

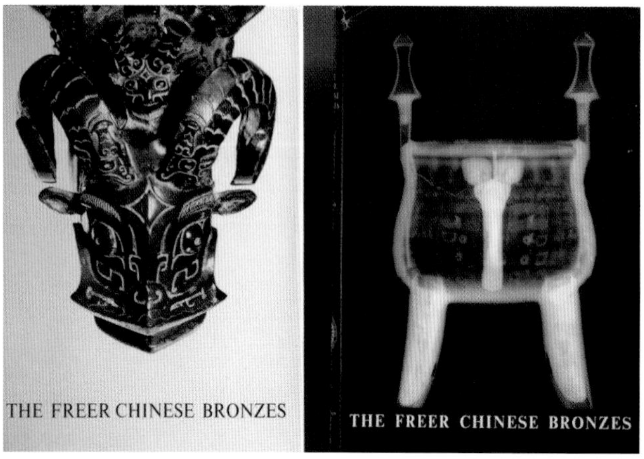

Figure 1. *The Freer Chinese Bronzes, volume 1 (Pope et al. 1967) and volume 2 (Gettens 1969).*

Pope mentions the "sustained and intensive research program." Toward the end of the publication, in fact, I think just when we had put it to bed for the final press run (in a state of both exhaustion and euphoria), we attempted to calculate how much time it had taken, and we came up with the figure of forty man-years—now we would probably say people-years, which would be more appropriate. The publication wouldn't have happened without Elisabeth West FitzHugh and Ilona V. Bene, as well as a slew of secretaries—don't forget, this was BC, before computers! Gettens's preface is worth reading for its acknowledgments, which show the number of people who contributed to the book.

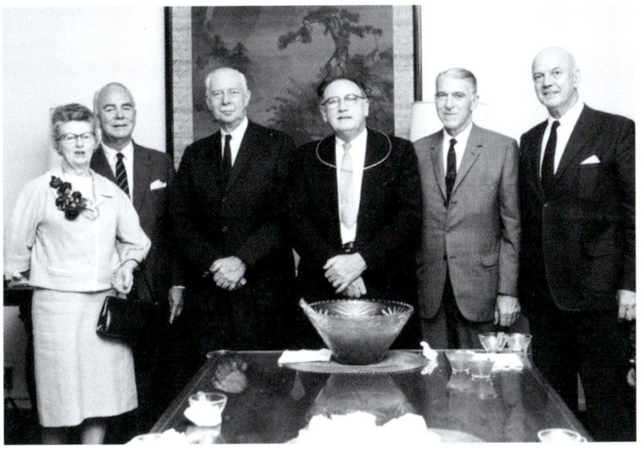

***Figure 2**. Rutherford J. Gettens's retirement party in 1968. From left, Katharine Gettens, Jack Thacher, Leonard Carmichael, R. J. Gettens, Frank Taylor, and John Pope. Photo by Raymond Schwartz, Freer Gallery of Art.*

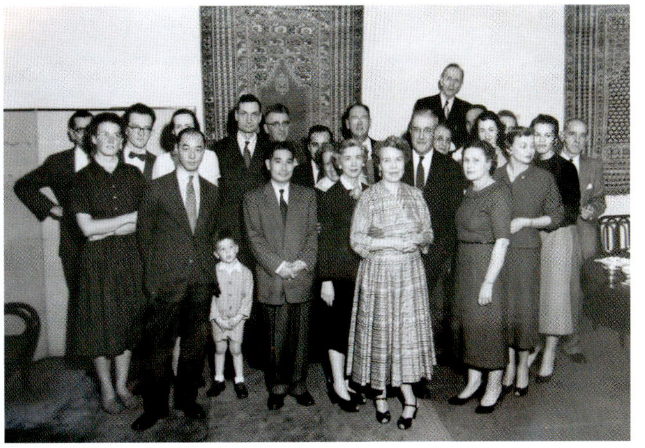

***Figure 3**. Burns Stubbs's retirement party, 1956. Smithsonian Archives. See text for some identifications.*

The Actors

Who were the principals involved in writing the bronze book? First, of course, was Gettens. Figure 2 shows Gettens's retirement party in 1968. Of course, Gettens is at center wearing a joke key ring—his retirement gift. This was a reference to the lab key ring, which was a piece of ½″ by ⅛″ brass and about 18″ long with all the lab keys on it. It hung in the closet. It was impossible to put in your pocket, so you couldn't walk home with it. From left to right are Mrs. Gettens (Katharine, or Kay), Jack Thacher (to whom we are indebted for his bequest establishing the Forbes Fund and this symposium), Leonard Carmichael (retired secretary of the Smithsonian), Gettens, Frank Taylor (director of the U.S. National Museum), and John Pope.

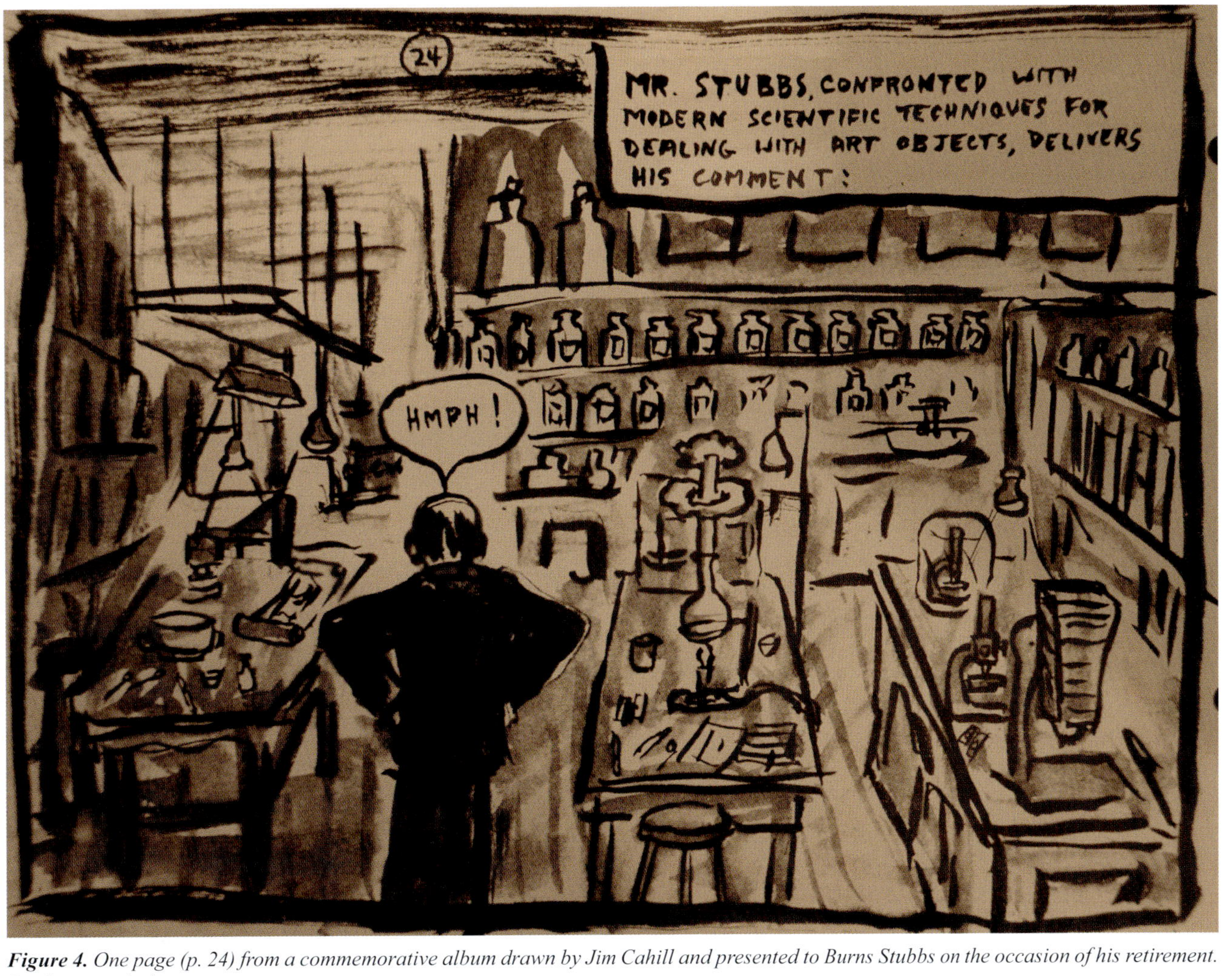

***Figure 4.** One page (p. 24) from a commemorative album drawn by Jim Cahill and presented to Burns Stubbs on the occasion of his retirement. Smithsonian Archives.*

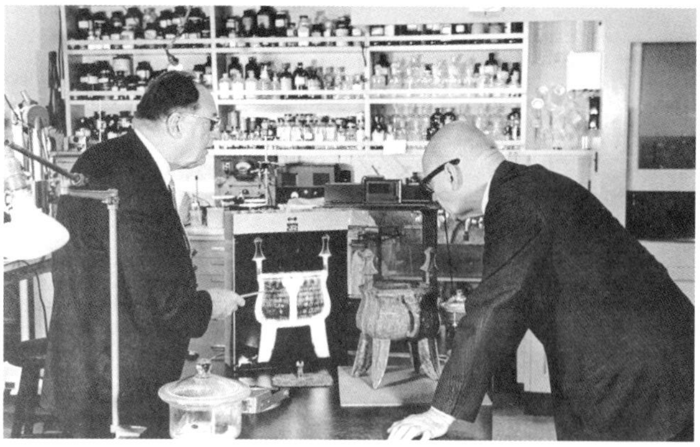

Figure 5. *The frontispiece to volume 2 (Gettens 1969, ii) showing R. J. Gettens and John A. Pope examining a Chinese bronze with its radiograph, which was used for the cover of volume 2. Photo by W. T. Chase.*

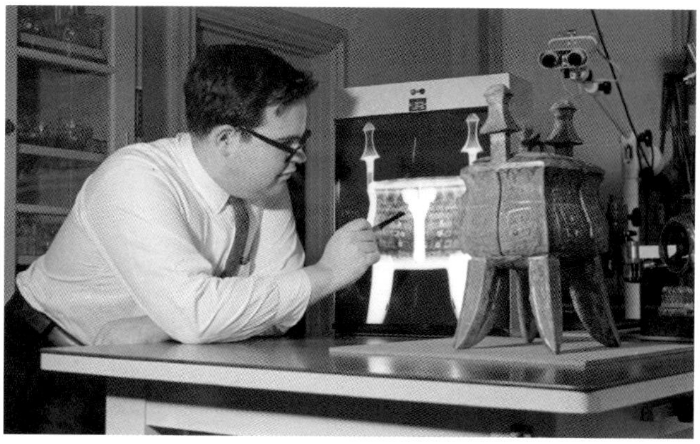

Figure 7. *W. T. Chase examining the same Chinese bronze as in figures 5 and 6. Photo by W. T. Chase (with a self-timer).*

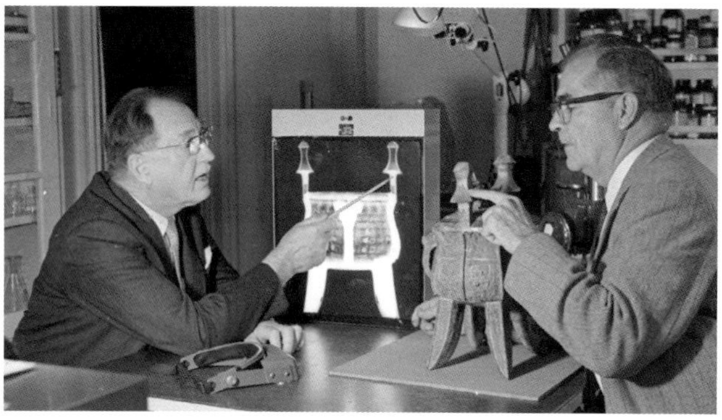

Figure 6. *R. J. Gettens and Lloyd Langford examining the same Chinese bronze as in figure 5 on the higher examining table. Photo by W. T. Chase.*

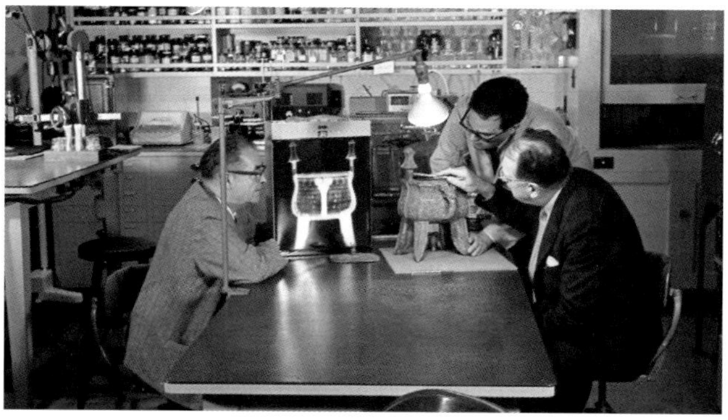

Figure 8. *R. J. Gettens, Lloyd Langford, and W. T. Chase examining the same Chinese bronze as in figures 5–7. Photo by W. T. Chase (with a self-timer).*

But the genesis of the bronze book began many years earlier. The best way to illustrate this is to show you a picture of Burns Stubbs's retirement party in 1956 (fig. 3). Stubbs was the general factotum, chauffer, and building superintendent from the beginning of the Freer Gallery. He is at the right rear of the photo, above the rest. Gettens is on Stubbs's right at the back, with Phil Stern on his right along with Russell Mielke (Stubbs's successor) at the back. Takashi Sugiura (then recently hired as *hyogushi*—restorer of Oriental paintings) is at the front on the left of the child, and Richard Ettinghausen is the tall figure behind him. Archibald Gibson Wenley, the director, is on Gettens's left. Just behind Wenley is Graziano Gagliardi, the carpenter who built the lab (and a lot of other beautiful furniture). And on the far left of the photograph is Lis FitzHugh, with Jim Cahill seen over her left shoulder.

At this party, Cahill presented Stubbs with a sketchbook illustrating his career. The illustration on page 24 (fig. 4) is quite an accurate depiction of the laboratory: you can see the examining table at left, the central table with the analytical balance at the center, and the microscope bench at right. Further actors came on the stage later.

Toward the end of the project we decided that we did not have an appropriate frontispiece for the book. We all thought that we should show people looking at a bronze in the laboratory, and we thought it would be most appropriate to have it be Gettens and Pope. I borrowed the Leica M3 from the photo lab, and we took a bunch of pictures. One of them became the frontispiece (fig. 5).

The unused shots are historically interesting. Figure 6 shows Gettens and Lloyd Langford, the editor, whom we got to know very well! In Pope's foreword, he mentions Gettens's illness. Gettens's daughter, Rebecca, thinks this was in the spring of 1966 (pers. comm., Rebecca Gettens Hayes, August 2010). Gettens was out for about three months, the first month of which he spent in the hospital. After this he went home, and Langford and I would go out and visit him with the manuscript. Langford gets kudos, even at this late date, for his patient help on the bronze book.

When I found some of the original negatives at home, I was surprised to see myself in the first test shot (fig. 7). Here I am, thinner and younger! I also found one of the few photographs showing me and Gettens together in the lab, along with Langford (fig. 8). I was glad to have found a picture showing us both.

The Stage

The Lab

Gettens published a plan in a description of the laboratory in 1959 (Gettens 1959b, fig. 1) (fig. 9). The top of the plan

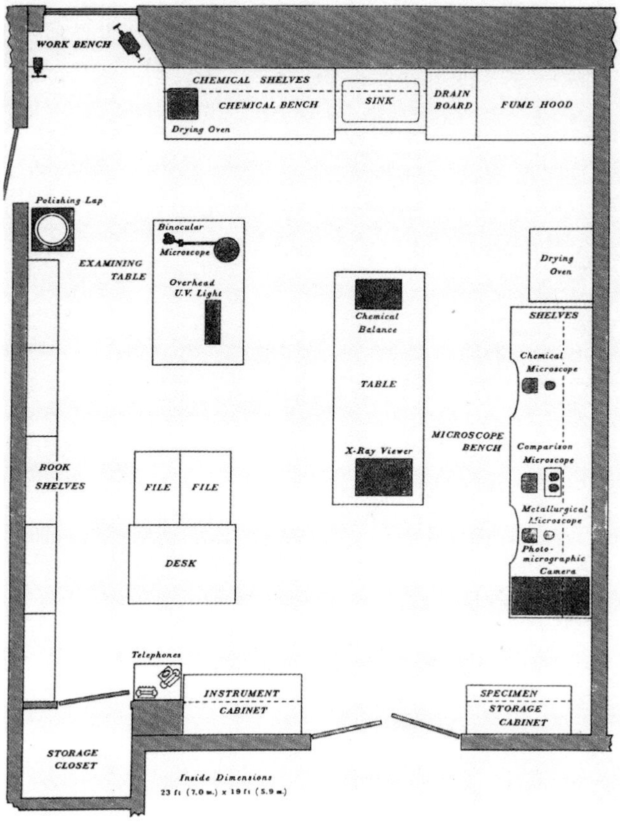

Figure 9. *Plan of the laboratory. Figure 1 in Gettens (1959b).*

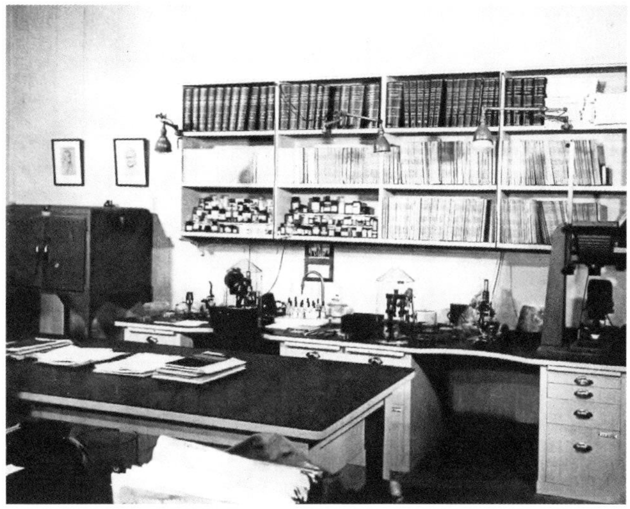

Figure 10. *The microscope bench. Figure 2 in Gettens (1959b).*

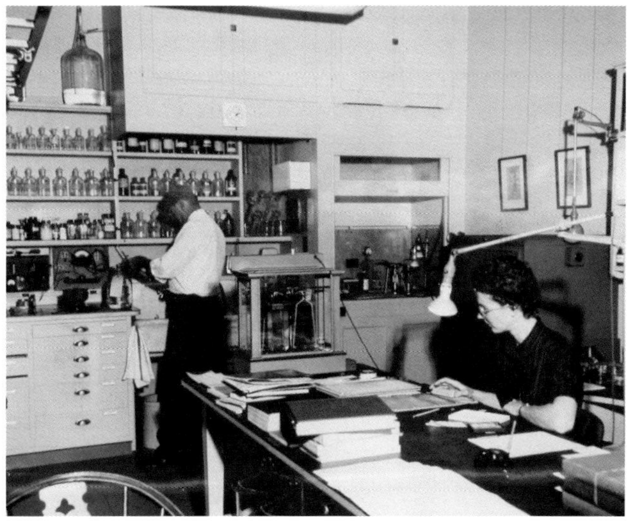

Figure 11. *R. J. Gettens and Elisabeth West FitzHugh in the laboratory, probably about 1959. Laboratory archives.*

is west, and the work bench was at our west window, with good daylight. The microscope bench is on the right. The central table is shown with the analytical balance on one end and the light box on the other. This is where we did a lot of our examination work and writing. It is at the center of figure 9.

The examining table, at the upper left, was the subject of another article by Gettens (1959a). Figures 7 and 8 show it prominently. It was high enough so that you could get your eyes close to an object easily, and three or four people could stand around and discuss what was on the table. Chemical operations took place on the chemical bench, shown at the top of the drawing, and in the fume hood. Samples were stored in the specimen storage cabinet. You could swivel around from the desk at left to pick up the telephone (the old intercom system, which was gone long before I arrived, also shows here). All the objects—except for particularly large ones—were locked up in the storage closet overnight along with the keys when we left.

The microscope bench was shown by Gettens in the 1959 article (1959b, 143, fig. 2) (fig. 10), and this is what it looked like while we were working on the bronze book. The chemical microscope is in the left-hand slot on the microscope bench, with reagents on its right and above. The comparison microscope is shown here, but we usually left it (and its illuminator) in its case; it's easier to take two photographs and put them next to each other. Dick Buck at Oberlin also had a comparison microscope that he used for student instruction. The metallurgical microscope is shown here at the right-hand slot on the bench, and the photomicrographic camera is at the far right.

Figure 11 shows the west wall and the northwest corner of the lab, with Gettens at the sink and FitzHugh writing on the table. The analytical balance shows prominently at the end of the table.

Office

When I came to the Freer in 1964, Gettens had a separate office carved out of the west side of the East Asian painting conservation studio to be used for writing the bronze book. I have no photographs of this. It was just wide enough for one desk, but it was a convenient and quiet place to write and to keep the bronze records.

Richard Ettinghausen had the office next to the laboratory. When Ettinghausen retired from the Freer and moved up to New York University, in 1969 or 1970, we got his office. Shortly after we moved in, Ursula Pariser, one of the staff photographers, took a series of photographs of Gettens for publicity purposes (fig. 12), which show the bookcases in our new office and the bronze book, open to the radiography chapter.

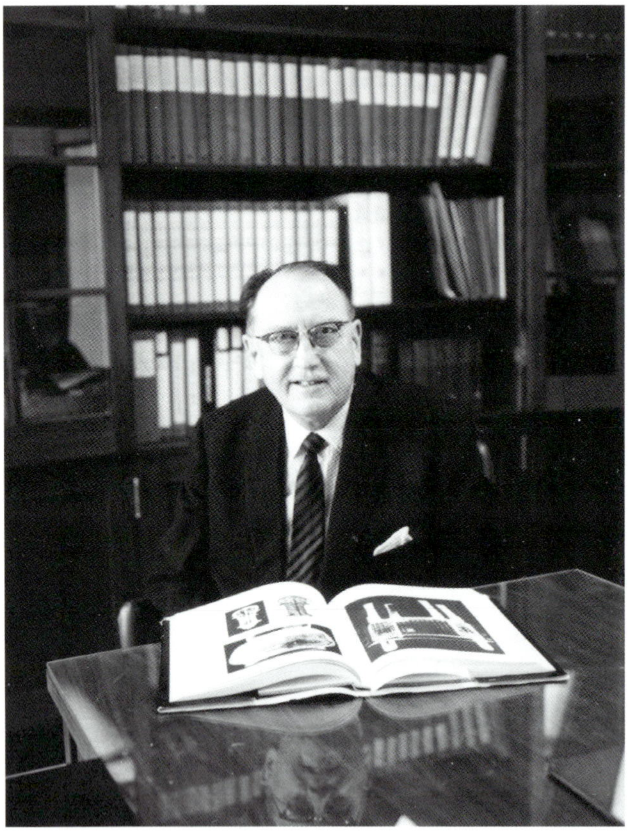

***Figure 12.** R. J. Gettens in our "new" office (see text). Taken in the early 1970s by Ursula Pariser. Laboratory archives.*

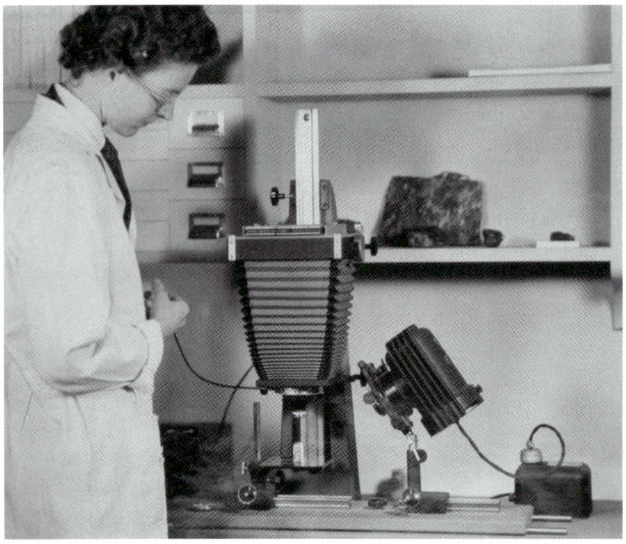

***Figure 13.** Elisabeth West FitzHugh using the Bausch & Lomb model L camera with a ribbon illuminator. Laboratory archives.*

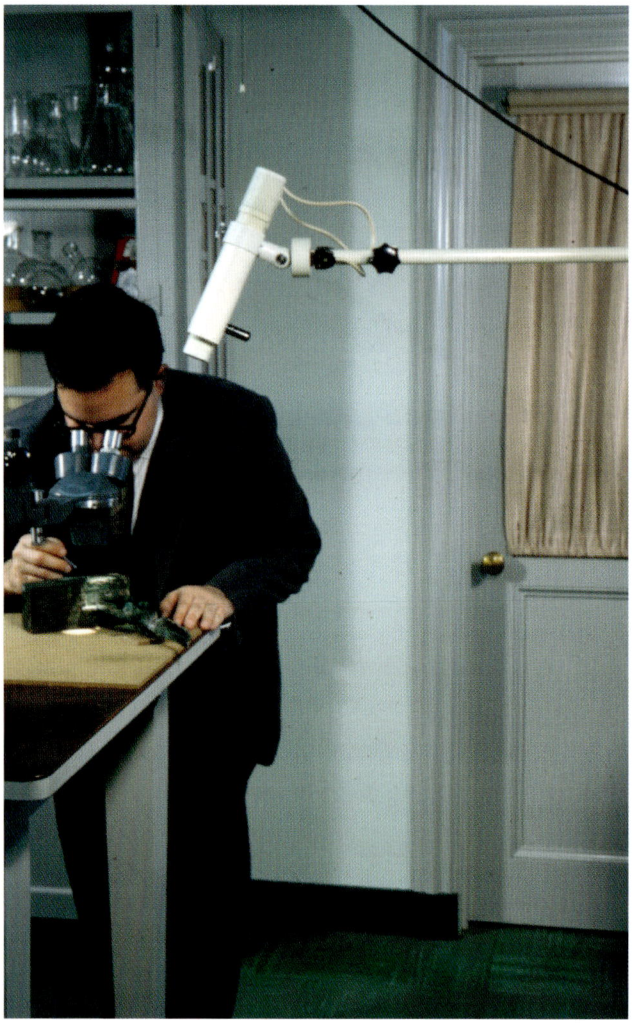

***Figure 14.** W. T. Chase mechanically cleaning an Egyptian statuette on the high examining table using a stereomicroscope and the Zeiss twin lamp illuminator. Laboratory archives.*

Equipment

Our equipment concentrated on microscopes, photographic apparatus for both macrographs and micrographs, chemicals, a fume hood, apparatus for chemical analysis, and hand tools. The photographic studio and the photo lab were directly across the hall from us. General photography of bronzes and details at low magnification was done by Ray Schwartz and Jim Riggs, and we took over at about 1:1 magnification. A lot of our work was done on the Bausch & Lomb model L camera, shown with FitzHugh in figure 13. It consisted of a bellows camera on a vertical counterweighted rail, a ribbon illuminator, and a specimen stage. It was very flexible, with four interchangeable macro lenses and a standard view camera shutter.

Figure 14 shows me working with the Bausch & Lomb stereomicroscope on the high examining table. We had a Zeiss twin lamp projection spotlight to use with this scope, and it gave a very fine sharply lit image. Figure 15 shows me working with what was then our new Zeiss operating stereomicroscope. It was particularly well suited for examining large objects. It had a second observer tube, camera, and three objectives, which allowed the working distance to be varied. Both of these microscopes were real workhorses!

We also had a Bausch & Lomb polarizing chemical microscope, seen just behind Gettens in figure 16. It had illumination both above and below the stage; the large round lamp just behind Gettens's nose is the lamp housing for the above-stage illumination. And we had a Bausch & Lomb Trivert metallographic microscope, with both bright and dark field illumination. Figure 17 shows FitzHugh using it; the illuminator housing protrudes to the left. This was quite a compact unit and had its own carrying case. Gettens brought it up to New York University to teach metallography when I was there in 1963.

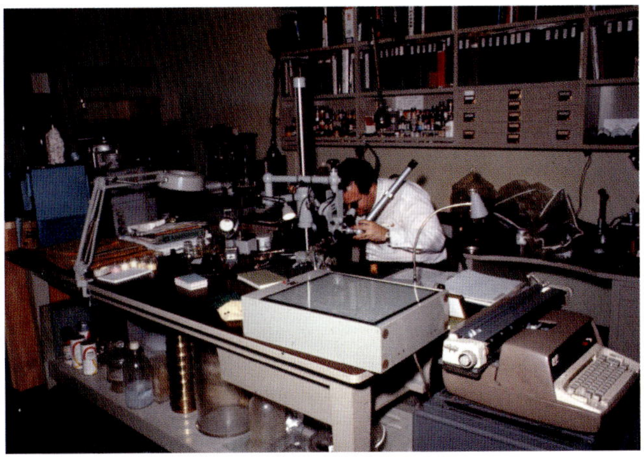

Figure 15. *W. T. Chase examining a Chinese bronze chimera using the Zeiss stereomicroscope. Laboratory archives.*

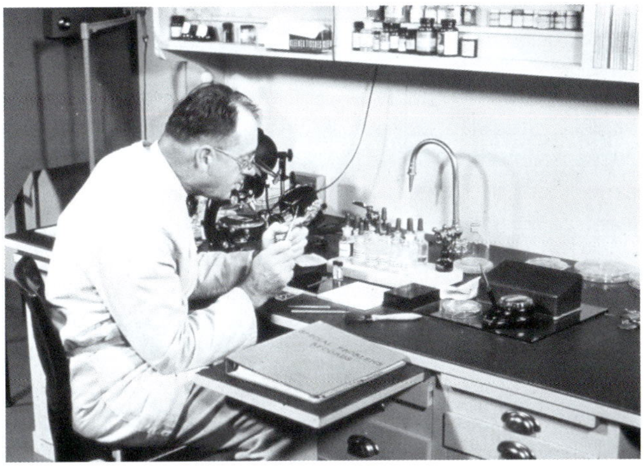

Figure 16. *R. J. Gettens preparing for a microchemical test using the chemical microscope. Laboratory archives.*

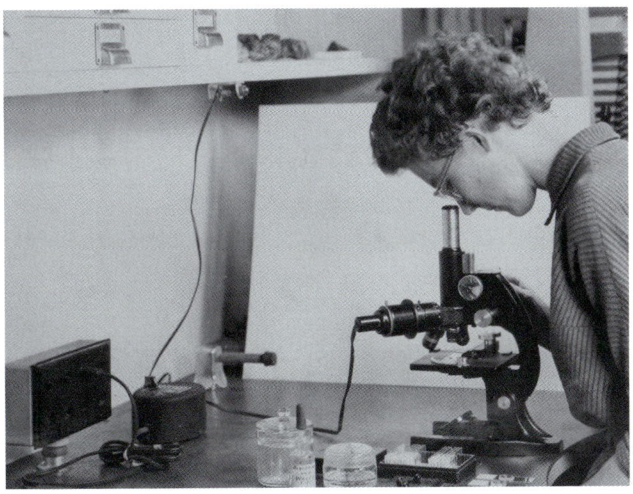

Figure 17. *Elisabeth West FitzHugh using the Bausch & Lomb Trivert metallographic microscope with integral light source. Laboratory archives.*

The Bronze Book

Chemical Analysis of the Bronzes

In 1946, a catalogue of Chinese bronzes acquired during the administration of John Ellerton Lodge was published by the Freer (Lodge et al. 1946). It included fifty-eight bronzes of all types—ceremonial vessels, weapons, mirrors, etc. The illustrations were done in collotype, and they were marvelously sharp.

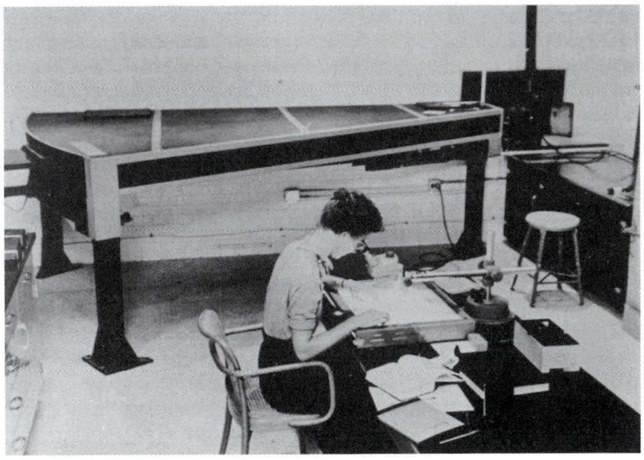

Figure 18. *Elisabeth West FitzHugh examining a plate from the 3 meter Jarrell-Ash spectrograph in the Conservation Center of the Institute of Fine Arts of New York University. Figure 17 in BBII (Gettens 1969, 37).*

In the Smithsonian Archives, I found a card from a file on publications that listed the details of the book (total cost was $10,250.54, and it sold for $5.00) and said, in red ink at the top, "Out of print in August, 1955." To quote again from John Pope (Pope et al. 1967, xii): "By 1956 that book was out of print, and Mr. Wenley began to think in terms of a second bronze catalog to include the many important pieces that had been added during his directorship. Unfortunately he died just as the project was getting well started, and the work has been carried on by the Galleries staff and has been expanded to include all the ceremonial vessels in the collection." In fact, the evolution of what was going to be included in the catalogue is an interesting study, but suffice it to say here that the final plan was to include all the ceremonial vessels and to publish them with art-historical and archaeological comparisons, inscription and epigraphic studies, and technical notes and to publish a separate volume of technical studies on the whole collection.

The newly formed Technical Laboratory, founded by Gettens when he came to the Freer in 1951 (see Bewer, in this vol.) was to take as one of its mandates analysis and examination of the Chinese bronzes, and to this end, FitzHugh was hired in 1956. With Gettens, she began to set up the chemical analyses of Chinese bronzes. "It was decided for practical reasons to carry out the analyses in two steps: wet chemical quantitative analysis for major constituents, and semiquantitative analysis by optical emission spectrometry for minor and trace constituents" (Gettens 1969, 34). Spectrometry was first carried out at Brookhaven National Laboratory and then at New York University. I first met FitzHugh in 1962 when she was working on the 3 meter Jarrell-Ash spectrograph in the basement of the James B. Duke house (fig. 18).

The major elements were analyzed "by classical gravimetric and electrolytic methods. Each 100 mg sample was dissolved in nitric acid; tin was separated as metastannic acid and determined as stannic oxide (fig.15) and from the filtrate copper and lead were separated and determined simultaneously by electrolysis (fig.16)" (Gettens 1969, 38). These

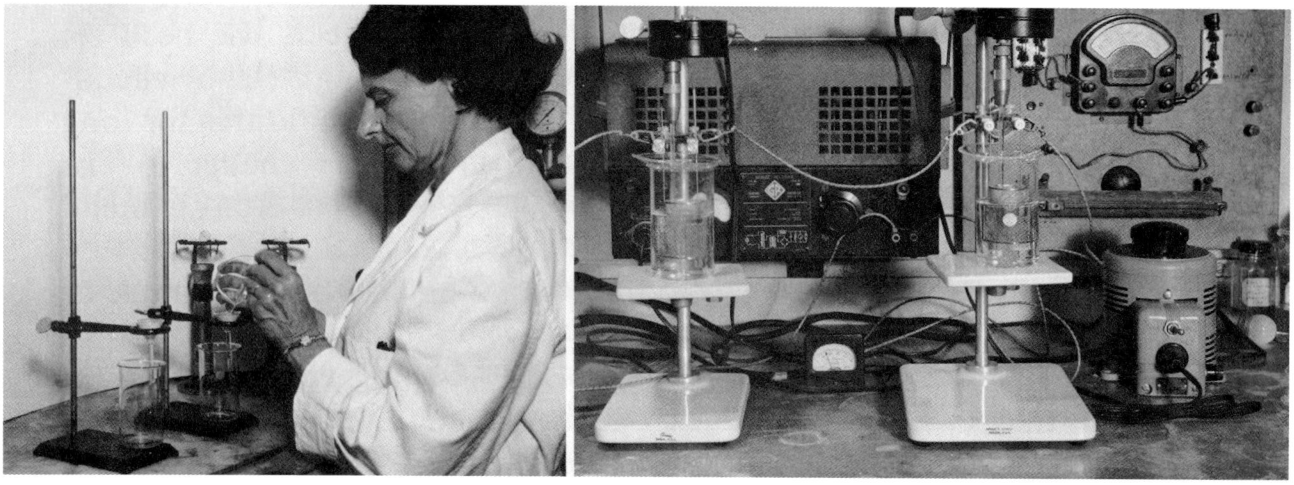

***Figure 19**. Ilona V. Bene doing chemical analysis, along with the equipment for electrolytic disposition of copper and lead. Figures 15 and 16 in BBII (Gettens 1969, 37).*

images are reproduced in this paper as figure 19. Ilona V. Bene is shown filtering the tin oxide in the image on the left, and the electrolytic setup is shown on the right. It was only when I was reading Francesca Bewer's new book on conservation at the Fogg (Bewer 2010) that I realized we know where this apparatus came from. Both the control board and the rectifier are clearly shown in Bewer's figure 4.11. The stirrers, stands, and platinum electrodes were bought after Gettens came to the Freer. When doing inventory, I saw the requisition for the platinum electrodes—which weren't cheap, even in 1956.

While the description of the technique is simple, the application isn't. To work on 100-mg samples you need to have an experienced analyst and a good balance. Analyses began with FitzHugh and continued with Bene. Eventually we had eighteen notebooks full of analyses—one page per bronze, one bronze per eight-hour day. The notebooks are kept in the lab archives. Page 1 of notebook 1 describes the quantification of the swings of the balance so that one could obtain accurate weights. Nowadays electronic microbalances simplify the process. I was amused to see on page 24 of notebook 1 that FitzHugh was doing the classical experiment of water of hydration of gypsum (an experiment done in high school chemistry classes), not because we didn't know the answer, but because if you get the right answer you know everything is working well with your balance and laboratory technique.

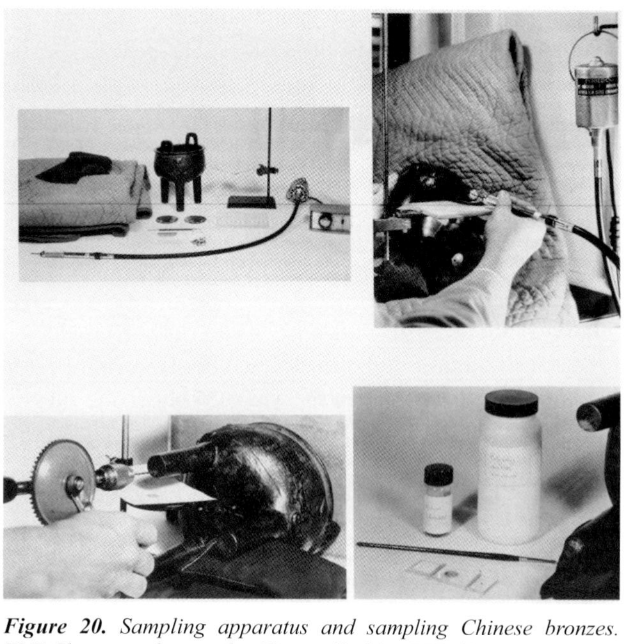

***Figure 20.** Sampling apparatus and sampling Chinese bronzes. W. T. Chase was the sampler. Figures 11–14 in BBII (Gettens 1969, 36).*

Sampling for bronze analysis was (and still remains) difficult. One of the first jobs I had when I arrived at the Freer as an intern was to take a photograph showing bronze sampling—we needed one for the book. I reproduce the images from the page about sampling as figure 20. We set up over in the window on the work bench, and those are my hands (without gloves!) sampling a *ding*. I think this is just about the first bronze that I ever drilled, and I remember sweating a lot by the end of it. I will point out that it's much easier to drill using a flexible shaft tool than a standard eggbeater drill. It's impossible to run an eggbeater drill without exerting a lot of sideways force on the drill bit as your hand moves around the crank.

Writing the Bronze Book

A lot of the bronze book had already been written when I arrived for my internship. This was particularly true of the fabrication chapter. A few years before, Gettens realized—when looking at a particular vessel, a *tsun* (F1954.122)—that what he thought were repairs were symmetrically placed between each of the four rising blades on the vessel (see Gettens 1969, 99, fig. 112). It would be hard to conceive of having repair spots in exactly the same place on each side. He came to realize that these were chaplets (or in Bob Bagley terms, spacers) used to hold the core and mold apart to maintain the casting space during casting. Chaplets are an interesting feature in most ancient Chinese bronzes.

We made an intensive study of chaplets where we could get samples. One of the items we sampled was the Kriger *p'an* (see Gettens 1969, 132, figs. 165 and 166) (fig. 21). The Kriger *p'an* sample shows a chilling effect where

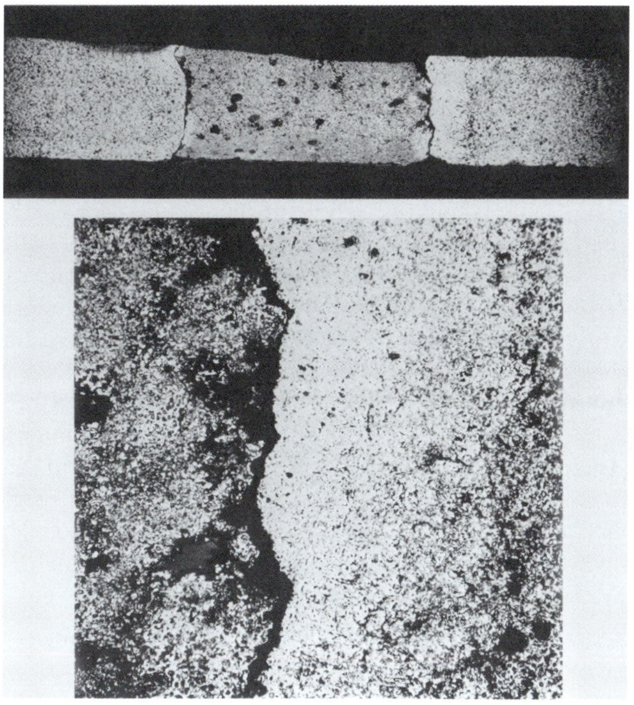

Figure 21*. A section through a chaplet in the Kriger* p'an *(FSC537). Figures 165 and 166 in BBII (Gettens 1969, 132).*

the molten metal of the bronze hit the chaplet. When we were making a similar photograph, we saw that the metal of the vessel was brighter next to the chaplet. We thought it was a photographic artifact, so we improved our lighting and shot it again and again. The bright effect wouldn't go away. Finally, we looked carefully at the metallography and realized that it wasn't a photographic artifact but a real difference in microstructure—and, in fact, a pretty important difference. It proves that at least this bronze was cast into a cold mold with a cold chaplet, and Kathy Ruhl's subsequent (and excellent) micrographs (done under Cyril Smith's direction) proved it beautifully.

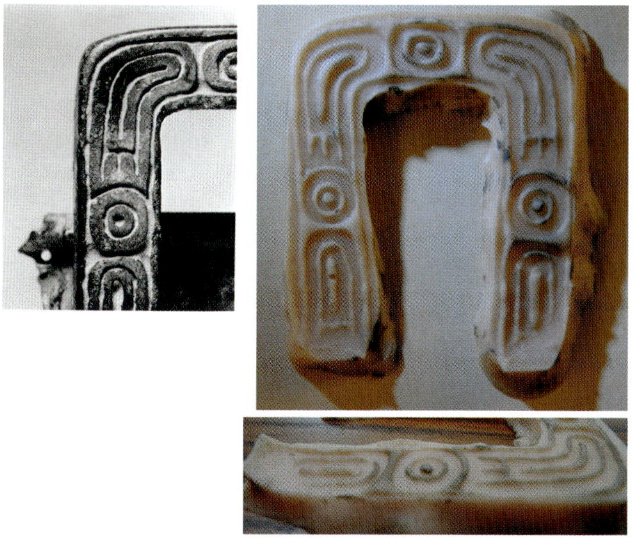

Figure 22*. A detail of the handle from the Kriger* p'an*, along with a silicone rubber impression showing that the decor appears to have been produced with balls and rolls of clay.*

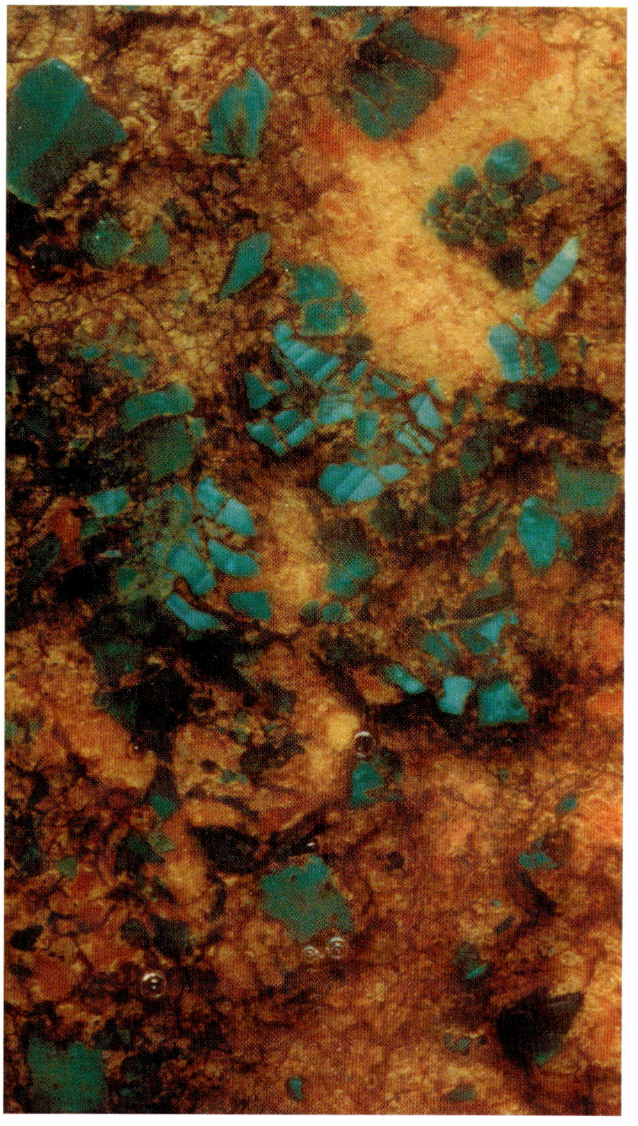

Figure 23*. Artificial patina made with crushed malachite. Note the bubbles in the immersion medium in the lower part of the photograph. Figure 290 in BBII (Gettens 1969, 215). Photo by W. T. Chase.*

The decor on the handle of the Kriger *p'an* was extremely undercut (Gettens 1969, 62, fig. 29). Since we were taking silicone rubber impressions of the inscription, I thought we should take a silicone rubber impression of the handle. This did not get published in the bronze book, but I am including it here in figure 22. It looks like another example of extruded lines used on the mold—*a la* Lukas Nickel's 2006 article in *Artibus Asiae*.

Most of our photomacrographic setups were done with the Bausch & Lomb model L camera with a ribbon illuminator (a great piece of gear) and a four-by-five camera back, starting with Polaroid film and then going to cut film, sometimes regular black-and-white, and sometimes full-size color transparencies. We would do anything to get a good photograph. The patina seen in figure 23 (Gettens 1969, 215, fig. 290), for instance, was difficult to photograph because the surface of the patina scattered light all over the place. You couldn't really see the banding in the malachite. Gettens was an excellent microscopist and I was an avid student, and we brainstormed how to get rid of those damn reflections. Finally I said, "What if we use

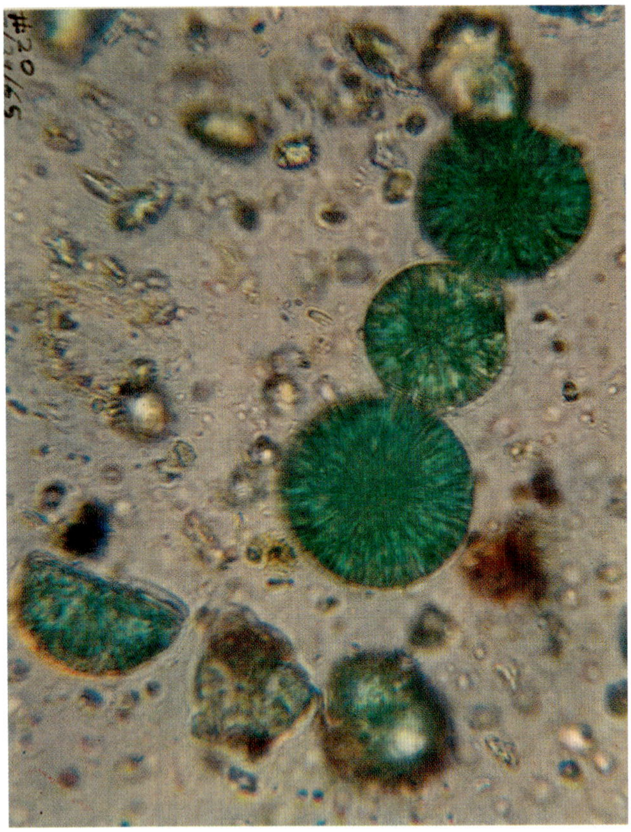

Figure 24. *Paris green in artificial patina showing the typical radiating needles. Figure 289a in BBII (Gettens 1969, 214). Photo by W. T. Chase.*

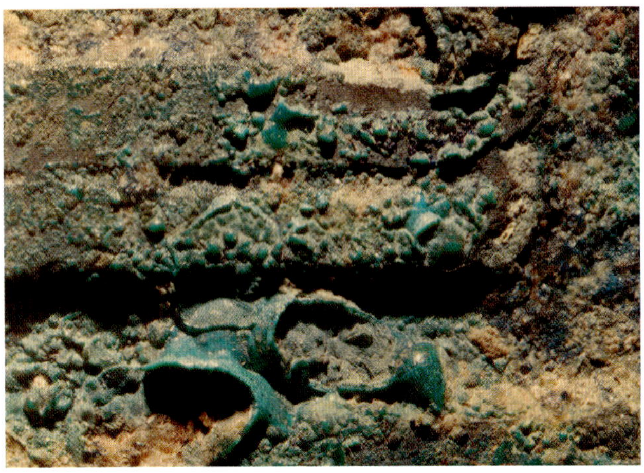

Figure 25. *Botryoidal malachite and azurite on a Chinese bronze. Figure 228 in BBII (Gettens 1969, 174). Photo by W. T. Chase.*

immersion oil and a cover glass?" The oil was inert enough that we could get it off the bronze later—the medium that this malachite is in is probably Oriental lacquer, which won't dissolve in anything short of aqua regia. We tried it, and it worked like a treat. If you look carefully toward the bottom of figure 23 you can see six bubbles in the immersion oil—of course we did not mention them in the caption!

We also took great care in manipulating aperture and field diaphragms to get the best picture possible—our photograph of Paris green is a good example (Gettens 1969, 214, fig. 289A) (fig. 24). We could probably find the old photographs taken with various settings of the aperture diaphragm as we attempted to bring out those nice needles in the Paris green. In the photograph, I see my handwriting in the upper left corner. (My favorite Gettens's quote is, "Sign and date everything.")

Sometimes we would take a day or more to get the right photograph. It could get a little frustrating: I'd take a shot, get it processed, and bring it into Gettens, and he would say, "It's not red enough." Back I'd go to change lights or filters or something and try again. Actually, this happened more with the micrographs for the pigment series, which we did subsequently. The bronze disease photographs of the Kelley bronze (Gettens 1969, 174, fig. 229) didn't yield as much cupric chloride dihydrate as we wanted, but on the first try it didn't yield any bronze disease at all—we found we had to be extremely careful of oil or any other films on the surface, which render the cuprous chloride insensitive to high relative humidity. And, of course, I can't not include my favorite corrosion macrograph of all time, with botryoidal malachite *and azurite* in the same field (Gettens 1969, 173, fig. 228) (fig. 25).

Making the Book

Table IA, Compositions, was a nightmare. We had to retype it again and again and again—sometimes because of errors found in proofreading, sometimes because we were adding bronzes, and once because we decided to change the way the confidence limits were used. Every secretary we had hated this table (and we sure hated to proofread it!).

Magnifications also turned out to be a nightmare. We stated all of the linear magnifications of the photographs as they were taken. This was fine, except that when the pages were designed Langford and, I believe, someone at Meriden Gravure changed the sizes of the photographs to fit the page design. This threw all of the magnifications in a cocked hat, and we had to remeasure all of the original photographs and the new versions and derive the correct linear magnification, which could then be put in the captions. Actually, at the same time this was happening, I went to the first McCrone symposium in Chicago, where the first scanning electron microscope (SEM) was being debuted. One of the things that the SEM does easily is put a magnification bar on the photograph—what a good idea! I see that it is required in the instructions to authors for the Forbes symposium, but I'm hoping that photographs from BBII are grandfathered in.

When we finally got to the point where we had a close-to-final version of the book, Gettens suggested that we should read it to each other so that we would catch any errors or infelicitous expressions. If you are trying to read a text out loud, any wording that is difficult to read stands out—sometimes you simply can't read it the way it's written. We took about two hours every day for a week or so and went into a quiet room (pottery storage was always a good candidate, although I remember reading some of the book in stone storage) and read it to each other, taking turns so we wouldn't get tired. Of course we each read proof after the book was set. Gettens was always a much better proofreader

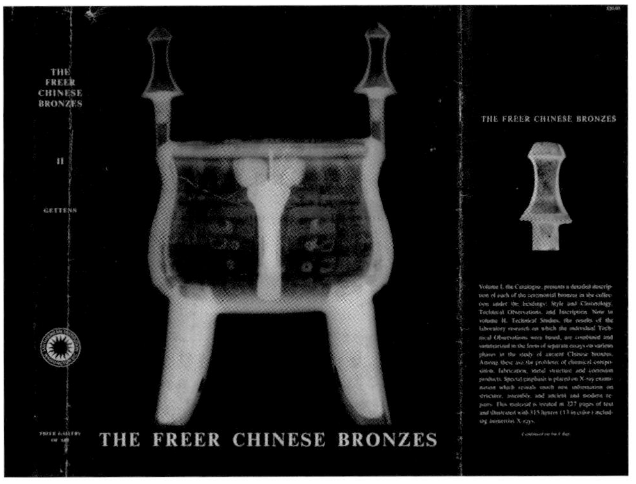

Figure 26. *Dust jacket for BBII (Gettens 1969).*

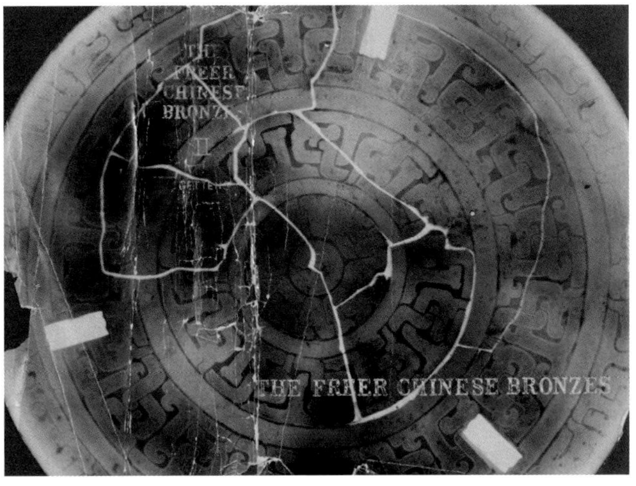

Figure 27. *Proposed dust jacket for BBII (Gettens 1969). Design by W. T. Chase and Frank Haentscke.*

than I ever was. I was one of the people reading proof on volume 1, and I took some fascicules of it with me to Sibley Hospital when I had a knee operation. The one error that I didn't catch was the lack of agreement of the catalogue number on the last page of the fascicule I had and the next fascicule, which was waiting back at the gallery. The error is preserved in the printed volume.

One of the final things we did was design a dust jacket for BBII. Gettens and Langford wanted to use a radiograph of the same bronze that we used in the frontispiece, and this was chosen as the final version (fig. 26). I tried to argue that an actual-size reproduction of the radiograph of the top of a big *ding* (with solder repairs) would be better, and I even had Frank Haentschke (our illustrator, who did beautiful work with line drawings and illustrations) letter it. My version—somewhat the worse for wear—is shown in figure 27. As you can see from the wear, I didn't give up easy, and this image is still on the outside of my copy of BBII.

Anyway, BBII was finally completed and sent off to Meriden Gravure, and eventually (I believe on a snowy day in 1969), the truck came in and we offloaded it at the back door of the Freer. I took some photographs (love that snow), but they seem to have been lost.

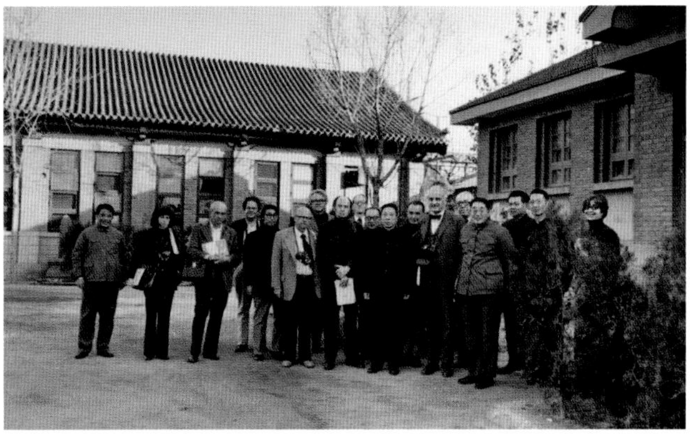

Figure 28. *The 1973 delegation on art and archaeology, sponsored by the Committee for Scholarly Communication with the People's Republic of China, standing in the yard of the Institute of Archaeology. Photo from W. T. Chase's collection.*

After BBII

There have been lots of developments, projects, collaborations, and publications since BBII—the wonderful Sackler bronze catalogue, lead isotope studies, micrography, studies on excavated Chinese bronzes, and more. We have gained a much better understanding of corrosion mechanisms than we had at the time of the bronze book.

When we wrote the bronze book, both Gettens and I understood that we were working from a selection of Chinese bronzes divorced from their archaeology. One of our major questions was how what we saw agreed with the genuine excavated bronzes in China. In 1973, I was fortunate enough to go on a trip sponsored by the Committee for Scholarly Communication with the People's Republic of China. Figure 28 shows our group at the Institute of Archaeology in Beijing. We had a great time, and we saw a lot. The Chinese seemed to be most interested in communicating about technical and scientific matters. But we saw lots of bronzes and excavations. I was glad to be able to come back and report to Gettens (in a Washington Conservation Guild meeting in February 1974, later printed as Chase 1974) that what I saw agreed with what we had published in the bronze book.

Barbara Keyser made a subsequent study of the Freer *jian* (F1939.5) and its Minneapolis counterpart, which we had shipped to the Freer so that they could be examined together (Keyser 1979). This was a fascinating study, and it made it even clearer that we would have to look at the evidence from the Houma foundry in detail. I thus entered into correspondence with our Chinese colleagues to make this possible. Jenny So helped set up the trip, and of course we invited Bob Bagley; Li Yung-ti also accompanied us. It was a very fruitful trip as witnessed by subsequent publications (Bagley 1996a, 1996b; Institute of Archaeology of Shanxi Province 1996). And things keep progressing in Chinese bronze studies (Bagley 1987; Rawson 1990; So 1995; Li Yung-ti 2003, 2007; Nickel 2006; Bagley 2009; Chen et al. 2011; etc.). It is still an interesting field to be involved in.

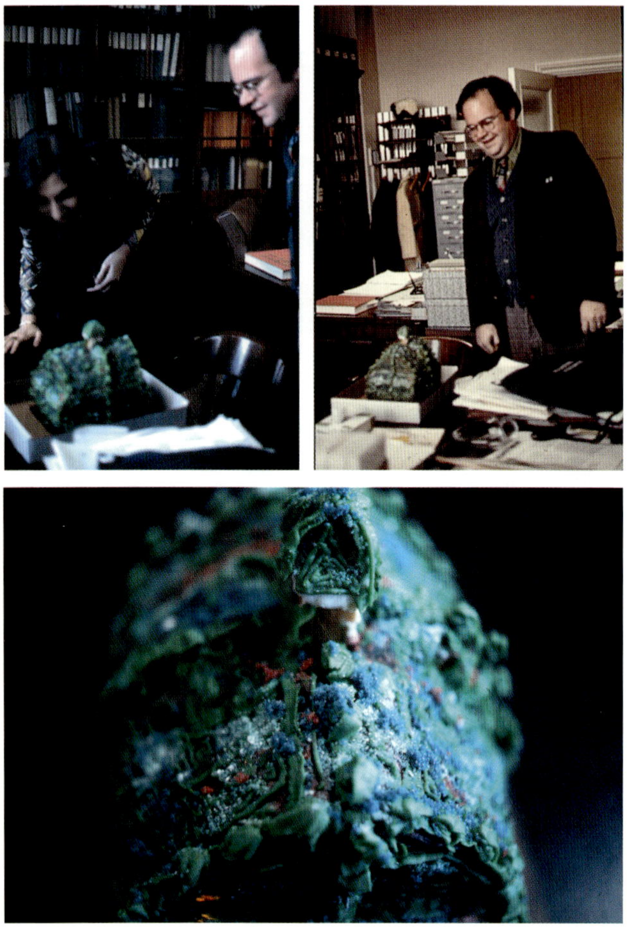

Figure 29. *Three photographs of a cake made in the shape of the Ling* fangyi *(F1930.54). W. T. Chase and Lynda Zycherman are pictured. Laboratory archives.*

Postscript

For even further amusement, I want you to know that for a party in the 1980s—I can't remember the occasion (although it might have been Linda Zycherman's farewell)—I produced a cake in the form of the Ling *fangyi* (fig. 29). I don't believe it was blue inside, but the top looked absolutely toxic—bronze disease at its most virulent. And with that I will close.

References

Bagley, R. W. 1987. *Shang ritual bronzes in the Arthur M. Sackler Collections*. Ancient Chinese Bronzes in the Arthur M. Sackler collections; v. 1. Washington, DC, and Cambridge, MA: The Arthur M. Sackler Foundation and the Arthur M. Sackler Museum.

———. 1996a. Debris from the Houma foundry. *Orientations* 27(9):50–58.

———. 1996b. What the bronzes from Hunyuan tell us about the foundry at Houma. *Orientations* 26(1):46–54.

———. 2009. Anyang mold-making and the decorated model. *Artibus Asiae* 69(1):39–90.

Bewer, F. G. 2010. *A laboratory for art: Harvard's Fogg Museum and the emergence of conservation in America, 1900–1950*. Cambridge, MA, and New Haven, CT: Harvard Art Museums and Yale University Press.

Chase, W. T. 1974. Conservation in the People's Republic of China. *Bulletin of the American Institute for Conservation of Historic and Artistic Works* 14(2):131–41.

Chen Jianming, Jay Xu, and Fu Juliang. 2011. *Along the Yangtze River: Regional culture of the Bronze Age from Hunan*. New York: China Institute.

Gettens, R. J. 1959a. Examining tables in use at the Freer Gallery of Art (laboratory notes). *Studies in Conservation* 4:23–27.

———. 1959b. The Freer Gallery Laboratory for Technical Studies in Oriental Art and Archaeology. *Studies in Conservation* 4:140–45.

———. 1969. *The Freer Chinese bronzes*, Vol. 2, *Technical studies*. Oriental Study Series 7. Washington, DC: Freer Gallery of Art, Smithsonian Institution.

Institute of Archaeology of Shanxi Province. 1996. *Art of the Houma foundry*. Princeton, NJ: Princeton University Press.

Keyser, B. W. 1979. Decor replication in two late Chou bronze *chien*. *Ars Orientalis* 11:127–62.

Li Yung-ti. 2003. The Anyang bronze foundries: Archaeological remains, casting technology, and production organization. PhD diss., Graduate School of Arts and Sciences—Anthropology, Harvard University.

———. 2007. Co-craft and multicraft: Section-mold casting and the organization of craft production at the Shang capital of Anyang. In *Craft production in complex societies: Multicraft and producer perspectives*, ed. I. Shimada, 184–223. Salt Lake City: University of Utah Press.

Lodge, J. E., A. Wenley, and J. A. Pope. 1946. *A descriptive and illustrated catalogue of Chinese bronzes acquired during the administration of John Ellerton Lodge*. Smithsonian Institution Publication 3805. Oriental Study Series 3. Washington, DC: Smithsonian Institution.

Nickel, L. 2006. Imperfect symmetry: Re-thinking bronze casting technology in ancient China. *Artibus Asiae* 66(1):5–39.

Pope, J. A., R. J. Gettens, J. Cahill, and N. Barnard. 1967. *The Freer Chinese bronzes*, Vol. 1, *Catalogue*. Oriental Study Series 7. Washington, DC: Freer Gallery of Art, Smithsonian Institution.

Rawson, J. 1990. *Western Zhou ritual bronzes from the Arthur M. Sackler collections*. Ancient Chinese bronzes in the Arthur M. Sackler collections; v. 2 a-b. Washington, DC and Cambridge, MA: The Arthur M. Sackler Foundation and the Arthur M. Sackler Museum.

So, J. 1995. *Eastern Zhou ritual bronzes from the Arthur M. Sackler collections*. Ancient Chinese bronzes in the Arthur M. Sackler collections; v. 3. Washington, DC and Cambridge, MA: The Arthur M. Sackler Foundation and the Arthur M. Sackler Museum.

Theories and Observations: Still Looking at the Freer Bronzes

Robert Bagley

ABSTRACT In the 1950s and 1960s R. J. Gettens carried out a comprehensive technical examination of the ancient Chinese ritual bronzes in the Freer collection. In the present paper I revisit several important Shang and Zhou castings about whose fabrication Gettens had interesting things to say. My concern is both the specific points he raised and, more generally, the relevance of technical studies to art history and of art history to the history of technology.

As an art historian I am something of an intruder at a symposium for scientists, but I am grateful to have been invited because I believe that even art historians should care about how things are made. And I am doubly grateful to be here because the person who led me to that conviction was John Gettens. I first read his volume on technical studies of the Freer bronzes when I was a graduate student. His long chapter on fabrication opened my eyes to the fact that my artists were actually making things, and perhaps even thinking about what they were doing, and I decided that an art historian would be unwise to ignore what his artists were doing and thinking. What I learned from Gettens has been a central

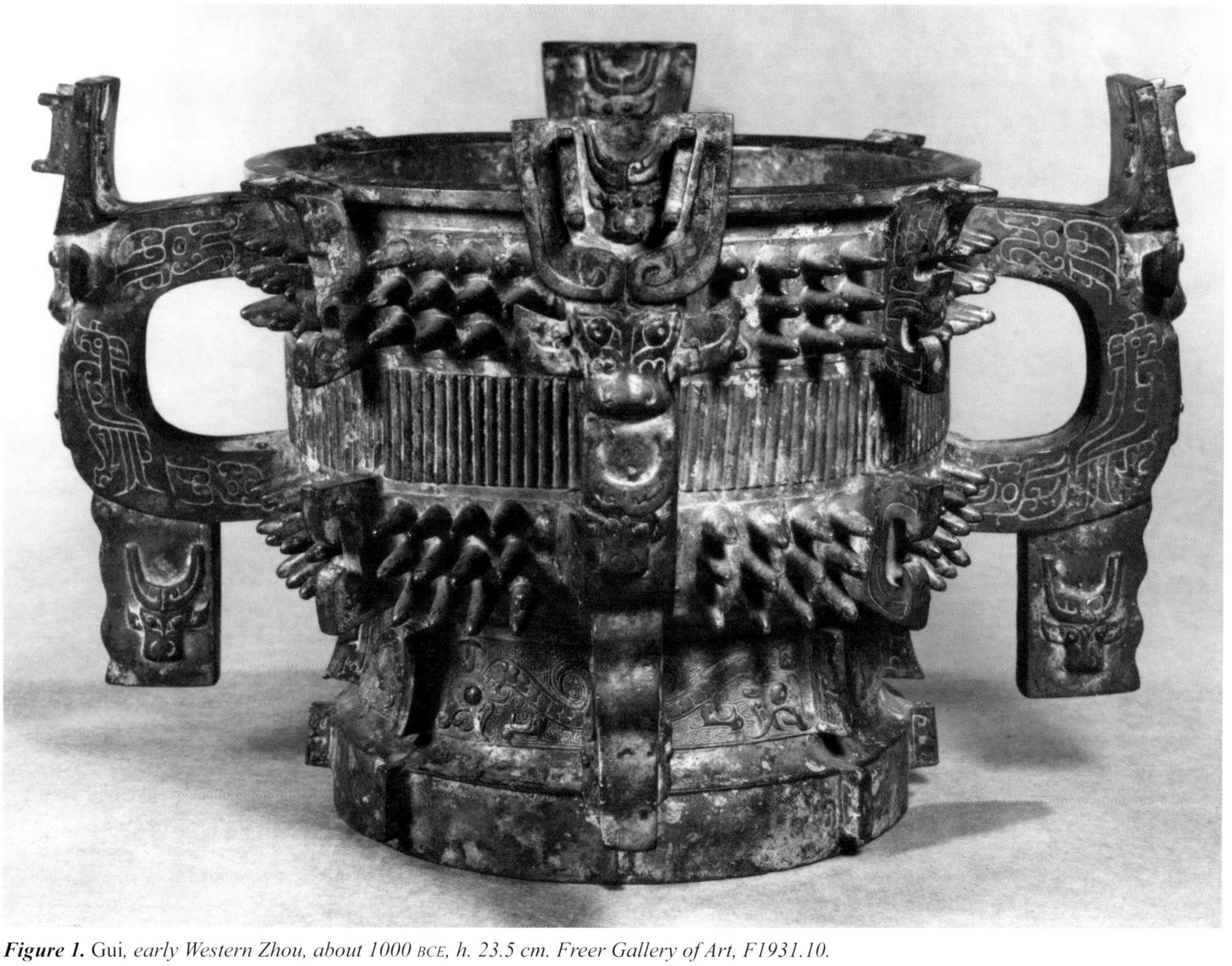

Figure 1. Gui, *early Western Zhou, about 1000* BCE, *h. 23.5 cm. Freer Gallery of Art, F1931.10.*

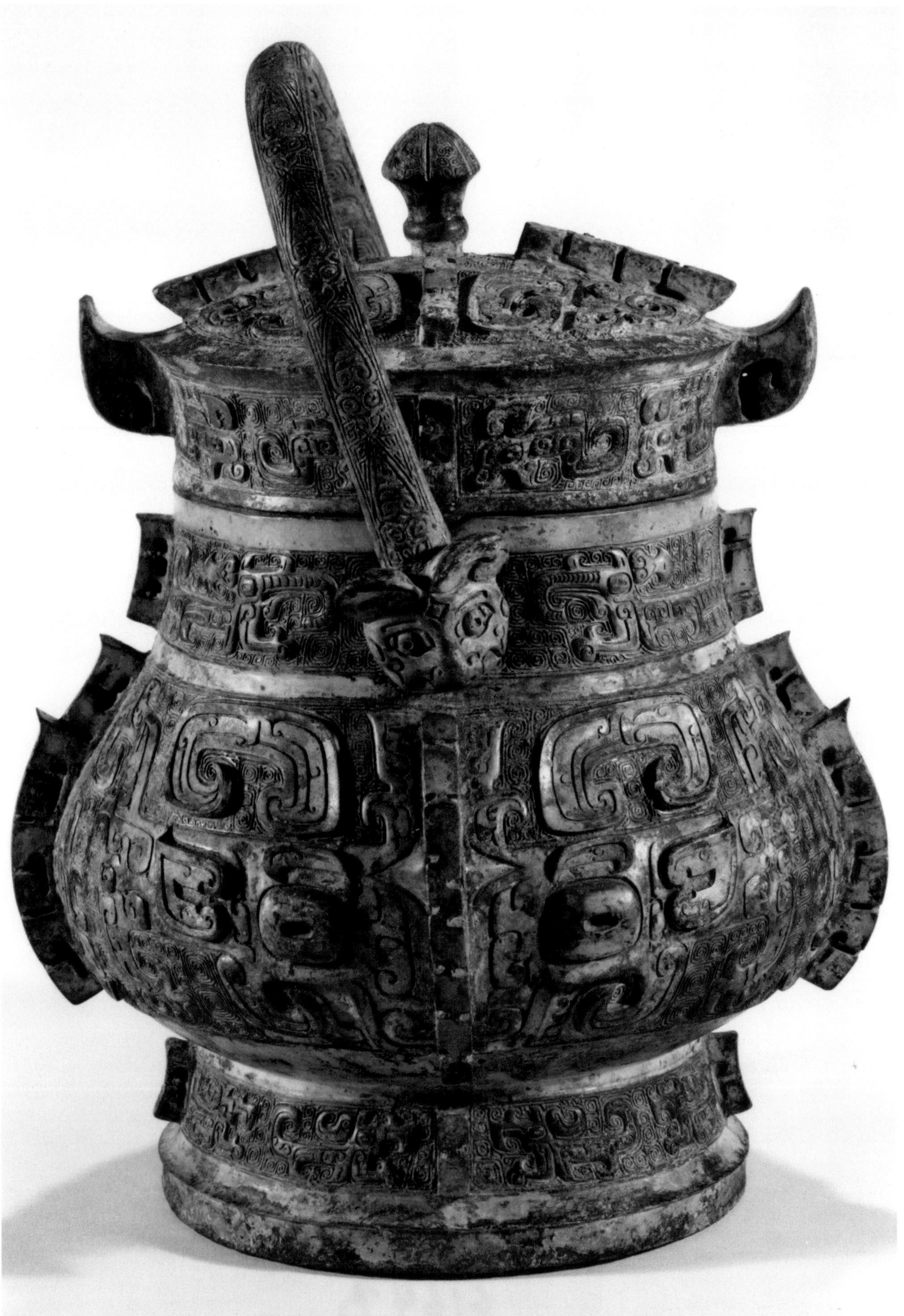

Figure 2. You, *Anyang period, twelfth or eleventh century* BCE, *h. 36.5 cm. Freer Gallery of Art, F1940.11.*

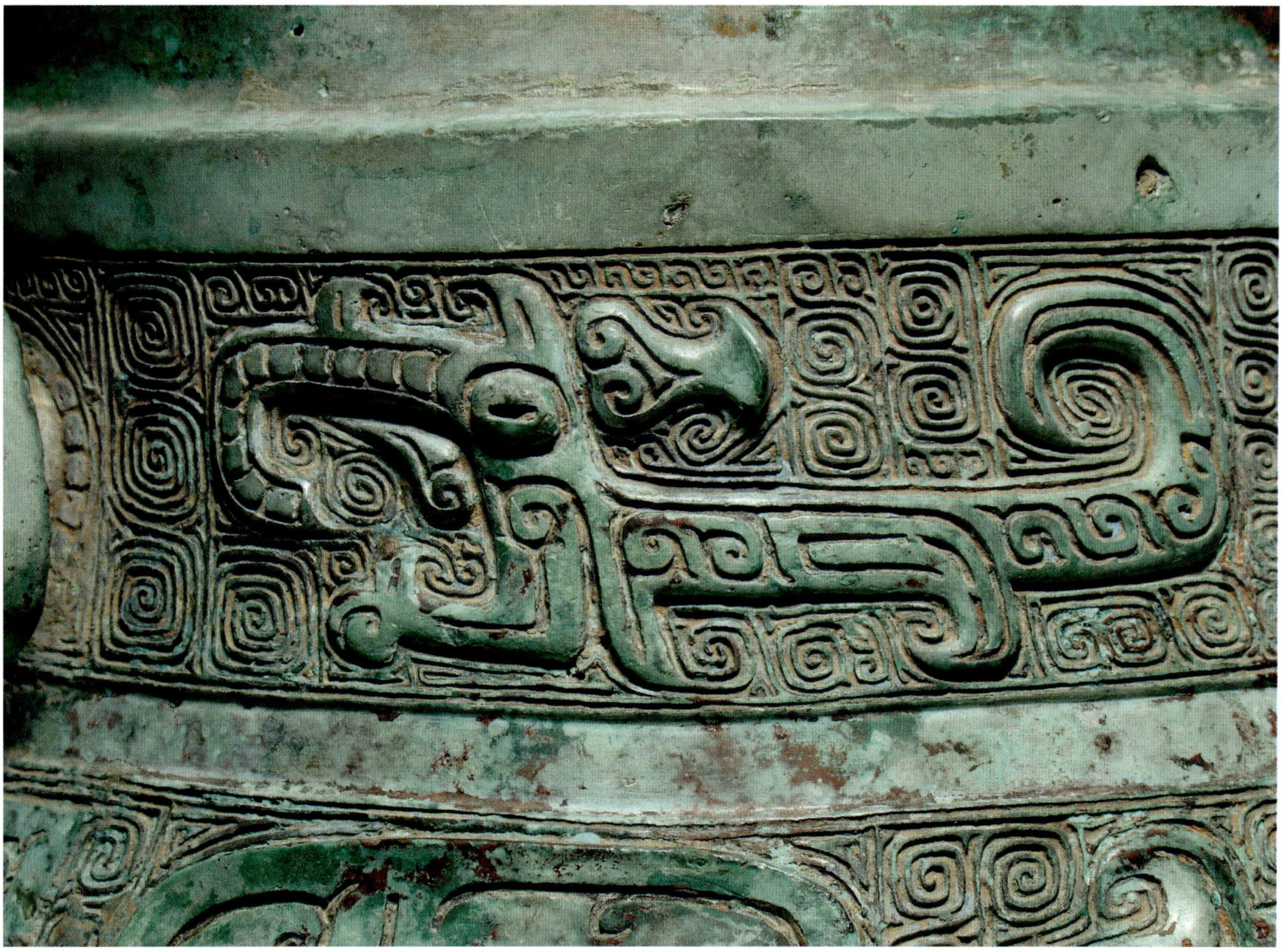

Figure 3. You *(same vessel as fig. 2), detail of neck register. Photograph by Kyle Steinke.*

part of my approach to art history ever since, so my debt to him is very great. I hope that my admiration for his work will be only the more obvious if I begin not by heaping him with praise he does not need but by analyzing one of his mistakes. When an observer as careful as Gettens makes a mistake, the reasons are worth thinking about.

In several of the Freer's most outstanding early bronzes Gettens saw evidence of mechanical replication of decor. I want to look at three of them, all from the last century or so of the second millennium BCE. Figure 1 shows the latest of the three, a rather ferocious *gui* dating from the very beginning of the Zhou period, in round figures 1000 BCE. Of this vessel Gettens writes: "The sunken decor on the sides of all four handles is identical possibly indicating the use of a stamp or stencil in their execution" (Pope et al. 1967, 373). He is saying that the little dragons and birds done in sunken line are replicated: the creatures on the left sides of all four handles match identically, and so do the ones on the right sides. He is hypothesizing a set of little stamps or stencils, maybe three stamps for one side and three for the other side.

The vessel shown in figure 2 is a *you*, probably an Anyang casting made about a century earlier than the *gui*: in round figures, 1100 BCE. Here Gettens believes that all of the high relief decoration was stamped but that the spiral background patterns, the so-called *leiwen* spirals, were executed by hand. He writes: "The principal elements in the high relief decor are identical, but the *lei-wen* which surround them are not thus implying that the high relief was stamped on the model or in the mold with dies, but the *lei-wen* were cut by hand in the mold sections" (Pope et al. 1967, 281). In other words, the *taotie* visible in figure 2, the staring animal in the main register, is identical to the *taotie* on the other side of the vessel and therefore must have been stamped, but the spirals that surround it do not match the spirals on the other side and therefore must have been carved by hand. Similarly the dragon visible in figure 3, just above the main register, is identical to the one diametrically opposite and must have been stamped, but the spirals surrounding it do not match the spirals surrounding the dragon on the other side. Gettens is telling us that this decoration was executed by a combination of techniques. The unit shown in figure 3, he says, is partly stamped and partly carved. If he is right, the combination is certainly seamless. The high relief misses no opportunity to spiral into the *leiwen* ground. The way these dragons relate to their boiling, seething surroundings makes this for me one of the most wonderful of the Freer bronzes.

My third vessel is another *you* (fig. 4). It is a little later than the previous one but earlier than the *gui* I began with; we might say Shang-Zhou borderline, eleventh century BCE. Gettens believed that on the previous vessel the high relief was stamped, the *leiwen* carved, but here he believes that everything was stamped: "On the vessel proper the main decor on diagonally opposed fields, and the *lei-wen* ground as well,

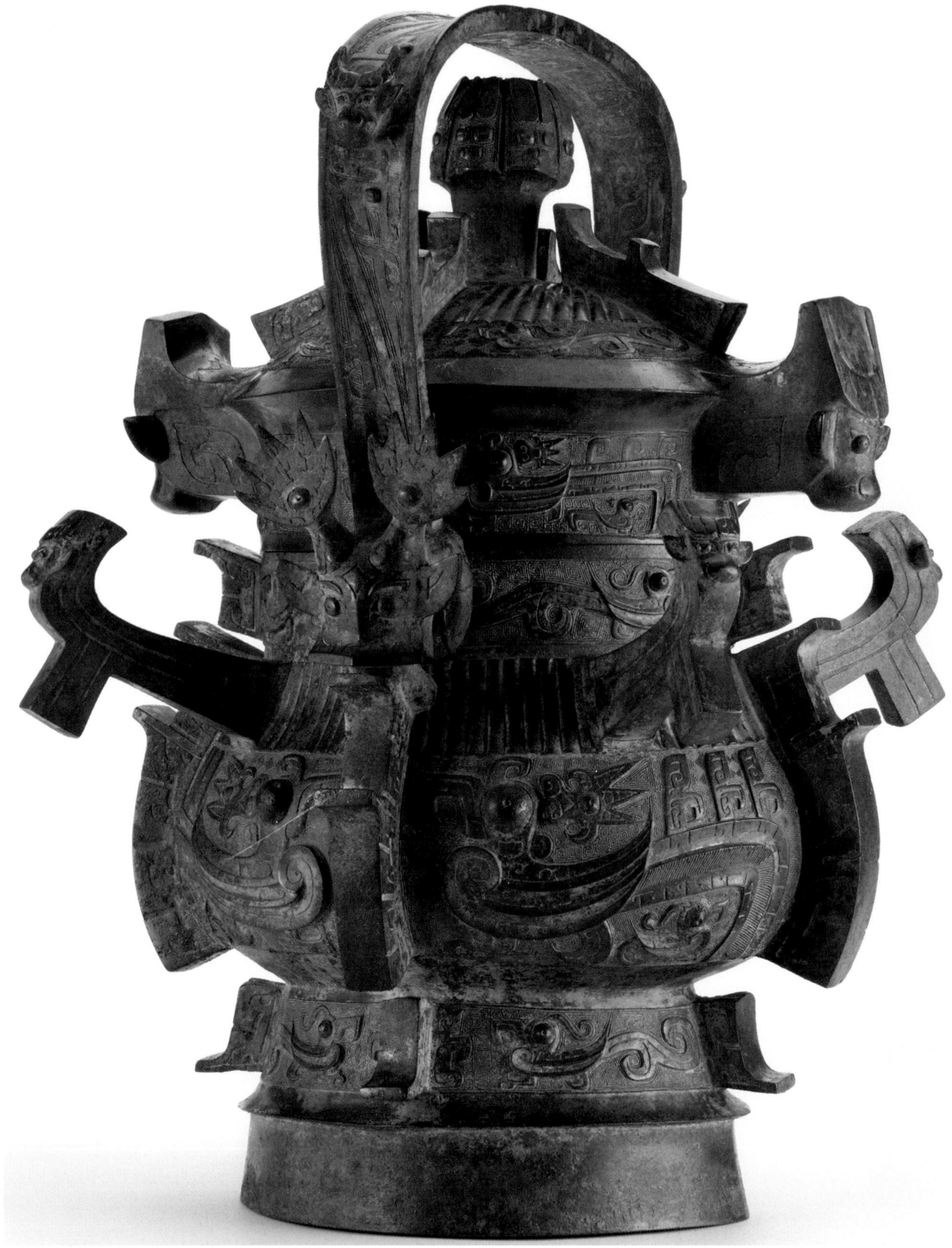

Figure 4. You, *late Anyang period or early Western Zhou, eleventh century* BCE, *h. 51.1 cm. Freer Gallery of Art, F1930.26.*

are identical suggesting that dies were impressed on the mold sections in the orthodox way" (Pope et al. 1967, 289).

Notice that phrase "the orthodox way." In his description of the previous vessel Gettens left it open where the stamping was done: he said "the high relief was stamped in the mold or on the model." But here he seems to be thinking that the caster's normal procedure was stamping in the mold. He believes that stamping was the normal way of executing high relief, and he believes that it is the way the *leiwen* spirals in figure 4 were executed too.[1]

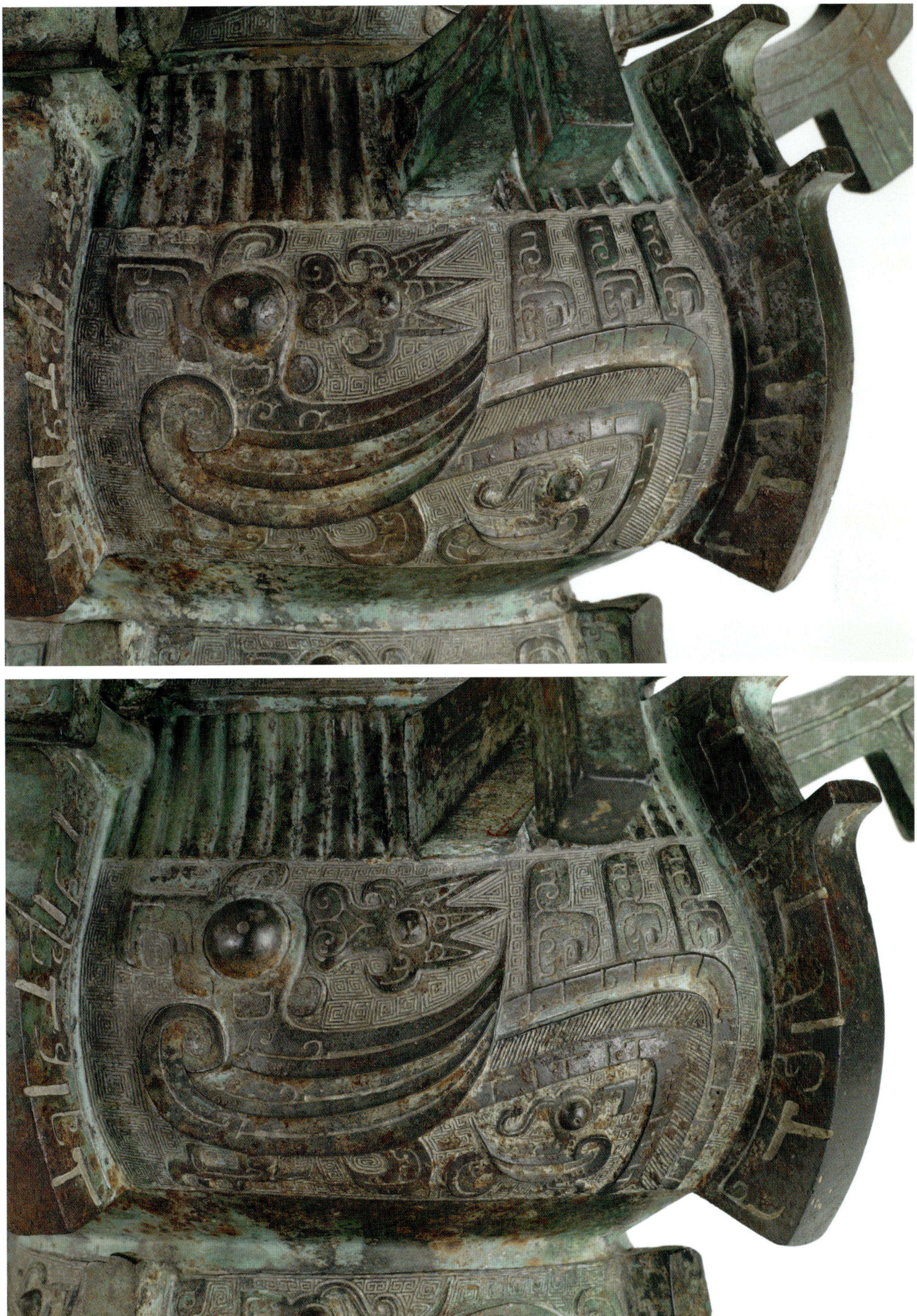

Figure 5. You *(same vessel as fig. 4), two views, showing diagonally opposed quadrants of the main register.*

Now, all of the observations I have quoted are incorrect. All of the elements that Gettens describes as identical are not identical. Everything that he says was stamped was executed by hand. If you look at the three bronzes in the Freer storeroom you will convince yourself of this in three minutes.

Let me show you what you would see in the storeroom if you looked closely at the *you* of figure 4. Gettens says that in the main register, diagonally opposed quadrants are identical: in other words, not only the big birds but also the tiny spirals that surround them match exactly. So, he concludes, there was a die for the right-hand quadrant in figure 4, and it was stamped into the mold section for that quadrant and also into the mold section for the corresponding quadrant on the other side. Since we cannot look at both sides of the vessel simultaneously, comparing the two quadrants is not very easy, but the Freer storeroom is equipped to deal with the difficulty because its wonderful tables have lazy susans built into them. The difficulty can also be overcome with photographs. Photographs of diagonally opposed quadrants are reproduced in figure 5, and a moment's inspection will reveal that neither the birds nor the spirals match. Nothing in these two quadrants is identical from one side of the vessel to the other.[2]

For the *gui* of figure 1 we do not need a lazy susan. To see that the designs on the handles were carved freehand we need only compare corresponding sides of any two. As figure 6 shows, only a glance is required to see that they are different. There is no evidence on any of my three vessels for any kind of mechanical replication, any impressing of stamps or dies. And that confronts us with a puzzling question: Why did Gettens believe there was? Why did he see things as identical that plainly are not?

To answer that question we must try to think ourselves back to the time when Gettens was working. Half a century ago very little was known about the fabrication of Chinese bronzes, and much of what passed for knowledge was fiction. The one solid piece of work before Gettens was a paper published by Orvar Karlbeck (1935). Karlbeck was the first observer to suggest that the bronzes were cast in piece molds, and he gave decisive evidence for his suggestion in the form of a meticulous study of mold fragments from the Anyang site. His demonstration opened up all sorts of possibilities. If the casters had access to the mold surfaces, they did not have to do their decorating entirely on the preliminary model; if they wished, they could do part of it, or even all of it, directly in the mold pieces. Gettens's ideas about the execution of the bronze decoration thus depended on Karlbeck's proof that the casters had access to the mold surface.[3] In 1958 he began a systematic, vessel-by-vessel examination of the Freer bronzes, in the course of which he more or less created the field of technical studies of Chinese bronzes. His aim was a thorough examination of each object, but he had specific questions in mind that gave him things to look for, ideas to test against the objects, and one of the things he looked for was mechanical rep-

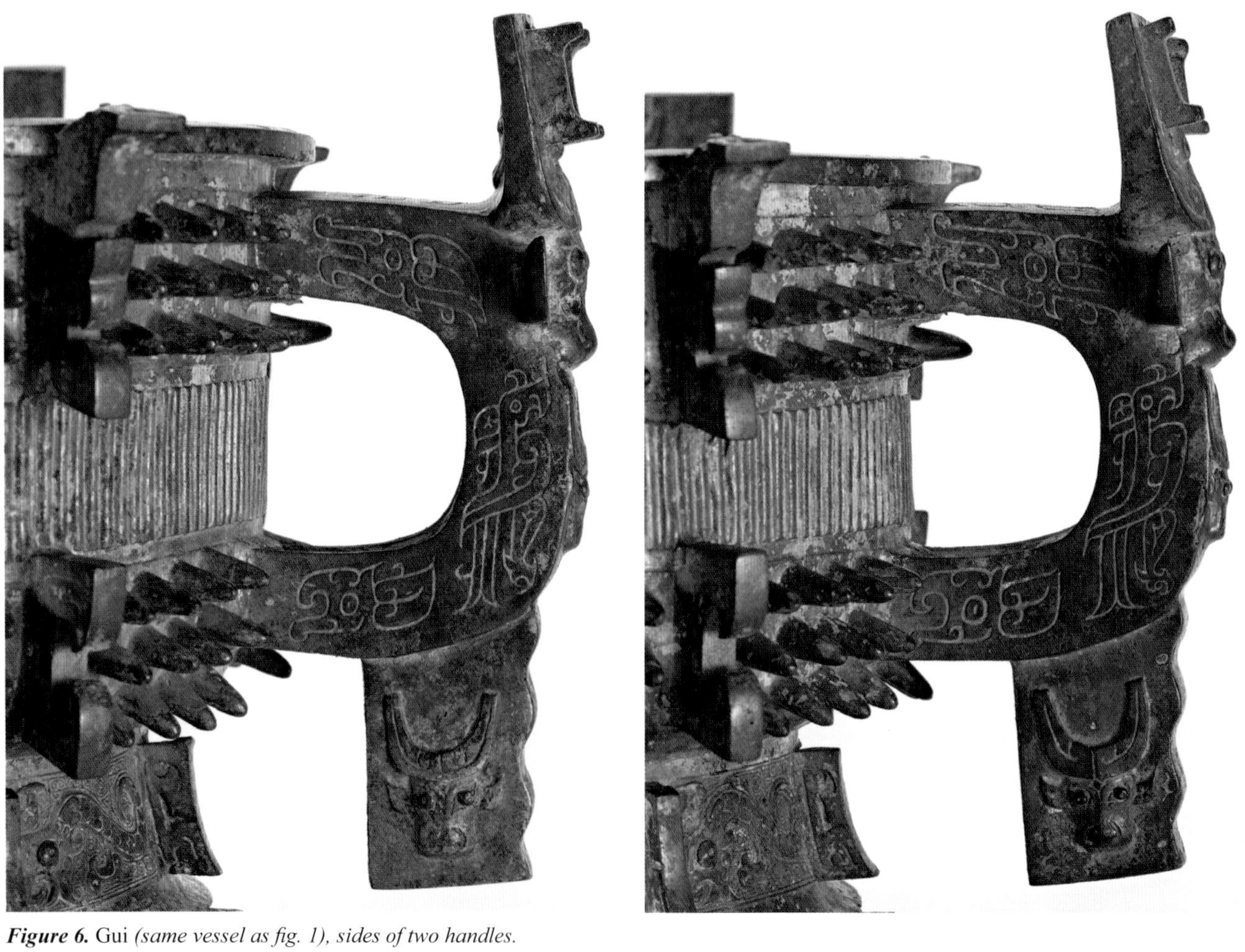

Figure 6. Gui *(same vessel as fig. 1), sides of two handles.*

lication. I can think of two or three reasons for him to be looking for that.

The first reason was archaeological. In the late 1950s Chinese archaeologists excavated the site of a fifth-century BCE foundry at a place called Houma, in Shanxi province, and between 1959 and 1962 they published several brief preliminary reports.[4] There was not much to be gleaned from the reports, but they did make it clear that the Houma foundry was using some sort of replication technique, for the foundry debris included baked pieces of decorated clay that were obviously positives for replicating single units of decoration, not broken fragments of decorated models. The Houma excavators were unable to figure out how these positives were used to make molds; that was done later, at the Freer Gallery, by Barbara Keyser. But the fifth-century Houma foundry debris gave Gettens every reason to be thinking about mechanical replication, and Houma replication techniques were not well enough understood for him to realize that they could not have produced the eleventh-century decoration I have been discussing.[5]

A second reason for Gettens to be thinking about replication was common sense—twentieth century common sense. Bilateral symmetry and repeating pattern units are everywhere in the early bronze decoration. When a modern observer looks at an Anyang bronze, he sees an object covered with opportunities for labor-saving shortcuts. In figure 5, why would anyone carve that decoration twice, when he could carve it once and stamp it forever? It is next to impossible for us to think ourselves back to a time when there were no minimum-wage laws. The shortcut we see here was surely just as obvious to the makers of the vessel as it is to us, and Gettens took it for granted that they would see it and take it.

Our instinct that stamping is the only sane way to make the decoration in figure 5 is probably reinforced by a belief that only machines can achieve this kind of precision; we take it for granted that craftsmen doing things freehand could never repeat themselves so exactly. But of course it is only since the Industrial Revolution that craftsmen have needed to be sloppy to convince their customers that their products are handmade. Before the Industrial Revolution, their patrons did not admire sloppiness. The finest craftsmen could achieve just about any desired level of precision, but what they achieved is sometimes invisible to us because we do not believe it possible.

Now so far I have been asking why Gettens did not see that the two quadrants in figure 5 are different. The inverse of that question is also important: Why did I see that they are? The reason is certainly not that I am a better observer than he was. That is very definitely not the case. In fact I suspect that I would never have looked closely at this decoration at all if I had not read Gettens's description of it—the sentences I quoted above—in the Freer catalogue. Gettens's description sent me straight to the bronze for the simple reason that it contradicted my own ideas. I believed that all the decoration on early bronzes was carved by hand, and that no stamps or dies were used. I cannot claim that my belief was well founded; I cannot even remember where I got it. But it gave me a motive for checking Gettens's description against the actual bronze.

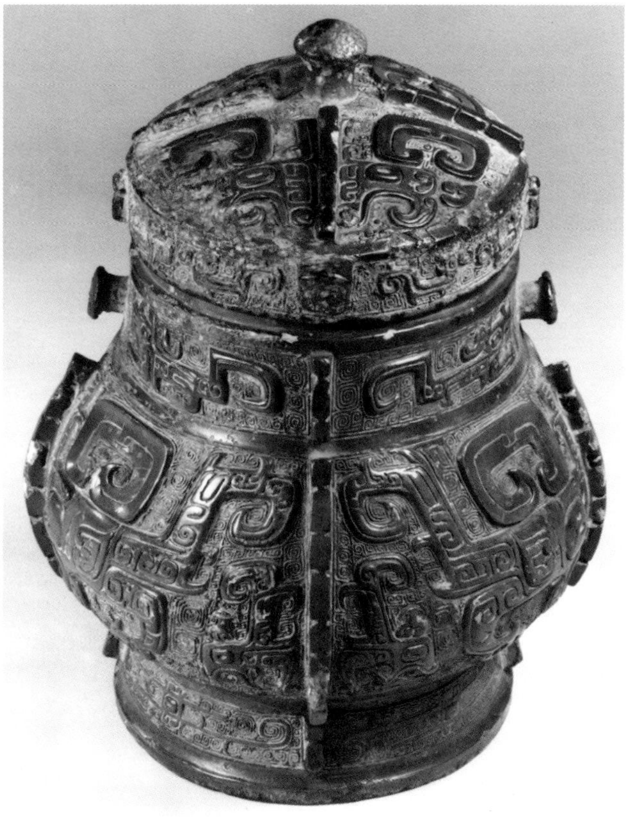

Figure 7. You, *side view (the* taotie *in the main register are facing the left and right edges of the photograph), Anyang period, twelfth century* BCE, *h. 22 cm. Arthur M. Sackler Gallery, S1987.968. Source: Bagley (1987, 373).*

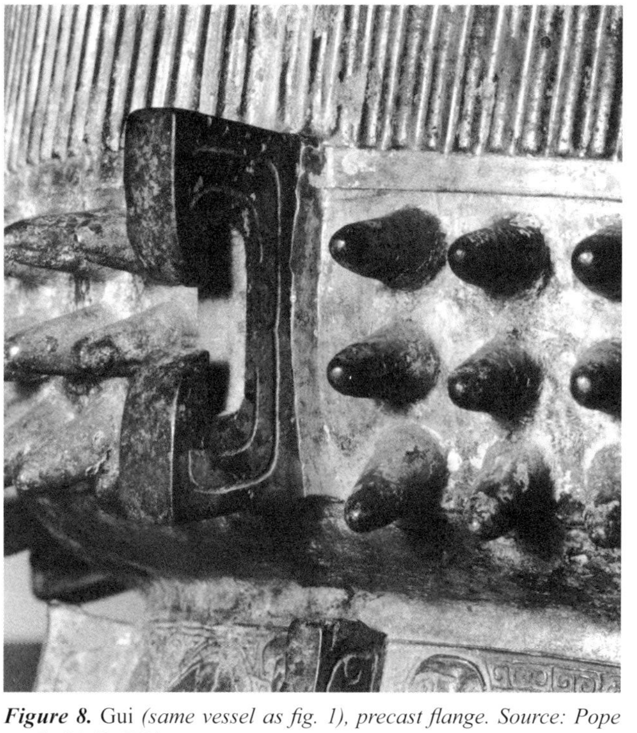

Figure 8. Gui *(same vessel as fig. 1), precast flange. Source: Pope et al. (1967, 372).*

The point here is the role of expectation in observation. When you see what you expect to see—when you see what your theory predicts—how much longer do you go on looking? How hard do you work to prove yourself wrong? When we see what we expect, most of us stop looking, and that is

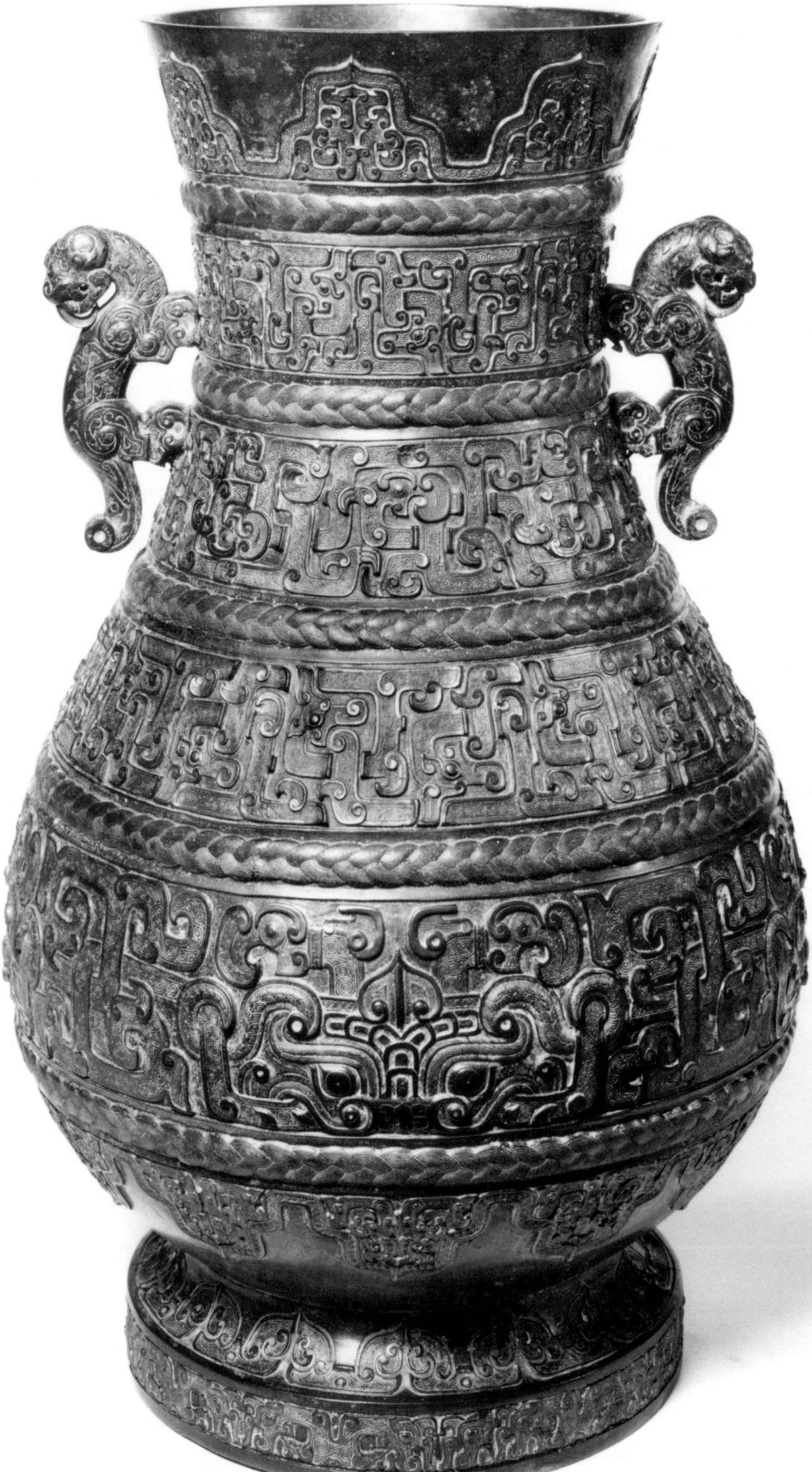

Figure 9. Hu, *fifth century* BCE, *h. 44.8 cm. Freer Gallery of Art, F1957.22.*

what Gettens did in this case. The fact that I had no good reason for believing what I believed did not really matter; the only thing that mattered about my theory of the manufacturing process was that it differed from his. As a result, an observation that was difficult for him was easy for me.

Would Gettens have been better off if he had approached the bronzes without preconceptions? Is the key to making accurate observations an empty mind? I do not think so. Without preconceptions, what would we look for? Without a theory, how would we decide what observations are worth making? An observer who brought nothing to the act of observation would see nothing. There is a famous remark to this effect in one of Darwin's letters. He writes: "About 30 years ago there was much talk that geologists ought to observe and not to theorize; and I well remember someone saying that at this rate a man might as well go into a gravel pit and count the pebbles and describe the colors."[6]

A theory does not make you infallible—alas, nothing does that—but it gives you something to look for, something to verify or refute. As I said a moment ago, Gettens was examining the Freer bronzes with a number of specific ideas in mind. Mechanical replication was one. Let me mention two more. These ideas are all, in essence, theories about the fabrication process that he set out to prove or disprove.

"Unitary Casting"

In the 1960s it was sometimes said that what is exceptional about Chinese bronzes is that they are one-piece castings—objects cast in one pour of metal. This is a theory that Gettens disproved by looking for, and finding, joins. He discovered for example that the beams projecting from the *you* in figure 4 were secured over tenons with hard solder.[7] In 1967 he published a long paper entitled "Joining Methods in the Fabrication of Ancient Chinese Ceremonial Vessels" (Gettens 1967). When I first read it I said to myself, "Why would anybody want to write a paper about joins? This is like writing a paper about Gothic mortar instead of Gothic cathedrals."

But of course the history of Gothic architecture would be rather different without mortar. No designer could have imagined a *you* with beams projecting from it if he had not already been a virtuoso in the making of joins; conversely, it was a taste for spectacular artistic gestures that tempted him to experiment with a new method. Anyang casting techniques were not developed on contract from the Shang royal air force and then borrowed by artists; they were developed by artists. The people doing cutting-edge technology in this period were artists.[8]

The joins on the *you* of figure 2 are trickier. The swing handle is free to move: its animal-head terminals are cup-shaped, and they fit over tenons that project from the vessel. (A *you* of the same type in the Sackler Gallery has lost its handle, exposing the tenons [fig. 7].) But the handle terminals cannot be removed from the tenons. The handle and vessel are permanently joined, and the join was made by casting. The vessel was cast with two bolt-shaped tenons projecting from its sides. Then the caster formed a wad of clay around each of the bolts. The clay wad served as the core for the handle terminal. The mold for the handle was assembled around the two cores, and the handle was cast in place, over the tenons. After casting, the clay inside the terminals could be pulverized by wiggling the handle.[9]

As for the *gui* of figure 1, Gettens showed that it was cast in thirteen pieces. Before Gettens, it had been claimed that Chinese casters never made joins; it had even been claimed that they did not know how to make joins. Gettens showed that they made many joins, they just did not make them the way a Western metalworker would. The Western metalworker joins two pieces of cold metal. That is how the maker of the *you* of figure 4 attached its projecting beams: he joined the already-cast beams to the already-cast vessel using solder. But the *you* is an isolated experiment, almost unique. Normally Chinese casters did not make joins between two pieces of cold metal, they made the join between two pieces of metal in the process of casting one of them. (The Freer *you*'s beams could perfectly well have been dealt with in this way.) Soldering and riveting are ways of joining pieces of cold metal; the early Chinese caster had no need of them. The caster of the *gui* cast four handles and eight flanges and embedded them in the mold for the rest of the vessel. When the thirteenth piece was cast, the twelve precast pieces were locked in place. Figure 8, which reproduces Gettens's illustration of one of the flanges, shows an overflow of metal from the vessel onto the flange: the flange was cast first.

All three of the vessels I have been discussing were designed by artists who thought in terms of their technology, artists who were, among other things, virtuoso makers of joins. If we do not understand their technology, we will be in no position to understand their thinking. When I first read Gettens's paper on joining techniques I was bored; now I think I might rank it as his single most important contribution to bronze studies.

Decalcomania

Another idea that was much discussed in the 1950s and 1960s was piece-mold or section-mold casting, the idea that Chinese bronzes were cast in molds made of fitted sections. As I said earlier, Orvar Karlbeck gave convincing arguments for this in 1935. Some of the Anyang mold fragments he examined had smooth edges provided with mortises and tenons, and some were scorched or even had traces of bronze caught in their sunken lines, proving that they had been used for casting bronze.[10] Since Gettens was interested above all in how the bronzes were made, he naturally went over the Freer bronzes keeping an eye out for mold marks, the raised lines formed wherever molten bronze leaked into the crevices between mold pieces. Again, a theory about fabrication gave him something specific to look for, and the mold marks he found told him a good deal about mold-building.

But on a few bronzes he found something he was not expecting. On the *hu* shown in figure 9, a fifth-century Houma casting, he found scattered lines that looked a little like mold marks but that did not run continuously from top to bottom of the vessel; they never ran through more than one or two registers. Figure 10 reproduces his illustration. The lines his three arrows point to do not continue into the register below; and anyway, why would the caster have divided the mold at two points an inch apart? As mold marks, these lines made no sense at all. Gettens seems to have had a pretty good idea of what they mean. In the Freer bronze catalogue he proposed a one-sentence explanation: "It is possible that clay master patterns in positive relief were employed in duplicating the repeated elements of the decor and that clay negative impressions from these were incorporated directly into the molds which were constructed around a plain core."[11] That is a promising first step toward the right answer, but the procedure was not elucidated in detail until shortly after his death by Barbara Keyser. Her paper, a worthy tribute, was published in the Gettens memorial volume of *Ars Orientalis* (Keyser 1979).

Now the lines in figure 10 look obvious in a photograph taken to show them—and with arrows inserted to help us find them—but on the vessel itself they are anything but obvious. In fact Gettens missed one: above his left arrow, there should be a fourth arrow. I think it is quite possible that he was the first person to notice these lines since the day the bronze left the foundry, and I suspect that he noticed them only because he was looking for something else, namely, mold marks. Here again, he made important observations because he had a theory about fabrication.

But what makes the observations important? Why do any of these minute details of fabrication matter? I believe that they matter a great deal, to the historian of technology and the art historian alike, because both of us want to get into the mind of the ancient maker. We want to understand the thinking—the inventiveness—of the person who made the object we are looking at. We want to fit him into a history of ancient artisans who century after century went on inventing new objects and new techniques.

A few of my fellow art historians would argue that this is impossible. It is trendy nowadays to declare that we have no access to the minds of people who died three thousand years ago. But if you think about it, pretty much every statement of any interest that we can make about these objects is a claim to know what was going on in the maker's mind. Every time we say "He did this *because*...," we are claiming to know what he was thinking. We can of course camouflage the claim by using the passive voice. But if we honestly try to eliminate such claims from what we write, I think we will quickly find ourselves counting pebbles and describing colors. The art historians who say that we have no access to the minds of ancient artists may talk themselves out of a job; at least, I hope so.

Seventy years ago Henri Focillon said very clearly what technical studies can do for the art historian. Focillon took technical matters seriously because he began his career in the one branch of art history that has always put technique front and center, printmaking. He said: "It has always seemed to me that in difficult studies of this sort...the obser-

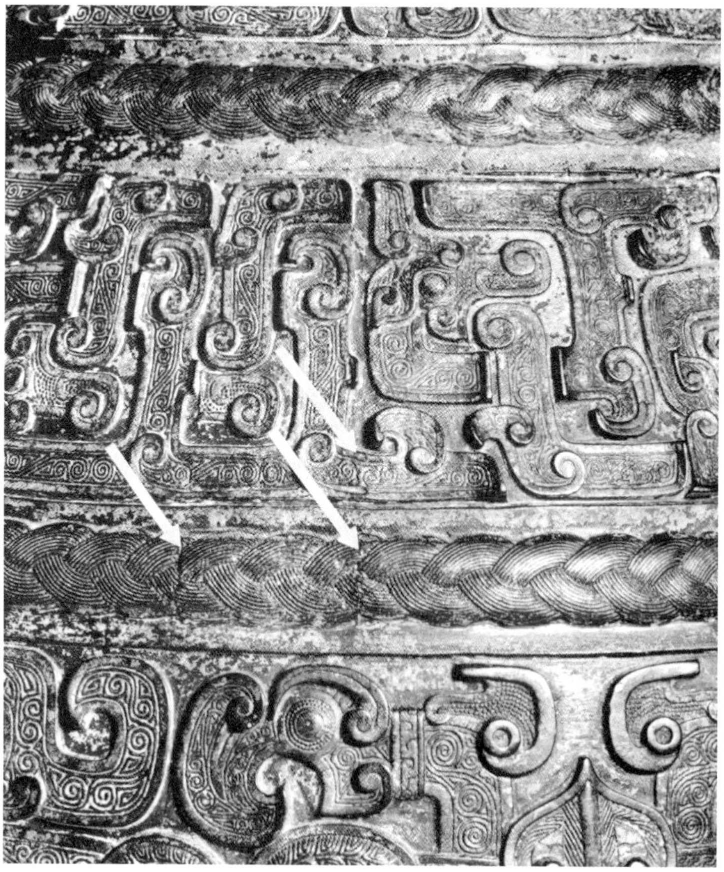

Figure 10. Hu *(same vessel as fig. 9), detail showing joins at the left and right edges of the truncated pattern unit in one of the bands of dragon interlace and in the braid below it. Source: Gettens (1969, 61).*

vation of technical phenomena not only guaranteed a certain controllable objectivity, but afforded an entrance into the very heart of the problem, *by presenting it to us in the same terms and from the same point of view as it is presented to the artist*" (Focillon 1942, 42 [his italics]).

Focillon is saying that the artist thinks about technique. But it is just as important to remember that the technician thinks about art—*because the technician and the artist are the same person.* Like a great engraver, the person who designed and executed the vessel in figure 2 was a master of both design and technique. He knew how to design *and* how to carve clay (fig. 3!) because he learned the two things together. He was no doubt the product of an apprenticeship during which his master set finished bronzes in front of him and made him copy them in clay, again and again (see Bagley 2008, 118). In the process of copying—a form of training nowadays greatly undervalued—he internalized both a technology and a system of design. Inevitably, therefore, the history of the technology he used is intertwined with the history of the designs he produced, and we cannot understand the one without the other. We must try to think like the ancient artisan, and in that effort art historians and technologists will need to collaborate, because the knowledge the artisan had is now split between two disciplines.

Now I am well aware that I am preaching to the converted. When I suggest that students of art and students of technology need to sit down and look at bronzes together, I am well aware that it is not always easy to get art historians to take an interest in technical studies. Some are too intimidated by science to feel comfortable around scientists. Too many still believe that the only function of technical studies is to identify fakes. But I believe that I can end on a hopeful note. Things are changing. The last few decades have seen an enormous increase in the number of art historians thinking about how objects are made; it is almost the only recent development in art history that I feel like applauding. Anything you can do to show art historians that you can help them answer the questions they care about will accelerate it. Certainly in the area I have been discussing, technical study has much to contribute to the solution of artistic problems. Anyone who helps us figure out how bronze inscriptions were written and cast is destined for fame in the annals of bronze studies. For the sake of my field, I solicit your help.

Notes

1. Compare his general statement of the point in Gettens (1969, 58). I should add that in the passage just quoted he was speaking only of "the vessel proper." He believes that the lid was done differently (Pope et al. 1967, 289): "On the lid, however, the main decor on two diagonally opposed sections on the vertical part of the lid is identical in respect to contours, but the *lei-wen* grounds are not. This indicates that the bird figures were made with a die but that the *lei-wen* was added later." In other words, he believes that the technique here is the same as on the previous *you*: high relief was stamped, *leiwen* was carved. He continues: "There are other inconsistencies as well. The legs as well as the positions of the small birds which are in rounded relief differ in detail." I am not sure what legs and birds he is referring to. Perhaps the ones on the domed part of the lid? On the domed part, of course, all the decoration can be seen at once, and differences are easier to detect.
2. I suppose that the bulging eyes of the birds could be, but before deciding that they were, I would want to take an impression from one of them and try fitting it over the others. Everything else was certainly carved by hand. Moreover if you examine the lines of the *leiwen* while thinking about the movements of the tool, I believe you will be persuaded that they were all carved on the model, in the positive, just as we see them on the finished vessel; nothing here was carved into the mold section. For full discussion see Bagley (2009).
3. Gettens seems to have been persuaded—perhaps by Karlbeck—that fine *leiwen* could not have been transferred from model to mold but had to be executed directly in the mold. I believe that he underestimated the sophistication of Anyang transfer techniques, and that virtually all the decoration of Anyang vessels was carved on the caster's preliminary model and transferred to the mold (Bagley 2009). As we have seen, in the case of the *you* of figure 4, Gettens suggested that the *leiwen* was stamped into the mold along with the rest of the design. If it could be transferred from a stamp, it could be transferred from a model: the stamp he is imagining is nothing but a section of a model.
4. For references see Institute of Archaeology of Shanxi Province (1996, 72, n. 7). This book illustrates about 1,300 pieces of foundry debris bearing decoration. Five of its illustrations are reproduced in Bagley (2009, figs. 18–22).
5. The very complex Houma replication procedure took shortcuts that entailed leaving incomplete units of decoration on the finished bronze. It could therefore be used only for executing intricate textured patterns in which incomplete units would be likely to pass unnoticed. Presumably it somehow lowered the cost or raised the speed of production, but it required a complete reorganization of the work of the foundry, and its advantage in efficiency must have depended on complicated tradeoffs. See Bagley (2001).
6. He continues: "How odd it is that anyone should not see that all observation must be for or against some view if it is to be of any service!" Charles Darwin, letter to Henry Fawcett, 18 September 1861 (Darwin and Seward 1903, 1:195).
7. At least two closely related vessels, a *you* in the Museum of Fine Arts, Boston (34.66), and a *fangyi* in the Harvard Art Museum (1944.57.37), have similar beams attached the same way. (For the *fangyi* see Murray [2001].) All three vessels are likely to be contemporary products of a single foundry. Their solder is the earliest yet known in China, and soldering was not used again, as far as I know, until more than four centuries later. The foundry that made the Freer *you* was experimenting with a technique it did not really need.
8. A point made eloquently by Cyril Stanley Smith, e.g., in Smith 1981 ch. 8, "Art, Technology, and Science: Notes on their Historical Interaction."
9. Gettens suggested that the handle was cast separately and "sprung" over the tenons (Pope et al. 1967, 281). This seems to me unlikely. For obsessive casters, which Anyang metalworkers certainly were, it would be out of character. As early as the fourteenth century BCE, pre-Anyang metalworkers were casting handles in place by procedures even trickier than the one I have described (see for example an Erligang-period *you* from Panlongcheng; Bagley 1987, fig. 219). In fifth-century bronzes I have seen joins that could not have been made in the way Gettens suggests because the head of the tenon is larger than the opening in the cup that encloses it.
10. Karlbeck cited the traces of bronze as evidence that the molds had not been used merely to make wax models. (The indirect lost-wax process forms a piece mold on a model that is not destroyed, uses the piece mold to make one or more wax

models, and then uses the wax model(s) to make mold(s) for casting one or more bronzes.)

11. The next sentence refers to the Houma foundry debris: "Examples of master patterns for this type of decor have been excavated from the Chan-kuo period bronze foundry site at Hou-ma." These remarks appear in Gettens's entry for the *Zhi Junzi jian* in the first volume of the bronze catalogue (Pope et al. 1967, 481). In the second volume he makes a different suggestion that is less prescient (Gettens 1969, 60): "On both these vessels [the *Zhi Jun Zi jian* 39.11 and the *hu* 57.22] the patterns have some peculiar characteristics such as the joins or stitches in the repeat patterns which are both uneven and asymmetrical (figs. 23–25). They resemble the fragmentary piecing together of odd bits of decalcomania, or transfer designs on modern chinaware. This is especially true on the braided rope patterns with fine curved strands which encircle the two vessels just mentioned. The joins in the braid are often quite unevenly spaced ranging from 5 cm to 19 cm apart. Some of these joins suggest that the design was impressed in the mold with a stamp; such a stamp with a positive design was found among the mold fragments at Hou-ma along with a negative mold. The stamps themselves may have been made of baked earth." The incomplete pattern units he has noticed played a key role in the Houma technique, and it has not occurred to him that stamps applied to the mold could not have produced them. See Bagley (2001).

References

Bagley, R. W. 1987. *Shang ritual bronzes in the Arthur M. Sackler Collections*. Ancient Chinese Bronzes in the Arthur M. Sackler collections; v. 1. Washington, DC, and Cambridge, MA: The Arthur M. Sackler Foundation and the Arthur M. Sackler Museum.

———. 2001. What the bronzes from Hunyuan tell us about the foundry at Houma. In *Chinese bronzes: Selected articles from* Orientations *1983–2000,* 214–22. Hong Kong: Orientations.

———. 2008. *Max Loehr and the study of Chinese bronzes: Style and classification in the history of art*. Ithaca, NY: Cornell East Asia Series.

———. 2009. Anyang mold-making and the decorated model. *Artibus Asiae* 69(1):39–90.

Darwin, F., and A. C. Seward, ed. 1903. *More letters of Charles Darwin*. New York: Appleton.

Focillon, H. 1942. *The life of forms in art*. New Haven, CT: Yale University Press.

Gettens, R. J. 1967. Joining methods in the fabrication of ancient Chinese ceremonial vessels. In *Application of science in examination of works of art,* 205–17. Boston: Museum of Fine Arts.

———. 1969. *The Freer Chinese bronzes,* Vol. 2, *Technical studies*. Oriental Study Series 7. Washington, DC: Freer Gallery of Art, Smithsonian Institution.

Institute of Archaeology of Shanxi Province. 1996. *Art of the Houma foundry*. Princeton, NJ: Princeton University Press.

Karlbeck, O. 1935. Anyang moulds. *Bulletin of the Museum of Far Eastern Antiquities* 7:39–60.

Keyser, B. 1979. Decor replication in two Late Chou bronze *chien*. *Ars Orientalis* 11:127–62.

Murray, J. K. 2001. Object of the month: A Western Zhou *fangyi*. In *Chinese bronzes: Selected articles from* Orientations *1983–2000,* 390–92. Hong Kong: Orientations.

Pope, J. A., R. J. Gettens, J. Cahill, and N. Barnard. 1967. *The Freer Chinese bronzes,* Vol. 1, *Catalogue*. Oriental Study Series 7. Washington, DC: Freer Gallery of Art, Smithsonian Institution.

Smith, C. S. 1981. *A search for structure: Selected essays on science, art, and history*. Cambridge, MA: MIT Press.

Chinese Bronzes

Recent Research on Early Bronze Metallurgy in Northwest China

Mei Jianjun, Xu Jianwei, Chen Kunlong, Shen Lu, and Wang Hui

ABSTRACT Over the past decade, several important archaeological discoveries have been made in Xinjiang and Gansu provinces in northwest China that throw new light on the early development of bronze metallurgy in the region. This paper first offers a brief review of studies made over the past several decades of early copper and bronze metallurgy in China. Then it highlights some recent archaeological discoveries of early metal objects in northwest China and presents preliminary examination results of some of these metal objects. Finally, it discusses a number of challenging issues in current studies, such as the significance of metal finds at the Xiaohe cemetery, the first identification of arsenical copper among the excavated Qijia metals, the discovery of early metallurgical production sites in Gansu, and the cultural connections between northwest China and the Eurasian steppe. It is argued that the region of Gansu-Qinghai played a crucial role in the early development of copper and bronze metallurgy in China.

Introduction

The past several decades have witnessed major transformations in studies of the beginnings of copper and bronze metallurgy in China. Magnificent bronze finds from the excavations at Anyang in the lower Yellow River valley in the 1930s stimulated scholarly interest in the origins of bronze metallurgy in China. In his book, *Origin of Far Eastern Civilizations,* Bishop (1942, 14) wrote that "an already well-developed Bronze Age civilization which had most (though not quite all) of the elements known to the Near East a thousand years or more earlier" must have appeared in northern China during the first half of the second millennium BCE. Loehr (1949, 129) also claimed that "primitive stages have, in fact, nowhere been discovered in China up to the present moment. Metallurgy seems to have been brought to China from outside." Such a diffusionist view was prevalent around the mid-twentieth century, when most scholars believed that "early Chinese civilization had developed through, and in large part because of, influence from West Asia" (Puett 1998, 699).

Archaeological finds since the 1950s, however, led Chang (1963, 138) to argue for the internal growth of Chinese civilization from a nucleus in the Yellow River valley. In his first major research monograph on bronze casting and bronze alloys in ancient China, Barnard (1961, 108) wrote that "bronze casting was almost certainly an art independently discovered in China, probably in the vicinity of the Cheng-chou area; it was a natural development from a highly advanced Neolithic pottery culture and because of its strong affinity thereunto techniques evolved which appear to contrast in many aspects with earlier or contemporaneous bronze cultures of the West." Chêng (1974, 226) also observed that "the investigations into the origin and processes of bronze casting in ancient China illustrate a progression of knowledge which has gradually replaced the theorizing speculations with concrete archaeological facts. Now the independent invention of the art in Shang China can no longer be dismissed as a mere possibility." On the basis of a systematic survey of metallurgical remains in ancient China, Barnard and Sato (1975, 68) stated that "the technical variations between Chinese and Western metallurgical industries in antiquity offer a substantial form of evidence in support of the thesis of independent discovery of metallurgy in China." Ho (1975, 221) noted the early metal finds belonging to the Qijia (Ch'i-chia) culture in Gansu, northwest China, and emphasized that "the primitive copper metallurgy of the loess highlands of China was of autochthonous local origins; and that the Shang bronze metallurgy, which by all technological criteria must be regarded as an indigenous development, owed its initial knowledge of metal to the loess highlands."

In the 1980s, scholars in mainland China began to join the discussion on the origins of bronze metallurgy in China, taking the debate of diffusion versus independent invention to a new level (An 1981; Yan 1984). Of particular importance is the paper published by Sun and Han (1981), which for the first time provided analytical results for thirty early copper and bronze objects excavated in China. These results indicated that tin bronze came into use in China as early as the early third millennium BCE, with copper, tin bronze, and leaded tin bronze all identified among early metal finds from the Qijia cultural sites in Gansu and Qinghai. Ko (1987, 225–43) also pointed out that the absence of arsenical copper among the early metals recovered in China exhibits a clear difference from what happened in the West.

While the view that copper and bronze metallurgy developed independently in China seemed to have become predominant during the 1970s and 1980s, some Western scholars remained in favor of the diffusion theory (Smith 1977, 81; Watson 1985, 335). Chase (1983, 106–7) preferred to see the early metals found in Gansu as one of the stages of the introduction of Western metallurgy into China. Muhly (1988, 13–16) believed that "what work during the past decade has accomplished is the solid documentation of the long-missing formative state in the development of Chinese copper and bronze metallurgy. The beginnings of metallurgy in China have now been shown to be not so much different from what can be seen elsewhere throughout the ancient world. This neither proves nor disproves the independent invention of metallurgy in east Asia, but it does move the argument into a more comprehensible frame of reference and should eliminate once and for all the argument that Chinese metallurgical technology developed in ways totally different (and thus apart from) anything known in the West."

It should be noted that the debate of the 1980s was mostly focused on early metal objects found in northwest China, especially those associated with the Qijia culture (ca. 2300–1700 BCE) in the Gansu-Qinghai region (Barnard 1983; Zhang 1987; Debaine-Francfort 1995). While scholars who favored the idea of independent invention took the metal finds from northwest China as signs of the initial stages of Chinese metallurgy (Ho 1975, 221; Sun and Han 1981), others preferred to see them as an indication of contact with the West (Jettmar 1981, 155; Chase 1983, 106–7). The reasons for the existence of diverse opinions concerning the issue were not just because of a lack of archaeological evidence or ambiguity about the evidence but were related to shifts in research paradigms, such as diffusionist theories around the 1950s and processual archaeology during the 1960s and 1970s (Puett 1998, 700–704), and even political passions (Wagner 1993, 28–33).

A significant change appeared when An Zhimin (1993), the leading Chinese archaeologist, remarked in a paper that "the appearance of early copper artefacts in China was relatively late; at least, there were still no copper objects produced during the Neolithic, five or six thousand years ago... Early copper artefacts quite possibly originated from or came into China through the prehistoric Silk Road." Fitzgerald-Huber (1995) also argued that a number of objects discovered at the Erlitou site near present-day Luoyang point to contact with neighboring and distant cultures of North and Central Asia, especially the Seima-Turbino cultural complex. French scholar Corinne Debaine-Francfort (1995, 324–26) also noticed the typological similarity between the Qijia socketed axes and those of the Seima-Turbino.

While discussions on early cultural contacts between northwest China and the Eurasian steppe have continued to attract considerable research interest, some new trends have emerged since the mid-1990s, including

- a fresh understanding of the early use of metals in the Gansu-Qinghai region based on a number of new archaeological finds and metallurgical analyses (Sun and Han 1997; Sun 1998; Li and Shui 2000; Sun et al. 2003);
- the initiation of systematic examinations of early copper and bronze objects in Xinjiang (Mei et al. 1998; Mei and Shell 1999; Mei 2000; IHMM 2001; Qian 2006);
- studies of metallurgical remains newly excavated at the Erlitou site of the first half of the second millennium BCE (Jin 2000; Liang and Sun 2004; Chen 2008);
- archaeometallurgical research on early metals of the Zhukaigou and Lower Xiajiadian cultures found in the northern border areas of China (Li and Han 2000; Lin 2002; Li et al. 2003); and
- systematic exploration of regional characteristics and interactions of early metallurgy in China (Linduff 1998; Bai 2002; Mei 2001, 2003a, 2003b; Li 2005; Liu and Li 2007).

It should be noted that new archaeological finds from eastern Xinjiang and western Gansu have revealed further evidence to support the existence of contacts with bronze cultures in the Eurasian steppe, leading some scholars to suggest the extension of "steppe metallurgy" into northwest China or to see early metallurgy in northwest China as "part of the Eurasian tradition" (Debaine-Francfort 2001, 57–70; Linduff 2004, 1–14).

In summary, our understanding of early metallurgy in China has undergone dramatic transformations over the past several decades. Although no definite consensus has been reached yet, archaeological and analytical evidence has been increasing steadily, throwing new light on the issue. Since the 1990s, a growing body of archaeological evidence indicating early cultural contacts between northwest China and the Eurasian steppe has amassed. Now the need for a re-evaluation of the role of outside influences on the early development of copper and bronze metallurgy in China is widely appreciated both in China and abroad, stimulating a new trend of research that focuses on early regional innovations and interactions within a wider Eurasian context (Linduff 2004; Li 2005; Liu and Li 2007; Mei 2009a, 2009b). On the basis of the most recent archaeological discoveries in northwest China, this paper first reports some new analytical results of early metal objects excavated in the region and then offers a discussion on the significance of these new discoveries, especially their implications for a fresh understanding of the early development of copper and bronze metallurgy in China.

Recent Archaeological and Archaeometallurgical Discoveries in Northwest China

Over the past ten years, there have been a number of significant archaeological discoveries revealing more evidence about metallurgical activities during the Early Bronze Age (ca. 2500–1500 BCE) in northwest China. At the same time, archaeometallurgical studies of the early copper and bronze objects excavated some decades ago have also been undertaken, revealing some significant new evidence for characterizing the early use of metals in the region. In this

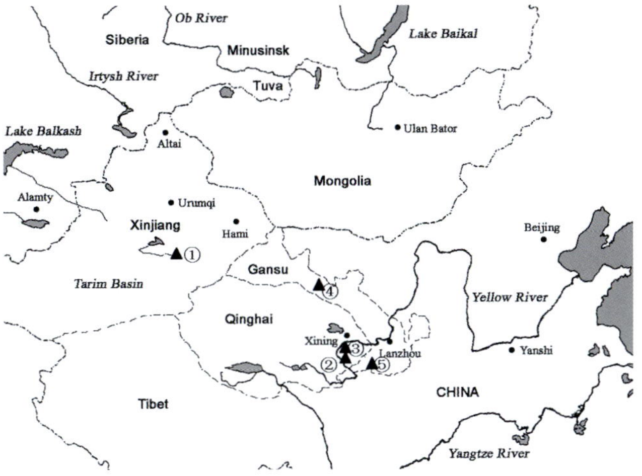

Figure 1. *Map of China showing the locations of the sites mentioned in the text. Key: 1 = Xiaohe cemetery; 2 = Zongri site; 3 = Gamatai cemetery; 4 = Heishuiguo site; 5 = Mogou site.*

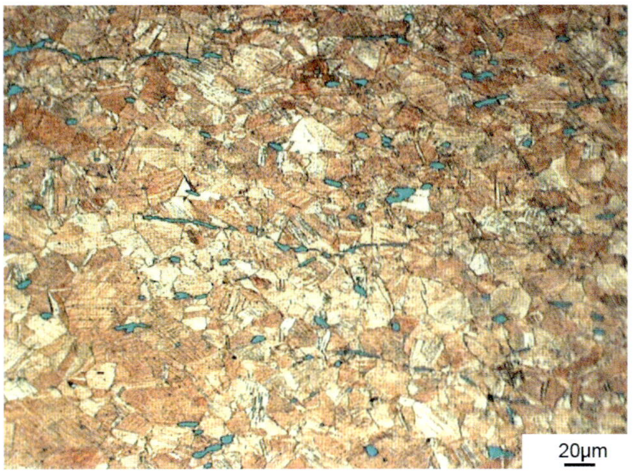

Figure 2. *Micrograph of the copper piece (XJ832) from the Xiaohe cemetery showing a hot-forged and cold-worked microstructure with some inclusions visible.*

paper, four major archaeological and archaeometallurgical discoveries will be highlighted: research on the early use of metals at the Xiaohe cemetery in Xinjiang, the examination of copper and bronze objects excavated at the Zongri and Gamatai sites in Qinghai, the discovery of a number of metallurgical production sites along the Hexi Corridor in Gansu, and the discovery and examination of a large number of copper and bronze objects at the Mogou site in Gansu (fig. 1).

Early Use of Metals at the Xiaohe Cemetery in the Tarim Basin in Xinjiang

Located at the eastern edge of the Taklamakan Desert and dated roughly to the first half of the second millennium BCE, the Xiaohe cemetery is one of the most important Early Bronze Age sites found so far in Xinjiang (see fig. 1). It was first reported in the 1930s (Bergman 1939), but a full-scale excavation of it only became possible in recent years. From 2002 to 2005, archaeologists from the Xinjiang Institute of Cultural Relics and Archaeology (XICRA) excavated the cemetery and uncovered hundreds of graves, which have yielded thousands of funerary goods, such as woollen blankets, felt hats, grass baskets, wooden implements, sculptures, and other objects. What is most extraordinary is that the majority of organic materials recovered from the graves are very well preserved due to the extremely dry environmental conditions. Neither ceramic nor metal implements have been found so far, though the cutting marks on the wooden coffins and poles unearthed at the cemetery strongly suggest the use of some sort of metal tools. The metal items recovered at the Xiaohe cemetery are mostly small, thin strips made of copper or copper alloys, measuring around 1 sq. cm. Some gold earrings or rings have also been found (XICRA 2003, 2007).

Table 1. *SEM-EDS analysis results of bronze pieces from the Xiaohe cemetery.*

Sample No.	Object No.	Chemical Compositions (wt.%)					Materials
		Cu	Sn	Pb	As	others	
XJ831	02XHM2:16-III	94.4	5.2	—	—	S: 0.3	Cu-Sn
XJ832	03XHT23C:55	88.5	10.1	—	1.1	—	Cu-Sn (As)
XJ833	04XHM129:27-IV	92.7	1.6	3.5	1.5	Cl: 0.7	Cu-Pb (Sn, As)

Table 2. *Results of portable XRF surface analysis of gold earrings from the Xiaohe cemetery.*

Lab. No.	Object No.	Chemical Compositions (wt.%)				Materials
		Au	Ag	Cu	Fe	
1464	04XHM1:22-1	81.16	17.77	0.84	—	Au-Ag
1465	04XHM1:22-2	90.15	9.25	0.14	0.20	Au-Ag
1466	04XHM130:18	84.82	14.04	0.73	0.20	Au-Ag
1467	04XHM100:11	83.71	14.93	1.09	0.21	Au-Ag (Cu)
1468	04XHM70:18	97.99	1.52	0.24	0.17	Au (Ag)
1469	04XHBM9:11-1	81.99	16.65	0.92	—	Au-Ag
1470	04XHBM9:11-2	81.09	17.73	1.01	—	Au-Ag (Cu)

The analysis was carried out on the surface of the gold earrings.

Table 3. *SEM-EDS analysis results of bronze objects from the Zongri site.*

Sample No.	Objects	Chemical Compositions (wt.%)						Materials	Preservation
		Cu	Sn	As	Pb	Fe	S		
ZR001	fragment	65.51	—	26.42	0.54	6.27	1.27	Cu-As (Pb)	heavily corroded
ZR002	fragment	87.90	7.73	4.37	—	—	—	Cu-Sn-As (?)	completely corroded
ZR003	ring	92.45	0.52	7.03	—	—	—	Cu-As (Sn)	lightly corroded

With the support of Xinjiang archaeologists, a preliminary examination of metal objects recovered at the Xiaohe cemetery has been carried out, including scanning electron microscopy–energy-dispersive spectrometry (SEM-EDS) analyses of three samples taken from three small pieces of copper or bronze as well as nondestructive x-ray fluorescence spectroscopy (XRF) surface tests of seven gold earrings. The examination results are given in tables 1 and 2. As we can see from table 1, the first three samples are all bronze, including two tin bronze (Cu-Sn) and one leaded bronze (Cu-Pb) that contains a small amount of tin and arsenic. Metallographic examination of these three samples has also been undertaken, showing that they were cast, hot-forged, and cold-worked to their final shape (fig. 2). As shown in table 2, among the seven gold earrings, six are made of Au-Ag alloys, with Ag contents mostly ranging between 9 and 18 wt.%, and only one is made of unalloyed gold (with Ag being less than 2 wt.%). All the gold earrings also contain a small amount of copper (Chen et al. 2007).

Examination of Bronze Objects Recovered at the Zongri and Gamatai Sites in Qinghai

Previous research has already shown that Qinghai is a major region where early copper and bronze objects have been recovered (Mei 2003a, 5–6). Among those sites yielding these objects in Qinghai, Zongri is the earliest and thus worthy of special attention (see fig. 1). Located at Bagou, Tongde county, and radiocarbon dated to 3600–2000 BCE, the site was excavated from 1994 to 1996, yielding thousands of pottery, stone, and bone objects as well as a small amount of copper and bronze ones, mostly rings and fragments. Its cultural characteristics are complicated and have been associated with the cultures of Majiayao, Zongri, and Qijia (QOCR 1998).

In cooperation with local archaeologists in Qinghai, three objects (two fragments and one ring) that are ascribed to the Qijia period were sampled and examined using SEM-EDS. The analysis results are given in table 3. It is important to note that among the three analyzed objects, two fragments (ZR001 and ZR002) were corroded heavily or completely and the ring (ZR003) was only slightly corroded. Therefore, while the result for the ring (ZR003) can be taken as quantitative—namely, 92.45 wt.% Cu, 7.03 wt.% As, and 0.52 wt.% Sn, revealing the use of Cu-As alloy—the data for the two fragments can only be taken as qualitative. The data suggest that they were probably made of Cu-As or Cu-Sn-As alloy, but their exact chemical compositions are unknown. Although the heavily corroded samples cannot offer quantitative data about their compositions, table 3 still reveals the crucial fact that all three objects contain arsenic (Xu et al. 2010a). This is a new phenomenon for metal objects of the Qijia culture found to date.

Another important site in Qinghai that has been ascribed to the Qijia culture is the Gamatai cemetery, which was reported to have yielded forty-nine copper and bronze objects (Li 2005, 243). Located in Guinan county (see fig. 1), the cemetery was excavated in 1977, but no excavation report has been published so far. It was briefly reported that more than forty graves had been excavated there, yielding hundreds of funerary objects of bone, stone, copper, and bronze. Among the forty-nine copper and bronze objects recovered, finger rings and buttons are most numerous (QOCR 1979). A mirror with a geometric pattern on the back was also reported, and scientific examination of it shows that it was made of tin bronze (Li 1980, 1981). Except for this mirror, however, no other copper and bronze objects from Gamatai had been subjected to scientific analysis until recently.

With the support of the Qinghai Institute of Cultural Relics and Archaeology (QICRA), we obtained eight bronze samples (seven buttons and one bracelet) unearthed

Table 4. *SEM-EDS analysis results of bronze objects from the Gamatai cemetery.*

Lab. No.	Objects	Chemical Compositions (wt.%)				Materials	Preservation
		Cu	Sn	As	S		
GM001	button	92.48	7.52	—	—	Cu-Sn	completely corroded
GM002	button	91.56	8.44	—	—	Cu-Sn	completely corroded
GM003	button	88.59	11.41	—	—	Cu-Sn	completely corroded
GM004	button	93.47	5.90	—	0.63	Cu-Sn	metal
GM005	button	87.49	12.51	—	—	Cu-Sn	completely corroded
GM006	button	91.45	8.55	—	—	Cu-Sn	completely corroded
GM007	button	85.71	14.29	—	—	Cu-Sn	completely corroded
GM008	bracelet	92.89	—	7.11	—	Cu-As	slightly corroded

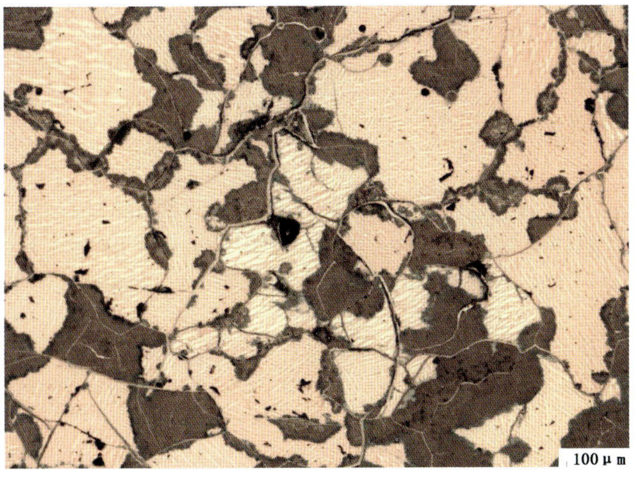

Figure 3. *Micrograph of a slag sample (HS013) from the Heishuiguo site showing some copper phases embedded in the slag matrix.*

Table 5. *SEM-EDS analysis results of slag samples from the Heishuiguo site.*

Lab. No.	Chemical Compositions (wt.%)					
	FeO	SiO_2	Al_2O_3	CaO	MgO	K_2O
HS002	56.13	27.53	11.9	0	3.86	0.59
HS005	37.18	50.37	4.58	4.97	2.22	0.68
HS006	59.99	28.65	4.42	3.25	1.76	1.94
HS007	61.21	21.81	13.93	1.24	1.8	0
HS008	46.55	38.84	9.17	1.28	3.39	0.77
HS010	59.07	29.86	4.65	4.43	0	1.98
HS011	51.95	37.83	7.16	1.01	1.78	0.27
HS013	52.76	33.9	9.58	0.35	2.84	0.57
HS015	33.35	45.09	11.23	3.1	4.7	2.53
HS016	51.73	36.33	7.64	1.33	2.98	0
HS017	66.82	26.25	6.39	0	0.54	0

at Gamatai. These samples were examined using SEM-EDS, and the results are given in table 4. Six of the buttons were completely corroded, one button was preserved as metal, and the bracelet was only slightly corroded. The data indicate that all the buttons are made of tin bronze. Once again, data for the completely corroded samples are qualitative only; the only button sample preserved as metal (GM004) contains 93.47 wt.% Cu and 5.90 wt.% Sn. Of greatest interest is the bracelet (GM008), which is made of Cu-As, containing 92.89 wt.% Cu and 7.11 wt.% As (Xu et al. 2010b). This analytical work not only confirms the preponderance of tin bronze among the metals found at Gamatai but also provides further concrete evidence for the use of Cu-As among the Qijia metals in Qinghai.

Discovery of Metallurgical Production Sites along the Hexi Corridor in Gansu

Over the past five years, several surveys of early metallurgical sites along the Hexi Corridor in Gansu have been carried out by a team led by Professor H. Wang from the Gansu Institute of Cultural Relics and Archaeology (GICRA) and Professor Y. Li from the Institute of Historical Metallurgy and Materials, University of Science and Technology, Beijing. These surveys have located a number of copper mining, smelting, and melting sites, such as the Heishuiguo site in Zhangye (see fig.1), the Gudongtan and Shaguoliang sites in Yumen, and the Ganggangwa, Huoshiliang, and Baishantang sites in Jinta. This pioneering work has been complemented by environmental research carried out by Dodson et al. (2009), which suggests that bronze manufacture becomes visible in the Huoshiliang sediments at about 2200 BCE. Although only some preliminary studies have been conducted on the slag or ore samples found on the surface at these sites, the evidence revealed so far is sufficient to show that metallurgical activities existed along the Hexi Corridor in Gansu no later than the first half of the second millennium BCE.

The Heishuiguo site in Zhangye county has been attracting considerable research interest recently because the fragments of painted pottery recovered at this site indicate its probable association with such cultures as Machang, Qijia, and Siba. The find of a small amount of smelting slag at this site further demonstrates its importance. Table 5 gives some preliminary results of the analysis of eleven slag samples collected at the Heishuiguo site. The samples consist of a $FeO-SiO_2-Al_2O_3$ system, with small amounts of CaO, MgO, and K_2O. Microscopic examination of many of these samples has revealed the existence of copper or copper sulfide particles or phases, suggesting that they must have been the products of a copper smelting process (fig. 3).

Discovery and Examination of Copper and Bronze Objects at the Mogou Site in Gansu

Located at Mogou village in Lintan county, Gansu province (see fig.1), the Mogou site is one of the most important archaeological discoveries made in northwest China in recent years. It has been under excavation since 2008 and has already revealed more than one thousand graves, most of which have been ascribed to the Qijia culture and roughly dated to the first half of the second millennium BCE. According to the brief reports available, the Qijia graves have yielded thousands of funerary goods, such as pottery vessels, bone ornaments and implements, stone implements, shells, and metal objects. The discovery of more than three hundred copper and bronze objects ascribed to the Qijia culture is totally unexpected, as previously no single Qijia site had yielded more than fifty. The copper and bronze objects recovered at the Mogou site are mostly implements and ornaments, such as buttons, earrings, tubes, torques, armbands, beads, and knives (fig. 4). Some types have been seen before at other Qijia sites, while others are new, such as trumpet-shaped earrings, torques, and armbands (GICRA 2009a, 2009b).

Preliminary analysis of eighteen copper and bronze objects unearthed at the Mogou site in 2008 has been carried out, and the results are given in table 6. Ten of the samples are completely corroded, so their analysis results can be taken as qualitative only, whereas the results for

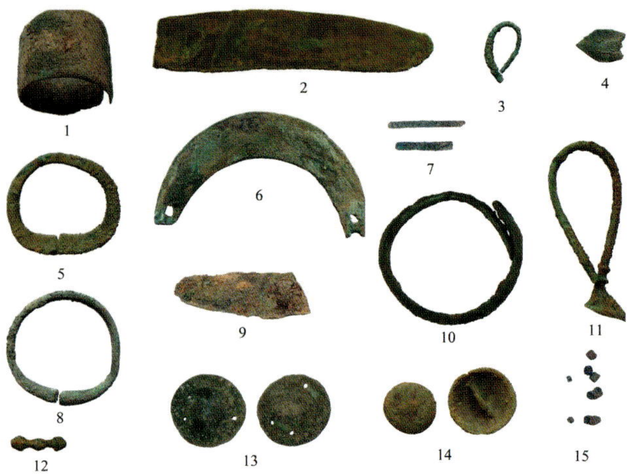

Figure 4. Some copper and bronze artifacts found at the Mogou site in eastern Gansu. Key: 1 = armband; 2, 9 = knives; 3, 5, 8 = earrings; 4, 12 = small ornaments; 6 = torque; 7 = tubes; 10 = bracelet; 11 = trumpet-earring; 13, 14 = buttons; 15 = beads.

the remaining eight metal samples can be used as quantitative. The results show that eight of the ten corroded samples are made of Cu-Sn and the remaining two are possibly Cu-As-Pb and Cu (Sn?), respectively. As for the eight metal samples, five are Cu-Sn, and the remaining three are Cu, Cu-Sn-Pb, and Cu-Sn-As, respectively. These results indicate clearly that tin bronze (Cu-Sn) was the most important alloy used at the Mogou site, although Cu and other copper alloys, such as Cu-Sn-Pb and Cu-Sn-As, were also in use. Six objects with metal cores remaining (including two tubes, one knife, one button, and two ornaments) were subjected to metallographic examination, and the results reveal that they were manufactured by casting and hot-forging, with cold-working as well at the final stage of shaping (figs. 5 and 6).

Discussion

In the preceding section, some new research results concerning early metallurgical activities and the use of metals and alloys in Xinjiang, Qinghai, and Gansu have been presented. To what extent can these new results contribute to our understanding of early bronze metallurgy in northwest China? To answer this question, we would like to place these new analytical results in a broader context and offer a brief discussion of the significance of the metal finds at Xiaohe, the identification of arsenical copper among the Qijia metals, the discovery of metallurgical production sites in Gansu, and cultural connections between northwest China and the Eurasian steppe.

The Significance of the Metal Finds at the Xiaohe Cemetery

Archaeological evidence for the early use of metals and alloys in Xinjiang is still rather limited. As far as published materials are concerned, the earliest metals recovered in Xinjiang thus far are those from the Tianshanbeilu cemetery in Hami, eastern Xinjiang, which are roughly dated to the first half of the second millennium BCE. Scientific examinations have demonstrated that the metals used by the Tianshanbeilu people include not just tin bronze and copper, but also arsenical copper (Mei 2000, 38–39; IHMM 2001, 83). It has also been reported that three small copper pieces were unearthed at the Gumugou cemetery near the eastern rim of the Tarim Basin, southern Xinjiang, and analysis suggests that they are made of unalloyed copper (Wang 1983, 121–22). The Xiaohe cemetery is also located at the eastern rim of the Tarim Basin, quite close to the Gumugou cemetery. Local archaeologists in Xinjiang now consider

Table 6. SEM-EDS analysis results of copper and bronze samples from the Mogou site.

Lab. No.	Objects	Chemical Compositions (wt.%)				Materials	Notes
		Cu	Sn	As	Pb		
CQ001	tubes	73.58	26.43	—	—	Cu-Sn	completely corroded
CQ002	button	51.74	—	36.10	12.17	Cu-As-Pb	completely corroded
CQ003	earring	88.87	11.13	—	—	Cu-Sn	metal
CQ004	fragment	89.12	10.88	—	—	Cu-Sn	metal
CQ005	tube	73.54	26.47	—	—	Cu-Sn	completely corroded
CQ006	tube	88.58	11.43	—	—	Cu-Sn	metal
CQ007	ornament	85.55	12.61	1.85	—	Cu-Sn (As)?	completely corroded
CQ008	ornament	92.67	4.95	2.39	—	Cu-Sn-As	metal
CQ009	ornament	84.24	15.73	—	—	Cu-Sn	completely corroded
CQ010	ornament	84.47	15.54	—	—	Cu-Sn	completely corroded
CQ011	ornament	77.85	22.15	—	—	Cu-Sn	completely corroded
CQ012	button	70.50	11.06	—	18.44	Cu-Sn-Pb	metal
CQ013	ornament	85.75	14.25	—	—	Cu-Sn	metal
CQ014	armband	79.00	21.00	—	—	Cu-Sn	completely corroded
CQ015	knife	100	—	—	—	Cu	metal
CQ016	tube	94.19	5.81	—	—	Cu-Sn	metal
CQ017	torque	98.67	1.34	—	—	Cu (Sn)?	completely corroded
CQ018	tube	86.75	13.26	—	—	Cu-Sn	completely corroded

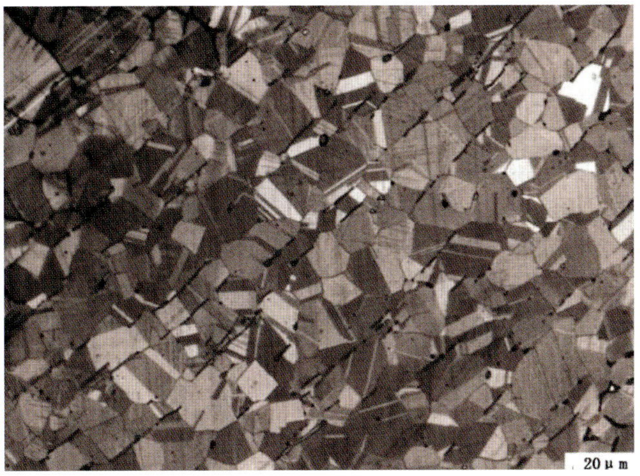

Figure 5. *Micrograph of the tin bronze tube (CQ006) from the Mogou site showing a hot-forged microstructure.*

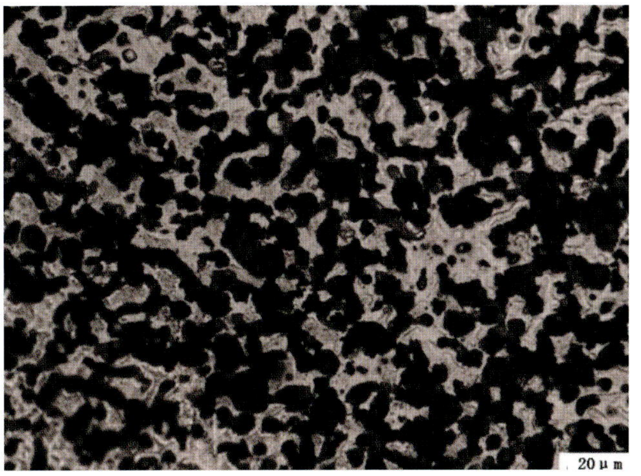

Figure 6. *Micrograph of the leaded tin bronze button (CQ012) from the Mogou site showing a casting microstructure.*

Gumugou and Xiaohe to belong to one archaeological culture and to date to the same period. The examination of three metal fragments from Xiaohe, however, showed that they are made of bronze rather than unalloyed copper. One of the Xiaohe samples even contains a small amount of arsenic. Therefore, from a broader perspective, the discovery of tin bronze at the Xiaohe cemetery undoubtedly provides crucial evidence not just for understanding the early use of metals in southern Xinjiang but also for exploring cultural connections between eastern Xinjiang and southern Xinjiang during the Early Bronze Age.

The examination of the seven gold earrings from Xiaohe is also significant because it provides the first analytical evidence for the earliest use of Au-Ag alloy in Xinjiang. A few early gold and silver objects have also been found at the Tianshanbeilu cemetery in eastern Xinjiang, the Huoshaogou cemetery in western Gansu, and the Mogou cemetery in eastern Gansu. Whether or not their appearance at these sites implies connections between them is difficult to conclude at this stage, but the possibility cannot be ignored. The identification of Au-Ag alloys at Xiaohe represents the earliest discovery of noble metals so far in present-day China. Furthermore, the gold earrings found at Xiaohe were most likely obtained through trade or exchange, as there is no evidence to indicate a local source of gold in the Tarim Basin. In other words, the discovery of the gold earrings suggests that the Xiaohe people were in active interaction with other contemporary cultures or groups of people.

The Identification of Arsenical Copper among the Qijia Metals

As we have seen in the first section of this paper, the importance of the Qijia metals to the study of the beginnings of copper and bronze metallurgy in China has been widely recognized since the mid-1970s (Ho 1975, 221; Sun and Han 1981, 277–79; Muhly 1988, 14; An 1993, 1118). Previous understandings of the Qijia metals may be summarized as follows: first, they can be roughly dated to 2200–1800 BCE, later than a few copper and bronze objects associated with the Majiayao and Machang cultures (ca. 3300–2000 BCE) but slightly earlier than those metals ascribed to the Siba culture (1950–1550 BCE) in western Gansu (Li 2005, 240–45); second, previous scientific analyses have shown that among the twenty-three analyzed Qijia metals, eighteen are unalloyed copper, three are tin bronze (Cu-Sn), one is leaded bronze (Cu-Pb), and one is leaded tin bronze (Cu-Sn-Pb) (Sun and Han 1981; 1997); third, the Qijia metals can be judged to have undergone a full developmental sequence from unalloyed copper to tin bronze because the early Qijia sites have yielded unalloyed copper while bronze came from the late sites (Zhang 1987, 173–74; Li 2005, 244).

The new analytical data presented in the preceding section are obviously significant in forming a new understanding of the use of metals and alloys in the Qijia culture. First, the examinations of metal samples from Zongri and Gamatai provide concrete evidence for the use of arsenical copper (Cu-As) among the Qijia metals, indicating that the previous assumption of a developmental sequence from unalloyed copper to tin bronze is over-simplified. It has thus become essential to take Cu-As alloy into consideration when discussing the developmental sequence of the Qijia metals. Second, the discovery and examination results of the Mogou metals demonstrate the relatively high level of bronze technology achieved by the Qijia people. While the use of tin bronze seems to be predominant among the metals found at Mogou, the existence of other alloys such as Cu-Sn-As, Cu-As-Pb, and Cu-Sn-Pb is also worthy of attention. Third, the new metallographic examinations further confirm that the Qijia tradition of metalworking involved a casting, hot-forging, and cold-working technology (Muhly 1988, 14).

The Discovery of Early Metallurgical Production Sites in Gansu

The discovery of a number of early metallurgical sites along the Hexi Corridor in Gansu has important implications for understanding the early development of copper and bronze metallurgy in northwest China and signals a major change in research strategy. Previous studies were mainly concerned with artifacts recovered during archaeological excavations, with little attention being paid to the production sites. Such an artifact-oriented approach faces great difficulties in tracing the sources of the artifacts concerned. Scholars have long assumed that mineral resources could have been exploited in Gansu during the Early Bronze Age because the region is well known for its rich deposits of nonferrous metal resources (Sun and Han 1981, 290). Now the finds of copper mining and smelting sites along the Hexi Corridor not only prove this assumption but also demonstrate that the Hexi Corridor acted as an important metallurgical center in the Early Bronze Age, one which could have supplied metal materials to Gansu as well as its adjacent regions.

Another important implication of the discovery of early metallurgical sites in Gansu is that it may help us to understand local technological innovations. Previous studies have argued that although some metal objects found in northwest China parallel those from the Eurasian steppe, others exhibit local features, such as the circular mirrors with geometrical decorations on the back excavated from the Qijia and Tianshanbeilu sites (Mei 2009b, 219). It has also been suggested that the long-lasting use of arsenical copper in Gansu and eastern Xinjiang implied the existence of local production based on mineral resources available in the Hexi Corridor (Mei 2009b, 219; Sun and Han 1997). There is, no doubt, a long way to go before we can reconstruct a metallurgical network linking archaeological finds of metals with the production sites as well as the mineral resources. However, the discovery of early metallurgical sites along the Hexi Corridor is still a major step forward in understanding the nature and extent of local metallurgical production in the region. We can expect that archaeological excavations and scientific examinations in the years to come will provide further evidence to clarify the role of local technological innovations in shaping the development of early bronze metallurgy in northwest China.

Cultural Connections between Northwest China and the Eurasian Steppe

As pointed out in the first section of this paper, some scholars in the mid-1990s suggested that cultural connections may have existed between northwest China and the Eurasian steppe during the Early Bronze Age (An 1993; Fitzgerald-Huber 1995; Debaine-Francfort 1995, 324–26). This suggestion has become widely accepted since then as new archaeological finds from Xinjiang and Gansu have revealed more supporting evidence (Mei 2003a, 2003b; Li 2005; Liu and Li 2007). Now, the new research results presented in this paper offer further evidence to confirm such connections. The extraordinarily rich finds of metals at the Mogou sites exhibit many cultural features that are comparable to discoveries in the Hexi Corridor, Xinjiang, and the Eurasian steppe, such as the predominant use of personal ornaments (earrings, armbands, torques, necklaces of beads or tubes, etc.). The discovery of trumpet-shaped earrings at the Mogou site is of special importance, as earrings of this type have been recovered not only in Gansu and Xinjiang but also in Inner Mongolia, Kazakhstan, and Siberia, all of which are associated with Early Bronze Age cultures such as Lower Xiajiadian, Zhukaigou, Siba, and Andronovo (Lin 1986; Bunker 1998; Mei 2003a). It would seem that a long-sought missing link has now been offered up by the copper and bronze finds from the Mogou site.

Conclusion

The past decade has witnessed significant progress in our understanding of early metallurgy in northwest China. The examination of the metals from the Xiaohe cemetery has found the earliest use of Au-Ag alloys and tin bronzes in the Tarim Basin in southern Xinjiang. The analysis of metals from the Zongri and Gamatai sites in Qinghai has shown, for the first time, the appearance or existence of arsenical copper (Cu-As) among the Qijia metals. The discovery of a number of metallurgical sites along the Hexi Corridor, especially the Heishuigou site in Zhangye, has provided further evidence to show that the Hexi Corridor acted as a regional metallurgical center in the Early Bronze Age. The discovery and examination of metals recovered at the Mogou site in Gansu have not only confirmed the dominant presence of tin bronze during the later stage of the Qijia culture but also revealed the uses of other copper alloys such as Cu-Sn-As, Cu-As-Pb, and Cu-Sn-Pb. It has become increasingly clear that the cultures of the Gansu-Qinghai region came into contact with those of the Eurasian steppe no later than the early second millennium BCE and played a crucial role in the early development of bronze metallurgy in China.

Acknowledgments

This research is supported by the National Natural Science Foundation of China (Project No. 51074026) and the Ministry of Science and Technology of China (Project No. 2006BAK21B03). Grateful thanks are due to Professors Abudu Idriss, Li Wenying, Mao Ruilin, Qian Yaopeng, Xu Xinguo, Chen Honghai, Li Shuicheng, Yuan Jing, Han Rubin, Sun Shuyun, and Jessica Rawson for their kind support and help. Professor Mei would like to thank Mr. Paul Jett and the Freer Gallery of Art and Arthur M. Sackler Gallery for inviting him to participate in the Fifth Forbes Symposium. He is also most grateful to Mr. John Moffett at the Needham Research Institute in Cambridge for his considerate encouragement and assistance during the preparation of this paper.

References

An, Z. 1981. Zhongguo zaoqi tongqi de jige wenti [Some issues concerning early copper and bronze objects in China]. *Kaogu xuebao* [Acta Archaeologica Sinica] 3:269–84.

———. 1993. Shi lun Zhongguo de zaoqi tongqi [On early copper and bronze objects in China]. *Kaogu* [Archaeology] 12:1110–19.

Bai, Y. 2002. Zhongguo de zaoqi tongqi yu qingtongqi de qiyuan [The origins of early Chinese copper and bronze objects]. *Dongnan wenhua* [Southeast Culture] 7:25–37.

Barnard, N. 1961. *Bronze casting and bronze alloys in ancient China*. Monumenta Serica Monograph XIV. Canberra: Australian National University.

———. 1983. Further evidence to support the hypothesis of indigenous origins of metallurgy in ancient China. In *The origins of Chinese civilization,* ed. D. N. Keightley, 237–77. Berkeley: University of California Press.

Barnard, N., and T. Sato. 1975. *Metallurgical remains of ancient China*. Tokyo: Nichiosha.

Bergman, F. 1939. *Archaeological researches in Sinkiang—Especially the Lop-Nor region*. The Sino-Swedish Expedition 7. Stockholm: Bokförlags Aktiebolaget Thule.

Bishop, C. W. 1942. *Origin of the Far Eastern civilizations: A brief handbook*. Washington, DC: Smithsonian Institution.

Bunker, E. C. 1998. Cultural diversity in the Tarim Basin vicinity and its impact on ancient Chinese culture. In *The Bronze Age and Early Iron Age peoples of eastern Central Asia,* ed. V. H. Mair, 604–18. Washington, DC: Institute for the Study of Man in collaboration with the University of Pennsylvania Museum Publications.

Chang, K. C. 1963. *The archaeology of ancient China*. New Haven, CT, and London: Yale University Press.

Chase, W. T. 1983. Bronze casting in China: A short technical history. In *The great Bronze Age of China: A symposium,* ed. G. Kuwayama, 100–123. Los Angeles: Los Angeles County Museum of Art.

Chen, G. 2008. Erlitou wenhua tongqi yanjiu [A study of copper and bronze objects of the Erlitou culture]. In *Zhongguo zaoqi qingtong wenhua* [Early bronze cultures in China], ed. The Institute of Archaeology, Chinese Academy of Social Sciences, 124–274. Beijing: Kexue Chubanshe.

Chen, K., Y. Ling, and J. Mei. 2007. Xiaohe mudi chutu sanjian tongpian de chubu fenxi [A preliminary analysis of three pieces of copper unearthed at the Xiaohe cemetery]. *Xinjiang wenwu* [Cultural Relics in Xinjiang] 2:125–28.

Chêng, T. K. 1974. Metallurgy in Shang China. *T'oung Pao* 60(4/5):209–29.

Debaine-Francfort, C. 1995. *Du néolithique à l'âge du bronze en Chine du nord-ouest: La culture de Qijia et ses connexions*. Paris: Éditions Recherche sur les Civilisations.

———. 2001. Xinjiang and northwestern China around 1000 BC: Cultural contacts and transmissions. In *Migration und kulturtransfer: Der vandel vorder-und zentralasiatischer kulturen im umbruch vom 2, um 1 vorchrislichen jahrtausend,* ed. R. Eichmann and H. Parzinger, 57–70. Bonn: Rudolf Habelt GmbH.

Dodson, J., X. Li, M. Ji, K. Zhao, X. Zhou, and V. Levchenko. 2009. Early bronze in two Holocene archaeological sites in Gansu, NW China. *Quaternary Research* 72(3):319–24.

Fitzgerald-Huber, L. G. 1995. Qijia and Erlitou: The question of contacts with distant cultures. *Early China* 20:17–67.

GICRA (Gansu Institute for Cultural Relics and Archaeology). 2009a. Gansu Lintan Mogou Qijia wenhua mudi fajue jianbao [A brief excavation report of the Qijia culture cemetery at Mogou, Lintan, Gansu]. *Wenwu* [Cultural Relics] 10:4–24.

———. 2009b. Gansu Lintan xian Mogou Qijia wenhua mudi [The Qijia culture cemetery at Mogou in Lintan county, Gansu]. *Kaogu* [Archaeology] 7:10–17.

Ho, P. T. 1975. *The cradle of the East: An inquiry into the indigenous origins of techniques and ideas of Neolithic and Early Historic China, 5000–1000* B.C. Hong Kong: Chinese University of Hong Kong; Chicago: University of Chicago Press.

IHMM (The Institute of Historical Metallurgy and Materials, University of Science and Technology Beijing). 2001. Xinjiang Hami Tianshanbeilu mudi chutu tongqi de chubu yanjiu [A preliminary study of copper and bronze objects excavated at the Tianshanbeilu cemetery in Hami, Xinjiang]. *Wenwu* [Cultural Relics] 6:79–89.

Jettmar, K. 1981. Cultural and ethnic groups west of China in the second and first millennia BC. *Asian Perspectives* 24(2):145–62.

Jin, Z. 2000. Erlitou qingtongqi de ziran kexue yanjiu yu Xia wenming tansuo [A scientific study of the Erlitou bronzes and an exploration of the Xia civilization]. *Wenwu* [Cultural Relics] 1:56–64.

Ko, T. 1987. The development of metal technology in ancient China. In *Science and technology in Chinese civilization,* ed. C. Y. Chen, 225–43. Singapore: World Scientific Publishing Co.

Li, H. 1980. Qijia wenhua tongjin de fei pohuai xing jianding [A nondestructive examination of a mirror of the Qijia culture–Analysis of fast neutron radiation]. *Kaogu* [Archaeology] 4:365–68.

———. 1981. Analysis of an ancient Chinese bronze mirror by fast neutron activation. *Archaeometry* 23(2):217–19.

Li, S. 2005. Xibei ji Zhongyuan zaoqi yetongye de quyu tezheng ji jiaohu zuoyong [The regional characteristics and interaction of early metallurgical industries in the northwest and the Central Plains]. *Kaogu xuebao* [Acta Archaeologica Sinica] 3:239–78.

Li, S., and T. Shui. 2000. Siba wenhua tongqi yanjiu [A study of the copper and bronze objects of the Siba culture]. *Wenwu* [Cultural Relics] 3:36–44.

Li, X., and R. Han. 2000. Zhukaigou yizhi chutu tongqi de jinxiang xue yanjiu [A micrographic study of copper and bronze objects excavated from the Zhukaigou site]. In *Zhukaigou—qingtong shidai zaoqi yizhi fajue baogao* [Zhukaigou—excavation report of an Early Bronze Age site], ed. Inner Mongolian Institute of Archaeology, 422–46. Beijing: Wenwu Chubanshe.

Li, Y., H. Jia, and Y. Zhu. 2003. Dadianzi mudi chutu tongqi chubu yanjiu [A preliminary study of copper and bronze objects excavated from the Dadianzi cemetery]. *Wenwu* [Cultural Relics] 7:78–84.

Liang, H., and S. Sun. 2004. Erlitou yizhi chutu tongqi yanjiu zongshu [A review of the studies of copper and bronze objects from the Erlitou site]. *Zhongyuan wenwu* [Cultural Relics from the Central Plains] 1:29–39.

Lin, Y. 1986. A re-examination of the relationship between bronzes of the Shang culture and of the northern zone. In *Studies of Shang archaeology,* ed. K. C. Chang, 237–73. New Haven, CT: Yale University Press.

———. 2002. Xiadai de Zhongguo beifangxi qingtongqi [The northern zone bronzes of the Xia dynasty in China]. *Bianjiang kaogu yanjiu* [Research of China's Frontier Archaeology] 1:1–12.

Linduff, K. M. 1998. The emergence and demise of bronze-producing cultures outside the Central Plains of China. In *The Bronze Age and Early Iron Age peoples of eastern Central Asia,* ed. V. H. Mair, 619–43. Washington, DC: Institute for the Study of Man in collaboration with the University of Pennsylvania Museum Publications.

———. 2004. How far does the Eurasian metallurgical tradition extend? In *Metallurgy in ancient eastern Eurasia from the Urals to the Yellow River,* ed. K. M. Linduff, 1–14. Lewiston, NY: Edwin Mellen Press.

Linduff, K. M., R. Han, and S. Sun, ed. 2000. *The beginnings of metallurgy in China*. Lewiston, NY: Edwin Mellen Press.

Liu, X., and W. Li. 2007. Zhongguo zaoqi qingtong wenhua de qiyuan jiqi xiangguan wenti xintan [A new exploration of the origins of early Chinese bronze cultures and relevant issues]. *Zangxue xuekan* [Journal of Tibetology] 3:1–63.

Loehr, M. 1949. Weapons and tools from Anyang and Siberian analogies. *American Journal of Archaeology* 53(2):126–44.

Mei, J. 2000. *Copper and bronze metallurgy in late prehistoric Xinjiang: Its cultural context and relationship with neighboring regions*. BAR International Series 865. Oxford: Archaeopress.

———. 2001. Guanyu Zhongguo yejin qiyuan ji zaoqi tongqi yanjiu de jige wenti [On the origins of metallurgy and the studies of early copper objects in China]. *Tulufan xue yanjiu* [Turfanological Research] 2:57–68.

———. 2003a. Cultural interaction between China and Central Asia during the Bronze Age. *Proceedings of the British Academy* 121:1–39.

———. 2003b. Qijia and Seima-Turbino: The question of early contacts between northwest China and the Eurasian steppe. *Bulletin of the Museum of Far Eastern Antiquities* 75:31–54.

———. 2009a. Early metallurgy in China: Some challenging issues in current studies. In *Metallurgy and civilisation: Eurasia and beyond,* ed. J. Mei and T. Rehren, 9–16. London: Archetype Publications.

———. 2009b. Early metallurgy and socio-cultural complexity: Archaeological discoveries in northwest China. In *Social complexity in prehistoric Eurasia: Monuments, metals and mobility,* ed. B. Hanks and K. Linduff, 215–32. Cambridge: Cambridge University Press.

Mei, J., S. H. Colin, X. Li, and B. Wang. 1998. A metallurgical study of early copper and bronze artefacts from Xinjiang, China. *Bulletin of the Metals Museum* 30:1–22.

Mei, J., and C. Shell 1999. The existence of Andronovo cultural influence in Xinjiang during the second millennium BC. *Antiquity* 73(281):570–78.

Muhly, J. D. 1988. The beginnings of metallurgy in the Old World. In *The beginning of the use of metals and alloys,* ed. R. Maddin, 2–20. Cambridge, MA: MIT Press.

Puett, M. 1998. China in early Eurasian history: A brief review of recent scholarship on the issue. In *The Bronze Age and Early Iron Age peoples of eastern Central Asia,* ed. V. H. Mair, 699–715. Washington, DC: Institute for the Study of Man in collaboration with the University of Pennsylvania Museum Publications.

Qian, W. 2006. *Xinjiang Hami shiqian shiqi qingtonqi jiqi yu zhoubian wenhua de guanxi* [Prehistoric copper and bronze objects from Hami, Xinjiang and their relationships with neighboring cultures]. Beijing: Zhishi Chanquan Chubanshe.

QOCR (Qinghai Office of Cultural Relics). 1979. Qinghai wenwu kaogu gongzuo sanshinian [Thirty years of the work of cultural relics and archaeology in Qinghai]. In *Wenwu Kaogu Gongzuo Sanshinian* [Thirty years of the work of cultural relics and archaeology], 160–68. Beijing: Wenwu Chubanshe.

———. 1998. Qinghai Tongde xian Zongri yizhi fajue jianbao [A brief report of the excavation at the Zongri site in Tongde county, Qinghai]. *Kaogu* [Archaeology] 5:1–14.

Smith, C. S. 1977. Review of Barnard & Sato 1975 and Ho 1975. *Technology and Culture* 18(1):80–86.

Sun, S. 1998. Donghuishan yizhi siba wenhua tongqi de jianding ji yanjiu [The examination and study of the copper objects from the Siba culture site at Donghuishan]. In *Minle Donghuishan kaogu: Siba wenhua mudi de jieshi yu yanjiu* [Archaeology of the Donghuishan site in Mingle: Excavation and study on the cemetery of Siba culture], ed. Gansu Provincial Institute of Cultural Relics and Archaeology and Jilin University, 191–95. Beijing: Kexue Chubanshe.

Sun, S., and R. Han. 1981. Zhongguo zaoqi tongqi de chubu yanjiu [A preliminary study of early Chinese copper and bronze artifacts]. *Kaogu xuebao* [Acta Archaeologica Sinica] 3:287–301. For an English translation, see Linduff et al. (2000, 129–51).

———. 1997. Gansu zaoqi tongqi de faxian yu yelian zhizuo jishu de yanjiu [Discovery of early copper and bronze artifacts in Gansu and studies of their smelting and manufacturing techniques]. *Wenwu* [Cultural Relics] 7:75–84. For an English translation, see Linduff et al. (2000, 175–93).

Sun, S., W. Qian, and H. Wang. 2003. Huoshaogou Siba wenhua tongqi chengfen fenxi ji zhizuo jishu de yanjiu [A study of the compositions and manufacturing techniques of copper objects of the Siba culture from the Huoshaogou site]. *Wenwu* [Cultural Relics] 8:86–96.

Wagner, D. B. 1993. *Iron and steel in ancient China.* Leiden: E. J. Brill.

Wang, B. 1983. Kongque he gumugou fajue jiqi chubu yanjiu [The excavation and preliminary study of the Gumugou cemetery near Kongque River]. *Xinjiang shehui kexue* [Social Sciences in Xinjiang] 1:117–27.

Watson, W. 1985. An interpenetration of opposites? Pre-Han bronze metallurgy in West China. *Proceedings of the British Academy* 70:327–58.

XICRA (Xinjiang Institute of Cultural Relics and Archaeology). 2003. 2002 nian Xiaohe mudi kaogu diaocha yu fajue baogao 2002 [The report of the archaeological investigation and excavation at the Xiaohe cemetery in 2002]. *Xinjiang wenwu* [Cultural Relics in Xinjiang] 2:8–64.

———. 2007. 2003 nian Luobupo Xiaohe mudi fajue jianbao 2003 [Brief report of the excavation at the Xiaohe cemetery in Luobupo in 2003]. *Xinjiang wenwu* [Cultural Relics in Xinjiang] 1:1–54.

Xu, J., J. Mei, S. Ge, and H. Chen. 2010a. Qinghai Tongde Zongri yizhi chutu tongqi de chubu kexue fenxi [A preliminary scientific analysis of copper objects unearthed at the Zongri site, Tongde, Qinghai]. *Xiyu yanjiu* [Studies of the Western Region] 2:31–7.

Xu, J., J. Mei, S. Sun, and X. Xu. 2010b. Qinghai Guinan Gamatai mudi chutu tongqi de chubu kexue fenxi [A preliminary scientific analysis of copper objects unearthed at the Gamatai cemetery, Guinan, Qinghai]. *Wenwu keji yanjiu* [Scientific and Technological Research on Cultural Heritage] 7:1–8.

Yan, W. 1984. Lun Zhongguo de tongshi bingyong shidai [On the Chalcolithic period in China]. *Shiqian yanjiu* [Studies in Prehistory] 1:36–44.

Zhang, Z. 1987. Qijia wenhua yanjiu [A study of the Qijia culture]. *Kaogu xuebao* [Acta Archaeologica Sinica] 2:153–76.

Recent Discoveries Concerning Metallurgy in Bronze Age Yunnan

Jin Zhengyao, Li Gong, Wang Haigang, Tian Jianhua, Yan Lifeng, and Jiang Zhilong

ABSTRACT Most Dian bronzes have been unearthed at the Shizhaishan site beside Lake Dianchi and at the Lijiashan site beside Lake Fuxianhu, both in central Yunnan province. Unfortunately, no smelting or casting facilities have been excavated, so the bronzes themselves have been the only sources of our knowledge on Dian bronze technology. During the last five years, another important archaeological site, Jinlianshan, located between Lake Dianchi and Lake Fuxianhu, was excavated. Among the objects unearthed from graves at this site was a lump of rare, freshly smelted copper that has been a rich source of information, aiding our understanding of the metal-producing capabilities of Bronze Age Dian smiths and indicating some technical details of the process of turning ore to usable metal, such as the copper ore type, furnace volume, and the control of furnace temperature.

Introduction

The period between 400 BCE and 200 CE saw the establishment, growth, and decline of the ancient Dian kingdom in the Yunnan area of southwestern China. Sites of the ancient Dian kingdom have been found mainly around Lake Dianchi and Lake Fuxianhu in central Yunnan (fig. 1). The two most important sites are the Shizhaishan site in Jinning county and the Lijiashan site in Jiangchuan county. The Shizhaishan site, near Lake Dianchi, is the cemetery of the kings and other nobles of ancient Dian and is where the famous golden seal of the Dian king was unearthed. The Lijiashan site, near Lake Fuxianhu, is the largest and richest of all ancient Dian kingdom sites other than the Shizhaishan site. These two sites are about 50 km apart. Because few relics directly related to copper or bronze production have been found in these sites up to now, research on the technology of copper or bronze production has depended on the study of the unearthed bronze artifacts. This limitation led to a superficial and incomplete understanding.

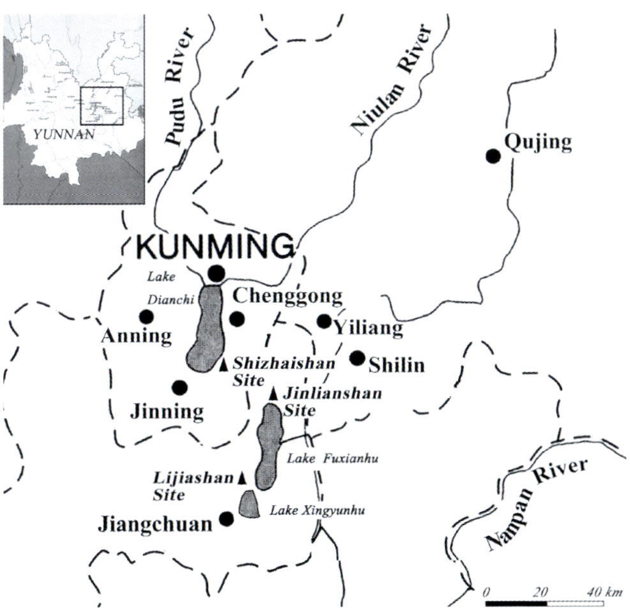

Figure 1. *Map of the sites of the ancient Dian kingdom.*

In the spring of 2006, another ancient Dian kingdom site, Jinlianshan, was found in Chengjiang county, 3 km north of Lake Fuxianhu (fig. 2). It is a cemetery site where 408 tombs were found in two excavations that occurred in 2006 and between 2008 and 2009, and where a huge number of human bones and artifacts were unearthed. The artifacts include weapons, farming tools, living goods, ornaments, and funerary objects. The bronze objects include swords (*jian*), axes, chisels, hoes, dagger-axes (*ge*), spears, cooking vessels (*fu*),

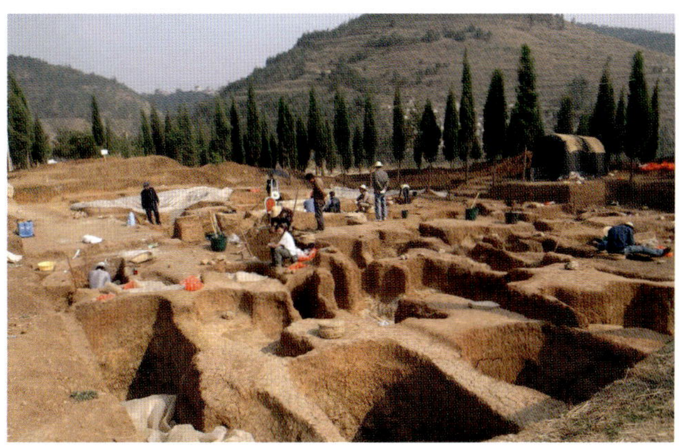

Figure 2. *The Jinlianshan site.*

pots, dishes, cups, arrows, armor, buckles, and belt hooks. The Jinlianshan site is about 30 km away from the Lijiashan site and about 20 km away from the Shizhaishan site and is located between the two sites. Furthermore, the ages of the three sites are roughly the same.

New Materials and Methods

Many cultural relics directly related to copper or bronze production have been unearthed at the Jinlianshan site. Two pairs of bivalve stone molds, excavated from tomb 2008CJM30, exhibit some casting features. Given their small size, it is likely that they were used for making funerary objects such as hoes, some analogues of which were found in other tombs at the Jinlianshan site. It is the first time that a complete pair of used molds has been archaeologically excavated at an ancient Dian site, and the molds provide precious new material for the study of copper and bronze production and use during that era.

A special piece of metal, which had the appearance of an ingot but with obvious layers, was found in tomb 2006CJM1. It is 27.2 cm in length (at the longest point), 4.4–19.2 cm in width, 0.7–1.2 cm in thickness, and 1059.3 g in weight. This is also the first time that archaeological material related to metal smelting or casting has been discovered in ancient Dian sites. This piece of metal, which has an undistinguished appearance, may be a treasure for research on the history of bronze technology in the ancient Dian kingdom. For convenience, we will refer to it as the "metal block" (fig. 3).

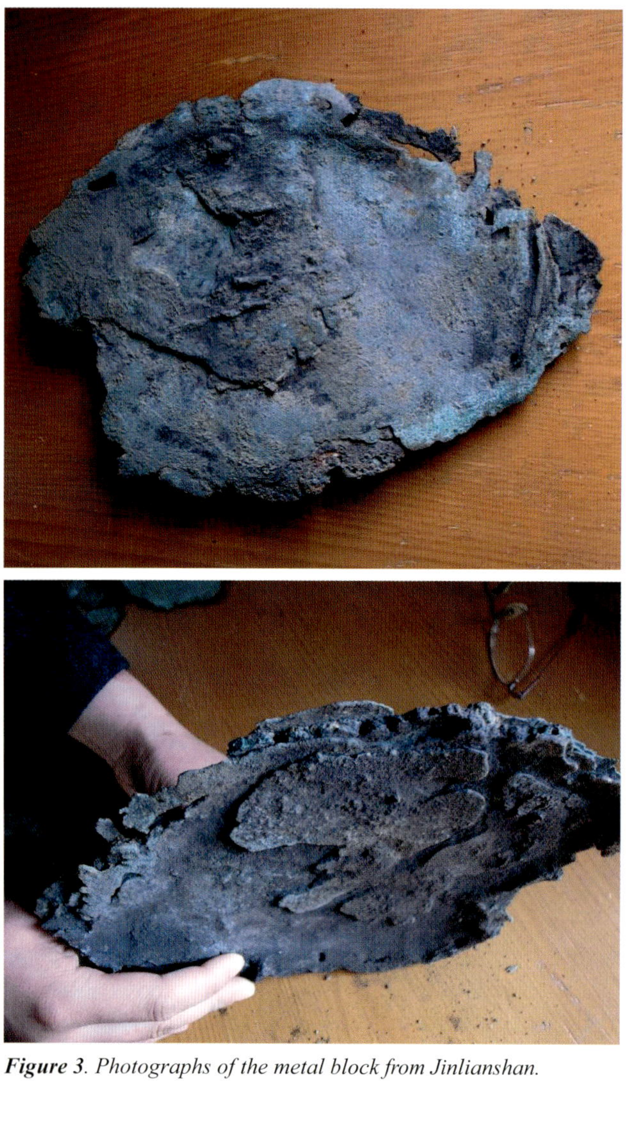

***Figure 3**. Photographs of the metal block from Jinlianshan.*

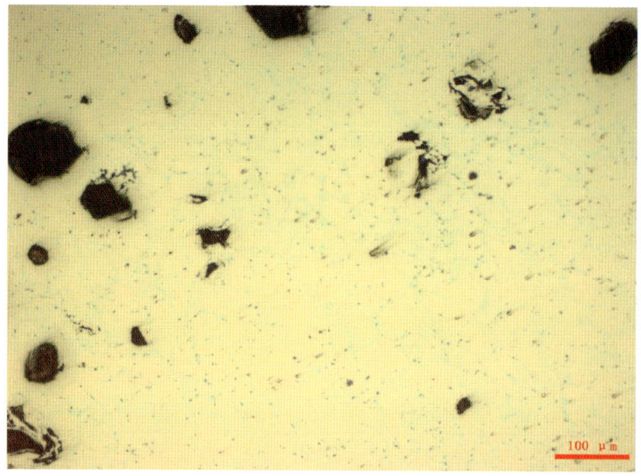

***Figure 4**. Micrograph of a fragment of the metal block from Jinlianshan.*

To analyze the metal and determine its manufacturing process, a sample of the metal block was removed, mounted, ground, and polished. The metallographic microstructure was then examined using optical microscopy. Backscatter images were also observed using scanning electron microscopy (SEM), and the basic elemental composition and inclusions were analyzed using scanning electron microscopy–energy-dispersive spectrometry (SEM-EDS). Additionally, a 92.8 mg sample was removed and dissolved in hydrochloric acid for analysis using inductively coupled plasma–atomic emission spectrometry/mass spectrometry (ICP-AES/MS) in order to determine the major and trace element compositions. The copper isotope ratios were analyzed using multiple collector–inductively coupled plasma–mass spectrometry (MC-ICP-MS) to learn more about the copper mineral information.

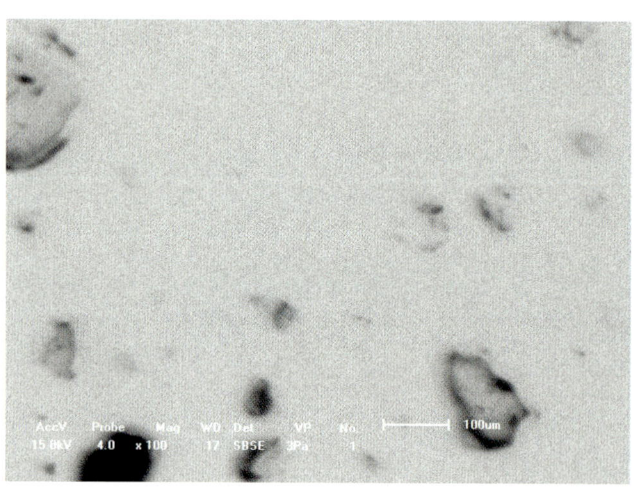

***Figure 5**. Backscattered electron image of the fragment from the metal block.*

Metallographic Microstructure

A small piece was cut from the metal block with a hacksaw. The specimen was mounted in epoxy resin. After grinding and polishing, it was examined using optical microscopy and SEM. The results are shown in figures 4 and 5.

Chemical Composition

A 92.8 mg sample was dissolved in hydrochloric acid to analyze the major and trace element compositions using ICP-AES/MS. The result shows that the metal block is more than 99 wt.% copper with no alloying elements present, thus showing that it is a high-purity copper (fig. 6). The so-called metal block should be called a "copper block." Trace element data are listed in table 1.

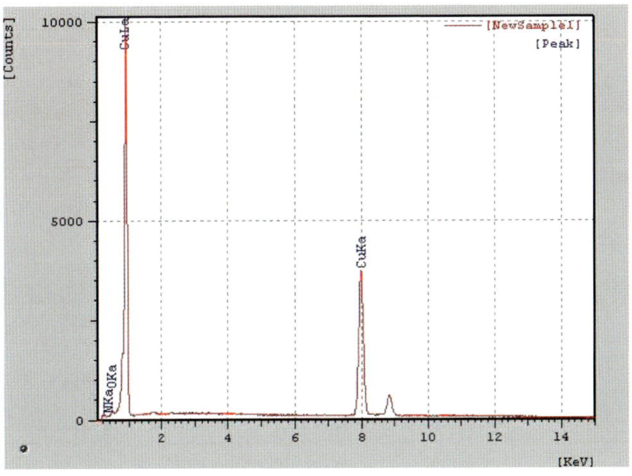

Figure 6. *SEM-EDS metal block data.*

Copper Isotope Analysis

Copper isotope analysis is a new method in archaeological study (Gale et al. 1999) that shows great promise. Although the regularity of copper isotope fractionation in natural copper minerals is still under study, recent research has already proven that copper isotopes do not show detectable fractionation in the processes of smelting or casting. On the other hand, copper isotope compositions vary in different types of copper minerals and, as a result, copper isotope analysis can be applied in archaeological research to the study of copper sources. Until now, the two successful applications of the method have been a study on turquoise (Hull et al. 2008) and a study on sources for copper coins in the United States in the nineteenth and twentieth centuries (Mather 2009).

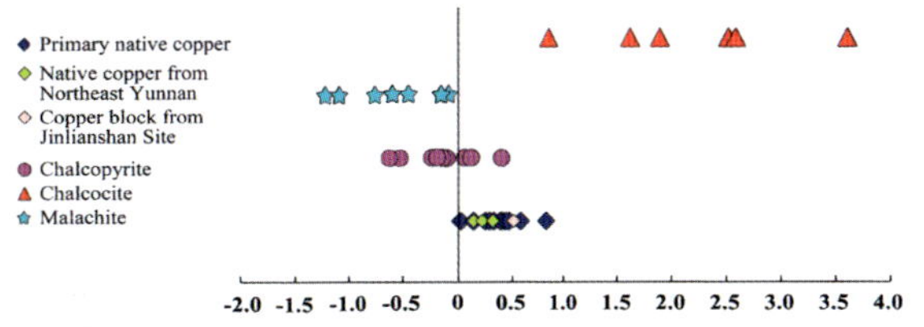

Figure 7. *Variation of δ^{65}Cu values of several copper minerals commonly used in the Bronze Age (Jiang 2003; Zhu et al. 2000; Maréchal et al. 1999) and the four samples in this study.*

Figure 7 shows the copper isotope variation of several copper minerals used in the Bronze Age. The copper isotope composition of chalcopyrite is in a narrow region because chalcopyrite is formed in high-temperature hydrothermal environments, while malachite and native copper can be easily distinguished. Research on the application of copper isotope analysis to archaeological studies began in our laboratory in 2006 with the help of Zhu Xiangkun. The experimental results for thirteen native copper samples from North America (Wang 2009) and the published

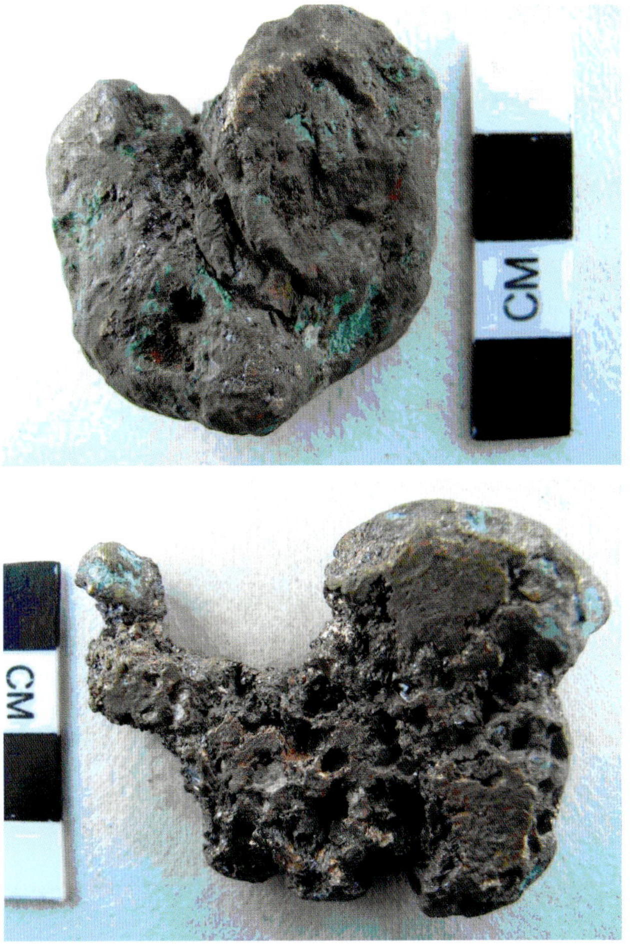

Figure 8. *Native copper from northeast Yunnan.*

Table 1. *Element composition of the "copper block" (sample no. ZY-6000).*

Cu	Pb	Sn	Mn	Fe	Ni	W	Mo	Th	Be	Ag	Co	Sr	Zn	Cd	Rb	Cr	Bi	Zr	Hf	As	Hg	Sb
>99	644	323	tr	tr	591	bdl	bdl	bdl	bdl	431	103	162	539	bdl	bdl	38	bdl	tr	bdl	tr	bdl	692

Cu in wt.% and analyzed by ICP-AES/MS, other elements in µg/g and analyzed by MC-ICP-MS. b.d.l. = below detection limit; tr. = trace and below the blank.

Table 2. *Copper isotope data of the metal block and native copper from northeast Yunnan.*

Sample (No.)	Metal Block (ZY-6000)	Samples of Native Copper from Northeast Yunnan		
		(ZY-2020)	(ZY-2021)	(ZY-2022)
δ^{65}Cu	0.51‰	0.22‰	0.14‰	0.32‰

The accuracy of δ^{65}Cu is 0.08‰ within 95% confidence limits.

data for seven other samples of native copper from North America show a narrow variation of δ^{65}Cu, within a range of 0‰–1‰.

Copper isotope analysis was carried out using Nu Plasma high resolution MC-ICP-MS; an anion exchange resin method was used for the separation of copper. A sample-standard bracketing method was used for mass fractionation correction, with standard sample NIST SRM-976. The data from the copper block are δ^{65}Cu = 0.51‰ (shown in table 2). In addition, three native copper samples from northeast Yunnan were also analyzed (fig. 8), and those results are also shown in table 2. δ^{65}Cu is defined as follows:

$$\delta^{65}Cu(\text{‰}) = [\frac{(^{65}Cu/^{63}Cu)_{sample}}{(^{65}Cu/^{63}Cu)_{standard}} - 1] \times 1000$$

As shown in figure 7, the result indicates that the mineral source of the copper block is not malachite. Based on the major and trace elements data, this copper block was determined to be high-purity copper. It is impossible to make such pure copper from chalcopyrite because of the limitations of ancient technology, so the possibility of chalcopyrite is excluded. The native copper from northeast Yunnan has a δ^{65}Cu variation of 0.14‰–0.32‰, which differs from the data of the copper block. Lacking more isotopic data from native copper, it is hard to determine if the source of the copper block is northeast Yunnan.

Pre-Ingot or Post-Ingot: The Metal Block and Its Technology

A series of processes—including mining, ore dressing, smelting, refining, and casting—are needed to make ore into metal artifacts. What was the process for production of the ingot-like copper block?

The multilayer structure of the metal block is obviously different from metal ingots found all over the world, such as the lead ingots of the Shang dynasty (ca. 1600–1050 BCE) found in Yinxu, the copper ingots of the Zhou dynasty (ca. 1050–221 BCE) found in the Zhouyuan area and the middle valley of the Yangtze River, and the oxhide ingots found in the Mediterranean area. It is neither like a copper ingot in its shape nor is it like a typical metal fragment.

As noted, the results of compositional analysis show that the metal block is high-purity copper. One can infer that it is probably a freshly smelted product from a furnace. Illustrations in the book *Tian Gong Kai Wu* (Exploitation of the Works of Nature) by Song Yingxing of the Ming dynasty (1368–1644) include copper furnace tappings, tin furnace tappings, and lead furnace tappings. They all have irregular shapes (fig. 9) and would need remelting to become the regular ingots used for transport. Judging from this, the Jinlianshan copper block appears to be a furnace tapping (i.e., the remains of the pre-ingot stage). However, it is still mere supposition because we lack any comparable materials.

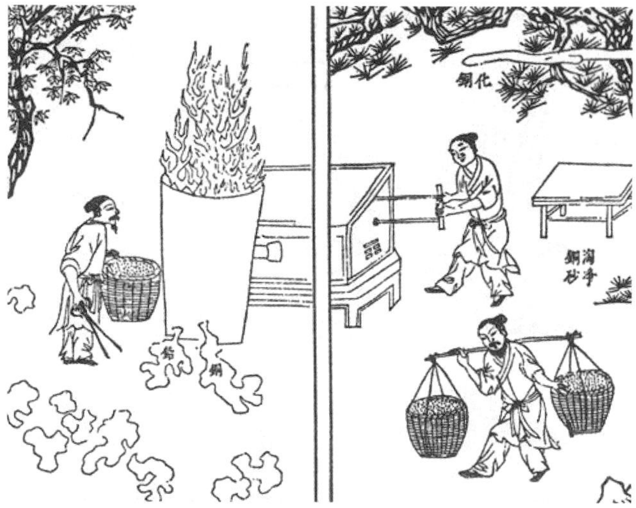

Figure 9. *Furnace tappings of copper and lead illustrated in* Tian Gong Kai Wu.

Discussion and Preliminary Conclusions

The metal block unearthed at M1 in the Jinlianshan site is probably a primary metal product from the furnace. This is not only the first discovery of such a metal block in the history of bronze technology from the ancient Dian kingdom but also the first discovery of its type from ancient China.

According to the results of metallographic microstructure, SEM, chemical composition, and copper isotope analyses, this primary metal product is made of native copper mineral and is high-purity copper. Native copper still exists in northeast Yunnan today, and it is possible that there were native copper mines in ancient times in the adjacent area of Lake Fuxianhu, given the richness of copper resources there. Thus, the copper block might have been made from the ancient native coppers in central Yunnan. This question needs to be studied further.

The relics related to melting or casting activity unearthed at the Jinlianshan site have provided precious material for studies on the bronze technology of the ancient Dian kingdom. They may also provide important clues that help to reveal the historical correlations among the sites at Jinlianshan, Shizhaishan, and Lijiashan.

Acknowledgments

This research was supported by National Natural Science Foundation of China grant 41073004. The authors are grateful to Professor Bingquan Zhu of Guangzhou Institute of Geochemistry, China Academy of Sciences, and Professor Xiangkun Zhu of Institute of Geology, China Academy of Geological Sciences, for their kind help and valuable suggestions on this research. Thanks are also due to Mr. W. T. Chase for his revision of the manuscript and professional suggestions, and to Mr. Peifang Zi of the Chengjiang County Administration of Culture Heritage for his great help in sampling.

References

Gale, N. H., A. P. Woodhead, Z. A. Stos-Gale, A. Walder, and I. Bowen. 1999. Natural variations detected in the isotopic composition of copper: Possible applications to archaeology and geochemistry. *International Journal of MASS Spectrometry* 184(1):1–10.

Hull, S., M. Fayek, F. J. Mathien, P. Shelley, and K. R. Durand. 2008. A new approach to determining the geological provenance of turquoise artifacts using hydrogen and copper stable isotopes. *Journal of Archaeological Science* 35(5):1355–69.

Jiang, S. Y. 2003. Transition metal isotopes: Analytical methods and geological applications. *Earth Science Frontiers* 10(2):269–78.

Maréchal, C. N., P. Télouk, and F. Albarède. 1999. Precise analysis of copper and zinc isotopic compositions by plasma-source mass spectrometry. *Chemical Geology* 156(1–4):251–73.

Wang, C. M. 2009. Geochemical analyses of native copper in North America: Implication for archaeological provenance studies. PhD diss., University of Science and Technology of China.

Zhu, X. K., R. K. O'Nions, Y. Guo, N. S. Belshaw, and D. Rickard. 2000. Determination of natural Cu isotope variation by plasma-source mass spectrometry: Implications for use as geochemical tracers. *Chemical Geology* 163(1–4):139–49.

Some Decorative Techniques and Corrosion on Chinese Bronze Mirrors in the Cotsen Collection

David A. Scott

ABSTRACT Technical examination of several Chinese bronze mirrors in the Cotsen collection revealed that several were decorated using techniques not previously described in the literature. A double-tier mirror was shown to have a silk interlayer between the front and the back. Another mirror illustrates the probable use of textiles to create designs that were then impressed into wax or clay to create cast decoration in bronze. A mirror from the Six Dynasties period was shown to be inlaid with intaglio glass decoration of both Chinese and imported glass, with one inlay shown to be of gold flakes in glass decoration contrasting with gold leaf and different colored glass inlay. Additionally, the problems of the reuse and redecoration of authentic, old mirrors are highlighted by one example from the Sui period with mineral paste decoration.

Introduction

As part of a much wider study of bronze mirrors in the Cotsen collection, Beverly Hills, California, an examination of several decorated examples from different dynastic periods were examined. The examinations revealed a number of interesting decorative techniques that employed glass, textiles, paint, and ceramic-related pastes to create designs on mirror backs. While many bronze mirrors have decorated backs that are cast into the bronze itself, mirrors that rely on additional components to create motifs are more unusual. The entire Cotsen collection, which is unusually rich in unique examples of decorated mirrors, has been published in two substantial volumes. These volumes, edited by Lothar von Falkenhausen and available from the Lloyd Cotsen Foundation, represent a substantial scholarly contribution to the subject of Chinese bronze mirrors. The first volume, written by Suzanne Cahill (2009), describes all of the mirrors from an art-historical perspective. Throughout this paper, reference is made to this volume for many sumptuous pictures of the mirrors.

Some Decorative Techniques

Most of the analyses of the Cotsen bronzes were conducted with a Bruker handheld x-ray fluorescence spectroscopy (XRF) instrument using an Al/Ti filter at 40 kV and 1.5 microamps for 180 seconds. Standards made from ternary alloys of copper, tin, and lead were used for comparison. The usual range of accuracy quoted for such analyses is +/- 10% for major elements and +/- 20% for minor and trace elements. If the mirror surfaces are heavily corroded, then the analytical results are not reliable at all (see table 1 for results). The standards file of alloys of known composition only provides reasonable results if the corrosion or patina is very thin or if bare metal is being analyzed. The nonreflecting rear surfaces of all mirrors in the Cotsen collection are decorated. Techniques used for decoration and coating include gilding, silvering, plating, lacquering, painting, and inlay. Several double-tier mirrors are also part of the collection. These are relatively heavy, complex mirrors with a plain reflecting surface riveted, soldered, or glued to the decorated back. This type of mirror first evolved, according to Dohrenwend (1964, 80), during the Spring and Autumn period (770–476 BCE), though the specimens in the Cotsen collection date no earlier than the Warring States period (475–221 BCE). The inlay materials used are malachite, copper, gold, turquoise, jade, shell, mother-of-pearl, a variety of other minerals, and glass paste. Glass arrived late in China, the earliest archaeological evidence coming from the second half of the first millennium BCE (Brill et al. 1991). Common adhesives used to attach inlay were animal glue and plant gums. For example, Fourier transform infrared spectroscopy (FTIR), applied by Costello to a double-tier mirror (HUAM 1943.52.155) and a Tang dynasty mirror (HUAM 1943.52.158) in the collections of the Harvard University Art Museums showed that the adhesive used on the double-tier mirror was a natural gum, while a sheet metal overlay on the Tang dynasty mirror was affixed with a mixture of animal glue and flour paste (Costello 2005).

An interesting example in the Cotsen collection is the double-tier mirror O-0360, dated to the Warring States period (fig. 1). The circular outer band of the decorated back is inlaid in small pieces of malachite and turquoise.

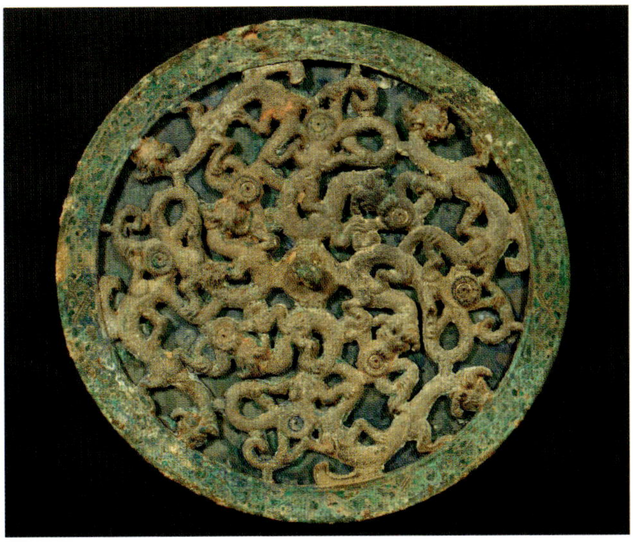

Figure 1. *Mirror O-0360, Warring States period. The alloy for the mirror part is 69.2 wt.% Cu, 22.8% Sn, 3.7 wt.% Pb. The alloy used for the openwork design of the back is of lower tin content and higher lead. Photograph courtesy of the UCLA Cotsen Institute of Archaeology.*

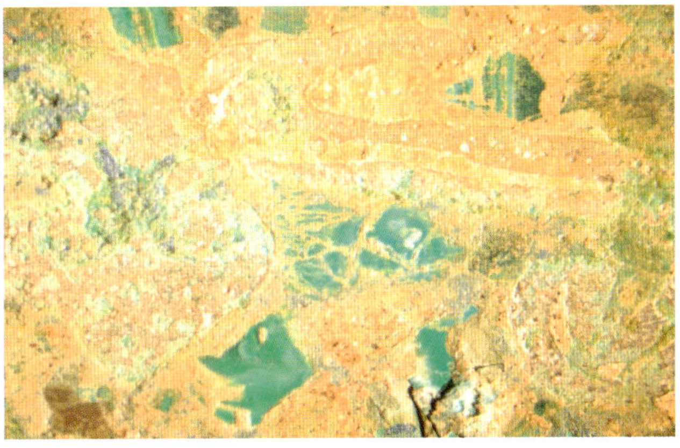

Figure 2. *Magnification (×35) of inlay on the outer rim of mirror O-0360 showing small pieces of malachite and turquoise decoration.*

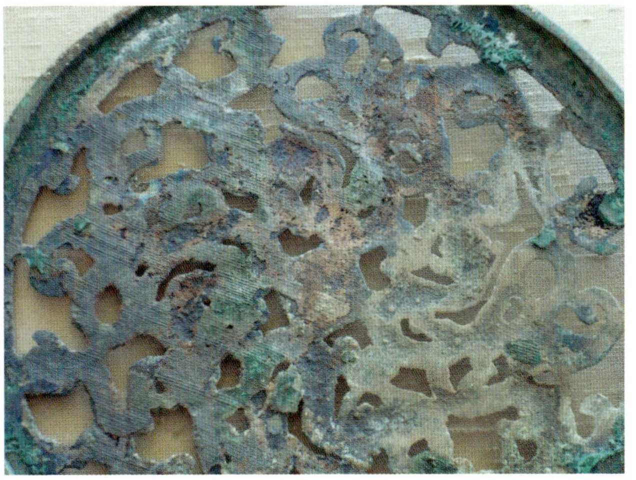

Figure 3. *Reverse of the back side of mirror O-0360 showing well-preserved textile pseudomorphs parallel across the back, indicating that originally it would have had a textile interlayer. Photograph by the author.*

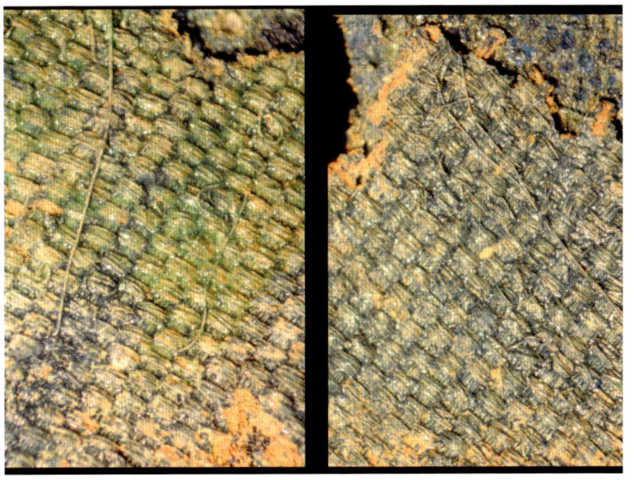

Figure 4. *Detail of mirror O-0360. Preserved textile of silk, multi-strand, plain-weave pseudomorphs on the rear of the back plate show that the original decorative scheme incorporated a textile interlayer. Photograph by the author.*

The inlaid minerals were partially obscured by subsequent corrosion of the bronze of the mirror itself, which produced cuprite, malachite, and in some places azurite. This visual confusion between the original inlaid mineral components and subsequent corrosion that produced minerals of essentially the same composition is the reason why the design is not immediately apparent. When the inlay is examined with a bench binocular microscope at ×20, then the random orientation of the inlaid minerals becomes evident, as shown in figure 2.

The flat polished disc of the front of this mirror can be detached from the back, which reveals excellently preserved textile pseudomorphs adhering to the rear surface of the decorated back plate. The direction of the weave of the remaining textile morphology is the same across the entire back side of the surface, showing that the decorated back had a silk textile inner surface between it and the flat front plate (fig. 3). In exposed areas of the design, the textile has completely degraded and been lost, while the copper corrosion products developing on the interior of the back have preserved the textile remnants from biodeterioration.

This textile was a plain weave with multiple strands of silk fibers forming both the warp and weft (fig. 4). The blue material seen in this figure is crystals of azurite, which often form in regions where the carbon dioxide content of the environment is elevated, such as regions poorly accessible and hidden under corroding bronze, as here. Because it is often not possible to separate the two parts of these double-tier mirrors, there may be other examples of textiles having been used as a decorative element to complement the bronze, but they have not yet been discovered. Most two-tier mirrors cannot be separated without damage, and some are riveted or soldered together to prevent the two parts from detaching, so it is entirely fortuitous that the mirror components examined here can be separated. To my knowledge, the present example of a two-tier mirror with a silk decorative layer sandwiched between is the first one to be published.

While on the subject of textiles, it is worth commenting on the extraordinary mirror O-0186 (see Cahill 2009, pl. 54), also from the Warring States period, which has been published by Mackenzie (1999) (fig. 5). Its decoration is in

Table 1. *Compositional and technical information (wt.%) regarding the Cotsen Mirror Collection. Most of the analyses were performed with a Bruker handheld XRF instrument. See the text for details. Some additional analyses were carried out with electron microprobe analysis on polished sections from seven of the mirrors. This analysis was performed on a JEOL SuperProbe at the Department of Geology, UCLA, with the assistance of Dr. Frank Kyte.*

Accession No.	Dynasty or Date	Cu	Sn	Pb	As	Fe	Ag	Au	Hg	Zn	Comment
O-0803	Qijia	75.2	18.6	0.3	nd	0.1	nd	nd	nd	nd	
O-0427	Shang	75.3	3.7	nd	7.1	nd	1.4	nd	nd	nd	high As content
O-0424	Eastern Zhou	53.3	10.5	29.9	5.9	0.4	nd	nd	tr	nd	high As content; some Hg
O-0648	Spring & Autumn	79	18.5	0.3	nd	0.3	nd	nd	nd	nd	
O-0200	Spring & Autumn	72	13.2	14.5	nd	nd	nd	nd	nd	nd	
O-0421	Eastern Zhou	61.2	24.3	1.7	nd	nd	nd	nd	nd	nd	
O-0201	Warring States	59.7	14.3	18.9	6	0.1	nd	tr	nd	nd	high As content; trace Au
O-0186	Warring States	75.1	20.2	3.7	0.3	0.3	nd	nd	nd	nd	
O-0833	Warring States	69.2	22.8	5.3	0.7	0.4	nd	nd	nd	nd	
O-0778	Warring States	84.2	10.2	6.2	nd	0.04	nd	nd	nd	nd	
O-0180	Eastern Zhou	59.6	17.1	5.4	0.7	0.3	nd	nd	nd	nd	
O-0279	Warring States	—	—	—	—	—	—	—	nd	—	
O-0129	Warring States	59	22.1	10.1	nd	0.1	nd	nd	tr	nd	reproduction
O-0096	Warring States	62.1	22.1	8.1	nd	0.3	nd	nd	tr	nd	reproduction
O-0127	Warring States	63.3	21.7	6.5	nd	0.01	nd	nd	tr	nd	reproduction
O-0398	Warring States	60.6	19.4	2.2	0.6	nd	tr	nd	nd	nd	
O-0128	Warring States	—	—	—	nd	—	—	—	tr		
O-0420	Warring States	—	—	—	tr	—	—	—	nd	—	high tin content surface layer
O-1504	Warring States	29.8	22.3	29	nd	nd	nd	nd	nd	nd	very mineralized
O-1506	Warring States	16.9	30.9	34.1	nd	nd	nd	nd	nd	nd	very mineralized
O-0646	Warring States	3.4	50.4	31.6	nd	nd	nd	nd	nd	nd	very mineralized
O-0457	Warring States	63.4	24.1	3.5	nd	nd	nd	nd	nd	nd	
O-0460 back	Warring States	71.8	21.4	8.2	nd	nd	nd	nd	nd	nd	
O-0460 front	Warring States	77.4	19	6.5	nd	nd	nd	nd	nd	nd	
O-0458 back	Warring States	66.3	17.9	17.4	nd	nd	nd	nd	nd	nd	
O-0131	Eastern Zhou	—	—	—	—	—	—	—	—	—	
O-0407	Warring States	72.8	24.5	8.6		nd	nd	nd	nd	nd	nd
O-0720 back	Warring States	74.5	15.4	9.6	0.4	nd	nd	nd	nd	nd	
O-0720 front	Warring States	74.5	19	3.3	0.4	nd	nd	nd	nd	nd	
O-0750	Warring States	49.3	19.6	25.8	nd	nd	nd	nd	nd	nd	surface very mineralized
O-0293	Warring States	61.7	19.1	21.9	nd	nd	nd	nd	nd	nd	highly leaded
O-0800 front	Warring States	48.2	22.9	27.2	nd	nd	nd	nd	tr	nd	
O-0800 back	Warring States	56.4	16.5	33.2	nd	nd	nd	nd	nd	nd	
O-0360 back	Warring States	25.6	13.2	59.3	6.9	1.0	nd	nd	nd	nd	high As
O-0360 front	Warring States	69.2	22.8	3.7	0.4	nd	nd	nd	tr	nd	
O-0295	Warring States	—	—	—	—	—	—	—	nd	—	
O-0185	Warring States	nd	nd	nd	nd	nd	nd	nd	nd	nd	surfaces covered in laquer
O-0245	Western Han	nd	nd	nd	nd	nd	nd	nd	nd	nd	
O-0779	Western Han	70.7	21.9	5.1	0.5	0.07	nd	nd	nd	nd	
O-0423	Western Han	—	—	—	—	—	—	—	nd	—	
O-0304	Western Han	—	—	—	—	—	—	—	nd	—	
O-0422	Western Han	69.4	22.9	0.5	0.5	0.6	nd	nd	nd	nd	trace of Mn
O-0278	Western Han	—	—	—	—	—	—	—	nd	—	
O-0856	Western Han	—	—	—	tr	—	—	—	nd	—	
O-0399	Eastern Han	61	23.9	3.3	1.2	0.4	nd	nd	nd	nd	0.05 wt.% Ni
O-0775	Eastern Han	—	—	—	—	—	—	—	nd	—	
O-0321	Western Han	—	—	—	—	—	—	—	nd	—	
O-0349	Eastern Han	—	—	—	tr	—	—	—	—	—	high tin content surface layer; trace Sb
O-0179	Eastern Han	—	—	—	—	—	—	—	nd	—	
O-0843	Eastern Han	—	—	—	tr	—	—	—	nd	—	high tin content surface layer
O-0233	Eastern Han	—	—	—	—	—	—	—	nd	—	high tin content surface layer
O-0744	Eastern Han	62.5	13.9	18.9	tr	0.3	nd	nd	nd	tr	gilded back
O-0133	Three Kingdoms	—	—	—	—	—	—	—	nd	—	high tin content surface layer
O-0246	Six Dynasties	—	—	—	tr	—	—	—	nd	—	high tin content surface layer
O-0226	Eastern Han	—	—	—	—	—	—	—	nd	—	
O-0832	Six Dynasties	79	22.3	0.5	0.25	0.2	nd	nd	nd	nd	
O-0815	Jin/Southern	59.4	22.7	5.7	1.2	0.2	nd	nd	nd	nd	
O-0774	Northern Wei/Sui	75.9	21.4	3.6	0.2	0.2	nd	nd	nd	nd	
O-0363	Sui/Early Tang	69.7	24.8	3.9	nd	nd	nd	nd	nd	nd	
O-0232	Sui/Early Tang	—	—	—	—	—	—	—	nd	—	
O-0231	Sui	70.1	24.9	3.5	nd	0.01	nd	nd	nd	nd	
O-0857	Sui	70.8	24.6	2.3	nd	0.4	nd	nd	nd	nd	
O-0193	Sui	70.5	23.4	3.7	0.5	0.09	nd	nd	nd	nd	

Accession No.	Dynasty or Date	Cu	Sn	Pb	As	Fe	Ag	Au	Hg	Zn	Comment
O-0426	Tang	69	23.1	2.3	1.4	0.05	nd	nd	nd	nd	mercury amalgam gilded
O-0308	Tang	61.7	23.2	3	0.7	0.6	nd	nd	nd	nd	silvered back: Ag, Pb, trace Fe, trace Cu
O-0792	Tang	66.4	23.9	1.8	0.7	0.12	nd	nd	nd	nd	0.01 wt.% Ni; back: high Ag, Au; trace Cu, Pb, Hg
O-0647	Tang	68.2	24.8	2.9	0.4	0.1	nd	nd	nd	nd	back: high Ag, little Au, trace Pb, Fe, Cu
O-0234	Tang	67.5	25	2.7	nd	0.05	nd	nd	nd	nd	
O-0742	Tang	70.1	25.2	2.2	0.6	0.4	nd	nd	nd	nd	0.01 wt.% Ni
O-0753	Tang	76.6	21	3.6	nd	0.3	nd	nd	nd	nd	
O-0874	Tang	70.3	25.3	0.2	nd	0.04	nd	nd	nd	nd	
O-0134	Tang	—	—	—	—	—	—	—	nd	—	
O-0312	Tang	70.9	20.9	7.3	1.4	0.3	nd	nd	nd	nd	
O-0418	Tang	65.6	16	18.5	2.4	0.7	nd	nd	nd	nd	
O-0649	Tang	69.6	20.2	3.2	0.2	2.4	nd	nd	nd	nd	
O-0781	Ordos	64.4	19.6	15	2.4	0.1	nd	nd	nd	nd	
O-0865	Tang	68.1	19.1	3	0.9	3.6	nd	nd	nd	nd	
O-0313	Tang	69.7	23.8	3.1	0.25	0.07	nd	nd	nd	nd	0.06 wt.% Ni
O-0425	Tang	63.5	23.9	3.2	tr	0.4	nd	nd	nd	nd	
O-0305	Tang	71.7	23	4.9	nd	0.09	nd	nd	nd	nd	
O-0135	Tang	—	—	—	—	—	—	—	nd	—	high tin content surface layer
O-0782	Tang	66.7	23.1	3.2	1.2	0.7	nd	nd	nd	nd	
O-0752	Tang	64	23.9	3.6	1	0.2	nd	nd	nd	nd	trace Au
O-0255	Tang	70.9	23.7	3.6	0.7	0.08	nd	nd	nd	nd	
O-0743	Tang	68.3	22.8	3.5	0.4	0.08	nd	nd	nd	nd	
O-0867	Tang	70.9	23.5	4.2	0.7	0.2	nd	nd	nd	nd	
O-0323	Tang	52.7	12.9	34.5	nd	0.13	nd	nd	high	nd	
O-0751	Jin	74.8	20.2	1.5	tr	0.03	nd	nd	nd	nd	
O-0322	unknown	69.4	23.2	3.7	0.7	0.01	nd	nd	nd	nd	0.01 wt.% Ni
O-0804	unknown	74.8	25.6	0.2	0.4	0.3	nd	nd	nd	nd	0.03 wt.% Ni
O-1505	unknown	55.8	18.1	17.8	3.7	0.1	nd	nd	nd	nd	
O-0291	unknown	54.5	18.9	16.1	2.4	0.06	nd	nd	tr	nd	
O-0292	unknown	76.3	2.5	7.3	nd	0.9	nd	nd	nd	7.6	note high zinc content; 0.1 wt.% Ni
O-0835	11th–13th centuries CE	—	—	—	tr	—	—	—	nd	—	high tin content surface layer

nd = not detected; tr = trace; blank = no data obtained.

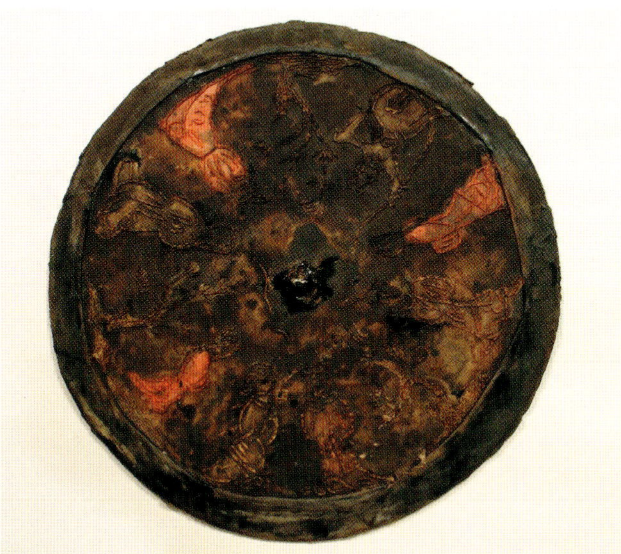

Figure 5. *Mirror O-0186 decorated with chain-stitched silk. The mirror itself is authentic judged on the basis of composition and corrosion, the latter being more decisive. XRF analysis gave 75.1 wt.% Cu, 20.2 wt.% Sn, 3.7 wt.% Pb, 0.3 wt.% As, and 0.3 wt.% Fe with no mercury detected. Photograph courtesy of the UCLA Cotsen Institute of Archaeology.*

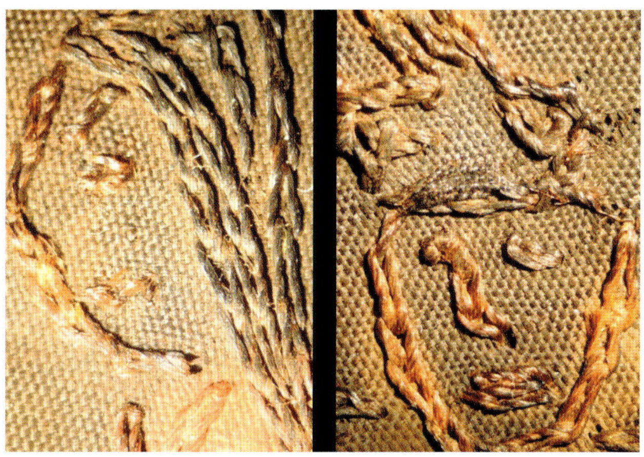

Figure 6. *Detail (×23) of mirror O-0186 showing part of the silk design in chain stitch with some traces of yellow pigment, especially to the left-hand side of the photograph. Photograph by the author.*

embroidered silk, which has apparently survived for 2,500 years! A coarse chain stitch is used to depict a figure in a red robe in juxtaposition with several other figures. Red and yellowish-gold pigments painted on the surface after the embroidery was completed make the design of the scene easy to read (fig. 6). Mackenzie (1999, 91) notes:

> [I]f the decoration because of its sophistication suggests a date as late as the Han period, the form of the mirror suggests an earlier one. The back has a simple loop in the center and a raised concentric ridge about one centimetre in from the edge. Plain backed mirrors like this are not uncommon in the fourth century BCE in Chu tombs and may originally have been painted,

or as this mirror suggests, embellished with a textile design.

Technical examination found nothing to contradict Mackenzie's assessment, and this mirror remains a marvellous illustration of another kind of surface design, which in most cases would have been subject to biodeterioration and dissolution. Radiocarbon dates gave an anomalous early date for this textile, which cannot be correct, but it is unlikely to be of modern manufacture.

Textiles may also have been used to create patterns in molds or wax models that were subsequently faithfully reproduced in the casting. Some of these fine textile- or feather-related patterns are to be seen in many Warring States-period mirror designs (Bulling 1955). The problem of how these patterns were created has been addressed by Maryon (1963), who examined an interwoven dragon motif on a mirror of Shouzhou (壽州) type from the Warring States period in the Royal Ontario Museum (933.12.18). Maryon suggests this pattern was made from spun threads, probably wool, slightly stiffened in gum. These threads were dipped in molten wax or lacquer to form the background design, which, when completed, was pressed into the clay form to produce raised textile-related designs on the mirror surface. Intricate surface designs could certainly have been made by such a technique. To illustrate this, we can look at mirror O-0460 shown in figure 7 and a photomicrograph of the design in figure 8 (see Cahill 2009, pl. 56). It is easy to visualize the animal figure as a cut-out in wax applied to the surface of a hardened textile backing, which was then used to create a ceramic mold. The protruding wax animal form would make a deeper depression in the clay, which would then be reproduced as an elevated feature on the mirror, as we see in this case.

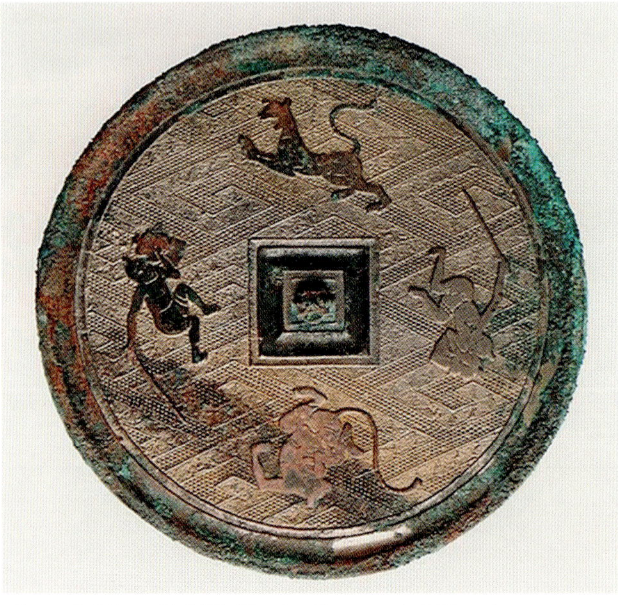

Figure 7. *Mirror O-0460, Warring States period. The mirror preserves a partially uncorroded back with animal designs. XRF analysis of the back gave 71.8 wt.% Cu, 21.4 wt.% Sn, and 8.2 wt.% Pb. XRF analysis of the front gave 77.4 wt.% Cu, 19.0 wt.% Sn, and 6.5 wt.% Pb. Photograph courtesy of the UCLA Cotsen Institute of Archaeology.*

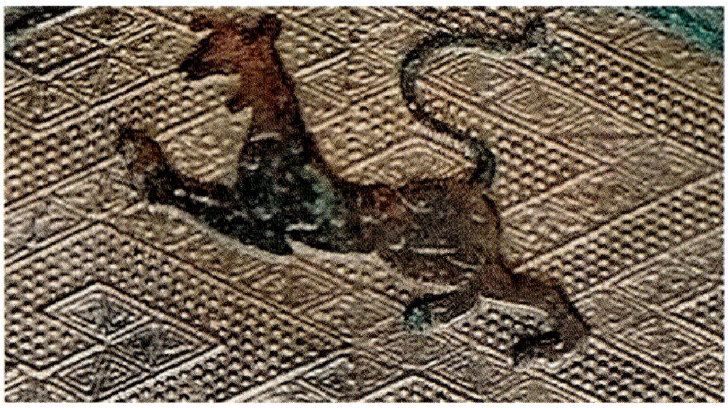

Figure 8. *Magnification (×25) of mirror O-0460 showing the probable use of a stiffened textile design over which wax or cut-out figures are then applied before creation of the ceramic mold. Photograph by the author.*

A finely painted mirror in the Cotsen collection, O-0322 (see Cahill 2009, pl. 131), has a fairly convincing metallic composition, although it is considered a modern piece in the catalogue. The decoration is similar in concept to square mirrors of the Tang dynasty (618–907 CE), decorated in the *pingtuo* (平脫; a term also known by its Japanese pronunciation as *heidatsu*) technique, in which delicate designs of birds and flowers are cut out and embedded in a lacquered surface (Yen 2005, 5). A detail is shown in figure 9.

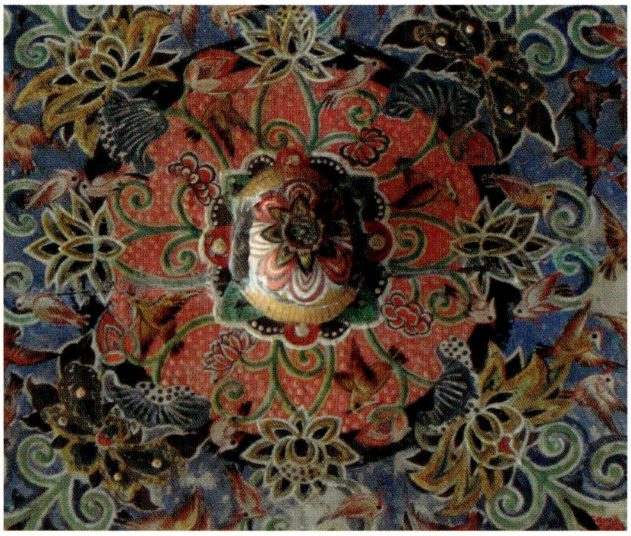

Figure 9. *Detail (×3) of mirror O-0322. Photograph courtesy of the UCLA Cotsen Institute of Archaeology.*

This mirror has also been repaired and then repainted with great skill, as can be seen from the central region of the design shown in figure 9. The composition of the mirror alloy is perfectly acceptable for age, although this can easily be faked, but more importantly, the corrosion on the front, shown in figure 10, cannot be so easily faked and looks to be authentic on microscopical study. The problem is, therefore, when this mirror was actually painted, and this is hard to ascertain.

Several mirrors are coated with lacquer. Coating samples from mirror O-0879 (fig. 11) were analyzed using gas chromatography–mass spectrometry (GC-MS), and the results indicated *urushi*.

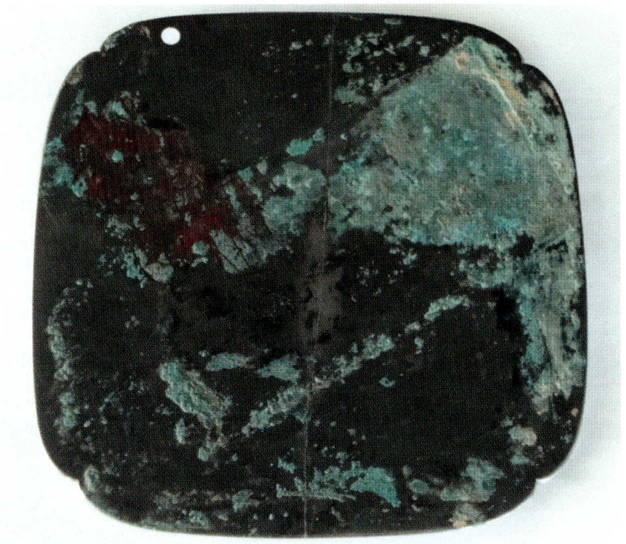

Figure 10. *Front view of mirror O-0322 showing patchy corrosion, although the mirror's morphology appears perfectly acceptable for age. XRF analysis gave 69.4 wt.% Cu, 23.2 wt.% Sn, 3.7 wt.% Pb, 0.7 wt.% As, and trace Ni and Fe. No mercury was detected, but the mirror does have arsenic content. Photograph by the author.*

Another interesting inlaid example, O-0832, shown in figure 12, dates from the Six Dynasties period (317–589 CE) or slightly later. It utilizes a number of different techniques and materials to achieve the inlaid design. Some of the inlays, in what could be described as an intaglio technique, are of minerals or glass inlay, such as the creamy-gray colored inlay made of glass (fig. 13, left half), while other inlaid designs are made of gold leaf on a decayed organic base (fig. 14).

One of the most interesting inlays is the glass paste material with flecks of gold randomly distributed within. The flecks are still vivid today and represent a sophisticated decorative use of gold for this time (see fig. 13, right

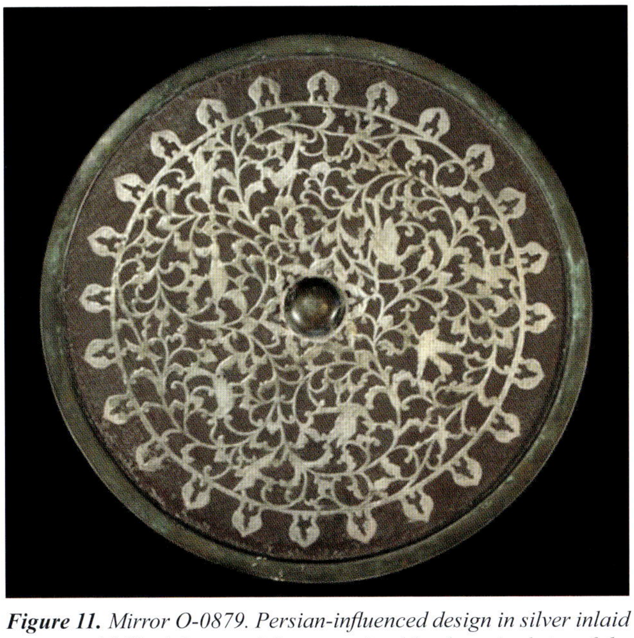

Figure 11. *Mirror O-0879. Persian-influenced design in silver inlaid in an* urushi *black lacquer inlay, contained by the raised rim of the mirror. XRF analysis of the mirror itself gave 71 wt.% Cu, 21 wt.% Sn, and 3 wt.% Pb. XRF showed that the lacquer is colored with iron oxides or iron sulfides, which are nonmagentic. Photograph courtesy of the UCLA Cotsen Institute of Archaeology.*

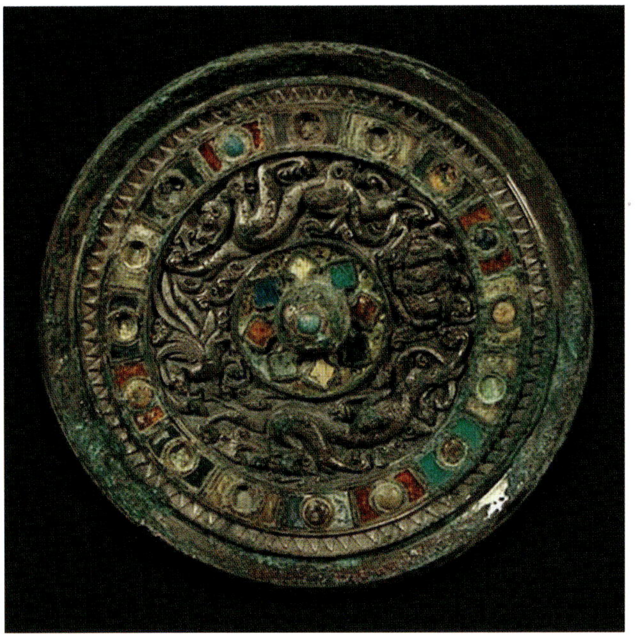

Figure 12. *Mirror O-0832, with the back extensively decorated with intaglio glass inlay work and an intricate animal design, Six Dynasties period. XRF showed that the alloy was essentially a copper-tin alloy with no other additions: 79.0 wt.% Cu, 22.3 wt.% Sn, 0.5 wt.% Pb, 0.25 wt.% As, and 0.05 wt.% Fe. Photograph courtesy of the UCLA Cotsen Institute of Archaeology.*

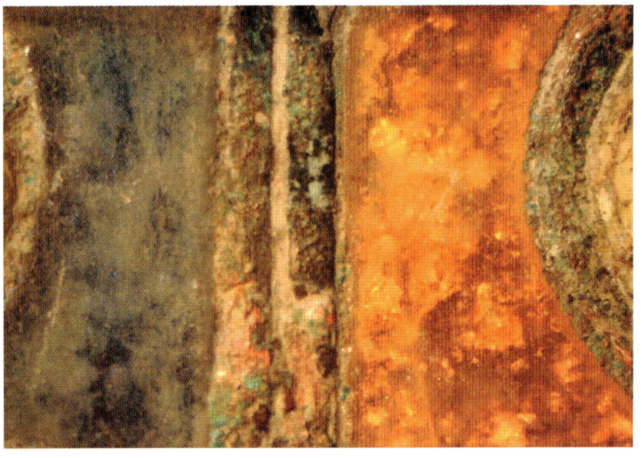

Figure 13. *Detail (×15) of mirror O-0832. Decayed glass inlay is visible on the left half of image. Flake gold in glass inlay is visible on the right half of image. Photograph by the author.*

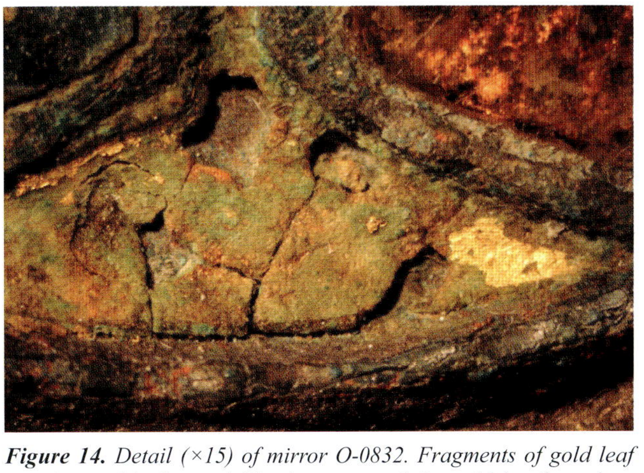

Figure 14. *Detail (×15) of mirror O-0832. Fragments of gold leaf remain on this decayed fill where most of the gold has been lost. The glass paste with gold leaf scattered inlay can be seen adjacent. Photograph by the author.*

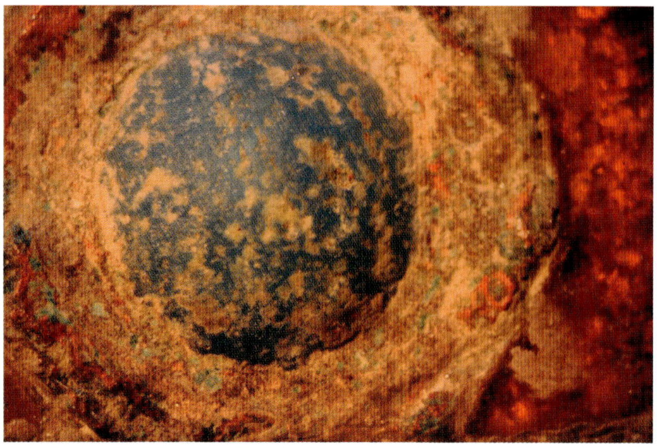

Figure 15. *Detail (×15) of mirror O-0832 showing the blue central glass boss and weathered surface. XRF analysis showed the presence of barium and probably a little lead, suggesting the use of a Chinese manufactured glass for this central boss. Photograph by the author.*

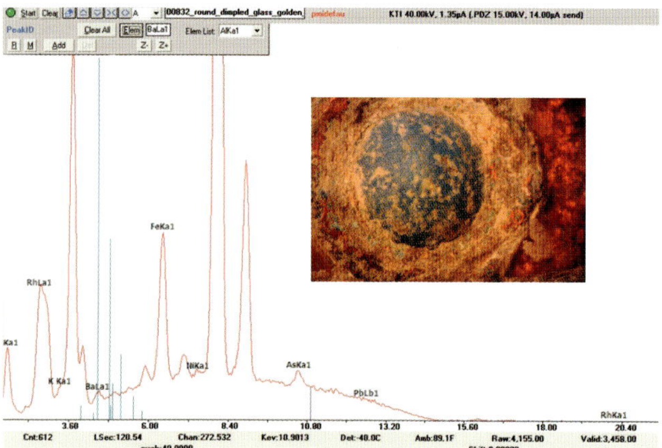

Figure 16. *Detail of mirror O-0832. The pale blue glass surrounding the central boss is of Chinese manufacture. X-ray fluorescence spectrum taken from the blue glass shows the presence of potassium, barium, iron, copper, arsenic, and lead.*

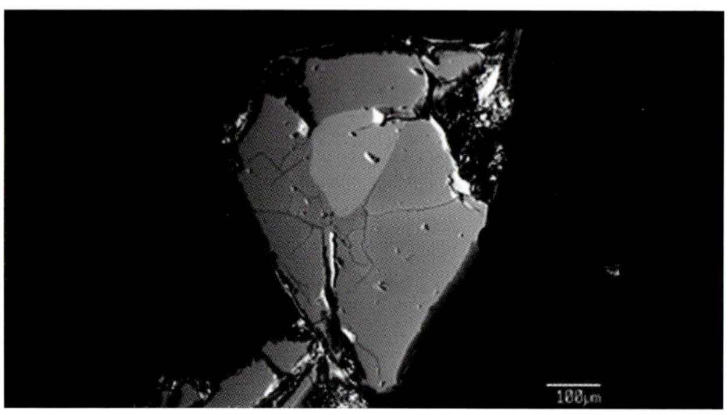

Figure 17. *Backscattered electron microprobe image for the light gray glass from mirror O-0832. The bright area in the center of the photomicrograph is uncorroded glass. Table 2 shows the elemental oxide concentrations obtained by electron microprobe analysis for four different areas of the glass compared with a typical eye bead.*

half). Conventionally, early Chinese techniques of decoration are said to have included *luodian* (螺鈿; inlaid shell; also known in Japanese pronunciation as *raden*) and the already-mentioned *pingtuo* (silver and gold sheet inlay), while the Japanese are credited with developing *makie* (蒔絵; "sprinkled illustrations" of gold filings) or *hiramakie* (平蒔絵), with the appearance of flat gold, often associated *urushi* (漆)). Mirror O-0832 has very well preserved sprinkled gold glass inlay, however, showing that this technique was already in use in China by the mid-first millennium CE. The use of gold foil in China goes back to the Shang period (ca. 1600–ca. 1050 BCE), but it is hard to find comparable examples of fine, sprinkled gold in glass decoration in other examples of Chinese art from the period during which this mirror was made, though there must be others.

Bronze mirrors with decorative glass intaglio are uncommon. From the late Warring States period, one specimen with ornamentation of hooks in the shape of the Chinese character *shan* (山) was excavated in 1992 from the site of the Eastern Zhou royal palace in Luoyang city (Luoyang Shi Wenwu Gongzuodui 1999, 26) and now resides in the collection of the Luoyang Museum. This mirror is 14.5 cm in diameter and has eighteen blue and white glass beads with dark blue eyes inlaid into the back surface. A recent study by Gao and Yang (2007) revealed another mirror from Luoyang containing inlaid glass beads with gold leaf wrapped around them. Examples of this kind of glass decoration have not been found outside of the Luoyang area. The example from the Cotsen collection is much later in date and may show that such techniques of decoration had been used in mirror production over a long period in different cultural settings.

XRF analysis performed *in situ* with a Bruker handheld instrument employing direct excitation at 40 kV and 1.35 microamps was used to study the composition of those glass inlays that could be positioned next to the head of the XRF instrument. The central bead of blue glass, shown in figure 15, gave a spectrum with peaks for copper, arsenic or lead, nickel, iron, barium, calcium, potassium, rhodium, and silicon. The presence of barium in this light blue, round glass bead shows that it is of Chinese fabrication, as lead and barium are typically found as major components only in ancient Chinese glass.

The flat, aquamarine-colored glass surrounding the central boss area, shown in figure 15, gave a clear XRF spectrum showing the presence of tin, rhodium, lead, copper, nickel, iron, barium, calcium, potassium, silicon, and aluminium. The rhodium content is from the XRF tube, and the tin and

Table 2. *Electron microprobe data (wt.%) for light gray glass from mirror O-0832 compared with typical data for a Chinese glass eye bead.*

Element Oxide	Area 1	Area 2	Area 3	Area 4	Eye Bead
SiO2	69.7	69.6	69.2	72.7	55.0
Al2O3	2.56	2.63	2.59	2.45	2.13
FeO	0.69	0.68	0.71	0.72	0.62
CaO	5.58	5.69	5.78	5.56	2.95
MgO	0.82	0.84	0.84	0.72	1.27
BaO	nd	nd	nd	nd	9.74
CuO	0.14	0.19	0.18	0.14	2.95
PbO	0.62	0.58	0.56	0.45	15.0
Na2O	19.52	19.03	19.28	0.11	7.53
K2O	0.62	0.59	0.60	0.25	4.05
Total	100.28	99.81	99.77	83.18	99.65

Areas 1, 2, and 3 represent the composition of the uncorroded glass, and area 4 is taken from the darker rim of corroded glass. nd = not detected.

copper are from the bronze. Because the lead content of the bronze alloy is so low, about 0.5 wt.%, the presence of larger lead peaks in the glass XRF spectrum together with the barium content argue for a typical Chinese glass composition in a leaded barium-containing glass used for the aquamarine blue of the flat glass elements surrounding the central boss (fig. 16).

Electron microprobe analysis was carried out on a tiny sliver of the light gray glass from one of the glass intaglios of mirror O-0832 (see fig. 13). The weathering of this glass sample was severe, as can be seen in the backscattered electron view of the polished surface shown in figure 17. The lighter area, which is uncrazed, is the only remaining region of sound glass. The exterior zone is heavily leached in alkalies, as the following analytical totals reveal. In table 2, areas 1, 2, and 3 represent the composition of the uncorroded glass, and area 4 is taken from the darker rim of corroded glass where extensive loss of sodium and some loss of lead and potassium have occurred. The sodium, lead, and potassium have been replaced by chlorine, due to the formation of chloride salts. Corrosion has not affected the brighter region shown in figure 17, and very good analytical totals were obtained from this glass. The composition of a typical Chinese eye bead from the late Warring States period is also shown for comparison in table 2. Chinese glasses are characterized by unique compositional characteristics, being high in both barium and lead. Interestingly, barium and lead are absent in the pale gray glass used on this mirror, meaning it is likely the bead was imported. During the centuries from the Warring States period to the Han dynasty, about the fifth century BCE to the early third century CE, glass was extensively imported to China from regions further to the west (Brill et al. 1991, 26). However, Mesopotamian glass usually has higher magnesium, calcium, or potassium content than that found in the glass on mirror O-0832: the composition would be atypical for a Mesopotamian origin (Moorey 1994, 211). Many early glasses roughly contemporaneous with the mirror O-0832, such as those of the Sasanian period (ca. 224–651 CE), were made with plant ash and have high levels of potassium and magnesium as a result (Brill 1999, 85–87). Consequently this glass is not from the Sasanian cultural area either. In fact, this low-magnesium and low-potassium glass prevailed in the Mediterranean area and continental Europe during the Roman and Byzantine periods (Shortland et al. 2006), whereas plant-ash glass is commonly found east of the Euphrates.

Another remarkable mirror in the collection, mirror O-0815 (see Cahill 2009, pl. 79), shown in figure 18, is thought by some to be a recent fake but is more probably an extraordinary example of reuse. The mirror was originally a plain-backed mirror, and the front of the mirror has a break across the entire surface. Examination of the corrosion on the front of this mirror suggested that it is perfectly acceptable for age (fig. 19). It is not a modern fabrication.

The mirror was repaired and the figures of the frolicking or boxing boys added. The presence of the heavily mineralized wood covering both the figures and the mirror back is evidence for the burial and natural corrosion of the mirror subsequent to the figural decoration being added to the back. Cahill (2009, 180) ascribes this painted scene to the Three Kingdoms period (220–280 CE) or the Six Dynasties period.

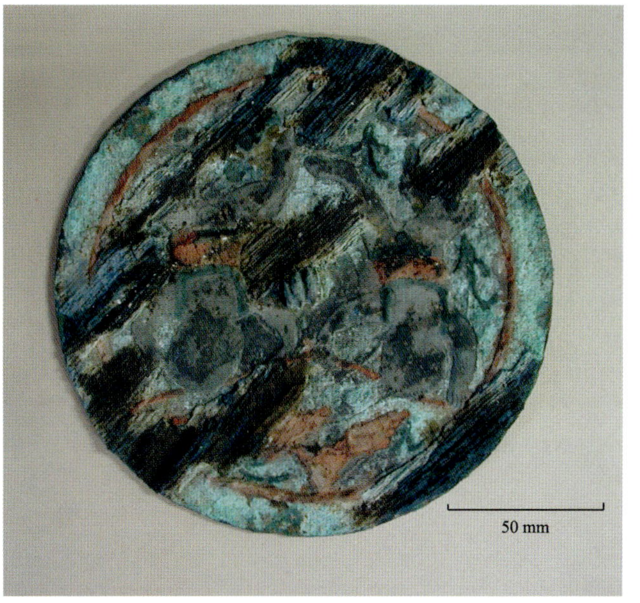

Figure 18. *Mirror O-0815, Jin/Southern dynasty. XRF gave 69.4 wt.% Cu, 22.7 wt.% Sn, 5.7 wt.% Pb, 1.2 wt.% As, 0.2 wt.% Fe, and no mercury, though the mirror does have arsenic content. Photograph courtesy of the UCLA Cotsen Institute of Archaeology.*

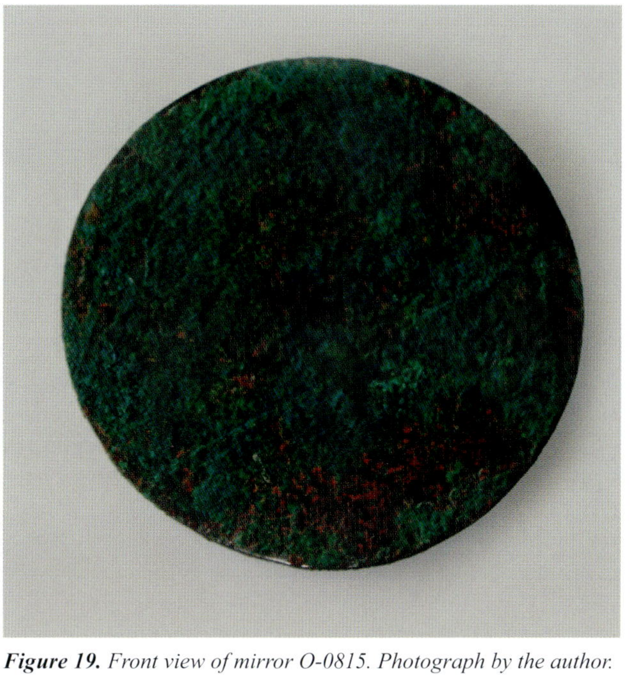

Figure 19. *Front view of mirror O-0815. Photograph by the author.*

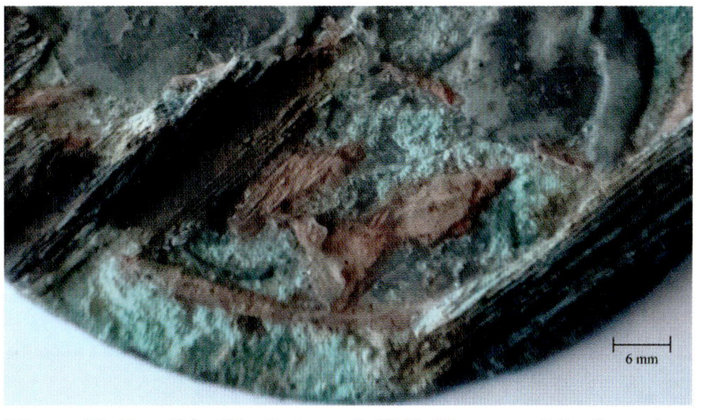

Figure 20. *Detail (×15) of mirror O-0815. Photograph by the author.*

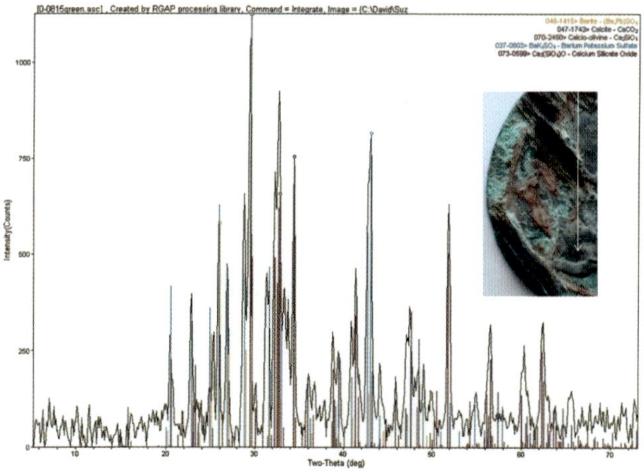

Figure 21. *X-ray diffraction scan of the green-gray mineral paste on mirror O-0815 showing the presence of a complex mineral assemblage of barite, calcite, calico-olivine, barium potassium sulfate, and calcium silicate oxide, the latter subsumed to be calcio-olivine.*

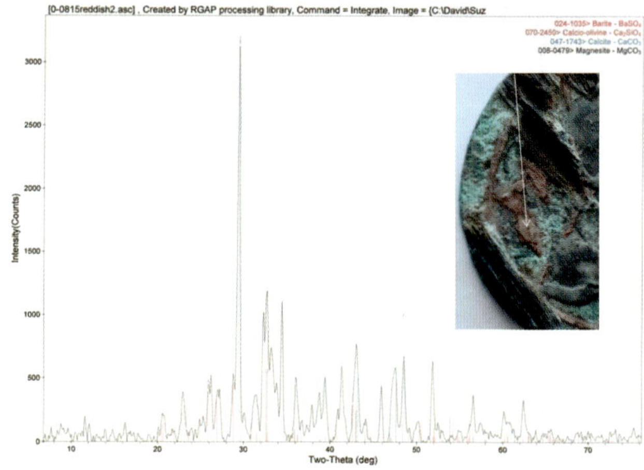

Figure 22. *X-ray diffraction scan for the reddish mineral paste on mirror O-0815 showing the presence of an iron compound, possibly marcasite, together with calcite, barium potassium sulfate, and probably calcio-olivine, here shown as calcium silicate oxide.*

The figures of the boys and the surrounding decoration are built up in a kind of mineral paste, a detail shown in figure 20, and are raised in relief on top of the background surface. They come in green-gray and brick-red colors. X-ray diffraction (XRD) studies showed that the green-gray paste was composed of calcite ($CaCO_3$), calico-olivine ($CaSiO_4$), and barium potassium sulfate (BaK_2SO_4) (fig. 21). XRD of another area gave barium potassium sulfate, periclase, MgO, zinc sulfide, ZnS, calcio-olivine, and calcite, while a third area showed the presence of a variety of barite, $BaSO_4$, which is a much more common mineral. These are complex results demanding further research. While gypsum, calcite, or natron were used in the West, barium compounds figure prominently in ancient Chinese glass and frits such as Han blue, Han purple, and early glass beads.

The presence of the calcio-olivine, or a mineral close to calcium silicate oxide, suggests the possibility that a ground-up ceramic or mineral paste was used in the fabrication of these figures. The barium potassium sulfate has never been reported before from archaeological contexts and may represent the weathering of a barium compound, such as barite, in the presence of potassium-containing groundwater, but at this stage it is not possible to say if this is true or not as, to my knowledge, we have no comparable material.

The reddish mineral mixture is similarly complex. XRD scans (fig. 22) suggested the presence of marcasite (FeS_2) along with the previously identified components of calcite, calico-olivine, and barium potassium sulfate. Although marcasite might be responsible for the reddish color of this pigment, it is unusual and may not be a satisfactory explanation for the color of this component. So what are we to make of this discovery? The fact that much of the design is obscured with mineralized wood remnants suggests that the design on this mirror is not new and that it was carried out before burial, but at what stage that might have been remains a mystery.

Finally, this mirror also displayed an unusual array of corrosion products on the bronze itself. The XRD chart in figure 23 shows malachite, tin sulfur (berndite, $4HSnS_2$), and connellite. Both tin sulfur and connellite are rarely reported as natural alteration products. The sources for the tin sulfur are numerous and may be related either to environmental factors or subsequent corrosion of the bronze in the indoor environment.

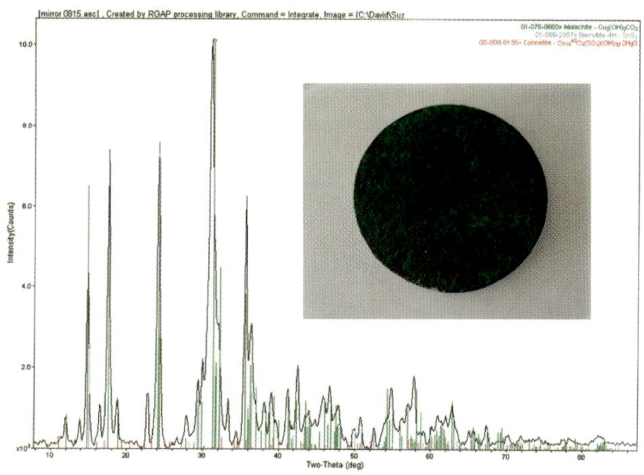

Figure 23. *X-ray diffraction data for the corrosion crust of mirror O-0815.*

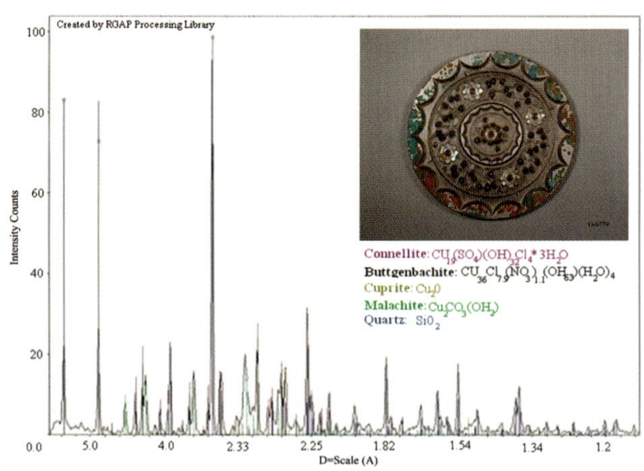

Figure 24. *X-ray diffraction data for the corrosion crust of mirror O-0779 showing the presence of connellite in the corrosion crust.*

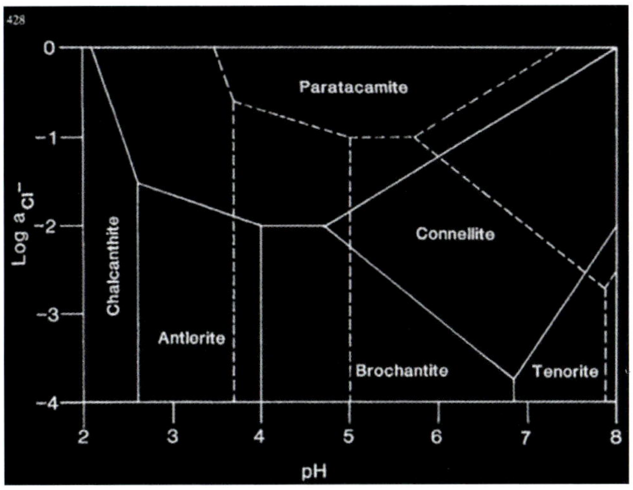

Figure 25. *Stability diagram for some of the copper sulfates and chlorides showing regions of stability. At fairly neutral pH values, anterlite and brochantite predominate at lower levels of chloride ion activity, but with the increasing presence of chloride together with sulfate, there is a substantial stability field for connellite at slightly alkaline pH levels.*

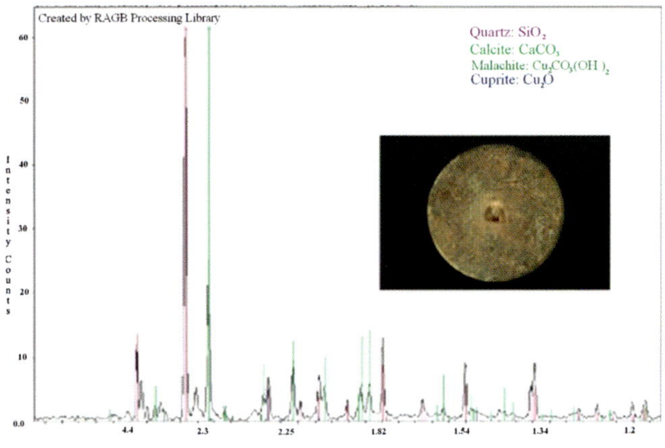

Figure 26. *X-ray diffraction spectrum for mirror O-0304.*

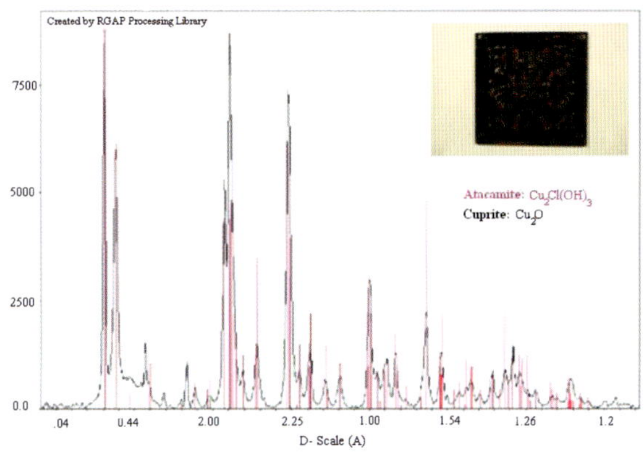

Figure 27. *X-ray diffraction data for mirror O-0185.*

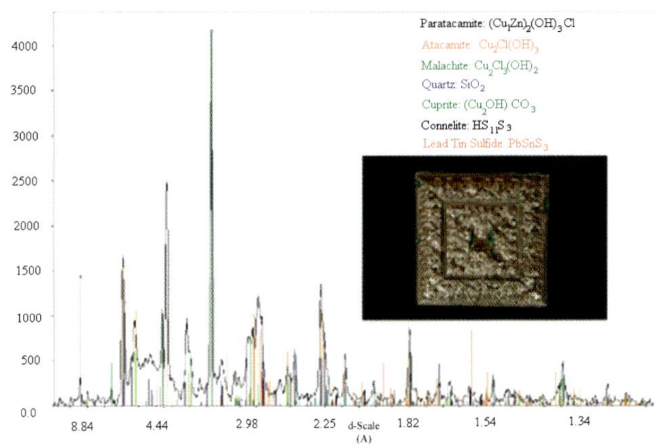

Figure 28. *X-ray diffraction data for mirror O-0742*

A curious discovery related to the corrosion of Chinese bronzes links together three very different mirrors: O-0779 from the Western Han dynasty (206 BCE–8 CE) (see Cahill 2009, pl. 59), O-0774 from the Sui dynasty (581–618 CE) (see Cahill 2009, pl. 81), and the mirror we are discussing here, O-0815. All three mirrors contain connellite ($Cu_{19}Cl_4(SO_4)(OH)_{32}.3H_2O$) as a patina constituent. Connellite is a complex mineral group isostructural with buttgenbachite ($Cu_{36}Cl_{7.9}(NO_3)_{1.1}(OH)_{63}(H_2O)_4$). The XRD data for mirror O-0779 are shown in figure 24.

How can we account for the occurrence of this apparently rare corrosion product in the patina of three widely disparate bronze mirrors? The principal factor is created by environmental variables rather than by composition. For connellite to form, there must be both chloride ions and some sulfate ions in the burial environment. The problem can be examined with the aid of a stability diagram based on the work of Pollard et al. (1990) (fig. 25). Crystallization of the connellite on the three mirrors must have taken place in a saline environment, but with sufficient sulfate ion content that the stability field of paratacamite was not reached. The presence of sulfate and chloride ions in groundwater is a common event. Connellite has been reported geologically from widely different milieu, which suggests that its occurrence in ancient bronze patinas may have been underestimated due to a lack of applied research directed toward patina characterization. Thus the apparently anomalous presence of connellite in these Chinese bronzes may, with time, come to be part of the corpus of alteration products routinely recognized as such.

The typical patina components seen in mirrors as a result of corrosion are malachite and cuprite, as seen in figure 26 for the Warring States-period mirror O-0304 (see Cahill 2009, pl. 39).

In mirrors with chloride-induced corrosion, atacamite and paratacamite are the most commonly encountered, with one example of botallackite. The presence of botallackite is suggestive of the rapid drying of the artifact from its burial environment, which helps to stabilize botallackite and prevent it from recrystallizing into atacamite (Scott 2002). The Warring States-period lacquered mirror O-0185 (see Cahill 2009, pl. 45), for example, shows the presence of atacamite on the front reflective surface (fig. 27). Atacamite is the most common chloride species on these mirrors and is generally recognized as the most thermodynamically stable of the isomers of the copper trihydroxychlorides (Scott 2002).

The Tang-period mirror O-0742 (see Cahill 2009, pl. 96) shows a complex patina (fig. 28) that incorporates paratacamite, malachite, atacamite, cuprite, and a lead-tin sulfide, $PbSnS_3$, which has probably arisen from conjoint corrosion of the lead and tin content with sulfur contaminants in previous storage or display, since it is not a part of a patina on this mirror in the sense of an overall film nor has it been reported before in Chinese contexts. The presence of malachite and paratacamite (or clinoatacamite) is a common occurrence in bronze patinas of this type (Scott 2002).

Conclusion

This brief survey of decorative techniques employed in some Chinese bronze mirrors from the Cotsen collection highlights some of the problems of examination of antiquities that have no excavated context. Many thousands of Chinese mirrors fall into this category, and the mirrors of the Cotsen collection are no exception. The most interesting mirrors are those that present previously unpublished decorative features, such as the double-tier mirror with a silk textile interlayer from the Warring States period, which is also inlaid with malachite and turquoise around the border. The gold flake in glass intaglio technique used for the mirror from the Six Dynasties period is an intriguing example of a technique not previously reported from Chinese art of this period, as far as I am aware. This mirror uses a combination of glasses of Chinese and foreign origin, as well as gold leaf adhered to a mineral base in juxtaposition to the gold flake glass, thus creating a stunning array of visual effects together with the silvery color of the entwined animal figures cast in the bronze itself. The intaglio casting of the mirror bronze reveals that the mirror was planned as an inlaid and decorated back from the very inception of the casting. It is difficult to comment further on the ceramic paste decoration used for the Sui-period mirror until further examples can be examined. Reuse and redecoration of Chinese mirrors in periods much later than the original casting of the bronze of the mirror itself continue to present challenges to the art historian and scientist alike.

Acknowledgments

I would like to thank Suzanne Cahill for the translations and Drew Pomatti and Seppi Lehner for assistance with the x-ray fluorescence analysis work and for some photomicrographs. Funds for one of the two sets of electron microprobe analyses carried out at UCLA Geology Department were kindly supplied by the Cotsen Institute Fund. I am grateful to Professor Lothar von Falkenhausen, Department of Art History, UCLA, for skillful editorial emendations of the long chapter for volume two of the Cotsen mirror compilation, from which this account is taken. The Lloyd Cotsen Foundation is thanked for permission to reproduce the images of the mirrors discussed here. Thanks are due to Lloyd Cotsen, Lyssa Stapleton, and Karen Hudson for facilitating this work.

References

Brill, R. H. 1999. *Chemical analyses of early glasses*. Vol. 1. Corning, NY: Corning Museum of Glass.

Brill, R. H., S. S. C. Tong, and D. Dohrenwend. 1991. Chemical analysis of some early chinese glasses. In *Scientific research in early Chinese glass*, ed. Robert Brill and John Martin, 31–58. Proceedings of the Archaeometry of Glass Sessions of the 1984 International Symposium of Glass, Beijing. Corning, NY: Corning Museum of Glass.

Bulling, A. G. 1955. The decoration of some mirrors of the Chou and Han periods. *Artibus Asiae* 18:20–45.

Cahill, S. 2009. *The Lloyd Cotsen study collection of Chinese bronze mirrors*, Vol. 1, *Catalogue*. Los Angeles: Cotsen Occasional Press and UCLA Cotsen Institute of Archaeology Press.

Costello, S. D. 2005. An investigation of early Chinese bronze mirrors at the Harvard University Art Museums. Paper presented at the 2005 Graduate Student Conference of the Association of North American Graduate Programs in the Conservation of Cultural Property, Buffalo, NY. Available online at http://www.ischool.utexas.edu/~anagpic/pdfs/Costello.pdf.

Dohrenwend, D. 1964. The early Chinese mirror. *Artibus Asiae* 27:79–98.

Gao, X., and G. Yang. 2007. Luoyang chutu Zhanguo tezhong gongyi tongjing ji xiangguan wenti yanjiu [Research on the bronze mirrors of Warring States periods made by special techniques unearthed in Luoyang]. *Zhongyuan wenwu* [Cultural Relics from the Central Plains] 2007.5:51–61.

Luoyang Shi Wenwu Gongzuodui. 1999. Luoyang shi Xigongqu C1M3943 Zhanguo mu [The Warring States-period tomb C1M3943 in the western industrial sector of Luoyang city]. *Wenwu* [Cultural Relics] 1999.8:4–13.

Mackenzie, C. 1999. The influence of textile designs on bronze lacquer and ceramic decoration styles during the Warring States period. *Orientations* 30(7):81–92.

Maryon, H. 1963. The making of a Chinese bronze mirror, pt. 2. *Archives of the Chinese Art Society of America* 17:23–25.

Moorey, P. R. S. 1994. *Ancient Mesopotamian materials and industries: The archaeological evidence*. Oxford: Clarendon Press.

Pollard, A. M., R. G. Thomas, and P. A. Williams. 1990. Connellite: Stability relationships with other secondary copper minerals. *Mineralogical Magazine* 54:425–30.

Scott, D. A. 2002. *Copper and bronze in art: Corrosion, colorants, conservation*. Los Angeles: Getty Conservation Institute.

Shortland, A., L. Schachner, I. Freestone, and M. Tite. 2006. Natron as a flux in the early vitreous materials industry: Sources, beginnings and reasons for decline. *Journal of Archaeological Science* 33:521–30.

Yen, C.-Y. 2005. The decorative motifs on Tang dynasty mirrors. In *Clarity and luster: New light on bronze mirrors in Tang and post-Tang dynasty China, 600–1300*, ed. C. Brown and J.-H. Chou, 1–10. Cleveland Studies in the History of Art, Vol. 9. Papers from a Symposium on the Carter Collection of Chinese Bronze Mirrors at the Cleveland Museum of Art. Cleveland: Cleveland Museum of Art.

Technical Studies of Three *Gui* Vessels of the Early Western Zhou Period in the British Museum Collection

Wang Quanyu

ABSTRACT Three Chinese bronze ritual *gui* vessels of the early Western Zhou period, including the Kang *gui* and Xing *gui*, were examined. The conditions of the vessels vary from being intact to extensively restored in modern times, although contemporary repairs, probably completed at the time of manufacture to rectify casting defects, were also identified. All vessels were found to have been made of ternary alloys of copper-tin-lead (Cu-Sn-Pb), with the Kang *gui* and Xing *gui* containing relatively smaller amounts of lead, which is probably reflective of the status of their owners. The Kang *gui* seems to have the handles attached by lock-on casting, while the other two appear to have loop-shaped handles cast as an integral part of the vessel. Squarish chaplets or spacers were found on the base of all the objects but not on the wall and were outside of the inscription area on the vessels with inscriptions. The materials for the chaplets were metals in two vessels but were nonmetal in the Xing *gui*. Nonmetal chaplets have not been previously reported and deserve particular attention. Mold-making in connection with the investigation of the vessels studied here is also briefly discussed.

Introduction

Ancient Chinese bronze technology has been attracting considerable research interest since the magnificent discovery of thousands of Shang bronzes in Anyang in the 1930s (Li 1933). Piece-mold casting technique was first mentioned by Karlbeck (1935) on the basis of clay casting molds collected in Anyang. Since the 1960s, major research has been carried out in the West as well as in China on the casting technology of ancient Chinese bronzes (Gettens 1969; Chase 1983; Meyers 1988; Barnard 1961; Li and Wan 1966, 1970, 1972; Hua et al. 1981; Tan and Huang 1996). A major contribution to the field was made by Bagley (1996) through his work on hundreds of casting molds recovered at the foundry site of the Warring States period (475–221 BCE) in Houma, Shanxi province. A review of the history of the study of bronze casting in ancient China can be found in a paper by Su (2003).

In recent years, new discoveries of a number of clay molds made in Anyang, Henan, and Zhouyuan, Shaanxi, have stimulated some important new studies on mold-making (Li et al. 2007; Liu 2009). Most interestingly, there have been debates on how ornaments were made in molds (Nickel 2006; Bagley 2009).

The British Museum collection contains about 160 Chinese bronze vessels dating to the Shang (ca. 1600–ca. 1050 BCE) and Zhou (ca. 1050–221 BCE) periods. These bronze vessels had undergone little technical analysis until recently when BBC Radio 4 and the British Museum commissioned the program *The History of the World in 100 Objects*, which featured one bronze vessel. This paper reports on the study of three ritual food vessels, *gui*, of the early Western Zhou period (ca. 1050–771 BCE). Aspects of the fabrication, including the way the decorative details were made on molds, the production methods for the handles, and the material used for chaplets, are also discussed.

Scientific Examination

The vessels were examined by optical microscopy and x-radiography. Compositional analysis was carried out using x-ray fluorescence analysis (XRF) on untreated surfaces (for fragile objects, e.g., the Xing *gui*) or small areas in inconspicuous places cleaned by abrading using a scalpel and silicon carbide paper. Ultraviolet (UV) imaging and Fourier transform infrared spectroscopy (FTIR) were used for investigating modern restoration of the surfaces. X-ray diffraction (XRD) was used to identify the surface crust and the nonmetal chaplets.

Results and Discussions

Kang Hou Gui (1977,0404.1)

The vessel with two handles is known as the Kang Hou *gui* after the brother of Wu, the previous Zhou king (fig. 1). The ruler Kang Hou (the duke of Kang), named in the

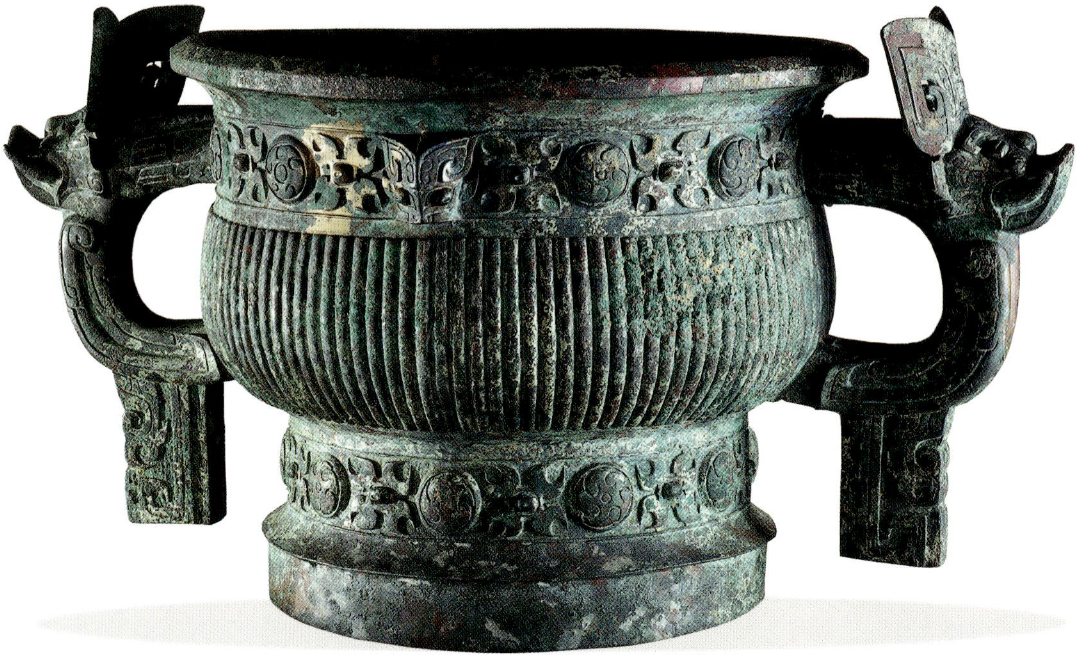

Figure 1. *The Kang* gui, *21 × 26.5 cm. British Museum, 1977,0404.1.*

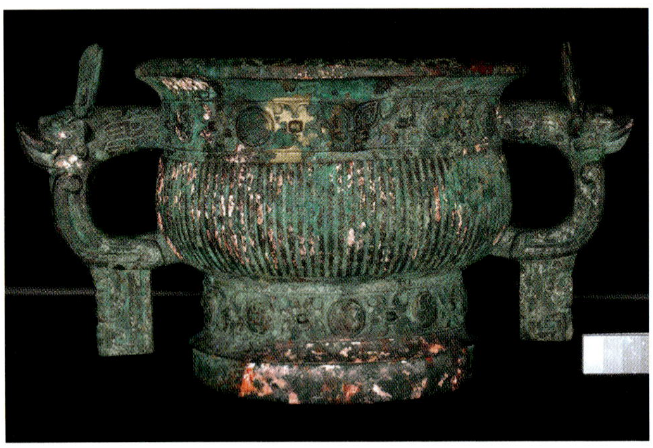

Figure 2. *The Kang* gui *with the UV image (fluorescing white and reddish) overlaid illustrating the surface retouching and repair at the foot ring.*

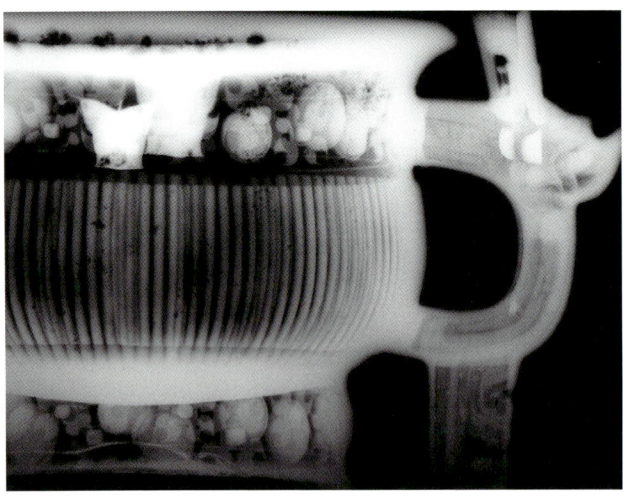

Figure 3. *X-radiograph of the Kang* gui *showing the separately cast handle and porosity (black globules) at the rim and the white line of soft solder at the repair to the foot ring. Note the dense areas (white) where the handle joins the vessel.*

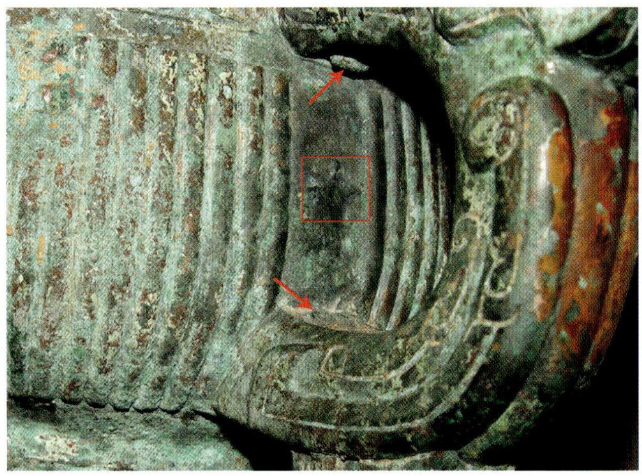

Figure 4. *Rivet-like "supporting nail" (arrowed) on the Kang* gui *at the vessel–handle joins and a cross-shaped chaplet (squared) used for positioning the handle.*

long inscription inside the vessel, was ordered by the king to establish the state of Wei, in what is present-day Henan province, after the defeat of a Shang rebellion during the reign of King Cheng (ca. 1042/35–1006 BCE). The vessel is decorated with dense vertical ribbing in its middle section and alternating roundels and quatrefoils in the narrow band on the neck and foot. The large handles are decorated with tusked animal heads with upright horns.

The vessel is in good condition except for a repair on the foot ring and surface retouching in some areas, indicated by strong fluorescence under UV light (fig. 2). An area of bare metal (see fig. 1) suggests that the vessel surface had been over-cleaned before it came to the museum. Tests with organic solvents indicated that organic materials were applied to some of the "crusts" on these areas. The white and greenish materials showing luminescence were respec-

Table 1. *Surface composition of the metal analyzed by XRF (wt.%).*

Object	Analyzed Area	Cu	Pb	Sn	Fe	As	Zn
1977,0404.1	bare metal near the rim	82.8	1.7	15.1	—	0.3	—
	original area on the foot ring	77.4	3.9	18.2	0.2	0.2	—
	repair on the foot ring	62.7	0.3	0.9	0.1	—	35.8
	handle lug 1 (without a seam)	70.4	9.7	19.2	0.1	0.3	0.3
	handle lug 2 (with a seam)	67.5	8.9	23.1	0.1	0.1	0.3
1936,1118.2	foot ring	78.8	2.7	18.2	0.3	—	—
	restoration (quarter 3)	0.7	26.5	0.1	3.1	—	69.6
1973,0726.11	foot ring	72.6	12.6	13.7	0.1	0.6	0.5
	hollow handle lug	70.9	12.3	15.5	0.1	0.7	0.6
	solid handle lug	73.4	7.7	17.8	0.1	0.7	0.3

tively identified as barium sulfate ($BaSO_4$) and Emerald green pigment (copper(II) acetoarsenite) or Scheele's green (copper arsenite) by XRD and FTIR (Watson and Hacke 2009). These pigments were not in use until the late eighteenth century, indicating that modern materials had been used to conceal the damage.

The handles were cast separately from the vessel, as revealed by x-radiography (fig. 3) and confirmed by XRF analysis. The vessel was made using a lock-on technique (Gettens 1969) by inserting the precast handles into the casting mold assemblage for the vessel; a slight spillover of metal from the vessel onto the handle stems at each vessel–handle junction occurred during the casting. A tenon projecting from the vessel wall at the junction was for positioning the handles. The vessel is not decorated behind the handles, where a cross on the vessel and a rivet-like "supporting nail" at each vessel–handle junction were probably used for keeping the handle in place (fig. 4). X-radiography also shows that the decorative lugs at the bottom of the handles are less dense than the vessel, although they are thicker than the rim of the vessel, indicating that the clay core extends right into the lugs. A mold-join mark is evident on one handle (but absent on the other) indicating that the handles were cast in a two-piece mold with an open core. Molds for handles with similar forms and designs to that on the Kang *gui* were found at the foundry site at Xiaomintun, which could provide information on the place of production of the Kang *gui* (Li et al. 2007).

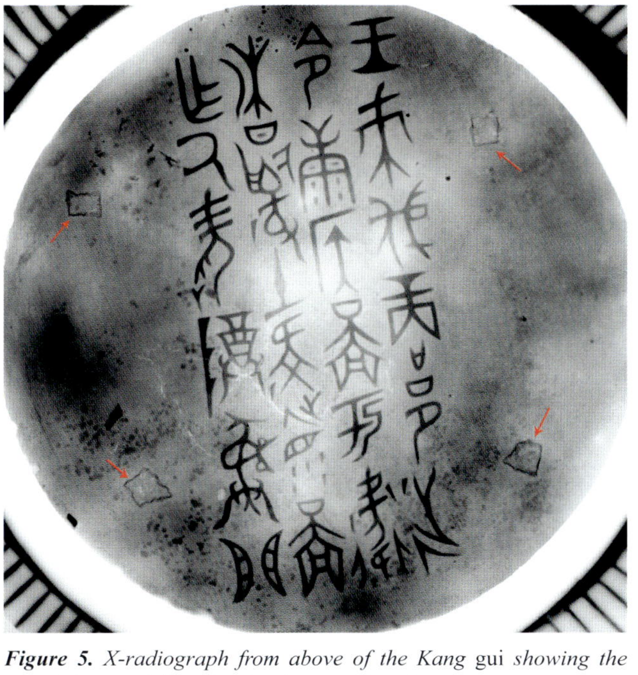

Figure 5. *X-radiograph from above of the Kang* gui *showing the inscriptions and four chaplets (arrowed) on the base.*

The vessel was cast in a piece mold that probably consisted of two or more parts (to be discussed later); the mold sections joined in line with the center of the two animal heads on the vessel, as indicated by the presence of a mold-join mark on the heads. Considering the size and shape of the handles, it is possible that a part of each handle rather than both entire handles was placed inside the mold prepared for the casting of the vessel. The mold construction for the Kang *gui* was so accurate that the vessel–handle joins were almost perfect. The spillover can be more extensive, as was reported on a similar vessel of the same period in the collection of the Freer Gallery of Art (Pope et al. 1967, 355, fig. 41).

Four squarish chaplets on the base of the vessel were revealed by x-radiography (fig. 5). The recessed inscription on the base was cast in the vessel, and the chaplets were placed carefully to avoid interrupting them. The quality of the casting is poor, as can be seen in the intensive porosity present in the upper region of the vessel near the rim and also in the middle of the vessel (see fig. 3). This suggests that the vessel was probably cast upright, as air bubbles rise during cooling. This is unlike most Chinese bronze vessels, especially those with legs, which are believed to have been cast upside down (Wan 1975).

XRF results on three areas of the vessel and on the bottom of each handle lug are listed in table 1. Because the analysis is of the surface and because minimal preparation was carried out to remove possible surface enrichment, such as patina, the results may not accurately represent the composition of the body of the object. Nevertheless, the difference between the handles and the vessel did confirm that they were not from one cast. The repair to the foot ring is made of brass and is therefore modern. The difference in lead content between the rim and the foot ring further confirms the hypothesis that the vessel was cast upright, as lead usually sinks to the bottom of the cast due to its greater density. The low content of lead may account for the porosity of the cast, as lead can improve the fluidity of molten metal, although it is immiscible in Cu-Sn alloys.

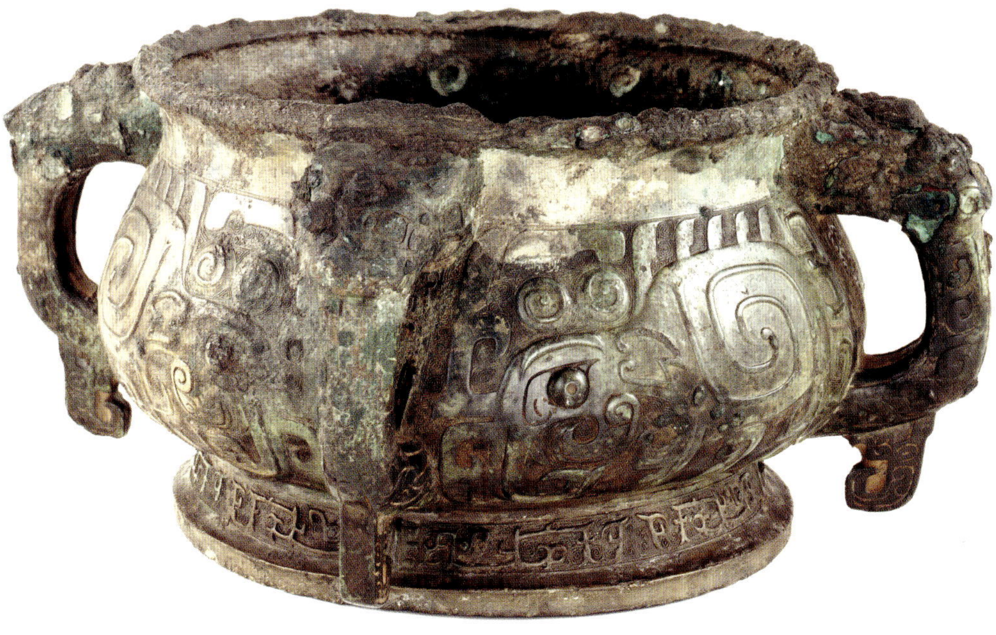

Figure 6. *The Xing* gui *showing pronounced nodular corrosion patches on the surface, 18.5 × 27.8 cm. British Museum, 1936,1118.2.*

Xing Hou Gui (1936,1118.2)

The vessel shown in figure 6 is called Xing Hou *gui* because of the inscription on its base. The lengthy inscription (figs. 7 and 8) records the grant of three groups of men to Xing Hou (ruler of Xing) and the dedication of the *gui* to the Zhou Gong (duke of Zhou, brother of King Wu). The vessel is divided into quarters by four handles and has an elephant monster motif on each quarter (see fig. 6). It is severely corroded with many nodular corrosion patches present on the surface, and corrosion has penetrated through the body

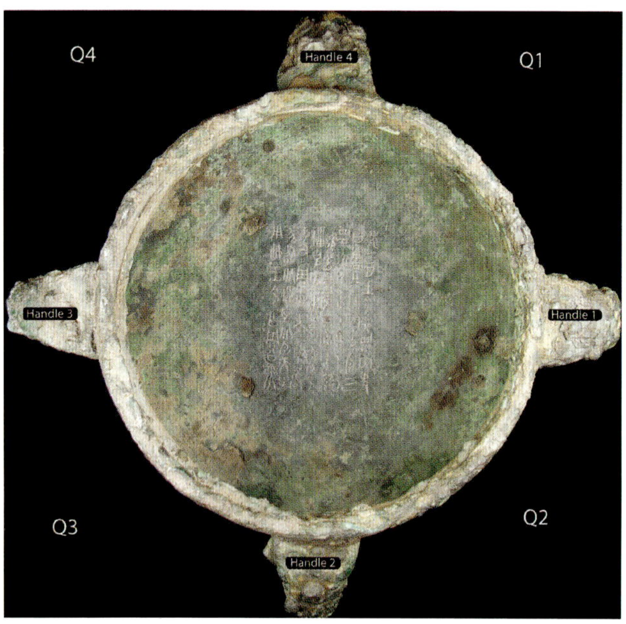

Figure 7. *The Xing* gui *from above showing the lengthy inscription on the inside bottom and four quarters divided by the handles (Q = quarter).*

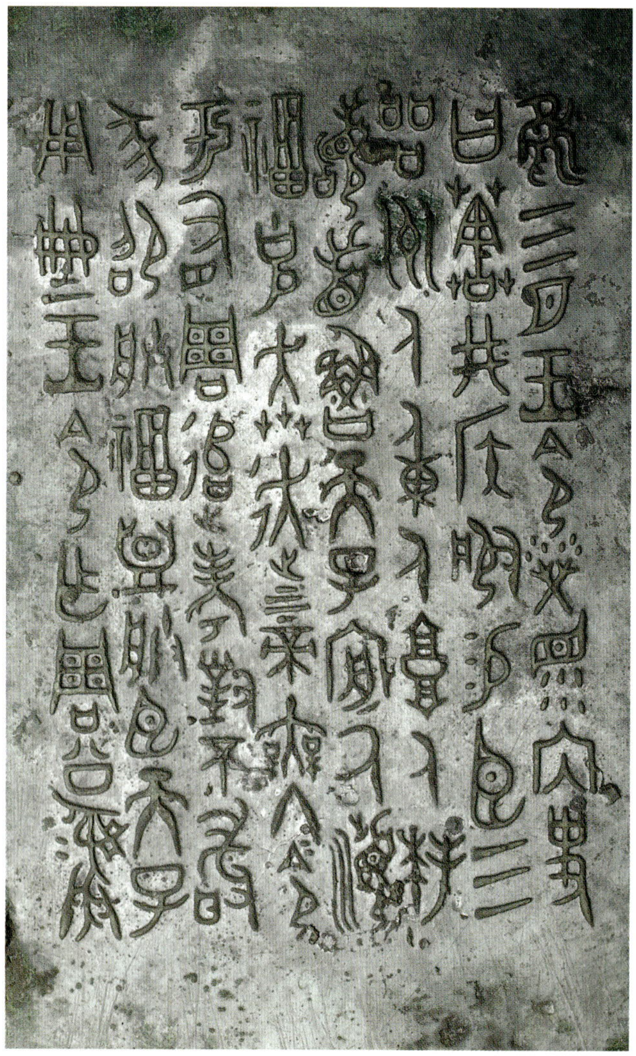

Figure 8. *The inscription on the base of the Xing* gui.

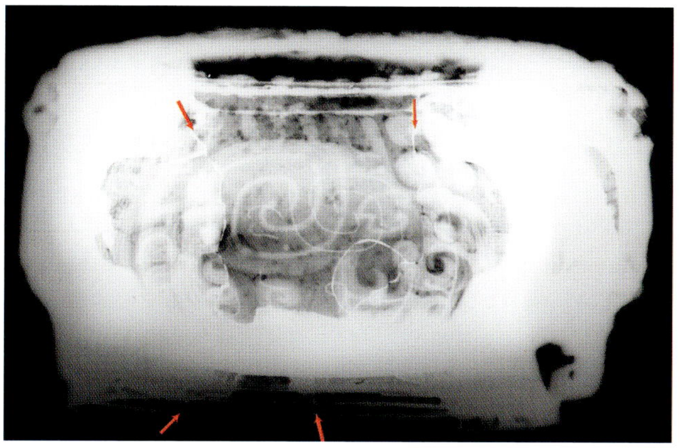

Figure 9. *X-radiograph of the Xing* gui *showing the restoration on the foot ring and repair on the wall. Solder joins are indicated by arrows. The area between the solder joins on quarter 3 is restoration.*

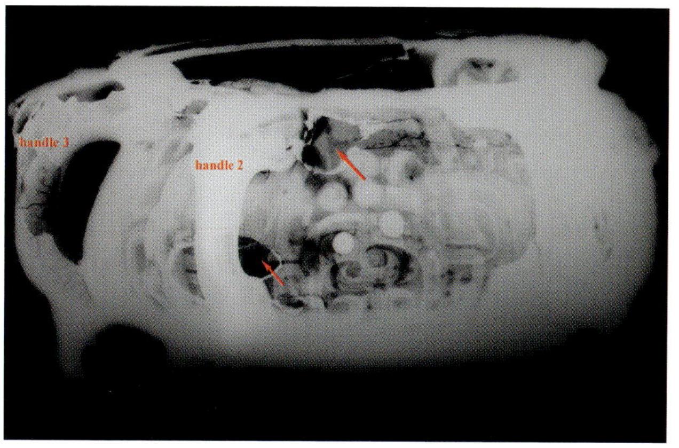

Figure 10. *X-radiograph of the Xing* gui *(with quarter 2 in front) showing the repair under the rim and beneath handle 2 (indicated by red arrows).*

wall, making small holes near the rim. Both x-radiography and UV light showed that it was badly damaged and heavily restored in many areas (figs. 9 and 10).

For this discussion, the handles of the vessel and the quadrant between the handles have been given the letter and number designations noted in figure 7. Area Q1 was severely corroded; the rim on this part seemed to have been retouched, and the foot ring is probably a restoration. Area Q2 appeared to be the best-preserved quadrant, although the foot ring on this part and the wall below the rim seemed to have been repaired. XRF analysis on the unabraded surface of this area showed that it is a high-tin bronze. Area Q3 fluoresced strongly under UV light, both inside and outside. XRF analysis showed that the major part of the wall of this quadrant contained mainly zinc instead of copper and tin, indicating a restoration. Although it is not possible to quantify the analysis due to surface corrosion, it provides convincing evidence of restoration. The metal used for the restoration is probably brass; lead white could have been applied to the surface to fake the appearance of natural patina. The bright line on the radiograph of area Q3 (see fig. 9) indicates that a denser material, possibly a lead-based soft solder, was used to join the restoration to the original metal. Area Q4 seemed from its surface appearance to be the worst-preserved quadrant. However, the majority of this area, including the foot ring and the wall, was shown to be genuine by XRF analysis.

The majority of the foot ring appeared to have been restored, and XRF analysis showed that the restoration contained mainly zinc (Zn) with lead (Pb), while the original was a high-tin bronze (see table 1). The base of the vessel appeared to be in good condition and was proved to be original by XRF analysis. The recessed inscription must have been formed by raised characters on the core for the upper body.

Three chaplets located outside the inscription area are visible on both sides of the base (fig. 11). These chaplets are very porous in texture, and x-radiography indicates that they may not be metal. A small sample of the chaplets was ana-

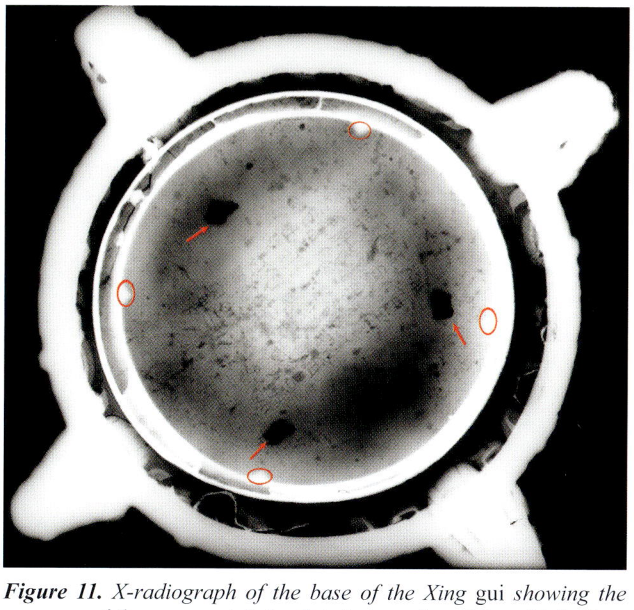

Figure 11. *X-radiograph of the base of the Xing* gui *showing the presence of three nonmetal chaplets (arrowed) and four thickenings (circled).*

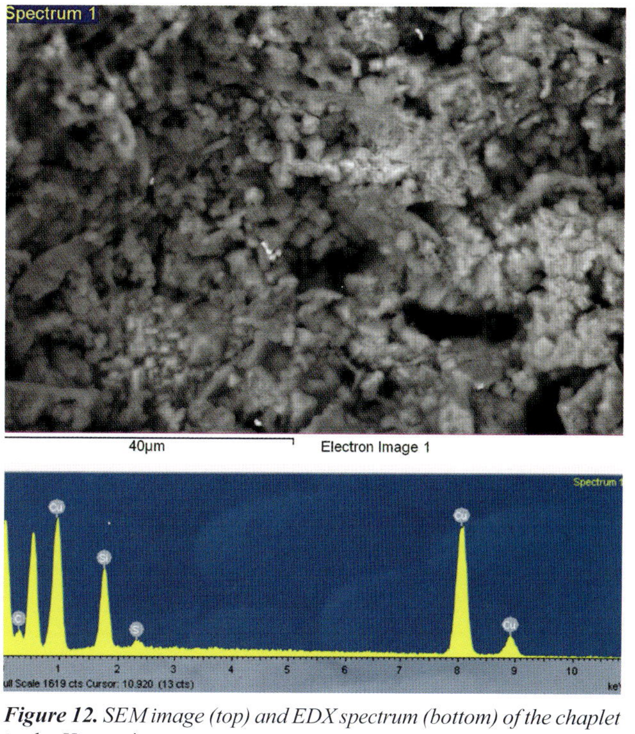

Figure 12. *SEM image (top) and EDX spectrum (bottom) of the chaplet in the Xing* gui.

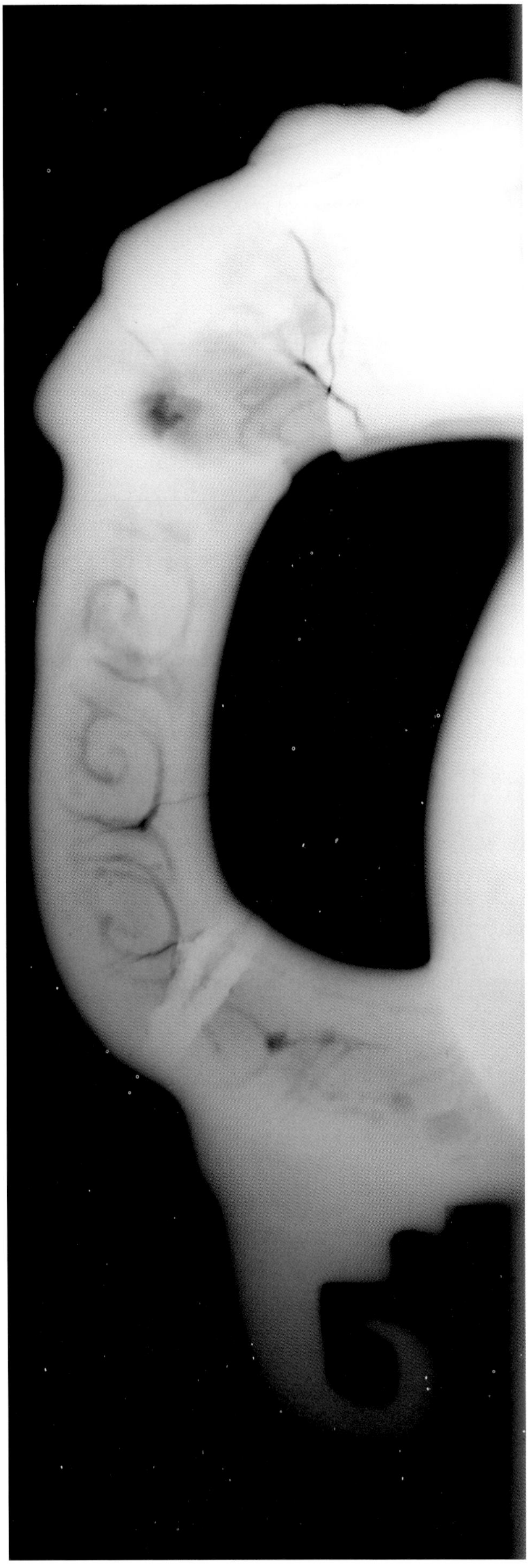

Figure 13. X-radiograph of handle 1 on the Xing gui *showing the cracks and repairs.*

lyzed by scanning electron microscopy–energy-dispersive x-ray spectrometry (SEM-EDX) and was found to contain mainly copper, silicon, and oxygen with sulfur (fig. 12). It was found to contain mainly malachite by XRD analysis. All the evidence points to a geological material, perhaps rock, rather than metal used for the chaplets. It is not known if the original material was malachite, as malachite begins to decompose at 200°C with slow heating (Simpson et al. 1964). Moreover, malachite is a common corrosion product, and it may be that the chaplets have become completely mineralized. Nevertheless, the use of nonmetal materials for chaplets has not been reported elsewhere and demands attention in future studies of Chinese bronzes. Four triangular thickenings at the junction of the base and foot (see fig. 11) were probably to strengthen the vessel wall and also to counteract "hot cracking" at the sharp angle at which the base and foot meet (Gettens 1969).

All the handles have clay core remnants and are covered with thick corrosion products. X-radiography showed that all have cracks and repairs (fig. 13) and even loss of surface on one (see fig. 10). X-radiography did not show any joins between the handles and the wall of the vessel. The XRF analysis suggests that the loop-shaped handles were probably cast integrally with the vessel. The mold assemblage was probably made in four pieces, joined in the middle of each handle, as evidenced by the presence of mold marks. It is not certain whether the pores, revealed by x-radiography, were from casting or have resulted from corrosion. It cannot be concluded from the x-radiography whether the vessel was cast upright or upside down.

The vessel is badly damaged and repaired extensively with pieces of metal containing zinc; the majority of the foot ring and one quarter of the vessel wall were modern restoration using this zinc-rich metal, which was soldered to the original metal. The restoration is highly skilled and was likely carried out in China before the vessel came to the West (Fleur Shearman, pers. comm.).

Gui (1973,0726.11)

This vessel is similar to the Kang *gui* in design but is smaller in size and has two loop-shaped handles (fig. 14). The underside of the base (fig. 15) is roughly decorated with raised diamond-shaped crisscross lines but is not well finished. Four triangular thickenings present at the junction of the base and foot were probably to strengthen the wall of the vessel, as detailed in the description of the previous vessel. Four barely visible squarish chaplets on the base, with a density similar to the metal of the vessel, were revealed by x-radiography.

The vessel is in good condition, and no artificial patina was detected by organic solvent tests or UV light. Residual mold joins were observed on the underside of the handles and the animal heads on the body. X-radiography (fig. 16) showed that the handles were probably cast integrally with the vessel. The solid metal seen on the radiograph of the lug of handle 1 (see fig. 16) was probably a contemporary repair by casting on to fill a failure in the original casting. The alloy used to make the lug is higher in tin but lower

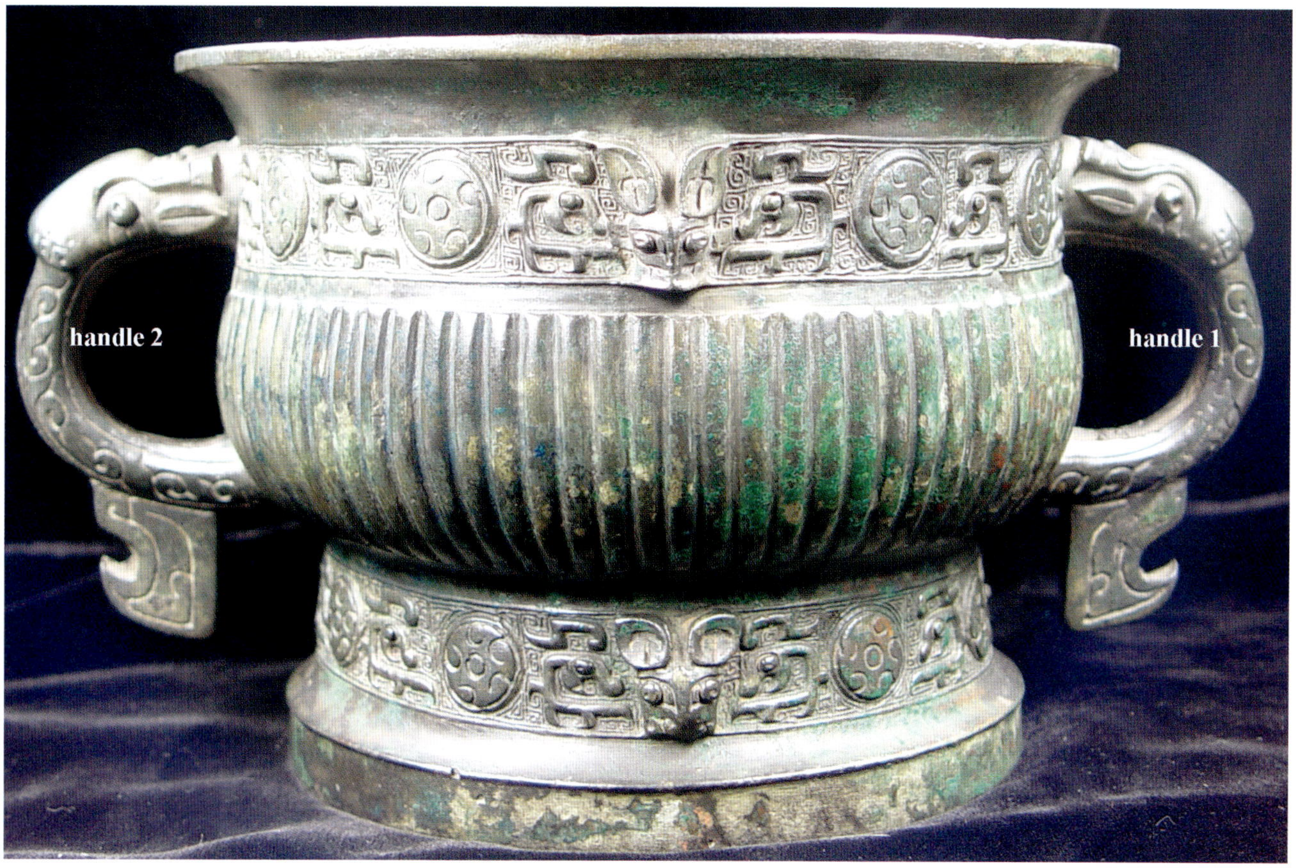

Figure 14. Gui *1973,0726.11 showing a similar design to the Kang* gui, *14 × 18.7 cm. British Museum.*

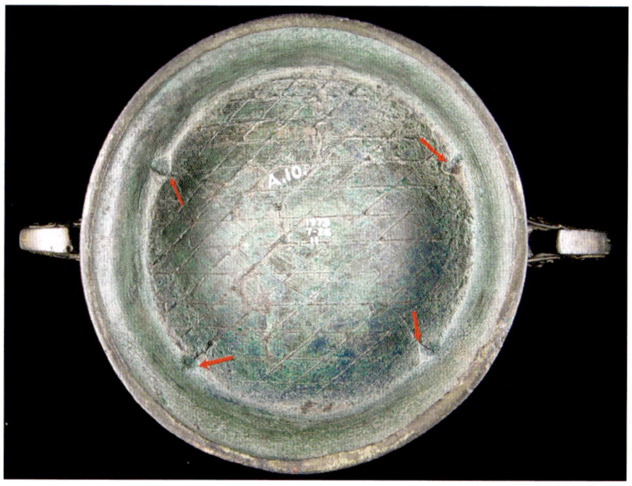

Figure 15. *The underside of the base on* gui *1973,0726.11 showing crisscross lines and four thickenings (arrowed).*

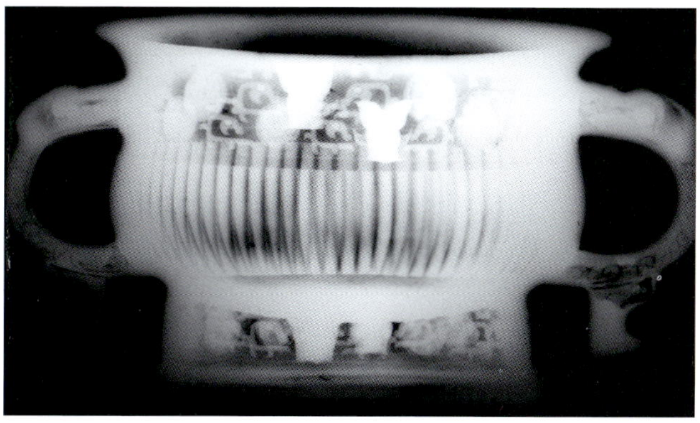

Figure 16. *X-radiograph of* gui *1973,0726.11 showing a solid handle lug (on the right).*

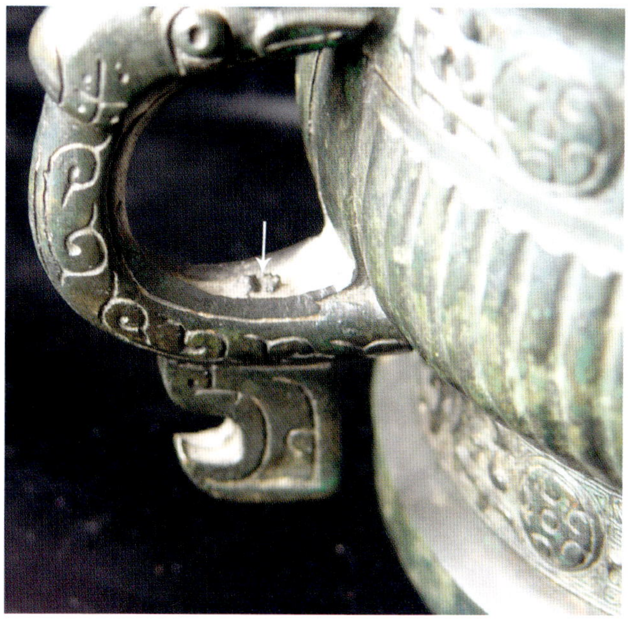

Figure 17. *Details of the handle–lug junction on* gui *1973,0726.11 showing the metal projections (arrowed) in the clay core of the repaired handle.*

in lead than the body metal (see table 1). Metal projections are visible in the clay core of the repaired handle (fig. 17), which could have been used for positioning the mold for the repaired handle lug. The lug of handle 2 from the original casting has a similar composition to the body, confirming that the handles were cast integrally with the body.

The cast is of high quality with a little porosity present in the lower part of the body wall near the handles (see fig. 16), which suggests that the object was probably cast upside down. The high tin and lead content of the alloy probably contributed to the good quality of the casting.

Discussion

There have been debates on how the decorative details were made on Chinese bronzes. It is generally believed that ornamentation was made on models and then transferred to molds, with the exceptions being the earliest bronzes decorated with simple thread relief that was executed by direct carving of the mold pieces. However, recent research by Nickel (2006) argues that all decorated patterns were executed on the molds rather than on the models. Nickel's argument is mainly based on imperfect symmetry in the ornamentation on bronze. This has been challenged by Bagley (2009) and Wang and Mei (2009), and a detailed analysis of Nickel's arguments can be found in Bagley (2009).

Recent experiments on mold-making carried out by Dong Yawei and his team are very interesting (Dong 2010). In their experiment, they did not adapt a traditional reproduction procedure (i.e., making a mold from a genuine bronze). They created models of clay in the shape of *gui, ding,* and *zun* on wheels, using wooden scrapers with the required geometry. They proposed that the model could either be a single piece or in sections, but they believed the latter to be more feasible so opted to use it in the experiments. In their experiment, a model for a *gui* vessel was made using a special scraper and then divided into four sections. Mold pieces were made from the model sections, and ornamentation was executed on the mold. Dong Yawei and his team claimed that execution of ornamentation on molds seemed to be fairly easy. However, execution of ornamentation on the model was not attempted, so consequently no comparison could be made. Furthermore, it was difficult to visualize the bronze to be cast by looking at the decorated mold, so they made copies of the ornamentation by pressing a thin layer of clay onto the decorated model to check.

One cannot be convinced that execution of ornamentation on the mold is easier than on the model. If the craftsman cannot visualize the appearance of the bronze to be cast, how would he make the ornamentation on the mold? Nevertheless, the idea of a piece model is rather interesting. Wood (1989) proposed that the mold was made in segments around the model rather than being made whole at once, in order to make a secure fit for the mold to be assembled for casting. It would be difficult to achieve equal shrinkage of each piece if the mold were made in this way, and a perfect registration of mold pieces would be unlikely, as experienced in my casting experiments (Wang and Ottaway 2004). If mold sections were made from model sections at the same time, this problem would not occur, and production would be more efficient. Furthermore, wheels had been used to make Chinese pottery since the Late Neolithic period (ca. 5000–2000 BCE) (Li 1998; Wang and Andrews 2002), so wheels could be used in the production of models for bronzes. A piece-model, in my opinion, could be more feasible than a one-piece model in the piece-mold process.

Horizontal mold joins were noted on foundry remains excavated from the site of Xiaomingun, although they were rarely observed on the bronze objects (Liu 2009). No traces of horizontal joins were observed on the objects studied, suggesting that either horizontal joins were not applied to these vessels or the vessels were finished too well to have any traces of joins remaining. It is also possible that some mold sections were joined in pairs (Bagley 2009) prior to use, and this could explain the lack of mold joins in the Kang *gui* and Xing *gui*.

Vessels with handles could have been made by three methods: (1) using engaged handles cast as an integral part of the vessel; (2) by lock-on casting of the handles, whereby the vessel was cast into a mold holding the precast handles; or (3) using brazed-on handles, in which the separately cast vessel and handles were joined with a filler metal made of Cu-Sn-Pb alloy with a lower melting point than that of the metals to be joined (Gettens 1969). Molds with a loop-shaped handle were found on the foundry site at Xiaomingtun, indicating the engaged handles were integrally cast with the vessels (Li et al. 2007). The large handles of tusked animal heads on the Kang *gui* seem to have been attached by lock-on casting, and the loop-shaped handles on the other two vessels appear to have been cast as an integral part of the vessel. Brazed-on handles were not observed on any of these objects, although this technique was in use before the Anyang period (Bagley 1987).

Metal chaplets used to hold the assemblage in position so as to maintain the wall thickness were usually found on the base and sides of vessels (Gettens 1969). For the *gui* vessels studied here, squarish chaplets were found on the base but not the wall. Nonmetal chaplets, of which there have been (to my knowledge) no previous report, used on the Xing *gui* are interesting and need further attention from archaeologists and researchers. The chaplets were found outside of the inscription area on the vessels with inscriptions, as Gettens (1969) described in his studies.

It was surprising to find that objects with sophisticated designs (i.e., the Kang *gui* and Xing *gui*) are poor quality castings, as evidenced by the presence of intensive porosity. This was probably due to the small content of lead in the alloy. Was this just accidental or indicative of the social status of the objects? It was reported that bronzes for the elite class seemed to have contained less lead than those for the lower classes (Wang 2002). It may have been the bronzesmith's intention to add less lead to the molten metal for these prestigious objects.

Conclusions

The conditions of the vessels vary from being intact to extensively restored in modern times, although contemporary repairs, probably made at the time of manufacture to rectify casting defects, were also identified. All vessels were found to have been made of ternary alloys of copper-tin-lead (Cu-Sn-Pb), with the Kang *gui* and Xing *gui* containing smaller amounts of lead, which is probably reflective of the status of their owners.

Two vessels seem to have engaged handles cast as an integral part of the vessel, and the Kang *gui* had the handles attached by lock-on casting. Chaplets were found on the base but not the wall in all the objects and were outside of the inscription area on the vessels with inscriptions. The use of nonmetal minerals for chaplets (in the Xing *gui*) has not been reported before and deserves special attention.

The inscriptions on the base of the Kang *gui* and Xing *gui* seem to have been cast with the vessel, although incised inscriptions could occur on bronzes from the late Shang period onward (Strahan and Fenn 2007). These vessels could have been cast upright to avoid air bubbles spoiling the inscription (Lai Guolong, pers. comm.). However, the weight of a clay core for a large object would have made upright casting difficult. This subject needs further study.

Acknowledgments

I am very grateful to Sascha Priewe, curator of the Chinese collection at the British Museum, for his support of my research and useful comments on this paper, and to Mr. Dong Yawei for allowing me to cite his publications in press and in progress. I would like to thank my colleagues Susan La Niece, for her help with x-radiography, Giovanni Verri, for his help with UV imaging, and Marei Hacke, for the identification of the conservation materials by FTIR.

References

Bagley, R. 1987. *Shang ritual bronzes in the Arthur M. Sackler collections*. Ancient Chinese Bronzes in the Arthur M. Sackler collections; v. 1. Washington, DC, and Cambridge, MA: The Arthur M. Sackler Foundation and the Arthur M. Sackler Museum.

———. 1996. Introduction to *Art of the Houma foundry*, by Institute of Archaeology of Shanxi Province, 69–75. Princeton, NJ: Princeton University Press.

———. 1999. Shang archaeology. In *The Cambridge history of ancient China: From the origins of civilization 221 B.C.*, ed. M. Loewe and E. I. Shaughnessy, 124–231. Cambridge: Cambridge University Press.

———. 2009. Anyang mold-making and the decorated model. *Artibus Asiae* 69(1):39–90.

Barnard, N. 1961. *Bronze casting and bronze alloys in ancient China*. Monumenta Serica Monograph XIV. Canberra: Australian National University.

Chase, W. T. 1983. Bronze casting in China: A short technical history. In *The great Bronze Age of China: A symposium*, ed. G. Kuwayama, 100–123. Los Angeles: Los Angeles County Museum of Art.

Dong, Y. 2010. Shang wan qi yuan xing ding de fan zhu mo ni shi yan yan jiu [Experimental production of round *ding* of the late Shang period]. *Sichuan wenwu* [Sichuan Cultural Relics] 5:73–87.

Gettens, R. J. 1969. *The Freer Chinese bronzes*, Vol. 2, *Technical studies*. Oriental Study Series 7. Washington, DC: Freer Gallery of Art, Smithsonian Institution.

Hua, J., F. Feng, Z. Wang, and R. Bai. 1981. Fuhao mu qing tong qi qun zhu zao ji shu de yan jiu [Studies of casting techniques of bronze objects from the Fuhao tomb]. *Kaogu xue jikan* [Archaeological Bulletin]1:244–72.

Karlbeck, O. 1935. Anyang moulds. *Bulletin of the Museum of Far Eastern Antiquities* 7:39–60.

Li, J. 1933. Yinxu tong qi wu zhong ji xiang guan wen ti [The types and some problems of the Yinxu bronzes]. *Qingzhu cai yuanpei xiansheng liushiwu sui lunwenji* [Collected papers on the occasion of Cai Yuanpei's 65th birthday] 1:73–104. Beijing: Zhongyang Yanjiuyan Lishi Yuyan Yanjiusuo.

Li, J. 1998. *Zhong guo ke xue ji shu shi: Tao ci juan* [The history of Chinese science and technology: Ceramics]. Beijing: Ke xue chu ban she.

Li, J., and J. Wan. 1966. Yinxu chu tu qing tong jue xing qi zhi yan jiu [A study of bronze *jue* from Yinxu]. *Kaogu xuebao* [Acta Archaeologica Sinica] 2.

———. 1970. Yinxu chu tu qing tong ding xing qi zhi yan jiu [A study of bronze *ding* from Yinxu]. *Kaogu xuebao* [Acta Archaeologica Sinica] 4.

———. 1972. Yinxu chu tu 53 jian qing tong rong qi zhi yan jiu [A study of 53 bronze vessels from Yinxu]. *Kaogu xuebao* [Acta Archaeologica Sinica] 5.

Li, Y., Z. Yue, and Y. Liu. 2007. Cong Xiaomintun Dongnandi chu tu tao fan tan dui Yinxu qing tong qi de ji dian xin ren shi [A few points in reconsideration of Yinxu bronzes based on the clay molds excavated at Xiaomingtun Locus Southeast]. *Kaogu* [Archaeology] 3:52–63.

Liu, Y. 2009. Origins and evolution of the casting technology of Anyang bronze ritual vessels: An exploratory survey. In *Metallurgy and civilisation: Eurasia and beyond*, ed. J. Mei and T. Rehren, 55–61. London: Archetype Publications.

Meyers, P. 1988. Characteristics of casting revealed by the study of ancient Chinese bronzes. In *The beginning of the use of metals and alloys*, ed. R. Maddin, 283–95. Cambridge, MA: MIT Press.

Nickel, L. 2006. Imperfect symmetry: Re-thinking bronze casting technology in ancient China. *Artibus Asiae* 66(1):5–39.

Pope, J. A., R. J. Gettens, J. Cahill, and N. Barnard. 1967. *The Freer Chinese bronzes*, Vol. 1, *Catalogue*. Oriental Studies Series 7. Washington, DC: Freer Gallery of Art, Smithsonian Institution.

Simpson, D. R., R. Fisher, and K. Libsch. 1964. Mineralogical notes: Thermal stability of azurite and malachite. *The American Mineralogist* 49(July–August):1111–14.

Strahan, D., and M. Fenn. 2007. A transfer of technology: Jade abrasive methods used to create inscriptions in ancient Chinese bronzes. In *Scientific research on the sculptural arts of Asia: Proceedings of the Third Forbes Symposium at the Freer Gallery of Art*, ed. J. G. Douglas, P. Jett, and J. Winter, 26–36. London: Archetype Publications.

Su, R. 2003. A history of the study of bronze casting in ancient China. In *Scientific research in the field of Asian art: Proceedings of the First Forbes Symposium at the Freer Gallery of Art*, ed. P. Jett with J.G. Douglas, B. McCarthy and J. Winter, 26–35. London: Archetype Publications.

Tan, D., and L. Huang. 1996. A study on treatment techniques about ancient clay mold materials. In *A collection of the Shanghai Museum papers on science of conservation*, ed. C. Ma, 47–70. Shanghai: The Shanghai Science and Technology Literature Press.

Wan, J. 1975. A comparative study of the casting of bronze *ting*-cauldrons from Anyang and Hui-hsien. In *Ancient Chinese bronzes and Southeast Asian metal and other archaeological artifacts*, ed. N. Barnard, 17–46. Melbourne, Australia: National Gallery of Victoria.

Wang, Q. 2002. *Metalworking technology and deterioration of Jin bronzes from the Tianma-Qucun site, Shanxi, China*. BAR International Series 1023. Oxford: Archaeopress.

Wang Q., and K. C. Andrews. 2002. Technological investigation of the decorative coatings on Yangshao pottery from Henan, China. *Archaeometry* 44(2):241–50.

Wang, Q., and J. Mei. 2009. Some observations on recent studies of bronze casting technology in ancient China. In *Metals and societies: Studies in honour of Barbara S. Ottaway*, ed. T. K. Kienlin and B. W. Roberts, 383–91. Bonn: Universitätsforschungen zur prähistorischen Archäologie.

Wang, Q., and B. S. Ottaway. 2004. *Casting experiments and microstructure of archaeologically relevant bronzes*. BAR International Series 1331. Oxford: Archaeopress.

Watson, K., and Hacke, M. 2009. Identification of green and white pigments on a Western Zhou (1050–770 BCE) bronze ritual food vessel (OA1977,0404.1), CSR analytical request no. AR2009/37. The British Museum, London.

Wood, N. 1989. Ceramic puzzles from China's Bronze Age. *New Scientist* 18(February):50–53.

An Enlightened Journey: Transitions in Casting of Chinese Buddhist Images

Donna Strahan

ABSTRACT When Buddhism began to arrive in northern China from India during the third century CE, travelers probably brought small lost-wax cast images with them. With the full-fledged arrival of Buddhism in the mid-fourth century CE these lost-wax cast images served as prototypes for locally produced sculptures; however, now they were cast by the traditional Chinese piece-mold method. At some point between the late sixth and early seventh century, Buddha sculptures in China transitioned from piece-mold casting to lost-wax casting. For a while, both methods were probably used contemporaneously. By the mid-seventh century, lost-wax casting became the norm, with only minor parts produced by the piece-mold method, and the lost-wax method continued as the preferred tradition into the twentieth century. In addition to addressing the question of the fabrication of Chinese Buddhist images using technical evidence, this paper examines how style reflects the casting technology.

Introduction

The arrival of Buddhism in China in the late third century CE brought with it an unprecedented need for a multitude of images, promoting a vast production of sculptural art. Central to Buddhist practice is the making of images, a process that serves the dual purpose of worship and attaining merit. As Buddhism traveled overland from India through Afghanistan and into western China, believers probably brought with them small images of the Buddha, which served as prototypes for local production. At this time, in the ancient region of Gandhara (roughly present-day Pakistan), Buddha figures were being cast by the lost-wax casting method from bronze and brass alloys containing varying amounts of copper, tin, zinc, and lead (fig. 1). During the initial period of Buddhism's arrival in China, it is unlikely that the technology for fabricating images or the metal-casters themselves traveled with the monks. When the local Chinese craftsmen were asked to faithfully copy the imported images, it was natural for them to employ their own traditional casting methods and alloys. The first locally produced, freestanding cast bronze icons appeared around the end of the third century CE (Leidy 2008).[1] These copper-alloy sculptures continued to be made intermittently up through the twentieth century, although their fabrication methods, alloy compositions, and styles changed through the centuries.

In northern China, the majority of early copper-alloy Buddha images identified thus far were produced around present-day Hebei (Leidy 2008). Stylistically they do not reflect the orthodox images imported from India and Gandharan Pakistan, but rather they are much stiffer and blockier in appearance, reflecting both Chinese technological traditions and tastes. Other factors may have contributed to the style of the Buddha images. The third and fourth centuries CE were a low point in the history of Chinese bronze manufacture as political upheavals and constant warfare greatly disrupted life, particularly in northern China. Although Hebei province had produced outstanding bronzes from the late Warring States period (475–221 BCE) to the Western Han dynasty (206 BCE–8 CE), its bronze industry greatly declined in the following periods. The stiff look of the Buddha images was also probably the result of the limitations of local workshops, as the bronze industry in the north did not begin to recover until the fifth century CE, when Hebei became a leading center for the production of Buddhist bronzes.

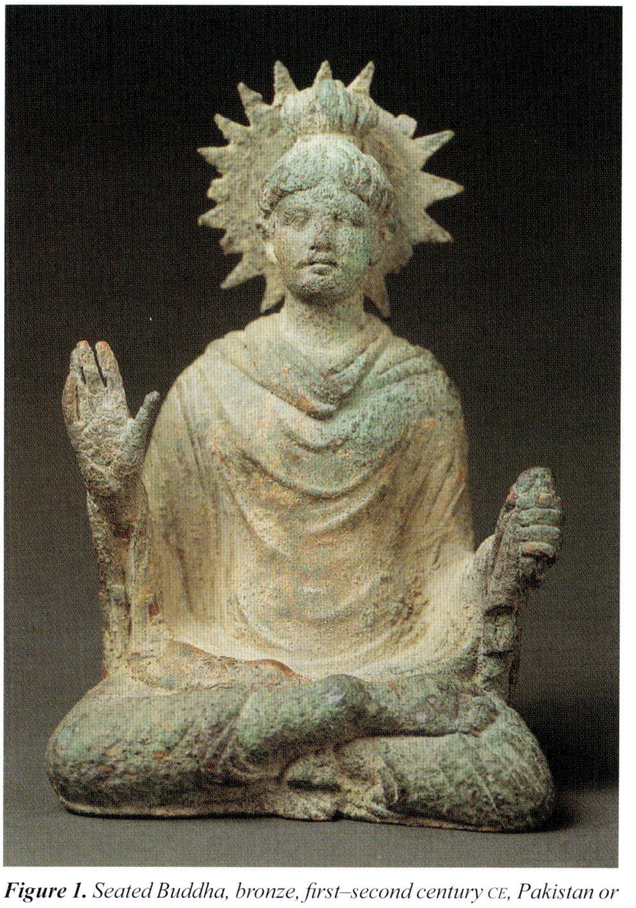

Figure 1. *Seated Buddha, bronze, first–second century CE, Pakistan or Afghanistan, ancient region of Gandhara, h. 16.8 cm. The Metropolitan Museum of Art, Gift of Muneichi Nitta, 2003, 2003.593.1. Photograph Studio, The Metropolitan Museum of Art.*

Because no ancient foundries for casting Buddhist images have ever been located, the exact method of production is speculative and can be deduced only by examining the existing works. Hundreds of small bronze Buddha sculptures have survived, most of which are less than 20 cm in height. The earliest known dated Chinese Buddha sculpture is a seated image inscribed with the date corresponding to 338 CE (Strahan 2010).

Piece-Mold Casting

Contrary to the findings of the few earlier technical studies, these early Buddhist bronzes were cast using a piece-mold method and not by the lost-wax technique (Chase 1991; Cowell et al. 2003; Jett and Douglas 1992; Tan 1989, 85–86).[2] A detailed technical study of three important sculptures from the fourth and fifth centuries CE confirms this (Strahan 2010). Indeed these three sculptures represent two distinct piece-mold casting strategies: one used for smaller sculptures and the other for large ones. For the smaller sculptures, metal spacers were used to hold the core in position during casting; for the larger sculpture, extensions from the core were used to support it within the mold assemblage.

These sculptures are representative of a large group of Buddha images of similar style, cast in piece molds during the fourth, fifth, and sixth centuries CE. During the sixth and seventh centuries, piece-mold casting gave way in favor of the lost-wax method. A study of this transition was undertaken during research for a catalogue of the Metropolitan Museum of Art's collection of Chinese Buddhist sculpture (Leidy and Strahan 2010). This paper provides further technical details from that research.

The piece-mold method used for smaller images will be illustrated by a fourth-century seated Buddha in the collection of the Harvard University Art Museums (fig. 2). This image is considered an important link between the Indian/Gandharan images and early Chinese works. Some scholars believe that it was made in the region of present-day Afghanistan and Pakistan. The trapezoidal base, lack of undercuts, flat hands, and simple shallow drapery are stylistic elements that point to piece-mold casting and not lost-wax casting. It is hypothesized that the image was cast in a three-piece mold: front, back, and bottom that included the core. The front mold section began at the top of the head and continued down the side of the neck, across the shoulders, down the outside of the arms, across the thighs, and down the front corners of the base. The back mold covered the entire back half of the sculpture, with the tang out the back of the head, and three sides of the base, including the figures. A separate mold was probably necessary for the proper right figure. There is no evidence of core pins, which are typically associated with lost-wax casts. Instead, metal spacers, characteristic of the piece-mold casting method, were used to hold the core and mold apart. Square and rectangular metal spacers were inserted between the core and

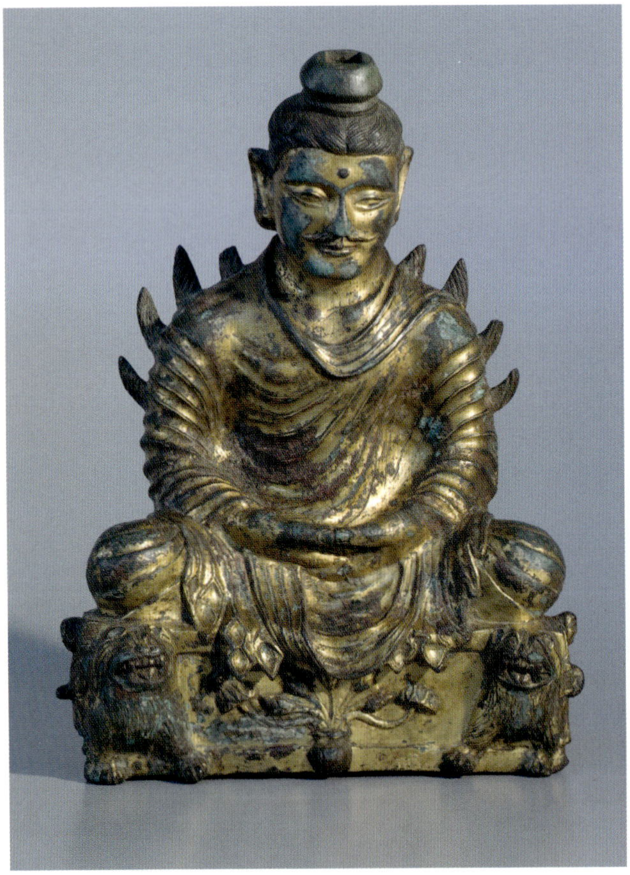

__Figure 2.__ Seated Buddha, gilt leaded bronze, Sixteen Kingdoms (304–439 CE), fourth century, h. 32 cm. Harvard University Art Museums, Grenville L. Winthrop Collection, 1943.53.80.

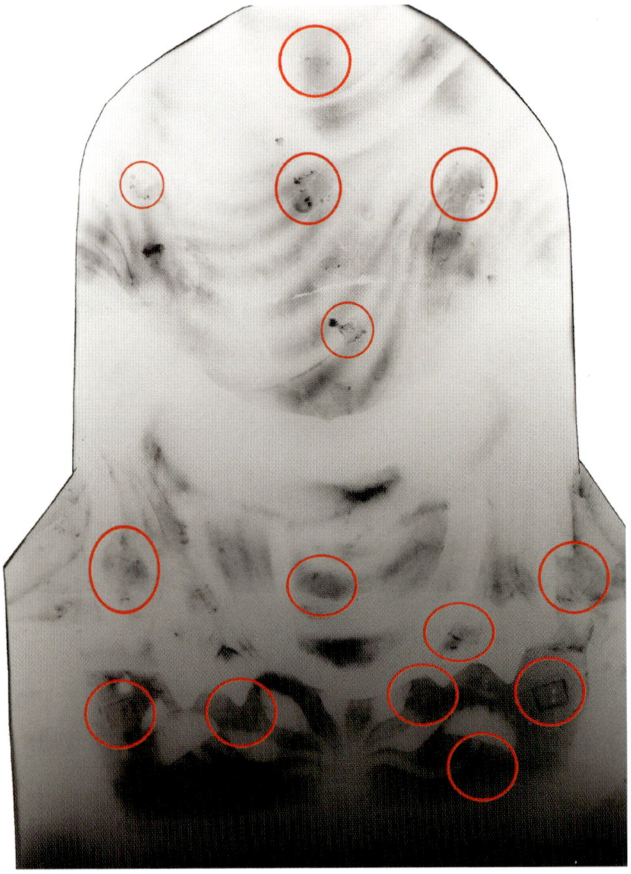

__Figure 3.__ Digitally enhanced x-ray radiograph revealing spacers, noted in red, in the front of the Harvard Buddha in figure 2. Department of Objects Conservation, The Metropolitan Museum of Art.

mold sections at strategic locations. Radiographs reveal the locations of at least twenty spacers that are not visible on the sculpture's surface as they have become incorporated in the metal walls. Fourteen spacers have been identified on the front of the sculpture alone (figs. 3 and 4).

After casting, the entire core was removed, and the exposed interior walls show the rough bronze texture typical of many "as cast" surfaces. They do not exhibit any evidence of tool marks, brush marks, finger prints or impressions, joins between wax sheets, or drip marks, as might be present if the sculptures had been cast using the indirect lost-wax method. After the sculptures were cast, the flashings located at mold junctures were removed mechanically, eliminating any evidence of piece-mold section joins. This study of the casting technology along with its alloy composition and core analysis provide clear evidence that the sculpture was made somewhere in northern present-day China.[3]

An example of piece-mold casting of larger sculptures is the monumental Buddha Maitreya in the Metropolitan dated 486 CE (fig. 5). Similar to the small Buddha above, its details are executed in low relief without undercuts; however, it is an order of magnitude larger in size and far more complicated. It was also cast by the piece-mold method, although the challenges of creating such a large sculpture necessitated a different strategy. This Buddha was probably cast in one piece, including the hands.[4]

The model of the Buddha—an "identical twin" of the finished metal sculpture—was probably created in loess (fine local clay). It would have had all the details that now appear on the final bronze except for fine finishing marks, and when it was finished, a piece mold would have been taken of the entire model. The mold sections would have been made one at a time, cutting mortises into the edges so that they could be keyed to their neighbors at each seam, and then allowed to dry *in situ*. The next section would contain tenons fitting the mortises next to it, guaranteeing a tight fit when the molds were reassembled for casting. Here the use of mortises and tenons is only conjecture since no Buddha sculpture foundries have ever been found.[5]

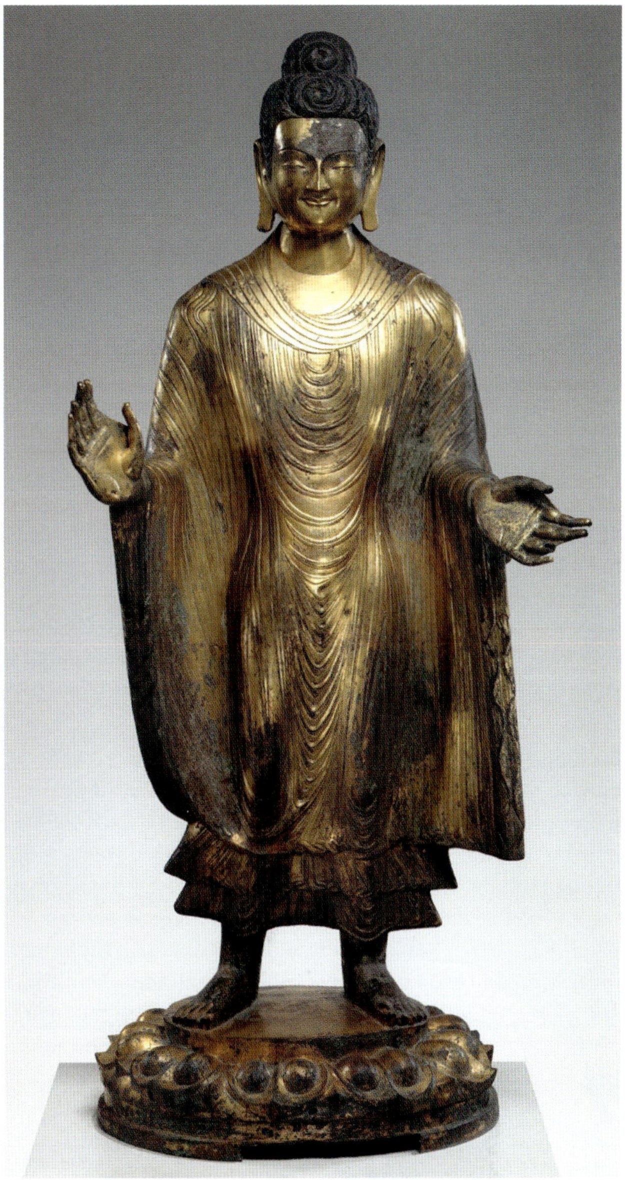

***Figure 5.** Buddha Maitreya, gilt leaded bronze with traces of pigment, dated 486 CE, Shanxi province, h. 140.5 cm. The Metropolitan Museum of Art, John Stewart Kennedy Fund, 1926, 26.123. Photograph Studio, The Metropolitan Museum of Art.*

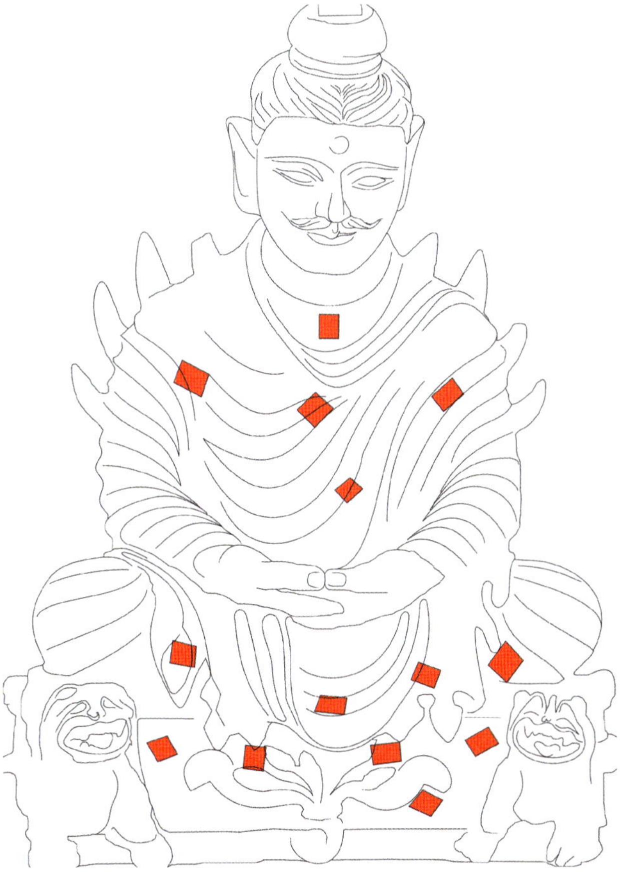

***Figure 4.** Harvard Buddha with locations of metal spacers on the front shown as red rectangles. Drawing by Federico Caró.*

These outer mold sections were probably given a light biscuit firing. However, the original model was probably not fired, making it easier to shave down to serve as the core as well as to remove after casting. A thickness of the loess corresponding to the intended wall thickness of the finished sculpture was removed. During this process, numerous rectangular protrusions were left intact to be used to separate the core from the mold sections at the level of the model's original surface. These core extensions were positioned at regular intervals over the entire sculpture (fig. 6). The large opening in the back of the sculpture would have accommodated what was effectively a very large core extension. Since most of the mold's seam flashings were removed during the finishing process, it is not clear how many mold sections were used or what their exact placement would have been. The positions of the core extensions on the standing Buddha Maitreya may help determine the basic mold construction.

Visible in the radiographs are periodic horizontal lines of bubbles indicating that the sculpture was cast upright using multiple crucibles of molten metal to fill the complex mold assemblage. The line of bubbles rising toward the surface as the metal cooled may correspond to the approximate level of a pour. Some of the gases were trapped in the metal, and the rest were most likely absorbed by the mold sections or along the mold seams. One of the virtues of fired loess is its fine particle size and high porosity, allowing the absorption of gases from the molten metal and thus avoiding the pores and flaws in the finished object or the need for a venting system necessary for materials high in platy clay minerals (Kerr and Wood 2004, 103).

After the assemblage cooled, the mold sections were broken off. At that time, the core extensions were cut out, and depressions that undercut the edges of the openings were gouged into the core. Before the surfaces were finished by grinding and polishing, molten bronze was poured into each hole, filling also the undercuts and flowing over the outer surfaces of the wall, thus helping to secure the patch. One patch has a thin circular depression in the center, indicating that the patch metal was poured into the middle and as it cooled, shrinkage drew the metal from the center toward the outer edges leaving a depression in the middle (fig. 7).

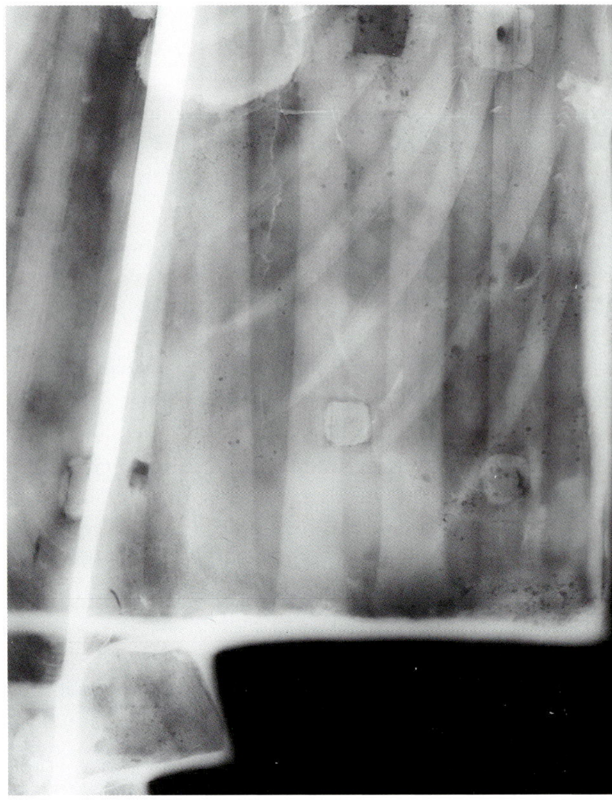

Figure 7. *X-ray radiograph of Buddha Maitraya's drapery on the left side revealing plugged core extensions. Department of Objects Conservation, The Metropolitan Museum of Art.*

Core extensions, or core extension spacers as Gettens calls them, are not a new development in Chinese piece-mold fabrication (Gettens 1969, 78–80). During the Shang dynasty (ca. 1600–1050 BCE), they were cruciform-shaped projections producing decorative holes in the cast vessel's

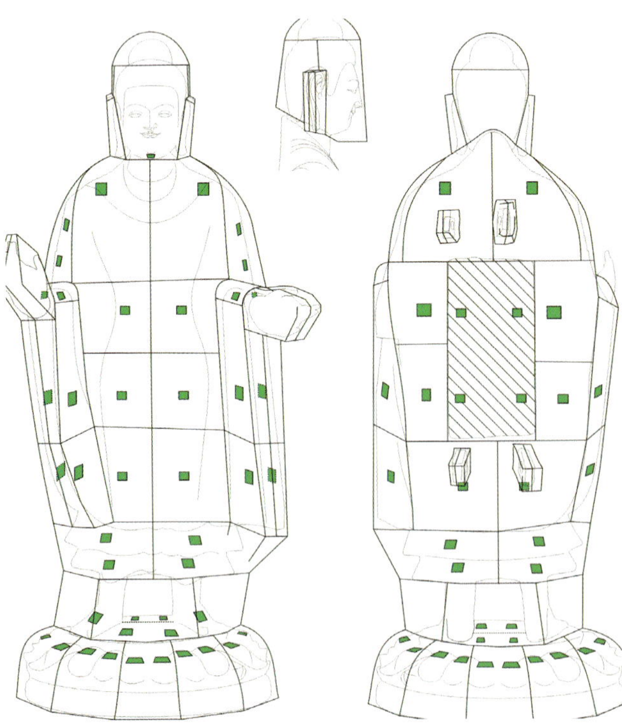

Figure 6. *Buddha Maitreya, conjectured piece-mold sections and core with core extensions shown as green rectangles. Drawing by Federico Caró.*

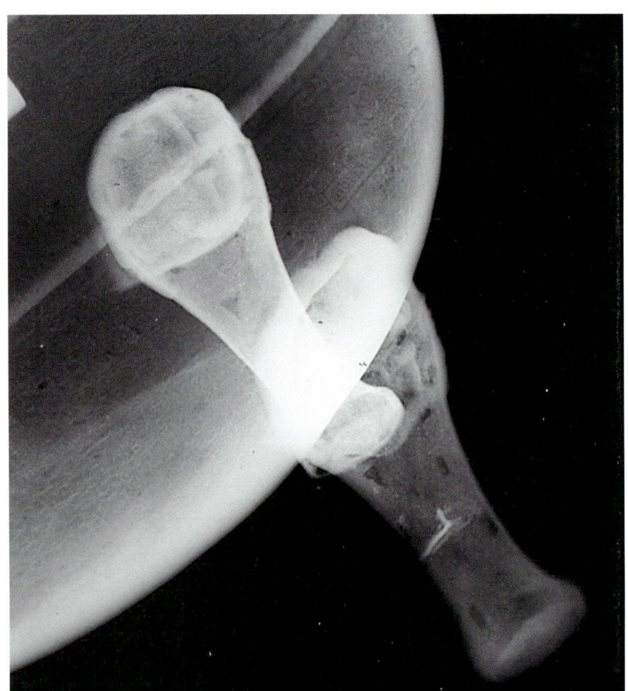

Figure 8. *X-ray radiograph of a* ding *showing pyramidal-shaped core extensions. The Metropolitan Museum of Art, 13.100.6a,b. Department of Objects Conservation, The Metropolitan Museum of Art.*

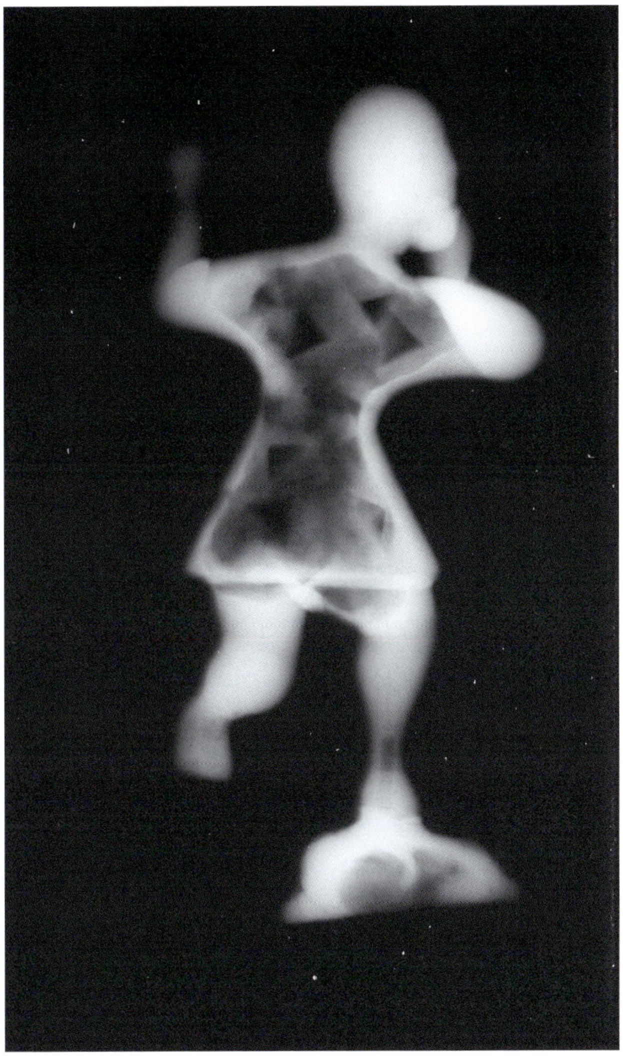

Figure 9. *X-ray radiograph of a standing figure of a nomad, bronze, Eastern Zhou dynasty (771–221* BCE*), h. 10.2 cm. The Metropolitan Museum of Art, Gift of Earnest Erickson Foundation, 1985.214.48. Department of Objects Conservation, The Metropolitan Museum of Art.*

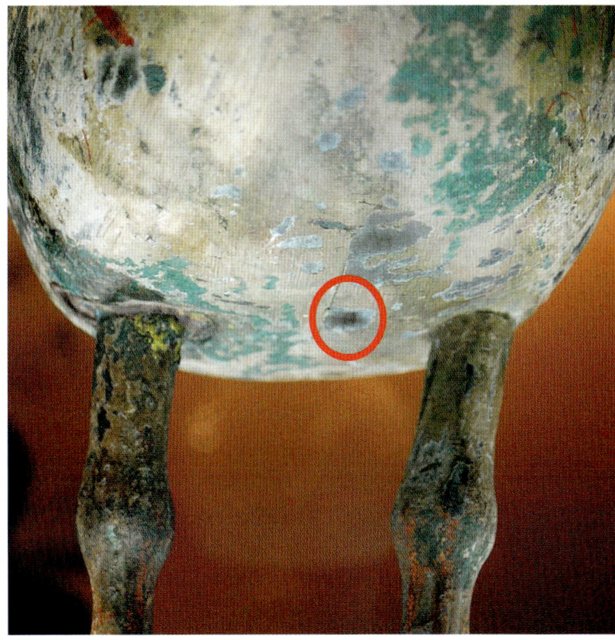

Figure 10. *Mold line and plugged core extension in the belly of a goose-shaped lamp, bronze with polychrome decor, Western Han dynasty (206 BCE–8 CE), excavated from Baoji City, h. approximately 70 cm. Shaanxi History Museum. Department of Objects Conservation, The Metropolitan Museum of Art.*

foot rim. At the Anyang foundry (Shang dynasty, 1200 BCE), independent cores were sealed inside vessel legs and supported by projections from the core (Bagley 1987, 44–45). During the early Western Zhou dynasty (ca.1050–771 BCE), they are present as pyramidal- or rectangular-shaped extensions projecting from the cores of legs and handles and serving to support the external mold sections by keeping the wall thickness even throughout and the core immobile (Meyers and Holmes 1983; Gettens 1969) (fig. 8). Core extensions continued to be used for small internal cores during the late Eastern Zhou dynasty (770–221 BCE), as can be seen in radiographs of a small figure of a nomad (fig. 9). Among the mold fragments at the Houma foundry (fifth century BCE) is an unused piece mold of a standing figure with multiple extensions on its core (Institute of Archaeology of Shanxi Province 1996). For larger objects, the core extensions developed into rectangular extensions. Some evidence of these may be seen in the large bronze birds from Qin Shihuang's mausoleum (Qin dynasty, 221–206 BCE) that were probably also piece-mold cast.[6] A mold line and a plugged core extension can be seen on the belly of a goose-shaped lamp from the late Western Han dynasty (fig. 10).

Lost-Wax Casting

Research indicates that piece-mold casting was the preferred technology and perhaps the only casting method used for the early Buddha images up until the late sixth and early seventh century. During that time, lost-wax casting began to take hold in northern China. Unfortunately, inscriptions were rarely used after this time period, making dating of the sculptures more difficult. Although it cannot be identified exactly where or how it occurred, it is clear when examining the images from this period that a dramatic change in both style and technology took place. It was not until the religion's hold strengthened during the Tang dynasty (618–907) and foreigners arrived that the technology of lost-wax casting was permanently adopted for images in a unified China. Imperial patronage promoted the growth of Buddhism by supporting foreign monks and by encouraging Chinese monks to travel abroad. They may have brought back lost-wax casting methods, or foreign metalworkers may have traveled to China to better imitate the imported sacred images. Lost-wax casting may have come in from the west (India/Afghanistan/Kashmir), the north (northern frontier), and/or even the south (Southeast Asia). The late sixth- to seventh-century Chinese sculptures found in the Cambodian Kampong Cham province cache provide clear evidence of the far-reaching trade with the south (Jett 2010).

It is likely that both the lost-wax and piece-mold methods were used at the same time. By the eighth century, sculptures were all made by the lost-wax method except for perhaps the occasional attachment. Founders were exposed to an array of styles and technologies during the Tang dynasty that broadened their perspective. This new technology was able to create stylistic changes, allowing for the production of more graceful figures that appear more natural than the

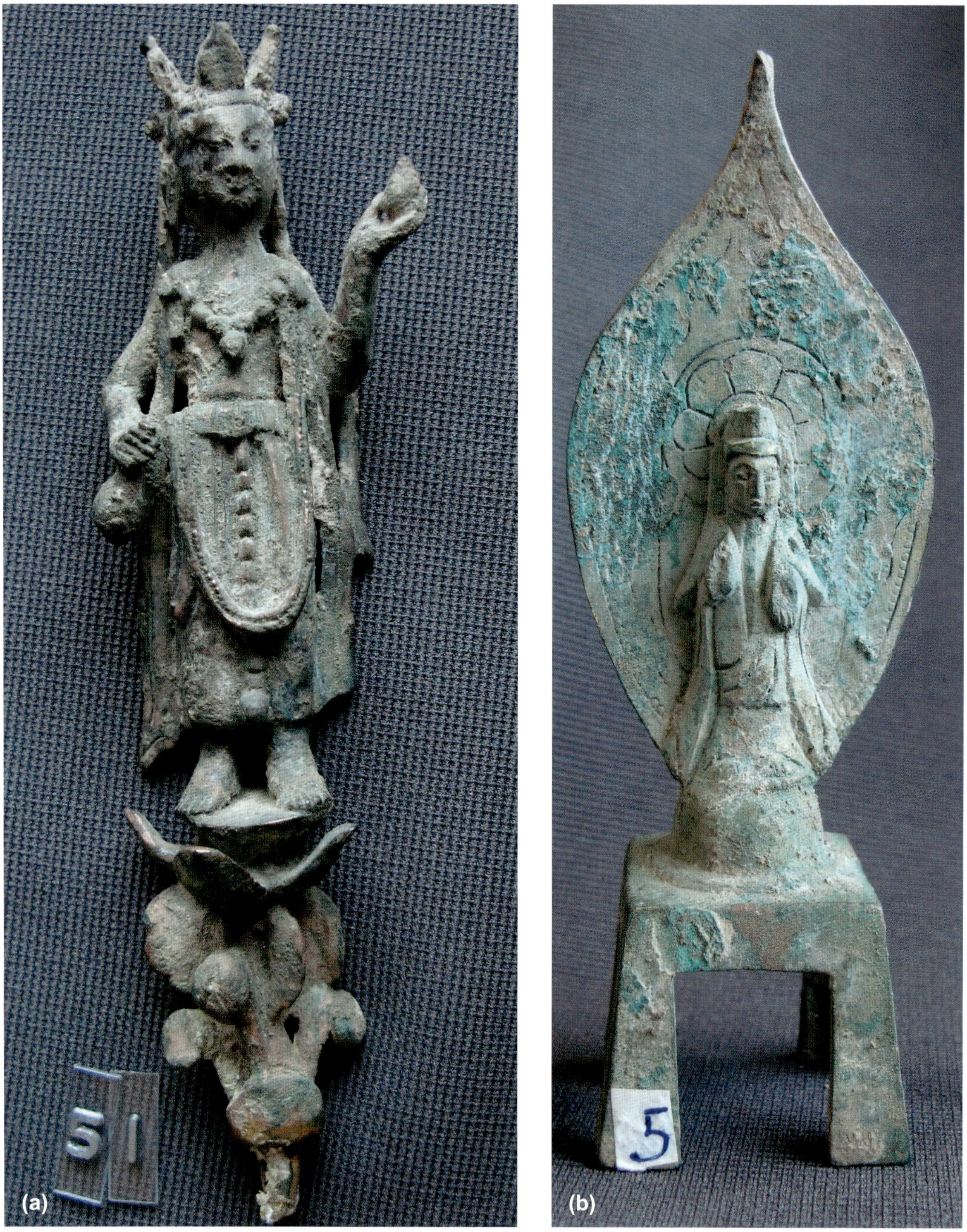

Figure 11. *Two sculptures from a cache excavated in Shouyang. The one on the left* ***(a)*** *was lost-wax cast, and the one on the right* ***(b)*** *was piece-mold cast. Photo by Donna Strahan.*

block-like piece-mold figures. Many of these changes probably resulted from the flexibility of the lost-wax method.

It is not always easy to recognize the difference between lost-wax and piece-mold casting, especially with solid sculptures. There were clearly times when both methods were being used, perhaps even on the same sculptures. Of the two solid cast Chinese figures in the sixth–seventh-century Kampong Cham province cache, one is clearly lost-wax cast (Ga 6943) but the other could be either lost-wax or piece-mold cast (Ga 6942) (Jett 2010). The images in the

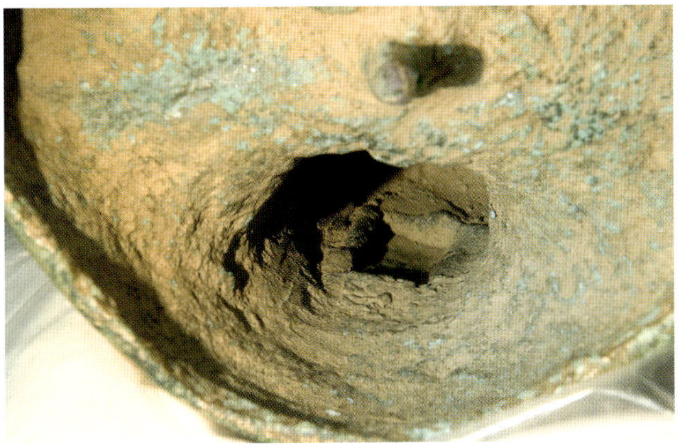

Figure 12. *Tool marks in the wax were reproduced in metal and can be seen, along with its thin sheet metal cover, on the interior of a Manjusri, the Bodhisattva of Wisdom, with Five Knots of Hair, mid-eleventh century, h. 15.9 cm. The Metropolitan Museum of Art purchase: Mary Livingstone Griggs and Mary Griggs Burke Foundation, Janet Neff Charitable Trust, and Christian Humann; Gifts: Mrs. Robert Herndon Fife Gift, in memory of her husband, and Seymour Fund, 1967, 67.228. Department of Objects Conservation, The Metropolitan Museum of Art.*

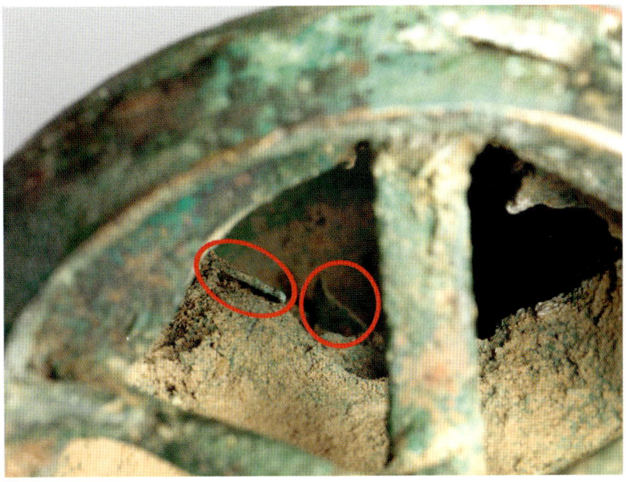

Figure 13. *Three flat triangular core pins are visible on the interior base of an early eighth-century seated Buddha Vairocana, h. 20 cm. The Metropolitan Museum of Art, Rogers Fund, 1943, 43.24.3. Photograph Studio, The Metropolitan Museum of Art.*

sixth–seventh-century Shouyang cache from near Taiyuan, China, contain both piece-mold and lost-wax cast images (fig. 11).

Lost-wax casting can be carried out by the direct method or the indirect method. Direct lost-wax casting produces a single image, because the original model is melted out and lost. Figures made by this method can be either block-like or naturalistic, with many projections and deep undercuts in all directions. Through the use of armatures, cores, and a highly malleable material such as wax, very expressive shapes can be created.

Using the indirect method requires a mold to be taken from a model. Once removed from the model, the mold is reassembled and lined with wax. Once the wax model is complete, the core is then introduced, the mold disassembled, and independently modeled wax components are appended. Before it is invested, surface detail can be scored into the wax model. Sometimes evidence of this process survives on the interior walls of the cast bronze. For example, if the wax was molten and poured or brushed into the mold, then drip or brush marks might be represented. If, instead, wax sheets were pressed in, then fingerprints, tool marks, or seam lines may be visible (fig. 12).

Both direct and indirect hollow lost-wax casting depend on pins to support the core in the investment once the wax has been melted out and until the molten metal is poured in. These core pins can be left in the finished sculpture or removed, leaving a small void that can be filled with plugs either hammered in position or, for larger holes, cast *in situ*.

Of those sculptures studied, beginning in the early eighth century, copper-alloy core pins, usually in a thin, flat, triangular shape, were used in hollow lost-wax cast sculptures (fig. 13). This core pin shape continued to be used into the seventeenth century and perhaps even later, regardless of the alloy composition of the sculpture.[7] Headless iron pins with a square cross-section have been found in Yuan dynasty (1279–1368) sculptures. During the Yuan dynasty, China was ruled by the Mongols, whose sculptures were influenced in turn by Kashmir, where iron core pins were common (Reedy 1997, 107; Leidy 2010). Iron- and copper-alloy core pins are the most common type found and continued to be used during the Ming (1368–1644) and Qing (1644–1911) dynasties. The Yongle emperor (1403–1424), a follower of Tibetan Buddhism, had great numbers of Buddhist sculptures produced at the court by Chinese and Tibetan artists who incorporated Indo-Himalayan styles that were transformed into Chinese expressions of belief (Howard et al. 2006). The core pins identified are predominantly iron, but small copper-alloy pins were also used. It is not possible to come to specific conclusions here on the type and geographic location of the use of core pins. Clearly along with the waxing and waning of the popularity of Buddhism and the enormous size and geographic variation of China, regional styles and fabrication innovations developed.

The presence of core pins appears to have eliminated the use of metal spacers, but core extensions may still occasionally have been used, even though they required more extensive repair and clean-up. It is possible they were used to create the rectangular openings in the backs, heads, and bases of sculptures. Interestingly some of these sculptures also have core pins. Once lost-wax casting dominated the sculpture industry, core extensions played a double role—that of holding the core in position during casting and, after casting, as an opening for accessing the interior of the image to insert consecratory deposits. Rectangular openings in the back and heads of sculptures are present in some of the earliest hollow lost-wax cast images (fig. 14). Although these openings could easily have been filled with cast-in repairs, they instead have thin sheet copper covers. Consecratory materials were usually composed of paper sutras, textiles, fragrant wood, semiprecious stones, and pearls. If consecratory material is deposited into the head or body of the sculpture, a cast-in cover would likely destroy the contents. The interior of a Manjusri exhibits tool marks in the wax and visual closure with a copper sheet cover (see fig. 12).

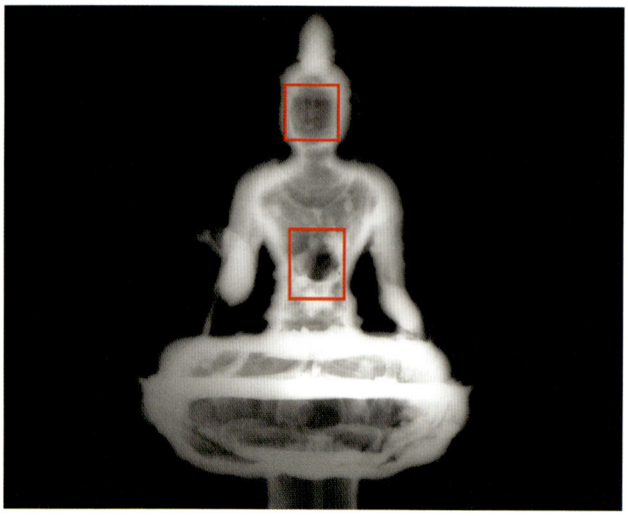

Figure 14. *X-ray radiograph revealing thin copper covers over a chamber in the head and in the back of a seated Bodhisattva (outlined in red), tenth century, h. 19.7 cm. The Metropolitan Museum of Art, Bequest of John M. Crawford Jr., 1988, 1989.363.220. Department of Objects Conservation, The Metropolitan Museum of Art.*

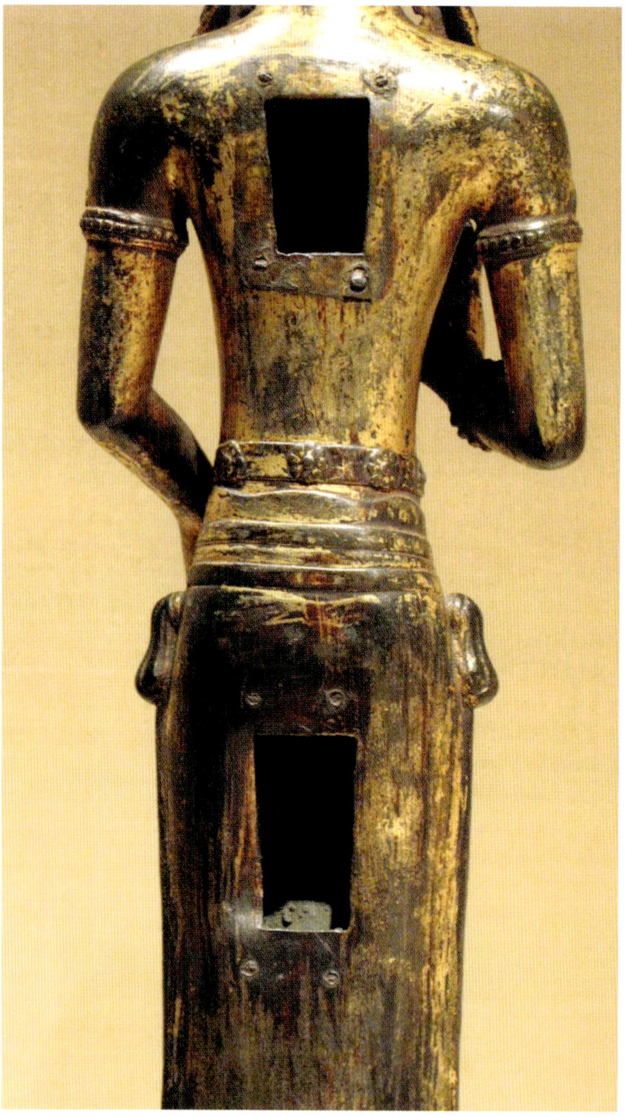

Figure 15. *The two openings in this Avalokiteshvara were once closed with sheet metal covers. Bodhisattva Avalokiteshvara, Yunnan province, twelfth century, h. 50.2 cm. The Metropolitan Museum of Art, Gift of Abby Aldrich Rockerfeller, 1942, 42.25.28. Department of Objects Conservation, The Metropolitan Museum of Art.*

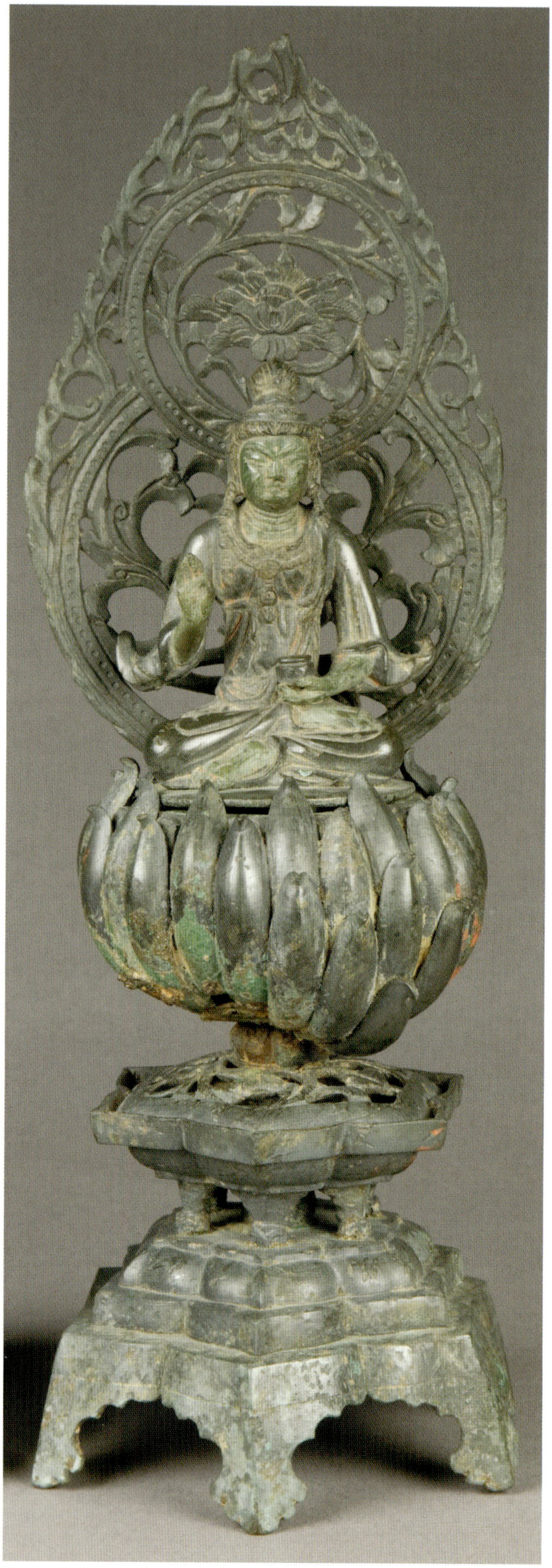

Figure 16. *Bodhisattva, high-leaded bronze, Wu Yue kingdom (893–978), h. 29.2 cm. The Metropolitan Museum of Art, Fletcher Fund, 1925, 25.217.3. Photograph Studio, The Metropolitan Museum of Art.*

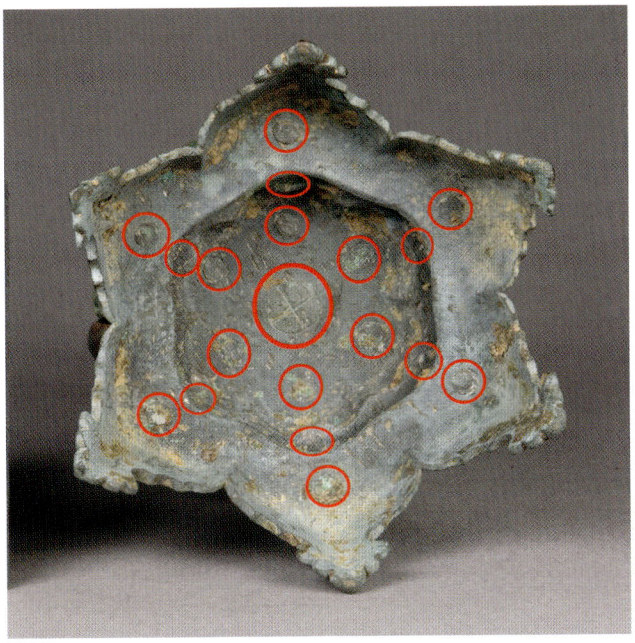

Figure 17. *Location of rivets holding multiple separately cast parts together on the underside of the Bodhisattva in figure 16. Department of Objects Conservation, The Metropolitan Museum of Art.*

A series of Bodhisattvas from the twelfth-century Dali kingdom in Yunnan province provides another example of the continuing use of core extensions in lost-wax casting but in a different regional style (Jett 1991) (fig. 15). These sculptures all have two rectangular openings in their backs, which were probably used as core extensions during fabrication. Aside from the core supports, metal armatures may have also been incorporated as core supports instead of core pins. Although the armatures were removed from those studied, the sculpture shown in figure 15 does have what may be the remains of an iron armature or core pin extending out the top of the head. There are no other core pins in the sculpture.

During this study, no identical sculptures have been identified. It is tempting to say that they were all made by the direct lost-wax casting method. Other regional foundries clearly existed, as can be seen with the Yunnan sculptures discussed above. Their differences are reflected in both their distinctive styles and technologies. Another example is a unique group of sculptures made in Zhejiang province, of which the Metropolitan has one sculpture and the Freer Gallery of Art has three (fig. 16).[8] This group of sculptures uses a complex system of rivets to pin together multiple separately cast parts. The small Metropolitan Bodhisattva is made up of nine different sections and twenty-six lotus petals, all riveted together to create a single sculpture (fig. 17). Perhaps parts of the petals, mandorla, and base were impressed in molds, and then the pieces were indirect lost-wax cast separately prior to riveting them together.

Conclusion

This study of copper-alloy Buddha images demonstrates that piece-mold casting, a tradition that harks back to the Shang dynasty, continued to be used in north China into the seventh century and perhaps later. Aside from the molds themselves, the main features of piece-mold casting are the use of metal spacers and core extensions to support the clay cores in hollow castings. Strong outside influences during the sixth and seventh centuries brought about a transfer of technology that resulted in the casting of sculptures by the lost-wax method. These two technologies were undoubtedly used at the same time, but by the eighth century the lost-wax method dominated. This new technology did incorporate a feature of piece-mold casting—the use of core extensions. Regional methods and styles also developed, but thus far only a few have been studied.

More images should be studied to confirm or refute these findings. Future study should re-examine images made between the Han and Tang dynasties in light of their casting method to learn more about the use of core extensions and the continuation of piece-mold casting. It is suspected that piece-mold casting will dominate.

Acknowledgments

Special appreciation is given to Federico Caró for his precise analysis of bronze core thin-sections and for his excellent illustrations. Henry Lie and his staff at the Harvard Art Museums were very generous with their time, radiography, and access to the Harvard Art Museums sculpture. Zhou Weirong is thanked for his lively discussions on casting. Wang Xiaoyi of the Archaeological Institute of Shanxi Province generously provided access to the Shouyang cache.

Notes

1. The earliest evidence of the new religion can be seen in Buddhist imagery on funerary pottery, bronze money trees, and tomb walls dating to the second and third centuries. The third-century Szechuan bronze money trees with Buddha images were piece-mold cast in sections (Steele et al. 1997).
2. The literature states that the lost-wax casting method was first used to create complex components of vessels in northern China during the Eastern Zhou dynasty (around the fifth century BCE) (see Bagley 1987; Mackenzie 1991; Chase 1991; Jett and Douglas 1992; Tan 1989). Recent debate has arisen over the date of the arrival of the lost-wax casting method into northern China. Zhou et al. (2009) propose that the lost-wax casting method did not arrive in north China until after Buddhism arrived sometime in the late sixth to early seventh century.
3. This technology and alloy composition are very similar to a seated Buddha dated 338 CE in the Asian Art Museum (B60B1034) discussed in Strahan (2010). Federico Caró, fellow, Department of Scientific Research, the Metropolitan Museum of Art, analyzed core material from the Metropolitan Museum of Art Maitreya (26.123) and the Harvard Art Museums Buddha (1943.53.80) and found that they are both consistent with Chinese loess and are of similar composition. There is no organic material added to the cores, and they have similar amounts of matrix, similar grain size, and texture. The grain size distribution and composition are also comparable.

The only difference is the presence of rounded calcite in the Harvard Buddha, but this is also found in Chinese loess.

4. The casting method of this image has been described in earlier publications by the author (Strahan 2010; Leidy and Strahan 2010). Both articles stated that the hands were cast separately, but further investigation provided evidence that they were made in the same pour as the rest of the figure. This was further confirmed by the alloy composition that is nearly identical to the figure, indicating that the hands were part of the original pour. All traces of mold marks have been ground off.
5. These steps are similar to those described by Bagley (2009, 44). Mold sections at the Anyang excavations have been found both with and without mortises and tenons (Li 2003, 109–11).
6. In 2000, a pit at the northeast corner of the Qin emperor's (d. 210 BCE) mausoleum unearthed forty-six life-sized bronze birds. Although the author has not examined them closely, mold seams and examples of plugged core extensions can be seen in the large bronze birds in the Xian Archaeological Institute and the Shaanxi History Museum. At the Xian Archaeological Institute, the author examined a broken bird's neck with the core still present. Two rectangular clay inclusions were present, projecting into the clay core from the sides of the neck. It is possible that the neck core (with wire armature) was made separately and inset into the bird body as the mold sections were assembled. Thanks are given to Shao An Ding for his assistance in the examination.
7. Other sculptures in the Metropolitan Museum of Art that have flat, triangular copper-alloy core pins include a seated tenth-century Laojun (42.25.19); a seated Daoist immortal dated 1438 (1997.139), and a seventeenth-century Guardian figure (2002.254). Cowell et al. (2003) illustrate a Ming example in plate 11.
8. Three other similarly fabricated examples of the same style, date, and alloy composition are in the collection of the Freer Gallery of Art (F1912.92, F1912.93, and F1912.94).

References

Bagley, R. 1987. *Shang ritual bronzes in the Arthur M. Sackler collections*. Ancient Chinese Bronzes in the Arthur M. Sackler collections; v. 1. Washington, DC, and Cambridge, MA: The Arthur M. Sackler Foundation and the Arthur M. Sackler Museum.

———. 2009. Anyang mold-making and the decorated model. *Artibus Asiae* 69(1):39–90.

Chase, W.T. 1991. *Ancient Chinese bronze art: Casting the precious sacral vessel*. New York: China Institute in America.

Cowell, M., S. La Niece, and J. Rawson. 2003. A study of later Chinese metalwork. In *Scientific research in the field of Asian Art: Proceedings of the First Forbes Symposium at the Freer Gallery of Art,* ed. Paul Jett with J.G. Douglas, B. McCarthy and J. Winter, 80–89. London: Archetype Publications.

Gettens, R. J. 1969. *The Freer Chinese bronzes,* Vol. 2, *Technical studies*. Oriental Study Series 7. Washington, DC: Freer Gallery of Art, Smithsonian Institution.

Howard, A. F., W. Hung, L. Song, and Y. Hong. 2006. *Chinese sculpture*. New Haven, CT: Yale University Press.

Institute of Archaeology of Shanxi Province. 1996. *Art of the Houma foundry*. Princeton, NJ: Princeton University Press.

Jett, P. 1991. Technologische Studie zu den vergoldeten Guanyin-figuren aus dem Dali-Königreich. In *Der goldschatz der drei pagoden: Buddhistische kunst des Nanzhao-Königreichs in Yunnan, China*, ed. A. Lutz, 68–74. Zürich: Museum Rietberg.

———. 2010. A technical study of the Kampong Cham figure group. In *Gods of Angkor: Bronzes from the National Museum of Cambodia,* ed. L. A. Cort and P. Jett, 78–87. Washington, DC: Smithsonian Institution.

Jett, P., and J. Douglas. 1992. Chinese Buddhist bronzes in the Freer Gallery of Art: Physical features and elemental composition. In *Materials issues in art and archaeology III,* ed. P. B. Vandiver, J. R. Druzik, G. S. Wheeler, and I. C. Freestone, 205–23. Material Research Society Symposium Proceedings 27 April–1 May 1992. Pittsburgh: Material Research Society.

Kerr, R., and N. Wood. 2004. *Science and civilization in China,* Vol. 5, *Chemistry and chemical technology,* Part 12, *Ceramic technology*. Cambridge: Cambridge University Press.

Leidy, D. P. 2008. *The art of Buddhism: An introduction to its history and meaning*. Boston: Shambala.

———. 2010. Buddhism and other "foreign" practices in Yuan China. In *The world of Khubilai Khan: Chinese art of the Yuan dynasty,* ed. J. C. Y. Watt, 87–122. New Haven, CT: Yale University Press.

Leidy, D. P., and D. Strahan. 2010. *Wisdom embodied: Chinese Buddhist and Daoist sculpture in The Metropolitan Museum of Art*. New York: The Metropolitan Museum of Art.

Li Yung-ti. 2003. The Anyang bronze foundries: Archaeological remains, casting technology, and production organization. PhD diss., Graduate School of Arts and Sciences—Anthropology, Harvard University.

Mackenzie, C. 1991. Chu bronze work: A unilinear tradition, or a synthesis of diverse sources? In *New perspectives on Chu culture during the Eastern Zhou period,* ed. T. Lawton, 107–57. Washington, DC: Arthur M. Sackler Gallery, Smithsonian Institution.

Meyers, P., and L. L. Holmes. 1983. Technical studies of ancient Chinese bronzes: Some observations. In *The great Bronze Age of China: A symposium,* ed. G. Kuwayama, 124–36. Los Angeles: Los Angeles County Museum of Art.

Reedy, C. L. 1997. *Himalayan bronzes: Technology, style, and choices*. Newark and London: University of Delaware Press.

Steele, J., L. Stodulski, and K. Trentleman. 1997. Deciphering the puzzle: The examination and analysis of an Eastern Han dynasty money tree. In *Objects Specialty Group Postprints,* Vol. 5, compiled by V. Greene and I. Neuman, 125–41. Washington, DC: American Institute for Conservation of Historic and Artistic Works.

Strahan, D. 2010. Piece-mold casting: A Chinese tradition for fourth- and fifth-century bronze Buddha images. *Metropolitan Museum Studies in Art, Science and Technology* 1:133–53.

Tan, D. R. 1989. *Canlan de zhounguo gudai shila zhuzao* [The splendid craft of lost-wax casting in ancient China]. Shanghai: Shanghai Scientific and Technological Literature Publishing House.

Zhou, W.R., Y. W. Dong, Q. W. Wan, and C. S. Wang. 2009. New research on lost-wax casting in ancient China. In *Metallurgy and civilisation: Eurasia and beyond,* ed. J. Mei and T. Rehren, 73–78. London: Archetype Publications.

Southeast Asian Bronzes

Looking East: The Coming of Tin Bronze in the Context of Bronze Age Trans-Eurasian Exchange—An Overview

Vincent C. Pigott

ABSTRACT While the origin and development of tin-bronze metallurgy in the Old World has for some time been the focus of considerable scholarly attention, this research has mostly proceeded on an area-by-area basis. This paper attempts to link current data for this all-important technological development across Eurasia. The technology appears first in Southwest Asia, ca. the early third millennium BCE, moving across the Eurasian steppe and China's borderlands, into China's Central Plain, and finally into Southeast Asia, where it appears ca. the late second millennium BCE, if not before. Thus, current evidence suggests that over time the technology of tin bronze was being gradually transmitted eastward between and among receptive cultures across this vast region through which major tin belts are known to be located. Questions continue to surround the arrival of tin bronze in regions of China ca. 2000 BCE and later. Contact with steppe and forest-steppe cultures (including perhaps the enigmatic Seima-Turbino) appears to have had a role in the transmission of bronze metallurgy into the Central Plain. However, their potential role in any ensuing dissemination southward into Southeast Asia remains open to question. Considerable debate as to the sources, paths, and chronologies of transmission into this region revolves around two contrasting models; these models are briefly reviewed. The technology and its arrival in Southeast Asia may be ultimately linked to earlier developments in Eurasia. In Thailand and Laos, fieldwork and research in the laboratory have documented prehistoric copper and tin-bronze metalworking activity (e.g., mining, smelting, casting, trade). An overview of this activity is provided, and evidence in support of possible steppe connections is presented.

Introduction

A recent article titled "Development of Metallurgy in Eurasia" (Roberts et al. 2009) makes the point that scholars now know enough about Eurasian metallurgical developments to speak of them in pan-Asiatic terms and as comprising a socio-technical continuum extending across this vast continent (e.g., Sherratt 2006). This is especially true with regard to the coming of tin bronze across the Asian Old World, as one can no longer interpret metallurgical developments in East and Southeast Asia without understanding what took place prior to this all-important technological innovation. A succinct overview of current thinking concerning what transpired from approximately the third millennium BCE onward is offered here.

Independent Invention–Dead or Alive?

Before proceeding, it is necessary to beat (yet again) the "dead horse" of independent invention (and I emphasize "dead"). It may well be the case that the discussion of independent invention will never end (e.g., Chernykh 2009, 3–5; Mei 2009b, 13; Radivojević et al. 2010), but, certainly with regard to the transmission of copper-base metallurgy out of greater Southwest Asia, the Balkans included, it is hard to imagine that this region did not share culturally defined, Neolithic/Chalcolithic aesthetic and technological traditions and it thus comprises the "heartland" of metallurgy. Moreover, and especially with regard to the transmission of tin bronze across the Asian Old World, this "horse" is well and truly dead. The debate concerning the origins of tin bronze, in the Eurasian context, is moribund.

An excellent starting point for the study of the origins of tin bronze is the seminal article by Muhly (1988) titled "The Beginnings of Metallurgy in the Old World." Muhly tackles such issues as "Migration versus Diffusion" and "China: Diffusion or Independent Invention" (see also Ciarla 1994, 74–75; Sherratt 2006, 32–35; Mei 2009b, 9). He overviews the shift in thinking that began in the 1960s with the advent of the "New Archaeology" and the ensuing post-processual approaches that brought about a dramatic sidelining of early "diffusionist" arguments. Proponents of diffusion, whom Muhly terms the "old school," offer arguments ranging from the extreme views held by Heine-Geldern (e.g., 1937, 1952, 1974; cf. Karlgren 1942; see also van Heekeren 1958, 92–98) to the more reasoned diffusionist worldviews of Bishop (1942, 14), Creel (1935, 43–69), and Loehr (1974, 300), to which we can also add Kroeber's (1963) "stimulus diffusion" model (cf. White and Hamilton 2009, 358) and Watson's (1960, 7–15; 1985) discussions.

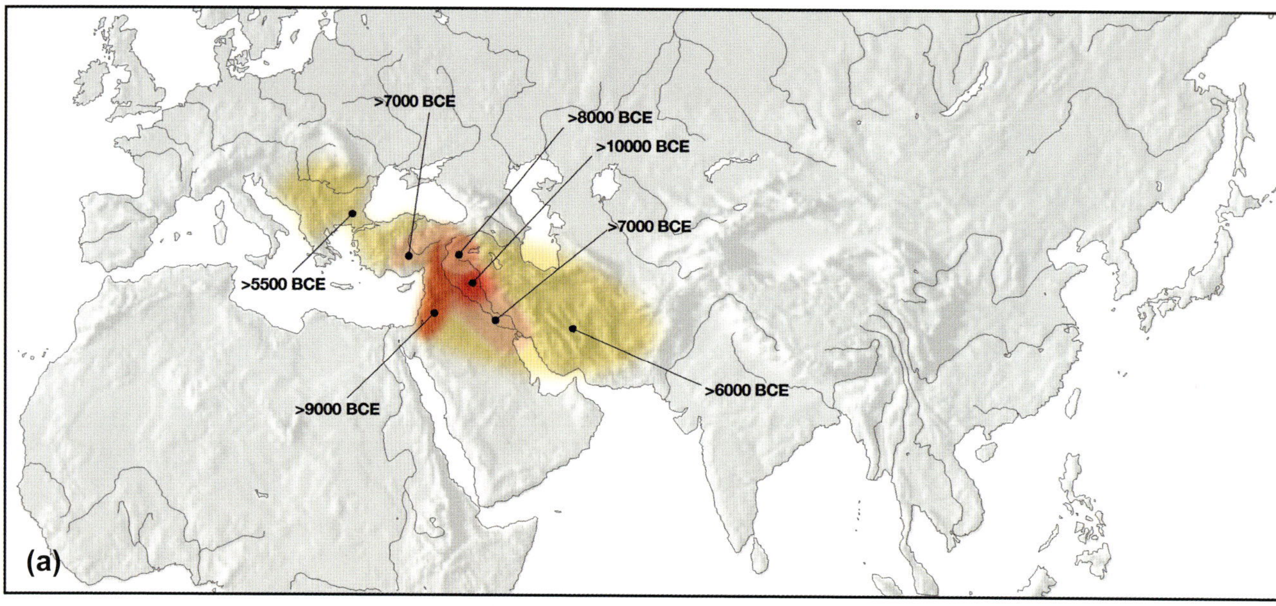

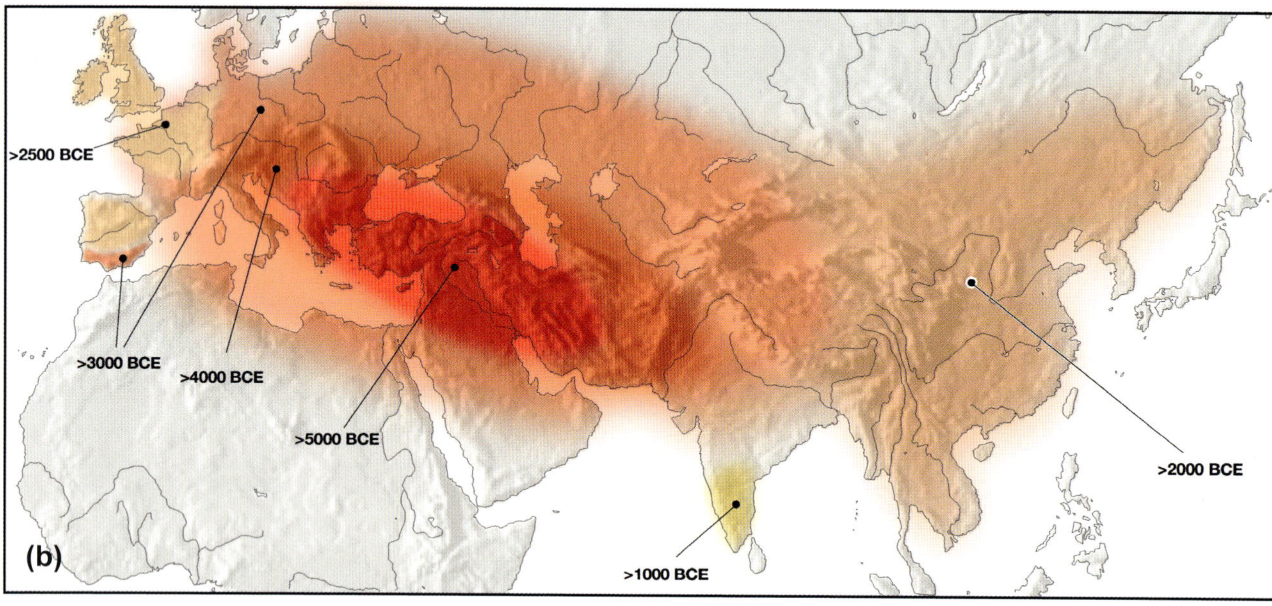

Figure 1. a. *Location of the earliest known exploitation of copper ores and native copper in greater Southwest Asia/Balkans.* ***b.*** *The spread of copper-base smelting technology, including that of tin bronze, eastward after ca. 3000* BCE *(from Roberts et al. 2009, 1014, fig. 1; © Trustees of the British Museum).*

Muhly indicates that from the 1960s onward the pendulum swung away from such "old school" approaches (mind you, much of it the product of experienced Asian specialists and anthropologists who had sensed pan-Asiatic connections) to the staunchly held views on the indigenous development of metallurgy (and culture), in particular on the Chinese Central Plain, by such eminent scholars as Barnard (e.g., 1983), Sullivan (1980, 11), Ho (1975), and Renfrew (1986, 15) (see Wertime 1973 vs. Renfrew 1973), among others.

In recent decades, on the strength of evidence pouring out of innumerable Chinese excavations, in concert with the gradual opening of China to foreign scholarly involvement, the pendulum has begun to retrace its path back and away from these isolationist approaches. Among the door-opening publications were those by the senior Chinese archaeologist An Zhimin (1982–83, 1993, 1998, 2000). He understood that there was an as-yet-to-be-defined connection between technological developments hailing from cultures to the West and what occurred over time in China proper. It was a courageous act on his part to step away, albeit cautiously, from thinking that saw no role for external influence in the development of the metallurgy in ancient China. What he offered was an open mind, and with it came the inching open of a conceptual door. This step, taken along with the collapse of the Iron Curtain, which opened the Eurasian steppe and Central Asia to extensive archaeological investigation, sealed the fate of those who saw no connection between developments in ancient China and the steppe/Central Asia.

Muhly (1988, 15) takes a pro-single source stance on origins that is very reminiscent of the stance of Wertime (1964, 1266), whose thinking had been strongly influenced by the Massachusetts Institute of Technology's eminent materials scientist, Cyril Stanley Smith (1981). Muhly

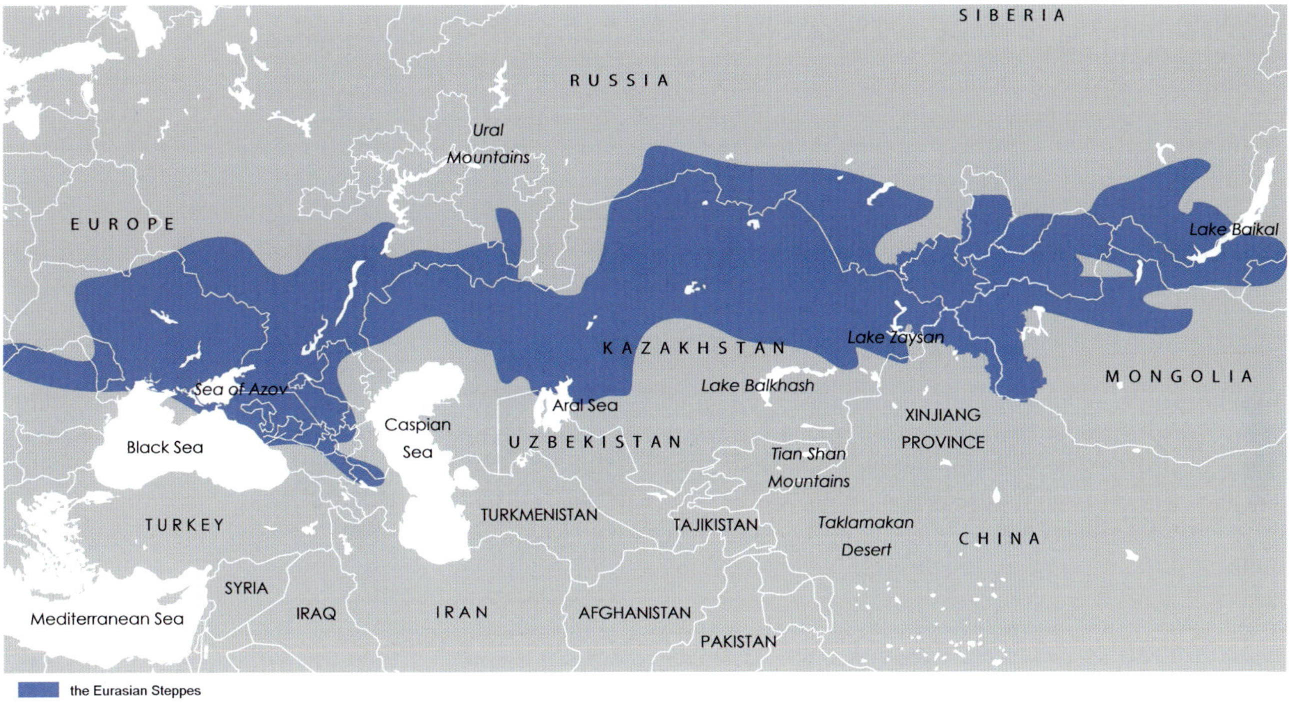

Figure 2. *The Eurasian steppe. Courtesy of Ivo Verhaeghe, Ghent University, 2006.*

argues that while precise moments of origin are not identifiable and first-ever artifacts not findable, there is true value in the study of origins. Significantly, with regard to metallurgy, he states: "I still believe that all these developments are, in ways, however indirect, somehow related. I believe that the discovery whereby a hard, intractable rock is turned into a soft, pliable and malleable metal, was a unique discovery, not one miraculously repeated in much the same way at different times in different parts of the world" (Muhly 1988, 3). Moreover, he concludes his article with a final statement indicating that research up to that point "should eliminate once and for all the argument that Chinese metallurgical technology developed in ways totally different (and thus apart from) anything known in the West" (Muhly 1988, 16). His article signaled a watershed moment in scholarly thinking that strongly suggested a need to explore the extent of what the late Andrew Sherratt (2006, 31) has more recently termed the "trans-Eurasian exchange." The nature of this process of exchange and the ever-increasing evidence for it are what now lie at the heart of demonstrating that discussions of "independent invention" are, in fact, passé.

In this trans-Eurasian context, the reader should refer to the map in figure 1, which shows the spread of copper-base metallurgy across Europe and Asia millennium by millennium. This map documents current locations for dated metallurgical finds with a clearly eastward sloping horizon. One could argue, however, that this simply documents what we now know of clusters of independently originated metalworking centers and does not necessarily demonstrate technological contiguity of any sort (e.g., Chernykh 2009). However, as we argued in our article "Development of Metallurgy in Eurasia," metallurgy developed initially in the metallogenic region of greater Southwest Asia in a context of pyrotechnological experimentation and with the "desire for new materials to serve as aesthetic visual displays of identity, whether of a social, cultural or ideological nature" (Roberts et al. 2009, 1019). For an additional excavated example strongly in support of this argument, see Molist et al. (2009).

Thus, from the outset it was a socially and culturally "contextualized" choice that exercised a profound influence not only on metals-related technological innovations (including the choice to begin to use metal) and their

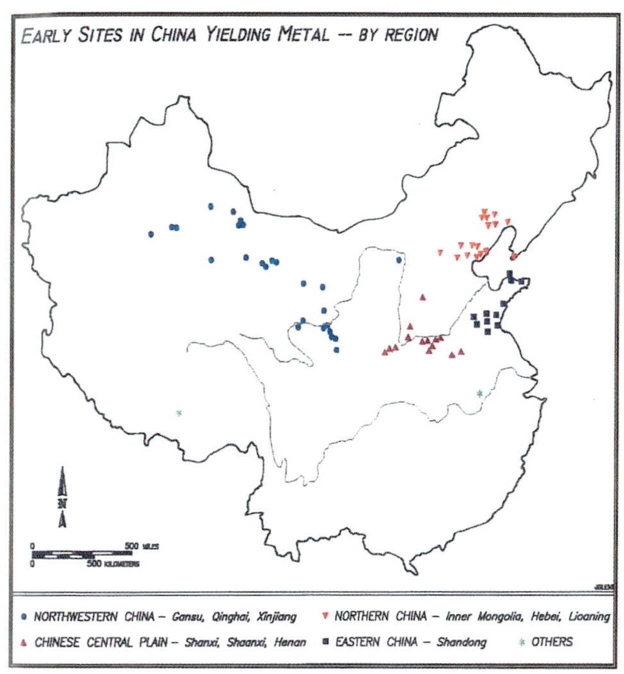

Figure 3. *Archaeological research has recorded a concentration of some of China's earliest copper-base artifacts in Xinjiang, Gansu, and Qinghai provinces, northwest China, where tin bronze, in any quantity, generally concentrates at ca. 2000* BCE *and later (after Linduff et al. 2000).*

subsequent acceptance but also on the process of their transmission across space and time, between and among cultures. Following the advent of metal use in Southwest Asia, "[t]he subsequent appearance of metals throughout Eurasia is due to the acquisition of metal objects by individuals and communities re-inventing traditions of adornment, even in regions hundreds of kilometers from the nearest sources of native metals or ores" (Roberts et al. 2009, 1019).

To the east of Southwest Asia, peoples were moving across the Eurasian steppe proper (fig. 2) as well as across the contiguous but more arid zones of Central Asia to the south. Among such peoples were groups who possessed metallurgical know-how that they transmitted through or in proximity to zones bearing the two critical ores (i.e., copper and tin).[1] The Eurasian steppe and Central Asia had relatively substantial, albeit not uniformly distributed, ore availability. In particular, a relative abundance of tin reserves allowed access to this essential raw material for bronze production.

In regions where ores were not readily available, current reconstructions suggest long-distance trade in metals and even ores—for example, from the massive mining complex at Kargaly in the Urals (Kohl 2007, 174; Zhang 2009 on trade in artifacts and ingots between the Sintashta culture and the Seimo-Turbino among others). Thus, from at least the later third millennium BCE, it is clear that tin bronze was being produced with gradually increasing regularity across the breadth of this region (e.g., Chernykh 1992; Anthony 2007; Frachetti 2008; Hanks and Doonan 2009; Hanks and Linduff 2009; Linduff 2004; Linduff et al. 2000; Linduff and Mei 2009; Kohl 2007; Kuzmina 2008; Mei and Rehren 2009).

Strong support documenting metallurgy's transmission across the Eurasian steppe and Central Asia can be found among this recent host of publications by Eurasian specialists. Brief mention can be made of the publications of two prominent members of the very active community of archaeologists working across the greater region, namely Anthony's *The Horse, the Wheel and Language: How Bronze Age Riders from the Eurasian Steppe Shaped the Modern World* (2007) and Kohl's *The Making of Bronze Age Eurasia* (2007). Kohl (2007, 168), for example, suggests that metal technology "diffused" across the steppe and forest steppe from west to east beginning in Chalcolithic times.[2] Intriguingly, with reference to the Kargaly mining complex, Kohl (2007, 178) offers a "gold rush" model that involves "prospectors occupying known or recently discovered ore sources, exploiting them with relatively primitive extraction procedures, and hauling away ore on their sturdy oxen-driven wagons to work elsewhere during other times of the year." He sees this model as an alternative to models of production that would call for either an "overarching social and political system, that is a Late Bronze Age despotic totalitarian state or a less centralized confederacy of related independent groups" (Kohl 2007, 178).

In Anthony's volume, part two, "The Opening of the Eurasian Steppe," explains that the nonlinear movements of steppe populations and the ensuing cultural contact fostered the spread of the knowledge and practice of copper-base metallurgy from at least the fourth millennium BCE onward (if not earlier). Indications are that tin-bronze technology and artifacts were transmitted relatively rapidly across the steppe beginning in the later third/early second millennium BCE.[3]

The final result of this burst of steppe-based research is burgeoning evidence that suggests the technology moved from "the steppe to the sown" (i.e., off the Eurasian steppe, into Xinjiang and northwest China and/or via southern Siberia into northwest China) and was flourishing in the Central Plain by the early centuries of the second millennium BCE. It has been widely argued that as tin-bronze technology reached the region of Xinjiang, it was transmitted further eastward through the Hexi Corridor of Gansu province leading to the Central Plain. Archaeological research has recorded a concentration of some of China's earliest copper-base artifacts in this region (fig. 3), whereas tin bronze, in any quantity, generally concentrates at ca. 2000 BCE and later. Thus, among the significant outcomes of what might be seen as a Bronze Age "gold rush," it can be argued that the almost overnight occurrence of copper-base metallurgy in Xinjiang/northwest China occurred in the final centuries preceding about 2000 BCE. A burst of remarkable and ever more deeply ritualized, technologically sophisticated bronze production from the Central Plain followed shortly thereafter. This cultural "event" is initially recognized at Erlitou and then at later sites, where it became a cultural marker of the Shang (ca. 1600–ca. 1050 BCE) and later dynasties (see Mei 2009b, 12–14 for a balanced discussion of these developments). Interestingly, Mei intimates that both arsenical copper and tin bronze appear to "arrive," though not exactly at the same time but with some rapidity, in Xinjiang/northwest China in the transitional centuries on either side of about 2000 BCE. Sherratt (2006, 34) characterized this period ca. 2000 BCE as an "Achsenzeit," an axial time of "ideological transformation" comprising "dimly perceivable changes of enormous magnitude."

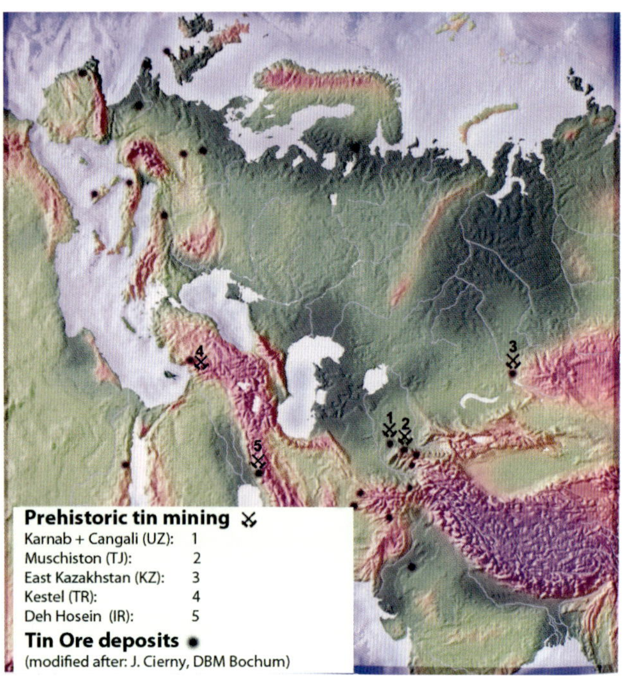

Figure 4. *Prehistoric tin mines and deposits in Southwest and Central Asia (after Cierny et al. 2005, fig. 1; Courtesy of the Deutsches Bergbau-Museum, Bochum).*

One might ask, given such a context of ensuing contact and culture change, did copper-alloy metallurgy arrive in Xinjiang from further west, where traditions of arsenical copper and tin-bronze use persisted among Andronovo cultures (e.g., Mei 2009b, 11–12, 14; Zhang 2009)? Or, at the same time, did tin-bronze metallurgy arrive in northwest China (Gansu) from the southern Siberian forest steppe, traveling with late third-millennium-BCE Seima-Turbino metalworkers (Anthony 2007, 443–48; Chernykh 1992, 215–33; Fitzgerald-Huber 1995; Frachetti 2008, 52–55, 173–74; Mei 2003b; White and Hamilton 2009, 381–84; Zhang 2009)? All of these events would have occurred just prior to the burst of metallurgical creativity on the Central Plain under the Shang. However, it remains difficult to directly link the metallurgy of northwest and northern China with that of the early second-millennium-BCE bronze-making Erligang-Erilitou cultures of the Central Plain. Here I fall back on the words of Smith (1973, 23), who intimated technological contiguity in saying, "Personally, I believe that the idea of making metals from rocks is not easily arrived at and second independent nucleation of metallurgy in China (or the New World either) seems highly improbable. This view by no means denies that there is something uniquely Chinese in the technique and beauty of the superb bronzes of Shang and Chou China, but this seems to lie in the making of moulds and the decoration of them, not necessarily in the metallurgy itself."[4] For now the jury remains out on how exactly this remarkable industry came into being and what the nature of its connections to points west truly is. One must, however, keep Smith's musings in the back of one's mind (see Linduff and Mei 2009; Mei 2009b).

Thus, while there is no question of the unique character of the metallurgy as practiced on the Central Plain and in neighboring regions by metalworkers of the Shang and Zhou (ca. 1050–221 BCE) dynasties, the relatively sudden appearance in Xinjiang/northwest China of tin-alloy metallurgy has all the hallmarks of a technology that was "introduced"—mostly probably, as increasing consensus would suggest, through contact with steppe and forest-steppe cultures.

Though controversial, particularly eloquent and cogent arguments in this regard have been put forth by Sherratt (2006), who saw the early second-millennium-BCE metallurgical developments in the Central Plain as the outcome of a long-term process of trans-Eurasian exchange, which, over time, linked western and eastern Eurasia.

Interestingly, there is something quite reminiscent, via this exchange process, of metallurgical "introductions" in northwest and northern China and on the Central Plain. Similar circumstances are presently being argued for in neighboring Southeast Asia. Tin bronze appears there from off the steppes and out of China, virtually "overnight" at the end of the second millennium BCE (Ciarla 2007b; Pigott and Ciarla 2007; Higham et al. 2011; see also Pryce 2011; Pryce et al. 2010; 2011).

Andrew Sherratt (2006, 43–44) was among those prescient scholars who recognized that over the course of the second millennium BCE, tin-bronze metallurgy off the steppe/forest steppe must have transited across China and arrived in Southeast Asia (see also Ciarla 2007b; Higham 2002, 2006, 2007, 2008, 2009; Higham and Higham 2009; White 1988, 179; 1997, 104; 2000; cf. White and Hamilton 2009; see also Ciarla 1994 for discussion of first-millennium-BCE steppe–Southeast Asian connections). While the direction and mechanisms of transmission are currently under dispute (e.g., Pigott and Ciarla 2007; Higham et al. 2011 vs. White and Hamilton 2009), the strength of typological similarities between, for example, socketed spearpoints and axes on the steppe and forest steppe and those in Thailand must be given some consideration (Sherratt 2006, 43–44; see also White and Hamilton 2009, 380ff. for the discussion of typological comparanda; cf. Ciarla 2007b; 315–19 on socketed axes and their connections). This debate will be touched upon again later in this paper.

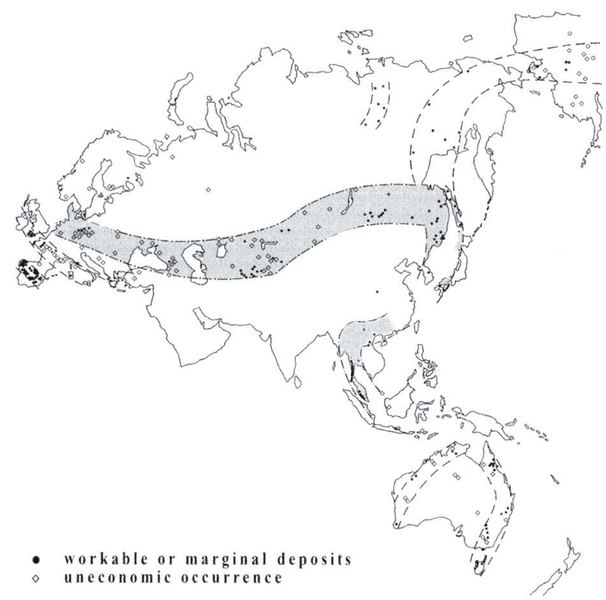

Figure 5. Map of Eurasian tin belts (after de Jesus 1978, 36, fig. 3).

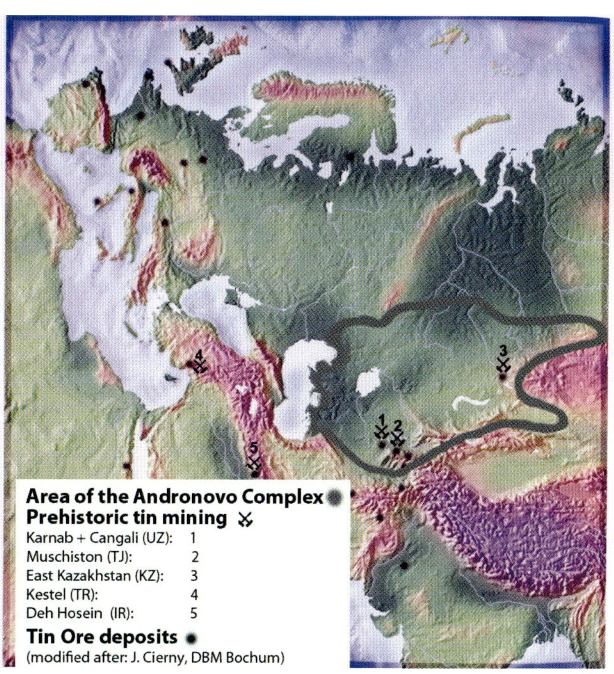

Figure 6. Map of the Andronovo culture's distribution in relation to known ancient tin mines and deposits (after Weisgerber and Cierny 2002, 180, fig. 1; Courtesy of the Deutsches Bergbau-Museum, Bochum).

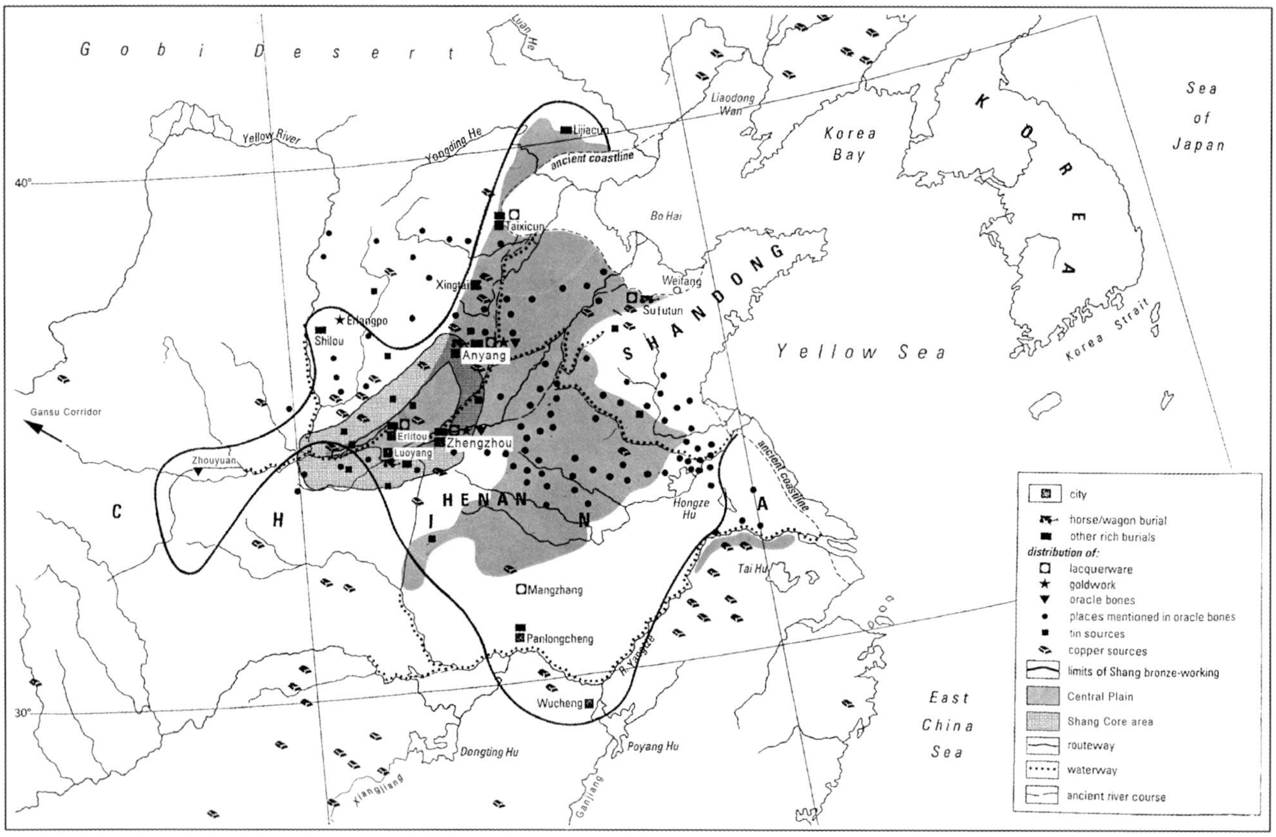

Figure 7. Map of the extent of the Shang empire as distributed on the Central Plain of China (© 1999 Chris Scarre. Permission to publish this map courtesy of Harper Collins Publishing Ltd. Copyright in the customized version is vested in Archetype Publications, London).

Tin Sources in the Asian Old World

Given the relative ubiquity of copper ores, the scarcity of tin ore deposits is the major critical variable in producing tin bronze. In Southwest Asia, where metallurgy was born, Anatolia and Mesopotamia are home to many of the world's earliest known occurrences of tin bronzes (e.g., Helwing 2009, 212). These date to the early third millennium BCE, some five millennia later than the earliest documented use of metal (native copper, e.g., at Çayönu in Turkey [e.g., Maddin et al. 1999]) and some two millennia after the earliest documented smelting of copper ores (e.g., at the Vinča culture site of Belovode in Serbia [Radivojević et al. 2010]).

Thus, based on current evidence, Southwest Asia continues to be seen as a, if not *the,* heartland of tin-bronze metallurgy, though one wonders, given the availability of Central Asian tin, including that in Afghanistan (e.g., Rossovsky et al. 1987; Shareq et al. 1977), why the alloy is first seen so far afield from this tin-belt concentration of major deposits (fig. 4). Further west, tin deposits are decidedly rarer and have been the source of much scholarly debate (e.g., Muhly 1973, 1976, 1985; Charles 1975; Cleuziou and Berthoud 1982; Crawford 1974; Franklin et al. 1978; Stech and Pigott 1986; Pigott, in press; Weeks 2003). A partial answer to the above question may, however, lie with deposits in Southwest Asia proper, namely the documented ancient tin mine at Kestel with its associated production settlement at Göltepe in southeastern Turkey (Yener 2000) and the tin deposit at Deh Hosein in western, central Iran (Momenzadeh 2004; Momenzadeh et al. 2002; Nezafati 2006; Nezafati et al. 2006, 2008, 2009). These deposits, most probably, initially were sources of tin ore for the early Near Eastern metalworkers. At a later date, sources further east in Afghanistan (Stech and Pigott 1986) and Central Asia (e.g., Boroffka et al. 2002; Cierny and Weisgerber 2003; Litvinskij 1950; Parzinger 2002; Parzinger and Boroffka 2003; Thornton and Giardino, in press) had their reserves tapped. However, prior to likely encounters with tin from sources east of Iran, it is hard to imagine that the two known Southwest Asian sources were sufficient to supply third-millennium-BCE tin-bronze production in the ancient Near East. The established sea link between Sumerian Mesopotamia and the Harappan Indus valley may well have played a role in bringing tin from the latter to the former, but this particular topic remains a source of debate at present (e.g., Muhly 1973; Pigott, in press; Possehl 1996, 1997; Weeks 2003; Weisgerber 1984). While the mechanisms of technological transmission across the steppe and Central Asia are beginning to be explored, how much we don't know is also readily apparent (e.g, Pryce et al. 2010, 12–13; 2011, 155–59; White and Hamilton 2009, 377–79), and while copper smelting dates at least as early as the fifth or fourth millennium BCE in this mid-Eurasian zone (e.g., Anthony 2007; Chernykh 1992; Kohl 2007; Salvatori et al. 2009), if not before, the *current* consensus suggests that tin-bronze know-how was transmitted from Southwest Asia into Central Asia and onto the steppe to the north at some point during the late fourth to early third millennium BCE.

Interestingly, the further east one travels the more common tin becomes. One can argue that the transmission of bronze technology across Central Asia and the steppes was facilitated by the presence of a major tin belt extending the length of Eurasia and spreading north–south in East Asia (Pigott and Ciarla 2007, 80, fig. 6) (fig. 5). Significantly, two second-millennium-BCE Central Asian tin mines along this belt, found at Karnab in Uzbekhistan and Muschiston in Tadjikistan, are known to have strong links to the Andronovo culture (Alimov et al. 1998; Boroffka et al. 2002; Cierny and Weisgerber 2003) (fig. 6). The Andronovo, a major, tin-bronze-using, second-millennium-BCE culture, is linked to the spread of bronze metallurgy across the steppe—perhaps as far east as Xinjiang, China (e.g., Mei 2000, 2003a, 2003b, 2003c, 2004, 2009a, 2009b; Linduff and Mei 2009; Zhang 2009).

Of noteworthy interest is the Muschiston mine that was a source of the tin- and iron-bearing copper sulfide ore stannite (Cu_2FeSnS_4). Zhang (2009, 18) alludes to mined, copper-tin deposits (possibly stannite-bearing) in the Eastern Urals at Elenovka and Ush-Katta, among more than thirty Bronze Age mines documented there by Chernykh (1970, 35–49). Deh Hosein in central western Iran is also stannite bearing (e.g., Nezafati 2006). Stannite, if properly co-smelted, would yield a natural tin bronze. Thus happenstance may have resulted in the initial recognition of this metal, but not necessarily in how to produce it (e.g., Rovira and Montero 2003). The role of this perhaps not so unusual ore in early tin-bronze metallurgy may well be underestimated and comprises a subject in need of a substantive review. As a matter for future discussion, given the presence of stannite in multiple deposits, should we be looking perhaps for the origins of tin bronze in central Eurasia as opposed to greater Southwest Asia?

As mentioned, the Asian tin belt bifurcates at its eastern end, heading north into Mongolia and Siberia and south through China proper and on through the Indo-China peninsula. Thus, eastern Asian metalworkers did not lack for tin in the same manner as did their counterparts in Southwest Asia. Prior to and during the third and early second millenium BCE, multitudinous ethnic groups, many of whose locations are seen mapped in Frachetti (2008, 40–41, figs. 4 and 5), were moving about in Central Asia and on the steppe. This was not a simple linear progression of peoples eastward, though by all indications the technology of tin bronze itself was gradually being transmitted in that direction by diverse cultural mechanisms only now beginning to be explored (e.g., Anthony 2007; Frachetti 2008; Hanks and Doonan 2009; Hanks and Linduff 2009; Kohl 2007; Linduff and Mei 2009; Pryce et al. 2010, 259–64; White and Hamilton 2009). Among the motivations for movements of peoples eastward across this region, one can include the search for new ore sources, while at the same time this search facilitated the transmission of metalworking expertise and metal items themselves into and through a succession of clearly receptive societies (Sherratt 2006; Kohl 2007; White and Hamilton 2009).

Jianjun Mei is among a group of eminent East Asian archaeologists who have argued that steppe/forest-steppe peoples were coming into contact with cultures in northwest and northern China by the late third/early second millennium BCE, their metalworking traditions melding in the process. By ca. 1600 BCE, with the rise of the Shang dynasty in the Central Plain (the extent of the Shang empire can be seen in figure 7), the sophisticated tradition of casting ritual vessels in bronze arose. Among the factors critical to the rise of this industry is the proximity of copper and tin reserves to major casting centers, such as that at Anyang. Mining expeditions must, of necessity, have spread out from the production centers on the Central Plain in search of all-important mineral reserves. Such efforts, at a minimum, could have served to indicate to populations in proximity just how highly desirable copper and tin ores were to the Shang elite consumers. Additionally, in figure 7, the dark, sinuous line marks the limits of Shang bronze casting, but the products of its industry—and most probably knowledgeable metalworkers—disseminated southward along the river valleys toward Lingnan. This transmission proceeded into receptive communities that adapted the new technology to suit their needs (Ciarla 2007b; Pigott and Ciarla 2007).

The real controversy regarding tin-bronze production begins here. As has been argued regarding this process of dissemination, accompanying knowledge of how to produce tin bronze ultimately may have reached Southeast Asia, depending on which hotly debated model of transmission one agrees with (Ciarla 2007b; Higham 1996; Higham et al. 2011; Pigott and Ciarla 2007; Pryce et al. 2010, 2011; cf. White and Hamilton 2009).

Tin Bronze and Competing Models Explaining Its Arrival in Southeast Asia

A model for the transmission of tin-bronze metallurgy from the steppe through the Central Plain and into northern Southeast Asia has been put forth by Higham (e.g., Higham 2002, 2006, 2007, 2008; Higham and Higham

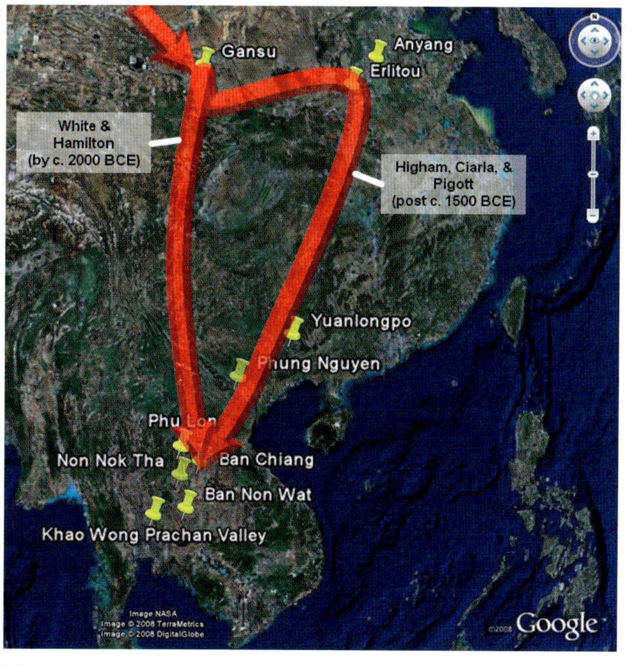

Figure 8. *Map indicating proposed paths of transmission of tin-bronze metallurgy as argued by Higham, Pigott, and Ciarla on the right and White and Hamilton on the left. Map by T. O. Pryce. Base map courtesy of GoogleEarth™.*

2009), by Ciarla (2000, 2007b) alone and in collaboration with the author (Pigott and Ciarla 2007), and most recently by Higham et al. (2011). This model argues that once metallurgy was in place in the Central Plain, transmission from the Shang and later production in this zone moved via southern-trending regional riverine conduits of communication that linked sites and bronze artifacts with the Lingnan region of southeast China. These riverine conduits play a critical role in tracking the potential route of the technology moving south from the Central Plain and into northern Southeast Asia (e.g., Vietnam and Thailand) by the close of the second millennium BCE.

However, as scholars love a good controversy, in late 2009, Joyce C. White and Elizabeth G. Hamilton, colleagues at the University of Pennsylvania Museum, proposed a competing model (White and Hamilton 2009). The route proposed separately by the aforementioned models is shown on the right in figure 8, while the path of the so-called Rapid Eurasian Technological Expansion Model (RETEM), as proposed by White and Hamilton, is shown on the left.

In RETEM, the source of the technology lies with late third-millennium-BCE, bronze-using forest-steppe peoples of southern Siberia, among whom it is argued are metalworkers of the Seima-Turbino cultural phenomenon knowledgeable in the production of tin bronze and who make distinctive bronze artifacts—in particular spearpoints with cast sockets and thin-walled, deep, hollow-socketed axes (celts). White and Hamilton (2009, 382) emphasize the important point made by Childe (1954) and Sherratt (2006) in this regard—namely, that this type of socketed casting was a technologically sophisticated innovation in its own right. White and Hamilton also point out that Childe (1954, 19) argued, “a supply of tin bronze is probably a prerequisite for the postulated advance in core-casting.” The method of core casting allowed for the production of implements that were lightweight (hence portable) and thin walled (economizes on bronze use, especially scarcer tin). Such casting was possible due to the fluidity of molten tin bronze and the relative ease with which it could be cast into bivalve molds used to produce such implements. Molten copper can prove far more difficult a casting medium when used in bivalve molds. In Southeast Asia, tin sources were concentrated in certain tin belts and zones, rather than being more evenly distributed. Interestingly, the arrival of tin-bronze metallurgy in Thailand is first noted in the northeast, where tin can be found immediately north in Laos. The arrival of this metallurgical tradition introduced these socketed implement types that were well suited to production via a tin-bronze technology. Childe (1954) also noted that the socketed, and hence well-hafted, axes were particularly good for woodworking given their long working edge (well defined in the typical Southeast Asian splayed edge ax/adzes). In terms of the process of socketed ax casting using a suspended core, no indications of the use of spacers or chaplets to hold the core in place are found as Childe (1954) suggested. However, by the late first millennium BCE in the Khao Wong Prachan valley (KWPV) of central Thailand, there is artifactual evidence for the use of a tongue-shaped, ceramic mold plug (a core of sorts) that fits snugly, without need of spacers, between the halves of a bivalve mold (for a description, see Pigott, in press).

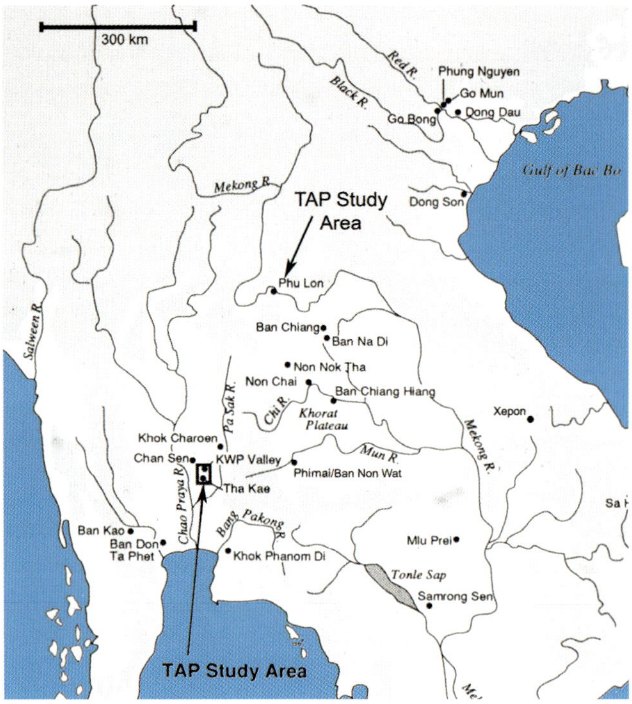

Figure 9. *Map indicating two TAP study areas as well as important prehistoric sites in Southeast Asia. Courtesy of the Thailand Archaeometallurgy Project.*

Thus, it is these socketed spearpoint and ax types linked to the Seima-Turbino casting tradition (e.g., Chernykh 1992, 191) that White and Hamilton (2009, 389) argue are to be found in northeast Thailand at the site of Ban Chiang beginning about 2000 BCE (White 2008). Hence they see a rapid transmission of bronze casting technology from southern Siberia to Southeast Asia along the eastern edge of the Tibetan plateau, from Gansu, south through Qinghai, Sichuan, and Yunnan, and into Thailand, prior to the coming of the Shang dynasty and bypassing the Central Plain altogether. However, despite White and Hamilton’s arguments to the contrary, current dates for metallurgy in Yunnan do not support such an early arrival of the technology in southwest China (e.g. Chiou-Peng 2009). Thus, while we have two distinct and competing models to consider, linking such theoretical models with evidence for early metallurgical evidence in Thailand, as we shall see, is by no means easy.

The Evidence for Prehistoric Copper Mining, Smelting, Alloying, and Casting in Southeast Asia: A Summary View from Thailand[5]

In regards to the evidence for prehistoric copper and bronze exploitation in Southeast Asia, much of the archaeological material related to bronze casting has been excavated in Thailand since the 1980s and most of that by the Thailand Archaeometallurgy Project (TAP), co-directed by Surapol Natapintu (Silpakorn University, Bangkok) and the author. TAP had two study areas, one in northeastern and one in central Thailand, and together these areas offer the current best detailed example of how copper ores were being extracted and processed during the prehistoric period in one region of Southeast Asia (fig. 9). It is, in part, upon this evidence, in concert with that from other parts of Thailand and Southeast Asia, that an external source for metallurgy in this region has been postulated (e.g., White 1988). Due to the frequently cited lack of evidence for a prior period of metallurgical experimentation, tin-bronze metallurgy, it has been argued, appears to have arrived virtually overnight from outside Southeast Asia and as a full-blown technology (e.g. Pryce et al. 2011). When this occurred is the subject of debate.

Other suggestions of external connections come from certain classes of artifacts that typify metallurgical production in Thailand in particular and that have links ultimately, it has also been argued, to the Eurasian steppe/forest steppe. Such artifacts are found among the markers of what White (1988) has termed the Southern Metallurgical Tradition and include the use of ceramic crucibles for smelting and the casting of socketed, tin-bronze artifacts (e.g., axes, spearpoints) in bivalve molds (in stone or ceramic) by means of a suspended core technology (e.g, Chernykh 1992; Sherratt 2006; White and Hamilton 2009; see also Pigott and Ciarla 2007). A ceramic core or plug is sandwiched between the mold halves in such a way that molten metal can be poured into the mold over and around the core, thus causing to form, once cooled, a deep, hollow socket in the cast implement suitable for hafting. One wonders if the use of casting core technology may be of considerable antiquity. Its use in the third-millennium-BCE steppe and forest-steppe regions may be traceable to the initial casting of shaft-hole axes in the fifth-millennium-BCE Balkans and the adjacent Pontic-Caspian region, home to the Cucuteni-Tripolye culture (e.g., Anthony 2007, 164–74) with links to the eastern end of the steppe/forest steppe.

The following discussion overviews the archaeological evidence for metallurgical activity from TAP fieldwork, including the evidence for possible markers of the Southern Metallurgical Tradition that share affinities with Eurasian artifacts as well as artifacts excavated at sites in China.

The Copper Mining Complex at Phu Lon, Northeast Thailand

TAP Survey and excavations in 1984–85 documented prehistoric copper mining activity dating to the first millennium BCE at the site of Phu Lon on the south bank of the Mekong River in northeast Thailand (e.g., Natapintu 1988; Pigott and Natapintu 1988, 1996–97; Pigott et al. 1992; Pigott and Weisgerber 1998) (see fig. 9). Two large adjacent hills, Phu Lon I and II, bore evidence of ancient mining, with the majority of the evidence apparent at Phu Lon I closest to the river (fig. 10). The green stains of copper salts were obvious on the exposed hillside hostrock as were numerous adits, shafts, and gallery openings (Pigott and Weisgerber 1998). Mine tailings spread down the hill to the west, and a tall, monolithic remnant of hostrock, termed "the Pinnacle," stands in mute testimony to countless years of mining activity. Considerable effort, most probably on a seasonal basis, had to have been expended extracting the oxidic copper ores (primarily malachite) from quartz-lined veins shot through the skarn hostrock matrix of this weathered copper sulfide ore body (see e.g., Kamvong and Zaw 2009 on Phu Lon geology; also Pryce 2009). Mining was conducted by hammering away by hand at the friable skarn hostrock with locally collected, river-worn, igneous cobble mining mauls that were found scattered over the surface of the mining area. Vestiges of mining range from the grubbing in small open pits of surface-deposited ores to the more sophisticated evidence of shaft and gallery mining that followed rich veins deep into the hill (see plans in Pigott and Weisgerber 1998). This activity ultimately riddled the hill to such an extent that it imploded on itself. The entire center of Phu Lon I at some point collapsed, an event well known from historical evidence of mining elsewhere in the world (Pigott and Weisgerber 1998, 364).

On the northern side of Phu Lon I, not far from the actual mining area, an ore dressing activity area known as the "Pottery Flat" was excavated. Here in an area of some 200 square meters, mined ores (malachite rich lumps of skarn) were brought to be beneficiated by hand crushing using hammerstones on stone anvils. This resulted in a shallow deposit about 50 cm thick of finely crushed hostrock from which the malachite was handpicked for further processing. Interestingly, this deposit of crushed ores closely resembles that found some 400 km to the south at the late first-millennium-BCE copper smelting settlement of Nil Kham Haeng in the KWPV of central Thailand.

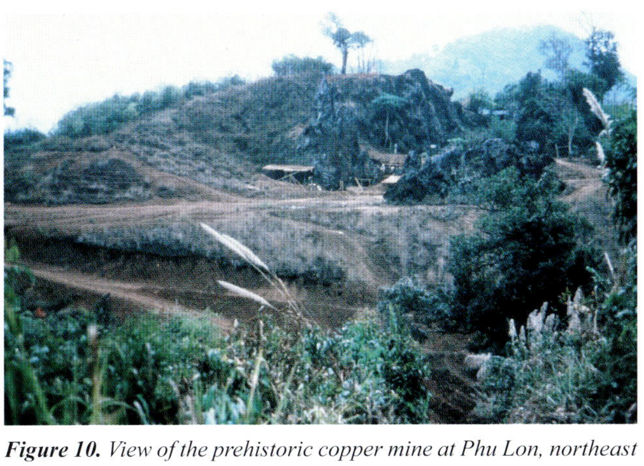

Figure 10. *View of the prehistoric copper mine at Phu Lon, northeast Thailand. Courtesy of the Thailand Archaeometallurgy Project.*

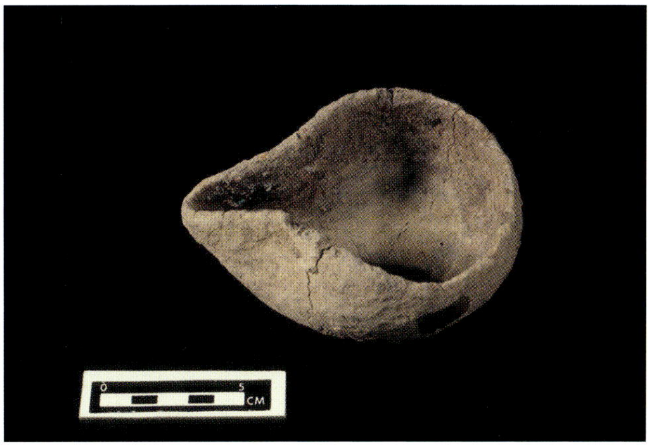

Figure 11. *Common Southeast Asian crucible from the prehistoric site of Ban Chiang, northeast Thailand. This crucible type also occurs at Phu Lon and other culturally related sites in the region. Courtesy of the Ban Chiang Project.*

Mixed in the Pottery Flat crushed ore deposit were steppe-linked markers of the Southern Tradition, namely bivalve molds and ceramic crucibles. Two fragmentary bivalve casting mold halves, one in ceramic with an "X" etched in its dorsal surface and one in sandstone, were found. Interestingly, the marked mold provides an additional link to the copper production center in central Thailand to be discussed shortly. It must also be noted that contemporaneous marked molds are known from Lingnan in southeast China (Ciarla 2007b, 315–19). These molds were for the casting of some type of socketed implement, most probably an ax not unlike that excavated elsewhere on the grounds of the mine with a parallel typological example known from Ban Chiang on the Khorat plateau (White 1982, 41, 78, fig. 94).

Additionally, about ninety-six rice chaff-tempered ceramic crucible sherds, remains of small vessels used to process metal and ores (see White and Hamilton 2009, 366, fig. 3, for a comparable, intact example), were found. An analysis of the Phu Lon crucibles by TAP geologist William W. Vernon indicated that high-temperature activities were conducted in these crucibles, including the processing of copper, tin, and tin bronze (Vernon 1997). Such residues could indicate simply the melting of metal, and while definitive evidence of ore smelting is lacking, recent research offers additional evidence suggesting that malachite was being smelted on the spot in these rather small vessels (Pryce et al. 2011). Thus, copper metal and/or baskets of sorted malachite could have been transported overland or up and down the Mekong River.

Downstream from Phu Lon, the Mekong flows along the northern edge of the Khorat plateau that comprises northeast Thailand, home to numerous culturally related, prehistoric settlements, including Ban Chiang, Non Nok Tha, Ban Na Di, and Ban Non Wat. These sites have all yielded evidence of tin-bronze founding activities and a predominance of artifacts (especially bangles and some socketed implements) in this alloy (e.g., Higham 1996, 2002, 2004; White and Hamilton 2009; White and Pigott 1996). Given this abundance of personal ornaments in tin bronze in northeast Thailand, one wonders if this reflects the very early stages of metal use, at which point its initial impact and appeal lay in the realm of the aesthetic, a theme intrinsic to arguments advanced by my co-authors and I in Roberts et al. (2009).

Mining expeditions from these sites, it has been argued, were dispatched overland or along the Mekong to Phu Lon to obtain copper (ore and/or metal) to bring back for possible smelting, melting, and casting (Pigott 1998; White and Pigott 1996). At the same time, many of these sites are linked technologically, as several share examples of the Phu Lon-style crucible, which has been termed the "common Southeast Asian crucible" (White and Hamilton 2009, 365–67) (fig. 11). Such crucibles have in common a tradition of manufacture that includes, after shaping, the lagging of the crucible interior with a quartz-rich slurry. This lagged layer serves to protect the crucible interior from erosion during high-temperature activity, keeping the molten slag and metal from penetrating the crucible fabric (Vernon 1996–97, 1997).

Current arguments, built around a sequence of recent ^{14}C dates, favor a very late second-millennium-BCE arrival of tin-bronze metallurgy in northeast Thailand (e.g., Higham and Higham 2009; Higham et al. 2011). Phu Lon, about this time, may well have enjoyed a "frontier" location as peoples and their knowledge moved along the Mekong. The copper mine at Phu Lon provides us with an important example of how copper was being obtained in and about the time of metallurgy's appearance in Southeast Asia. Finally, the links to the Khorat plateau argue for the probable role of Phu Lon as a major provider of ore and metal to consumers across a substantial region during the first millennium BCE, if not later (e.g., Pigott 1998; Pigott and White 1996).

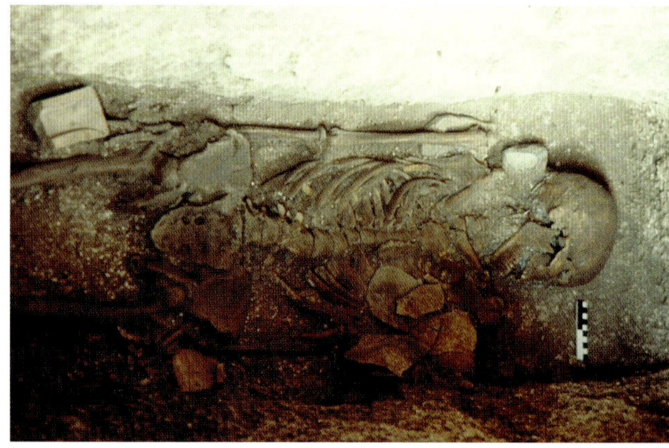

Figure 12. *The late second-millennium-BCE "Grave of the Metalworker" from basal Non Pa Wai in the Khao Wong Prachan valley of central Thailand. A pair of ceramic, bivalve molds for the casting of a large, socketed ax were broken up and distributed around the body, with some fragments placed in the hands, at burial. Courtesy of the Thailand Archaeometallurgy Project.*

A Recently Discovered Late Prehistoric Mining Center at Xepon, Laos

A greater understanding of prehistoric Southeast Asian mining is now becoming a reality as recent fieldwork at Xepon in Laos, due east of Phu Lon, by a Laotian–Australian team has revealed an ancient complex of multitudinous, deep, vertical mine shafts ("timbered" with bamboo laths to prevent collapse). The mining is preliminarily dated to ca. 200 BCE–200 CE and adds a new and distinct example to the archaeological record of the exploitation of copper in northern Southeast Asia (Sayavongkhamdy et al. 2009).

The evidence for prehistoric metallurgical activity is not limited to sites in northeastern Thailand and Laos. Moving south to central Thailand, two sites among the largest such sites documented anywhere in Asia offer a detailed look at Southeast Asia's most substantive evidence for premodern copper smelting.

Prehistoric Copper Smelting in the Khao Wong Prachan Valley, Central Thailand: Evidence from a Major Regional Production Center

Building on previous fieldwork conducted by Natapintu (e.g., 1984, 1988) in and around the modern town of Lopburi, TAP fieldwork between 1986 and 1994 documented the presence of human occupation and burials in association

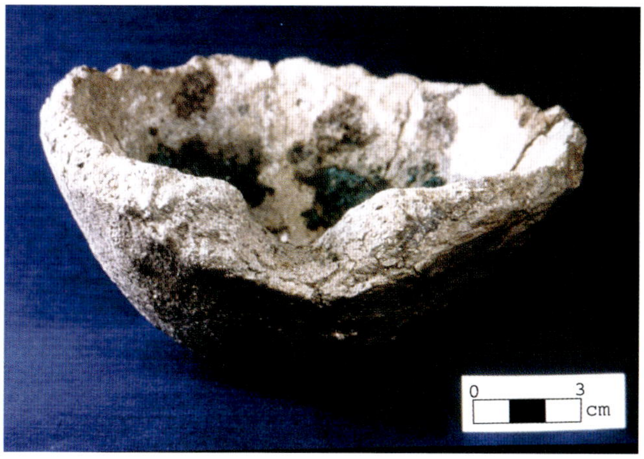

Figure 13. The only relatively intact example of a typical smelting crucible from Non Pa Wai in the Khao Wong Prachan valley of central Thailand. This vessel, which has had its rim broken off and pouring spout created, has been repurposed as a melting crucible. Courtesy of the Thailand Archaeometallurgy Project.

with large-scale copper smelting activity at two sites, Non Pa Wai (NPW) and Nil Kham Haeng (NKH), in the copper-rich KWPV (e.g., Natapintu 1988, 119–22; Pigott et al. 1997; White and Pigott 1996). A third, large, contemporaneous site, Non Mak La, lacking massive production evidence but replete with burials, was also excavated (e.g., Natapintu 1988, 116, 119; Pigott et al. 1997, 133–34). NPW, at five hectares, and neighboring NKH, at three-plus hectares, revealed copper processing contexts spanning the period from the very end of the second millennium BCE into the early centuries of the historic period. The most intensive period of production spans ca. 500 BCE to 500 CE.

In the neighboring region just south of the KWPV, recent research of the Thai–Italian Lopburi Regional Archaeological Project (LoRAP) at the sites of Khao Sai On and Noen Din has added new and contemporaneous data significant to our understanding of local area occupation and a shared tradition of copper production (Ciarla 2007a, 2008; Pryce et al, in press). In addition, the chronology for the greater Lopburi region has recently been revised (F. Rispoli, R. Ciarla, and V. C. Pigott, in preparation). An earlier version of the revised chronology was used in recent laboratory-based research on the comparative assessment of smelting remains from the two KWPV sites (Pryce 2009; Pryce et al. 2010, 2011).

Based on the latest research, not only is copper smelting now documented at Ban Non Wat (BNW) ca. 1000–900 BCE, but also lead isotope analyses strongly suggest that the KWPV was supplying copper to BNW metal founders (Higham 2008) and consumers (Pryce 2011). At NPW, the earliest documented copper processing activities have been dated to the late second millennium BCE (e.g., Pryce et al. 2010). Importantly, this new BNW evidence sits in support of contemporaneous smelting occurring at NPW. The KWPV processing evidence is manifested in the contents of two so-called "founder's burials," in which the dead were interred with the tools of their trade (fig. 12). Each burial contained a pair of ceramic, bivalve casting molds for large, socketed axes (e.g., Pigott et al. 1997, 147–48; for widespread typological connections see Ciarla 2007b, 315–19; Pigott and Ciarla 2007). Burials containing metallurgical artifacts continued in the KWPV in the late first millennium BCE at NKH and immediately to the south at Khao Sai On–Noen Din (Ciarla 2007a, 2008) and have, as termed by White and Hamilton (2009, 374), distinct "behavioral parallels" with probable founders' burials in northeast Thailand at sites including Non Nok Tha (Bayard 1984) and BNW (Higham 2008). Significantly such behavioral parallels can be compared to multiple occurrences on the Eurasian steppe (e.g., Chernykh 1992, 80–81, 83, pls. 7, 8, 135, pl. 12, 222, fig. 76; see also Pigott and Ciarla 2007, 82, 84) and further fuels the notion of steppe-related influences having spread into Thailand, if not across Southeast Asia. However, by what route remains at issue.

At NPW, while processing activity continued into the first millennium BCE, the exact timing of the advent of smelting has proven difficult to document due to the complex and compact stratigraphic context of the early basal deposit at NPW. However, the recent evidence from BNW offers some support to the argument that smelting was occurring at NPW in the late second millennium BCE. Ultimately, this led to the boom in Iron Age copper smelting in the valley that occurred over centuries following ca. 500 BCE when demand for copper was accelerating (Pryce et al. 2010, 260–62; 2011, 149, 158).

With this production boom, the site is seen to expand to its five hectare size and is comprised of an almost 2 m thick, ashy, powdery industrial layer filled with massive amounts of ore, slag, bivalve implement and ornament molds, cup and conical ingots molds, and large, smelting crucible sherds. The whole crucibles were distinctly shaped and larger than those known in northeast Thailand (fig. 13). This industrial deposit comprises one of two KWPV assemblages unique in the archaeometallurgical record of the Asian Old World. The sheer volume of later first-millennium-BCE production remains in the KWPV offers strong evidence of production meant for consumers well beyond the valley's immediate interaction sphere (e.g., Mudar and Pigott 2003; White and Pigott 1996).

As previously mentioned, Pryce undertook a comparative laboratory study of slag and technical ceramics from the two neighboring and partially contemporaneous production

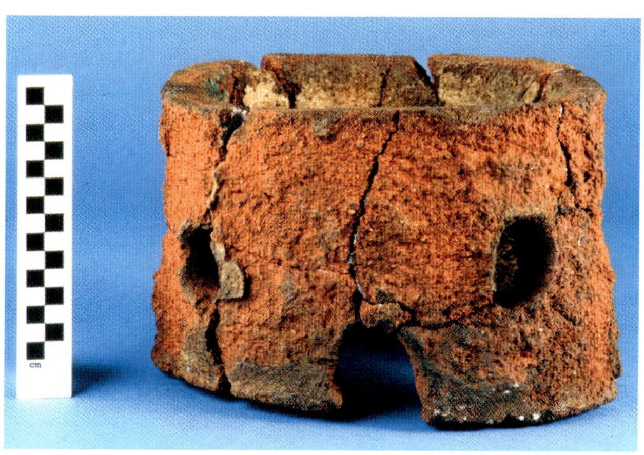

Figure 14. A reconstructed example of a typical ceramic chimney, several of which were excavated from burials at later first-millennium-BCE Nil Kham Haeng and Khao Sai On in and near the Khao Wong Prachan valley of central Thailand, respectively. Courtesy of the Thailand Archaeometallurgy Project.

sites, an approach not frequently seen in archaeometallurgical investigations. He characterized Iron Age smelting at NPW as "inefficient and non-standardised crucible-based reduction of oxidic and some sulfidic copper ores by 'serendipitous co-smelting.'" Whereas about two centuries later in the first millennium BCE at nearby NKH, the process is seen to be a "more efficient and standardised crucible-based reduction of sulfidic and some oxidic copper ores by 'deliberate co-smelting'" (Pryce et al. 2010, 261; see also Pryce 2009). With a gradual shift in production and what Pryce terms the valley's "metallurgical ethos," a marked shift in how ores were being beneficiated also occurred at NKH (Pryce et al. 2010, 259). This in turn created a unique site matrix as countless years of human effort went into crushing the increasingly sulfidic ore to mostly pea-sized gravel, presumably to improve hand separation of ore bits from gangue. NKH's three-plus hectares are composed predominantly of multitudinous, almost varve-like layers of gravel stacked up to six meters deep in places. The layering of the gravel presumably stems from piles of crushed material being flattened *in situ* underfoot over time as well as by seasonal heavy rains. By all indications, people at NKH lived, smelted, and were buried within the context of this massive production site. As mentioned earlier, a comparable, but far smaller, crushed hostrock deposit is known from the mining site at Phu Lon, another example suggestive of a shared technological tradition linking central with northeast Thailand.

While crucible smelting characterizes production at both KWPV sites, the air delivery mechanism that drove the smelting processes remains "archaeologically invisible" (Pryce et al. 2010, 261). How exactly these smelts were being driven is, at present, unknown. Were typical ethnographically documented Southeast Asian bamboo piston-bellows in use? If so, being organic, they could easily have not survived over time. No evidence of tuyère-like artifacts has been recovered, but enigmatic, ceramic chimneys known from NKH's matrix and from inclusion in several burials there and at Khao Sai On may have played a role (Pigott et al. 1997, 150, fig. 10; Ciarla 2008, 331, fig. 26) (fig. 14). Pryce (2010, 247–49) has argued cogently that, based in part on his smelting replication experiments, these chimneys are not directly related to the act of wind-driven smelting and hence were not related to a potential air delivery mechanism. He suggests instead that they may be foundry related, whereas, Ciarla, arguing from the Khao Sai On evidence, maintains support for the original contention that the chimneys are smelting related (Ciarla 2008, 330–36).

Pryce et al. (2010, 259) argue, "Whilst NKH copper production was certainly more effective in reducing metal loss, this 'efficiency' probably came at the cost of a much increased labor input." At the same time with the coming of massive production at NKH in the centuries following ca. 300 BCE, Pryce's perceived shift in local "metallurgical ethos" is best interpreted in terms of cultural response to the documented rise in regional demand for copper for social display, frequently funerary, in late prehistory across the greater region (Pryce et al. 2010, 237, 259; Higham and Higham 2009).

Conclusion

In closing, information gleaned from the new fieldwork at the mining site of Xepon, Laos, added to what we already know from Phu Lon and the KWPV in Thailand, is providing a detailed characterization of prehistoric copper mining and production in these areas of Southeast Asia. In addition, Pryce's Oxford-based Southeast Asian Lead Isotope Project (SEALIP) (in this vol.; also Pryce 2011; Pryce et al, in press) is an innovative analytical program that will be helpful in understanding from what sources copper was coming and to what consumers copper (e.g., ores, metal) was being distributed. Additionally this program can focus on RETEM, which, on present evidence, appears chronologically at odds with prevailing ^{14}C dates for prehistoric Thailand. Thus SEALIP offers the hope that we can move beyond the forty-year-old controversy over metallurgical origins in Southeast Asia. It may also shed new light on the compelling new evidence for technological links to Eurasian steppe cultures. Future efforts may offer new insights into potential technological links to Eurasian steppe cultures as well as to those within China itself that may also have played a part in the transmission of metallurgy southward.

In the end, the future of archaeometallurgy in East and Southeast Asia lies with increased excavation and, specifically, with projects purposefully designed to seek resolution to important regional archaeological problems. Moreover, hope also lies with the laboratory-based analysis of artifacts, including the combined insights of technological and geochemical approaches that should and must be an essential part of a holistic approach to the archaeology of the greater region. With every such research initiative, it becomes clearer that Southeast Asian prehistoric cultures and economies formed part of what, from as early as 1954, the eminent Cambridge sinologist Joseph Needham (1954; Sherratt 2006, 33–34, fig. 11) foresaw as a community of loosely connected Bronze Age cultures stretching from the Atlantic to the Pacific. In Sherratt's terms (2006, 34), this community, of necessity, formed through contact and interaction, which over time resulted in the cultural phenomenon of trans-Eurasian exchange. By all indications, the coming of metallurgy, especially that of tin bronze, played an integral role in this remarkable process connecting Europe and Southwest Asia, across the steppe/forest steppe and Central Asia into the realm of ancient China and ultimately into mainland Southeast Asia, and then out into the islands beyond. The door is only now beginning to open on understanding, interpreting, and explaining just how this connection came about.

Acknowledgments

This paper is dedicated to the memory of the late Gerd Weisgerber, a friend and colleague, who led a long and distinguished career researching the archaeometallurgy and archaeology of mining across Europe and Asia. His contributions to these fields, including his work at Phu Lon, from his base at the Deutsches Bergbau-Museum in Bochum,

Germany, set the highest standards for research into ancient metals and mining and their significance in antiquity.

The author wishes to thank his colleagues T. Oliver Pryce and Christopher P. Thornton for their insightful and helpful comments on this manuscript. Error and omissions that persist remain the responsibility of the author. A vote of thanks is due also to the Freer and Sackler Galleries staff who made the Fifth Forbes Symposium such a success and, in particular, my personal thanks to Janet G. Douglas for her remarkable patience and continued assistance and to Stacy Bowe for her computer graphics contribution.

Notes

1. To get a sense of ore body locations and availability, see discussions and maps in Anthony (2007, 290, fig. 15.9, 413, fig 16.1), with documented Bronze Age mines, and compare to his distribution map of steppe populations across the same region ca. 1900–1500 BCE (2007, 436, fig. 16.10). Compare also culture distribution maps in Frachetti (2008, 40–41, figs. 4 and 5). See also map of the major Eurasian tin belts in Pigott and Ciarla (2007, 80, fig. 6) and the ore and culture distribution maps in Zhang (2009, 18, figs. 1 and 2). See also Chernykh (1992).
2. See Anthony (2007, 165, fig. 9.1) for a map demarcating the forest steppe in western Eurasia.
3. For an example of how such transmission may have proceeded, see Zhang's (2009) discussion of the Sintashta (Andronovo) culture trade in copper ingots and artifacts to the tin-bronze-using Seima-Turbino of the forest steppe.
4. Compare Muhly (1988, 15), who cautions it as a "persuasive but dangerous argument," apt to lead one off the interpretive rails.
5. Images of all artifacts discussed in this section can be found in the publications by Pigott and colleagues, Pryce and colleagues, and White and Hamilton (2009), listed in the References.

References

Alimov, K., N. Boroffka, M. Bubnova, J. Burjakov, J. Cierny, J. Jakubov, J. Lutz, H. Parzinger, E. Pernicka, V. Radililovsky, V. Ruzanov, T. Sirinov, and G. Weisgerber. 1998. Prähistorischer zinnbergbau in Mittelasien. Vorbericht der ersten kampagne 1997. [Prehistoric tin mining in Central Asia: Preliminary report of the 1997 campaign]. *Eurasia Antiqua* 4:137–99.

An, Z. 1982–83. Some problems concerning China's early copper and bronze artifacts. *Early China* 5:53–75.

———. 1993. Shi lun Zhongguo de zaoqi tongqi [On early copper and bronze objects in China]. *Kaogu* [Archaeology] 12:1110–19.

———. 1998. Cultural complexes of the Bronze Age in the Tarim Basin and surrounding areas. In *The Bronze Age and early Iron Age peoples of eastern Central Asia,* Vol. 1, ed. V. H. Mair, 45–62. Washington, DC, and Philadelphia: Institute for the Study of Man in collaboration with the University of Pennsylvania Museum Publications.

———. 2000. On early copper and bronze objects in ancient China. In *The beginnings of metallurgy in China,* ed. K. M. Linduff, R. Han, and S. Sun, 29–49. Lewiston, NY: Edwin Mellen Press.

Anthony, D. W. 2007. *The horse, the wheel, and language: How Bronze-Age riders from the Eurasian steppes shaped the modern world.* Princeton, NJ: Princeton University Press.

Barnard, N. 1983. Further evidence to support the hypothesis of indigenous origins of metallurgy in China. In *The origins of Chinese civilization,* ed. D. N. Keightley, 237–77. Berkeley: University of California Press.

Bayard, D. 1984. Rank and wealth at Non Nok Tha: The mortuary evidence. In *Southeast Asian archaeology at the 15th Pacific Science Congress,* ed. D. Bayard, 87–122. Dunedin: Department of Anthropology, University of Otago.

Bazin, D., and H. Hübner. 1969. *Copper deposits in Iran.* Report 13. Tehran: Geological Survey of Iran.

Bishop, C. W. 1942. *Origin of the Far Eastern civilizations: A brief handbook.* Washington, DC: Smithsonian Institution.

Boroffka, N., J. Cierny, J. Lutz, J., H. Parzinger, E. Pernicka, and G. Weisgerber. 2002. Bronze Age tin from Central Asia: Preliminary notes. In *Ancient interactions: East and West in Eurasia,* ed. K. Boyle, C. Renfrew, and M. Levine, 135–60. McDonald Institute Monographs. Cambridge: McDonald Institute of Archaeological Research.

Charles, J. A. 1975. Where is the tin? *Antiquity* 49(193):19–24.

Chernykh, E. N. 1970. Drevneishaya metallurgiya Urala I Povolzh'ya. Moscow: Nauka.

———. 1992. *Ancient metallurgy in the USSR: The early metal age.* Cambridge: Cambridge University Press.

———. 2009. Ancient metallurgy in the Eurasian steppes and China: Problems of interactions. In *Metallurgy and civilisation: Eurasia and beyond,* ed. J. Mei and T. Rehren, 3–8. London: Archetype Publications.

Childe, V. G. 1954. The socketed celt in Upper Eurasia. Tenth Annual Report. London: London University Institute of Archaeology, 11–25.

Chiou-Peng, T. 2009. Incipient metallurgy in Yunnan: New data for old debates. In *Metallurgy and civilisation: Eurasia and beyond,* ed. J. Mei and T. Rehren, 79–84. London: Archetype Publications.

Ciarla, R. 1994. A long debated question: Pastoralism in south-western China. A different approach. In *The archaeology of the steppes: Methods and strategies,* ed. B. Genito, 73–85. Series Minor XLIV. Naples: Istituto Universitario Orientale, Dipartimento di Studi Asiatici.

———. 2000. Rethinking Yuanlongpo: The case for technological links between the Lingnan (PRC) and central Thailand in the Bronze Age. Paper presented at the 8th International Conference of the European Association of Southeast Asian Archaeologists, 2–6 October, Sarteano, Italy.

———. 2007a. A preliminary report on LoRAP archaeological excavations at prehistoric Khao Sai On, Lopburi province, central Thailand: *East and West* 57(1–4):395–401.

———. 2007b. Rethinking Yuanlongpo: The case for technological links between the Lingnan (PRC) and central Thailand in the Bronze Age. *East and West* 57(1–4):305–28.

———. 2008. The Thai–Italian "Lopburi Regional Archaeological Project" (LoRAP). Excavation at Khao Sai On-Noen Din 2008. Preliminary report. *East and West* 58:313–36.

Cierny, J., T. Stöllner, and G. Weisgerber. 2005. Zinn in und aus Mittelasien. In *Das schiff von Uluburun: Welthandel vor 3000 jahren,* ed. Ü. Yalçin, C. Pulak, and R. Slotta, 431–48. Bochum: Deutsches Bergbau-Museum.

Cierny, J., and G. Weisgerber. 2003. Bronze Age tin mines in Central Asia. In *Le problème de l'étain à l'origine de la métallurgie,* ed. A. Giumlia-Mair and F. Lo Schiavo, 23–31. Acts of the XIVth UISPP Congress, University of Liège, Belgium, 2–8 September 2001. BAR International Series 1199. Oxford: Archaeopress.

Cleuziou, S. and T. Berthoud. 1982. Early tin in the Near East: A reassessment in the light of new evidence from western Afghanistan. *Expedition* 25(1):14–19.

Crawford, H. E. W. 1974. The problem of tin in Mesopotamian bronzes. *World Archaeology* 6(2):242–47.

Creel, H. C. 1935. On the origins of the manufacture and decoration of bronze in the Shang period. *Monumenta Serica* 1:39–69.

de Jesus, P. S. 1978. Considerations on the occurrence and exploitation of tin sources in the ancient Near East. In *The search for ancient tin,* ed. A.D. Franklin, J. S. Olin, and T. A. Wertime, 33–38. Washington, DC: Smithsonian Institution.

Fitzgerald-Huber, L. G. 1995. Qijia and Erlitou: The question of contacts with distant cultures. *Early China* 20:17–67.

Frachetti, M. 2008. *Pastoralist landscapes and social interaction in Bronze Age Eurasia.* Berkeley: University of California Press.

Franklin, A. D., J. S. Olin, and T. A. Wertime. 1978. *The search for ancient tin.* Washington, DC: Smithsonian Institution.

Hanks, B. K., and R. Doonan. 2009. From scale to practice: A new agenda for the study of early metallurgy on the Eurasian steppe. *Journal of World Prehistory* 22(4):329–56.

Hanks, B. K., and K. M. Linduff, ed. 2009. *Social complexity in prehistoric Eurasia: Monuments, metals, and mobility.* Cambridge: Cambridge University Press.

Heine-Geldern, R. 1937. L'art prebouddhique de la Chine et de l'Asie du Sud-Est et son influence en Oceanie. *Revue des Arts Asiatiques* 11(4):177–206.

———. 1952. Some problems of migration in the Pacific. *Wiener Beiträge zur Kulturgeschicte und Linguistic* 9:311–62.

———. 1974. American metallurgy and the Old World. In *Early Chinese art and its possible influence in the Pacific Basin,* ed. N. Barnard, 787–822. Taiwan: Interculture Arts Press.

Helwing, B. 2009. Rethinking the tin mountains: Patterns of usage and circulation of tin in greater Iran from the 4th to the 1st millennium BC. *TÜBA-AR* 12:209–21.

Higham, C. F. W. 1996. *The Bronze Age of Southeast Asia.* Cambridge: Cambridge University Press.

———. 2002. *Early cultures of mainland Southeast Asia.* Bangkok: River Books.

———. 2004. Mainland Southeast Asia from the Neolithic to the Iron Age. In *Southeast Asia: From prehistory to history,* ed. I. Glover and P. Bellwood, 41–67. London: RoutledgeCurzon.

———. 2006. Crossing national boundaries: Southern China and Southeast Asia in prehistory. In *Uncovering Southeast Asia's past: Selected papers from the 10th International Conference of the European Association of Southeast Asian Archaeologists,* ed. E. A. Bacus, I. C. Glover, and V. C. Pigott, 13–21. Singapore: National University of Singapore Press.

———. 2007. Summary and conclusions. In *The excavation of Noen U-Loke and Non Muang Kao: The origins of the civilization of Angkor,* Vol. 2, ed. C. F. W. Higham, A. Kijingam, et al., 595–609. Bangkok: Thai Fine Arts Department.

———. 2008. Recasting Thailand: New discoveries at Ban Non Wat. *Current World Archaeology* 31:38–41.

———. 2009. Thailand's Bronze Age superburials: The aristocrats of Ban Non Wat. *Current World Archaeology* 35:18–23.

Higham, C. F. W., T. Higham, R. Ciarla, K. Douka, A. Kijngam, and F. Rispoli. 2011. The origins of the Bronze Age of Southeast Asia. *Journal of World Prehistory* 24: 227–274.

Higham, C. F. W., and T. Higham. 2009. A new chronological framework for prehistoric Southeast Asia, based on a Bayesian model from Ban Non Wat. *Antiquity* 83:125–44.

Higham, C. F. W., and R. Thosarat. 2006. Ban Non Wat: The first three seasons. In *Uncovering Southeast Asia's past: Selected papers from the 10th International Conference of the European Association of Southeast Asian Archaeologists,* ed. E. A. Bacus, I. C. Glover, and V. C. Pigott, 98–104. Singapore: National University of Singapore Press.

Ho, P. T. 1975. *The cradle of the East: An inquiry into the indigenous origins of techniques and ideas of Neolithic and Early Historic China, 5000–1000* B.C. Hong Kong: Chinese University of Hong Kong; Chicago: University of Chicago Press.

Kamvong, T., and K. Zaw. 2009. The origin and evolution of skarn forming fluids from the Phu Lon Deposit, Northern Loei Fold Belt, Thailand: Evidence from fluid inclusion and sulfur isotope studies. *Journal of Asian Earth Sciences* 34(5):624–33.

Karlgren, B. 1942. The date of the early Dong-son culture. *Bulletin of the Museum of Far Eastern Antiquities* 14:1–28.

Kars, H., and E. Burke, ed. 2005. *Proceedings of the 33rd International Symposium on Archaeometry, 22–26 April 2002, Amsterdam.* Geoarchaeological and Bioarchaeological Studies 3. Amsterdam: VU University of Amsterdam.

Kohl, P. L. 2007. *The making of Bronze Age Eurasia.* New York: Cambridge University Press.

Kroeber, A. L. 1963. *Anthropology: Culture patterns and processes.* New York: Harbinger Books.

Kuzmina, E. E. H. 2008. *The prehistory of the Silk Road.* Ed. V. H. Mair. Philadelphia: University of Pennsylvania Press.

La Niece, S., D. Hook, and P. Craddock, ed. 2007. *Metals and mines: Studies in archaeometallurgy.* London: Archetype Publications in association with the British Museum.

Linduff, K. M., ed. 2004. *Metallurgy in ancient eastern Eurasia from the Urals to the Yellow River.* Lewiston, NY: Edwin Mellen Press.

Linduff, K. M., R. Han, and S. Sun, ed. 2000. *The beginnings of metallurgy in China.* Lewiston, NY: Edwin Mellen Press.

Linduff, K. M., and J. Mei. 2009. Metallurgy in ancient Eastern Asia: Retrospect and prospects. *Journal of World Prehistory* 22(43):265–81.

Litvinskij, B.A. 1950. K istorii dobytschi olova v Uzbekistane. *Trudy Sredneaziatskogo Gosudarstvennogo Universiteta (Archeologija srednej Azii)* n.s. 9:51–97.

Loehr, M. 1974. China: Prehistory and archaeology. In *Encyclopaedia Britannica, Macropaedia,* 15th ed., 297–301. Chicago: William Benton.

Maddin, R., J. D. Muhly, and T. Stech. 1999. Early metalworking at Çayönü. In *The beginnings of metallurgy,* ed. A. Hauptmann, E. Pernicka, T. Rehren, and Ü. Yalçin, 37–44. Der Anschnitt, Beiheft 9. Bochum: Deutsches Bergbau-Museum.

Mair, V. H., ed. 1998. *The Bronze Age and early Iron Age peoples of eastern Central Asia.* Vols. 1–2. Ed. V. H. Mair. Washington, DC: Institute for the Study of Man in collaboration with the University of Pennsylvania Museum Publications.

Mei, J. 2000. *Copper and bronze metallurgy in late prehistoric Xinjiang: Its cultural context and relationship with neighboring regions.* BAR International Series 865. Oxford: Archaeopress.

———. 2003a. Cultural interaction between China and Central Asia during the Bronze Age. *Proceedings of the British Academy* 121:1–39.

———. 2003b. Qijia and Seima-Turbino: The question of early contacts between northwest China and the Eurasian steppe. *Bulletin of the Museum of Far Eastern Antiquities* 75:31–54.

———. 2003c. Early copper technology in Xinjiang, China: The evidence so far. In *Mining and metal production through the ages,* ed. P. Craddock and J. Lang, 111–21. London: British Museum Press.

———. 2004. Metallurgy in Bronze Age Xinjiang and its cultural context. In *Metallurgy in ancient Eastern Eurasia from the Urals to the Yellow River,* ed. K. M. Linduff, 173–88. Lewiston, NY: Edwin Mellen Press.

———. 2009a. Early metallurgy and socio-cultural complexity: Archaeological discoveries in northwest China. In *Social complexity in prehistoric Eurasia: Monuments, metals, and mobility,* ed. B. K. Hanks and K. M. Linduff, 215–32. Cambridge: Cambridge University Press.

———. 2009b. Early metallurgy in China: Some challenging issues in current studies. In *Metallurgy and civilisation. Eurasia and beyond,* ed. J. Mei and T. Rehren, 9–16. London: Archetype Publications.

Mei, J., and T. Rehren, ed. 2009. *Metallurgy and civilisation: Eurasia and beyond.* London: Archetype Publications.

Molist, M., I. Montero-Ruiz, X. Clop, S. Rovira, E. Guerrero, and J. Anfruns. 2009. New metallurgic finds from the pre-pottery Neolithic: Tell Halula (Euphrates valley, Syria). *Paléorient* 35(2):33–48.

Momenzadeh, M. 2004. Metallische bodenschätze in Iran in antiker zeit. Ein kurzer überblick. In *Persiens antike pracht: Bergbau, handwerk, archäeologie: Katalog der ausstellung des Deutschen Bergbau-Museums Bochum vom 28 November 2004 bis 29 Mai 2005*, ed. T. Stöllner, R. Slotta, and A. Vatandoost, 28–43. Bochum: Deutsches Bergbau-Museum.

Momenzadeh, M., N. Nezafati, and E. Pernicka. 2002. First indication of tin at the ancient mining site near Deh Hosein. In *Proceedings of the 33rd International Symposium on Archaeometry, 22–26 April 2002, Amsterdam,* ed. H. Kars and E. Burke, 116–17. Geoarchaeological and Bioarchaeological Studies 3. Amsterdam: VU University of Amsterdam.

Mudar, K. M., and V. C. Pigott. 2003. Subsistence changes and community-based craft production in prehistoric Thailand. In *Fishbones and glittering emblems,* ed. A. Källén and A. Karlström, 149–60. Stockholm: Östasiatiska Museet.

Muhly, J. D. 1973. *Copper and tin: The distribution of mineral resources and the nature of the metals trade in the Bronze Age.* Transactions of the Connecticut Academy of Arts 43. Hamden, CT: Archon Books.

———. 1976. *Supplement to copper and tin: The distribution of mineral resources and the nature of the metals trade in the Bronze Age.* Transactions of the Connecticut Academy of Arts 46. Hamden, CT: Archon Books.

———. 1985. Sources of tin and the beginnings of the bronze metallurgy. *American Journal of Archaeology* 89:275–91.

———. 1988. The beginnings of metallurgy in the Old World. In *The beginning of the use of metals and alloys,* ed. R. Maddin, 2–20. Cambridge, MA: MIT Press.

Natapintu, S. 1984. Ancient settlement at Ban Tha Khae in Lopburi (in Thai). *Muang Boran* 10:26–40.

———. 1988. Current research on ancient copper-base metallurgy in Thailand. In *Prehistoric studies: The Stone and Metal Ages in Thailand,* ed. P. Charoenwongsa and B. Bronson, 107–24. Papers in Thai Antiquity 1. Bangkok: Thai Antiquity Working Group.

Needham, J. 1954. *Science and civilization in China,* Vol. 1, *Introductory orientations.* Cambridge: Cambridge University Press.

Nezafati, N. 2006. Au-Sn-W-Cu-mineralization in the Astaneh-Sarband area, west central Iran: Including a comparison of the ores with ancient bronze artifacts from Western Asia. PhD diss., University of Tübingen. Available at: http://tobias-lib.ub.unituebingen.de/volltexte/2006/2533/.

Nezafati, N., E. Pernicka, and M. Momenzadeh. 2006. Ancient tin: Old question and a new answer. *Antiquity* 80(308):308–11.

———. 2008. Iranian ore deposits and their role in the development of the ancient cultures. In *Anatolian Metal IV*, ed. Ü. Yalçin, 77–91. Der Anschnitt, Beiheft 21. Bochum: Deutsches Bergbau-Museum.

———. 2009. Introduction of the Deh Hosein ancient tin-copper mine, western Iran: Evidence from geology, archaeology, geochemistry and lead isotope data. *TÜBA-AR* 12:223–36.

Parzinger, H. 2002. Das zinn in der Bronzezeit Eurasiens. In *Anatolian Metal II,* ed. Ü. Yalçin, 159–77. Bochum: Deutsches Bergbau-Museum.

Parzinger, H., and Boroffka, N. 2003. *Das zinn der Bronzezeit in Mittelasien I.* Archäologie in Iran und Turan, Bd. 5. Mainz am Rhein: Verlag Philipp von Zabern.

Pigott, V. C. 1998. Prehistoric copper mining in the context of emerging community craft specialization in northeast Thailand. In *Social aspects of an industrial past: The archaeology and anthropology of mining,* ed. A. B. Knapp, V. C. Pigott, and E. W. Herbert, 205–25. London: Routledge.

———. 2011. Sources of tin and the tin trade in Southwest Asia: The results of recent research and their importance to current understanding. In *Metallurgy: Understanding how, learning why. Studies in honor of James D. Muhly,* ed. P. P. Betancourt and S. C. Ferrence, 267–85. Prehistory Monographs 29. Philadelphia: INSTAP Academic Press.

———. In press. On ancient tin, its sources and trade: Further comments. In *Eastern Mediterranean metallurgy in the second millennium* BC, ed. V. Kassianidou and G. Papasavvas. Oxford: Oxbow Books.

Pigott, V. C., and R. Ciarla. 2007. On the origins of metallurgy in prehistoric Southeast Asia: The view from Thailand. In *Metals and mines: Studies in archaeometallurgy,* ed. S. La Niece, D. Hook, and P. Craddock, 76–88. London: Archetype Publications in association with the British Museum.

Pigott, V. C., and S. Natapintu. 1988. Archaeological investigations into prehistoric copper production: The Thailand Archaeometallurgy Project 1984–1986. In *The beginning of the use of metals and alloys,* ed. R. Maddin, 156–62. Cambridge, MA: MIT Press.

———. 1996–97. Investigating the origins of metal use in prehistoric Thailand. In *Ancient Chinese and Southeast Asian Bronze Age cultures,* Vol. 2, ed. F. D. Bulbeck and N. Barnard, 787–808. The proceedings of a conference held at the Edith and Joy London Foundation Property, Kioloa, NSW, Australia, 8–12 February 1988. Taipei: SMC Publishing.

Pigott, V. C., S. Natapintu, and U. Theetiparivatra. 1992. The Thailand Archaeometallurgy Project 1984–85: Research in the development of prehistoric metal use in northeast Thailand. In *Early metallurgy, trade and urban centres in Thailand and Southeast Asia,* ed. I. C. Glover, P. Suchitta, and J. Villers, 47–62. Bangkok: White Lotus.

Pigott, V. C., and G. Weisgerber. 1998. Mining archaeology in geological context: The prehistoric copper mining complex at Phu Lon, Nong Khai province, northeast Thailand. In *Metallurgica Antiqua. In honour of Hans-Gert Bachmann and Robert Maddin*, ed. T. Rehren, A. Hauptmann, and J. D. Muhly, 135–62. Der Anschnitt, Beiheft 8. Bochum: Deutsches Bergbau-Museum.

Pigott, V. C., A. D. Weiss, and S. Natapintu. 1997. The archaeology of copper production: Excavations in the Khao Wong Prachan valley, central Thailand. In *South-East Asian archaeology 1992,* ed. R. Ciarla and F. Rispoli, 119–57. Proceedings of the Fourth International Conference of the European Association of South-East Asian Archaeologists, Rome, 28 September–4 October 1992. Serie Orientale Roma, Vol. 77. Rome: Istituto Italiano per L'Africa e L'Oriente.

Possehl, G. L. 1996. Meluhha. In *The Indian Ocean in antiquity,* ed. J. Reade, 133–208. London: Kegan Paul International.

———. 1997. Seafaring merchants of Meluhha. In *South Asian archaeology 1995,* ed. F R. Allchin and B. Allchin, 87–100. Proceedings of the 13th Conference of the European Association of South Asian Archaeologists, Cambridge, 5–9 July 1995. Cambridge: Science Publishers, Inc.

Pryce, T. O. 2009. Prehistoric copper production and technological production in the Khao Wong Prachan valley of central Thailand. PhD diss., Institute of Archaeology, University College London. Available at http://sites.google.com/site/thomasoliverpryce/.

———. 2011. Technical analysis of Bronze Age Ban Non Wat copper-base artefacts. In *The excavation of Ban Non Wat: The Bronze Age,* ed. C. F. W. Higham and A. Kijngam, 489–97. The Origins of the Civilization of Angkor Vol. 5. Bangkok: The Thai Fine Arts Department.

Pryce, T. O., A. H. Bevan, R. Ciarla, F. Rispoli, J. L. Malakie, B. Hassett, and C. Castillo. In press. Intensive archaeological survey in Southeast Asia: Methodological and metallurgical insights from Khao Sai On, central Thailand. *Asian Perspectives.*

Pryce, T. O., M. Brauns, N. Chang, E. Pernicka, M. Pollard, C. Ramsey, T. Rehren, V. Souksavatdy and T. Sayavongkhamdy (2011). Isotopic and technological variation in prehistoric primary Southeast Asian copper production. *Journal of Archaeological Science* 38: 3309–22.

Pryce, T. O., and V. C. Pigott. 2008. Towards a definition of technological styles in prehistoric copper smelting in the Khao Wong Prachan valley of central Thailand. In *Archaeology in Southeast Asia: From* Homo erectus *to the living traditions,* ed. J.-P. Pautreau, A. Coupey, V. Zeitoun, and E. Rambault, 139–50. Choice of Papers from the 11th EurASEAA Conference, Bougon 25–29 September 2006. Chiang-Mai, Thailand: European Association of Southeast Asian Archaeologists.

Pryce, T. O., V. C. Pigott, M. Martinon-Torres, and T. Rehren. 2010. Prehistoric copper production and technological reproduction in the Khao Wong Prachan valley of central Thailand. *Archaeological and Anthropological Sciences* 2(4):237–64.

Pryce, T. O., M. Pollard, M. Martinon-Torres, V. C. Pigott, and E. Pernicka. 2011. Southeast Asia's first isotopically defined prehistoric copper production system: When did extractive metallurgy begin in the Khao Wong Prachan valley of central Thailand? *Archaeometry* 53(1):146–63.

Radivojević, M., T. Rehren, E. Pernicka, D. Šljivar, M. Brauns, and D. Borić. 2010. On the origins of extractive metallurgy: New evidence from Europe. *Journal of Archaeological Science* 37(11):2775–87.

Renfrew, C. 1973. Sitagroi and the independent invention of metallurgy in Europe. In *Actes du VIIIe Congrès international des sciences préhistoriques et protohistoriques, Rapports et corapports, Beograd 1971,* Vol. 2, ed. G. Novak, 473–81. Belgrade: University of Belgrade.

———. 1986. Varna and the emergence of wealth in prehistoric Europe. *In Social life of things: Commodities in cultural perspective,* ed. A. Appadurai, 141–68. Cambridge: Cambridge University Press.

Roberts, B. W., C. P. Thornton, and V. C. Pigott. 2009. Development of metallurgy in Eurasia. *Antiquity* 83:1012–22.

Rossovsky, L. N., V. V. Mogarovsky, and V. M. Chmirev. 1987. The metallogeny of tin and rare metals in eastern part of the Mediterranean folded belt. In *Mineral deposits of the Tethyan Eurasian metallogenic belt between the Alps and the Pamirs: Selected examples,* ed. S. Jankovic, 170–77. IGCP Project 169. Belgrade: UNESCO.

Rovira, S. Y., and I. Montero. 2003. Natural tin-bronze alloy in Iberian peninsula metallurgy: Potentiality and reality. In *Le problème de l'etain à l'origine de la métallurgie/The problem of early tin,* ed. A. Giumlia-Mair and F. Lo Schiavo, 15–22. BAR International Series 1199. Oxford: Archaeopress.

Salvatori, S., M. Vidale, G. Guida, and E. Masioli. 2009. Ilgynly-Depe (Turkmenistan) and the 4th millennium BC metallurgy of Central Asia. *Paléorient* 35(1):47–67.

Sayavongkhamdy, T., N. Chang, V. Souksavatady, and H. Cawte. 2009. The archaeology of Sepon, Lao PDR: Archaeometallurgy, unexploded bombs and collaborations. Paper presented at the 19th Congress of the Indo-Pacific Prehistory Association, 29 November–5 December 2009, Hanoi, Vietnam.

Shareq, A., V. M. Chmyriov, K. F. Stazhilo-Alexseev, V. I. Dronov, D. J. Gannon, G. K. Lubemov, A. K. Kafarshiy, and E. P. Malyarov. 1977. *Mineral resources of Afghanistan.* 2nd ed. Afghan Geological and Mines Survey, United Nations Development Support Project, AFG/74/012. Kabul: Ministry of Mines and Industries.

Sherratt, A. 2006. The trans-Eurasian exchange: The prehistory of Chinese relations with the West. In *Contact and exchange in the ancient world,* ed. V. H. Mair, 30–61. Honolulu: University of Hawaii Press.

Smith, C. S. 1973. Bronze technology in the East: A metallurgical study of early Thai bronzes, with some speculations on the cultural transmission of technology. In *Changing perspectives in the history of science, essays in honour of Joseph Needham,* ed. M. Teich and R. Young, 21–32. London: Heinemann.

———. 1981. *A search for structure: Selected essays on science, art, and history.* Cambridge, MA: MIT Press.

Stech, T., and V. C. Pigott. 1986. The metals trade in Southwest Asia in the third millennium BC. *Iraq* 48:39–64.

Sullivan, M. 1980. Review of *The Great Bronze Age of China*, ed. Wen Fong. *New York Times Book Review* (31 August 1980):10–11, 18.

Thornton, C. P., C. C. Lamberg-Karlovsky, M. Liezers, and S. M. M. Young. 2002. On pins and needles: Tracing the evolution of copper-base alloying at Tepe Yahya, Iran, via ICP-MS analysis of common-place items. *Journal of Archaeological Science* 29(12):1451–60.

Thornton, C. P., and C. Giardino, in press. Serge Cleuziou and the 'Tin Problem', in J. Giraud and G. Gernez (eds), *Aux Marges de l'archéologie: Hommage à Serge Cleuziou.* De Boccard, Travaux de la Maison René-Ginouvés, Paris.

Van Heerkeren, H. I. 1958. The Bronze Age of Indonesia. Verhandelingen van het Koninklijk Instituut voor taal-, land- en volkenkunde, Vol. 22. 's-Gravenhage, The Netherlands: Martinus Nijhoff.

Vernon, W. W. 1996–97. The crucible in copper-bronze production at prehistoric Phu Lon, northeast Thailand: Analyses and interpretation. In *Ancient Chinese and Southeast Asian Bronze Age cultures,* Vol. 2, ed. F. D. Bulbeck and N. Barnard, 809–20. Proceedings of a conference held at the Edith and Joy London Foundation Property, Kioloa, NSW, Australia, 8–12 February 1988. Taipei: SMC Publishing.

———. 1997. Chronological variation in crucible technology at Ban Chiang: A preliminary assessment. *Bulletin of the Indo-Pacific Prehistory Association* 16:107–10.

Watson, W. 1960. *Archaeology in China.* London: Max Parrish & Co.

———. 1985. An interpenetration of opposites? Pre-Han bronze metallurgy in West China. *Proceedings of the British Academy* 70:327–58.

Weeks, L. 2003. *Early metallurgy of the Persian Gulf: Technology, trade, and the Bronze Age world.* Boston: Brill Academic Publishers.

Weisgerber, G. 1984. Makan and Meluhha: Third millennium B.C. copper production in Oman and the evidence of contact with the Indus valley. In *South Asian Archaeology 1981,* ed. B. Allchin, 196–201. Proceedings of the Sixth International Conference of the Association of South Asian Archaeologists in Western Europe, Cambridge University, 5–10 July 1981, University of Cambridge Oriental Publications 34. Cambridge: Cambridge University Press.

Weisgerber, G., and J. Cierny. 2002. Tin for ancient Anatolia? In *Anatolian Metal II*, ed. Ü. Yalçin, 170–87. Der Anschnitt, Beiheft 15. Bochum: Deutsches Bergbau-Museum.

Wertime, T. A. 1964. Man's first encounters with metallurgy: Man's discovery of ores and metals helped to shape his sense of science, technology, and history. *Science* 146:1257–67.

———. 1973. The beginnings of metallurgy: A new look. *Science* 182:875–87.

White, J. C. 1982. *Ban Chiang: Discovery of a lost Bronze Age.* Philadelphia: The University Museum, University of Pennsylvania, and the Smithsonian Institution.

———. 1988. Early East Asian metallurgy: The southern tradition. In *The beginning of the use of metals and alloys,* ed. R. Maddin, 175–81. Cambridge, MA: MIT Press.

———. 1997. A brief note on new dates for the Ban Chiang cultural tradition. *Bulletin of the Indo-Pacific Prehistory Association* 16:103–6.

———. 2000. Review of *Prehistoric Thailand: From early settlement to Sukhothai* by Charles Higham and Rachanie Thorsarat. *Journal of Asian Studies* 59(4):1093–94.

———. 2008. Dating early bronze at Ban Chiang, Thailand. In *Archaeology in Southeast Asia: From* Homo erectus *to the living traditions,* ed. J. P. Pautreau, A. Coupey, V. Zeitoun, and E. Rambault, 91–104. Choice of Papers from the 11th EurASEAA Conference, Bougon, 25–29 September 2006. Chiang Mai, Thailand: European Association of Southeast Asian Archaeologists.

White, J. C., and E. G. Hamilton. 2009. The transmission of early bronze technology to Thailand: New perspectives. *Journal of World Prehistory* 22(4):357–97.

White, J. C., and Pigott, V. C. 1996. From community craft to regional specialization: Intensification of copper production in prestate Thailand. In *Craft specialization and social evolution: In memory of V. Gordon Childe,* ed. B. Wailes, 151–75. University Museum Monograph 93. Philadelphia: The University Museum of Archaeology and Anthropology, University of Pennsylvania.

Yalçin, Ü., C. Pulak, and R. Slotta, ed. 2005. *Das schiff von Uluburun. Welthandel vor 3000 jahren.* Bochum: Deutsches Bergbau-Museum.

Yener, K. A. 2000. *The domestication of metals: The rise of complex metal industries in Anatolia.* Boston: Brill.

Zhang, L. 2009. Metal trade in Bronze Age Central Asia. In *Metallurgy and civilisation: Eurasia and beyond,* ed. J. Mei and T. Rehren, 17–25. London: Archetype Publications.

The High-Tin Bronzes of Thailand

Ian Glover and Anna Bennett

The local people used brass that is cast, not the kind that is forged with the strange result that when they fall to the ground they break to pieces like pottery.[1]

ABSTRACT The term high tin, in the case of the objects we discuss and illustrate, is conventionally used for a binary copper alloy containing between about 20 and 28 wt.% tin. High-tin bronzes are rather few in number but found over an extensive region across West, South, Southeast and East Asia. The metal is brittle when cast, impossible to shape when cold, and can only be worked when hot. When freshly cast it has a yellowish color resembling gold, which is one reason that has been suggested for why this difficult alloy was used for special, perhaps ritual and ornamental, purposes. In this paper, we discuss excavations at Ban Don Ta Phet in Thailand, focusing on the high-tin bronze vessels found during these excavations. We also mention bronze vessels found elsewhere in Thailand and in Southeast and South Asia.

High-Tin Bronze Vessels and Ornaments in Thailand

As an archaeologist and a conservator/metallurgist, our interest in and knowledge of high tin comes primarily from studying objects excavated at the late prehistoric Iron Age cemetery of Ban Don Ta Phet (BDTP) in Kanchanaburi province, west-central Thailand (fig. 1), where many high-tin bronze vessels and some personal ornaments were excavated (Glover 1990) (fig. 2), and from a few surface finds found by alluvial tin miners at Khao Jamook near the Thai–Burmese border, by school children camping at Khao Kwark Cave in Ratchburi province in west central Thailand[2] (Bennett and Glover 1992), by villagers at Khao Sam Kaeo (KSK) in Chumphon province on the Thai–Malay peninsula (Pryce et al. 2006), and by archaeologists at Kok Khon and a number of other unpublished sites in northeast Thailand studied by Rajpitak (1983).

Excavations at Ban Don Ta Phet

BDTP was excavated on four occasions between 1975 and 2000 by teams from the Fine Arts Department of Thailand (FAD) and the Institute of Archaeology, University College London (UCL) (You-di 1976; Glover et al. 1983; Glover 1990; Dhepsuriyanon 2001), and over one hundred funerary deposits were found, richly equipped with iron tools and weapons, bronze vessels, and jewelry made from bronze, bone, ivory, glass, and semiprecious stones (fig. 3). Five radiocarbon dates calculated from organic temper, primarily rice grains, obtained from pottery from the 1980–81 season have dated the site to 390–360 BCE (table 1).

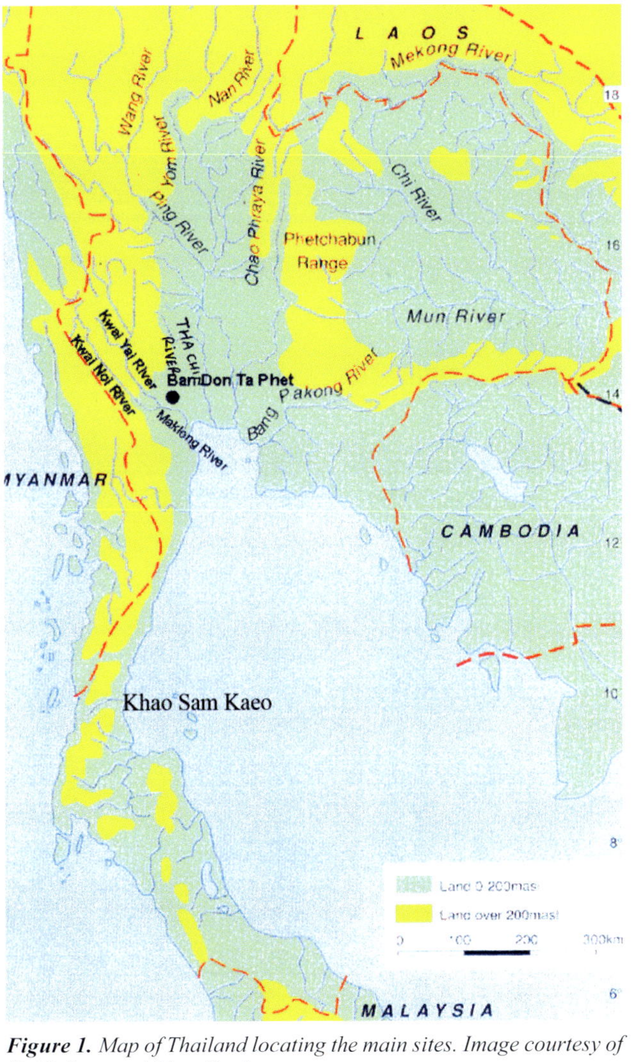

Figure 1. Map of Thailand locating the main sites. Image courtesy of C. Higham and River Books.

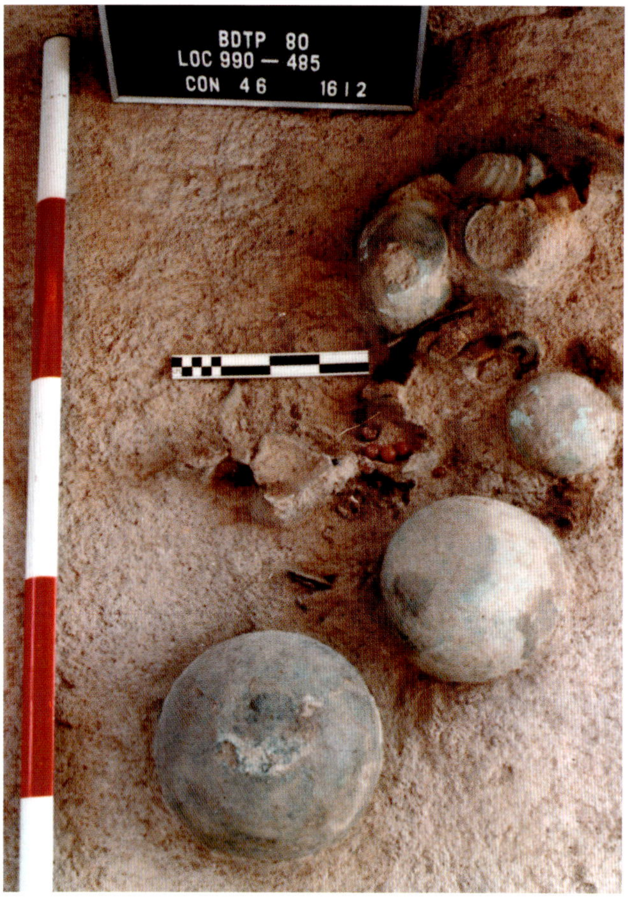

Figure 2. *Burial context 46 at Ban Don Ta Phet during excavation showing high-tin bowls turned rim down.*

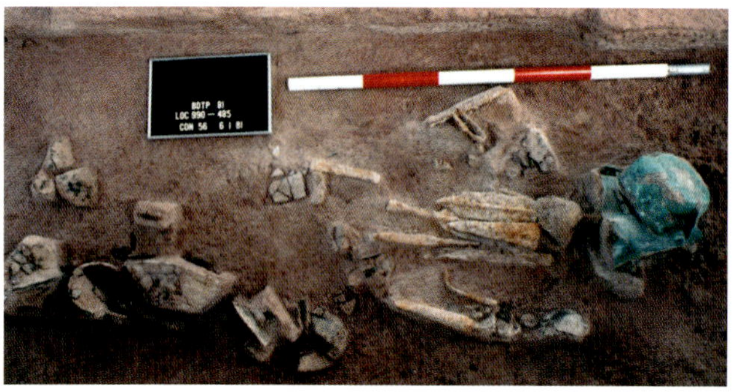

Figure 3. *Burial context 56 at Ban Don Ta Phet showing bronzes, iron weapons, pottery, and beads.*

Table 1. *Calibrated AMS radiocarbon dates from Ban Don Ta Phet.*

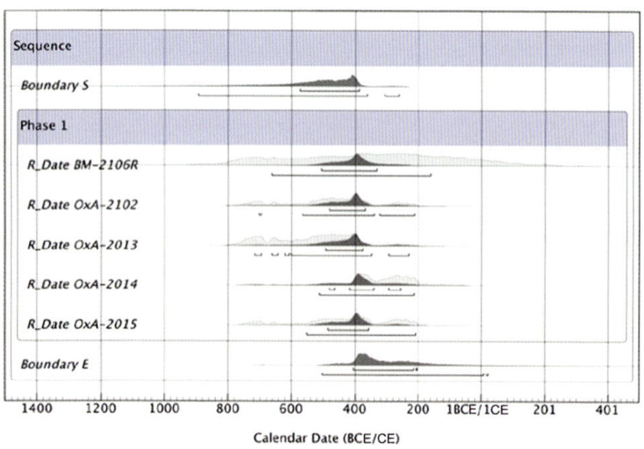

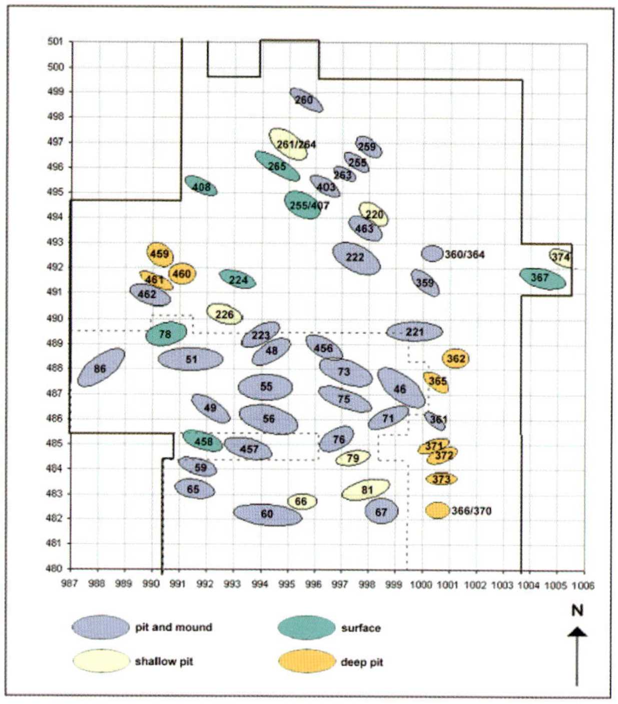

Figure 4. *Burial groups from the 1981 and 1985 excavations at Ban Don Ta Phet. Image courtesy of M. Woods.*

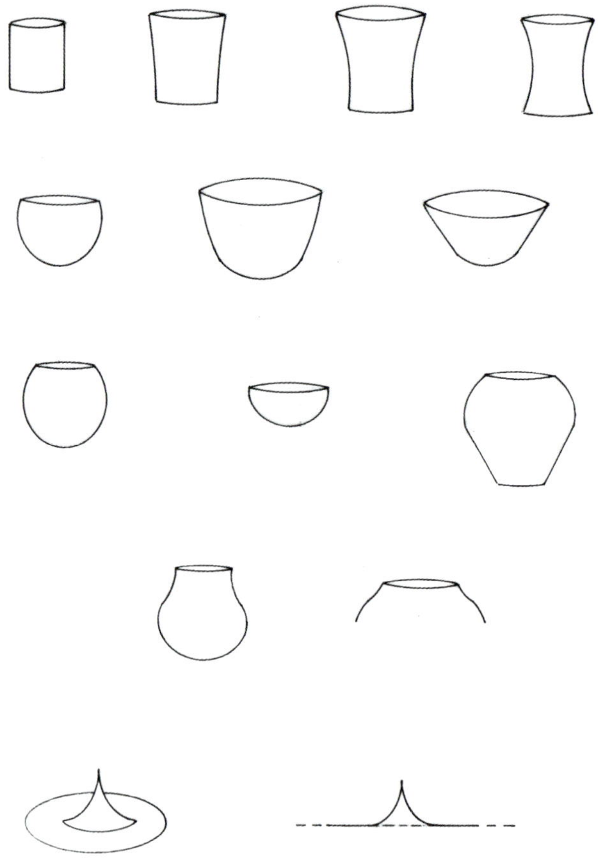

Figure 5. *The principal shapes of high-tin containers found at Ban Don Ta Phet during the Thai FAD excavations in 1975–76. Redrawn by B. Campbell Cole from Rajpitak (1979).*

Table 2. *Bronzes excavated over four seasons at Ban Don Ta Phet.*

Excavation Season	Containers	Bracelets, Rings, and Anklets	Bird Figurines	Small Bells	Ladles	Various
1975–76	163	61	3	2	1 or 2	—
1980–81	31	112	1	8	—	1 cage
1984–85	48	51	—	—	—	1 "stupa," 1 situla
2000	25	several	—	several	1	1 pendant
Totals	267	224+	4	10+	2+	

These are minimum numbers since many fragments cannot be attributed to any of the items listed in the table.

Table 3. *Bronzes examined by Rajpitak (1983).*

Sites	No. Examined	No. High Tin	No. Containers	No. Bracelets & Ornaments
Ban Don Ta Phet	28	14	7	7
Ban Na Di	81	5	1	4
Kok Khon	36	10	6	4
Non Chai	49	2	0	2
Ban Na Di	81	5	1	4
Nai Tong Soon	3	3	0	3
Ban Na-An	2	1	0	1
Totals	280	40	15	25

The total of 280 includes 200 fragments that could not be attributed to specific forms.

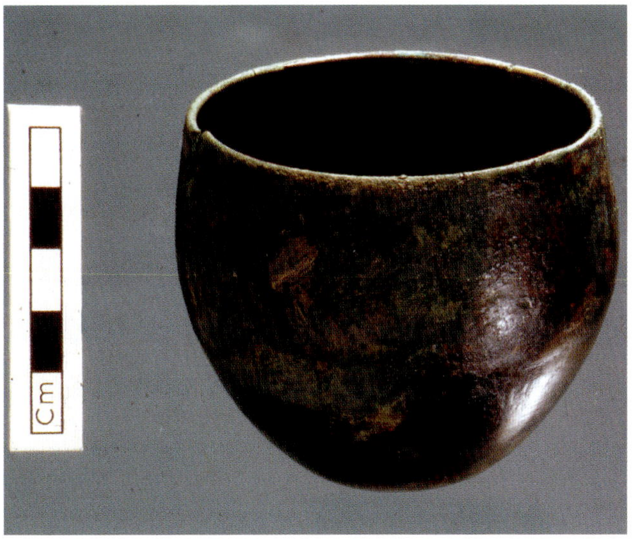

Figure 6. *Small beaker from context 59 showing the lustrous surface and traces of matting.*

The BDTP site contains secondary burials with some fragmentary bones and materials brought together and buried in regular rows some 50–70 cm apart (fig. 4), perhaps in one mass "grave," over a rather short time and within an area about 40 m across surrounded by a ditch and low bank. No traces of settlement rubbish or occupational layers were found, and no sites exactly comparable with BDTP have yet been excavated in Thailand, although the distinctive iron tools, beads, and bronze vessels have been recognized at a number of sites in central and western Thailand and more recently as far south as KSK—a settlement, trading, and manufacturing proto-urban site contemporary with BDTP and situated on the east coast of peninsular Thailand (Murillo-Barroso et al. 2010). During the first excavation, from November 1975 to 1976, Thai archaeologists with FAD (Rajpitak and Seely 1979, 26–27) excavated about

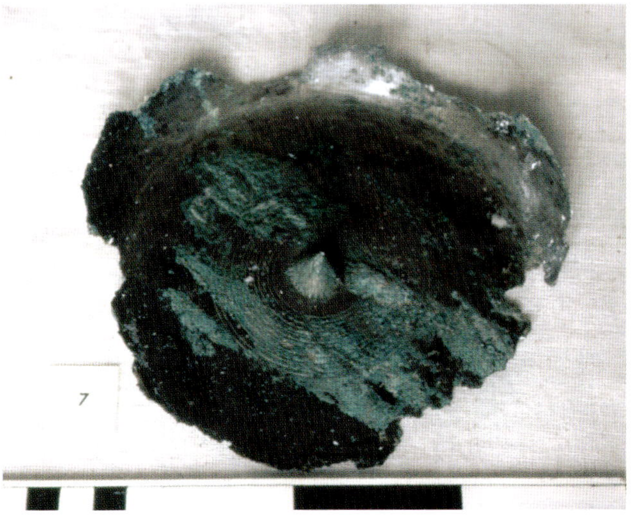

Figure 7. *Base of a bowl showing the central conical knob and surrounding circles.*

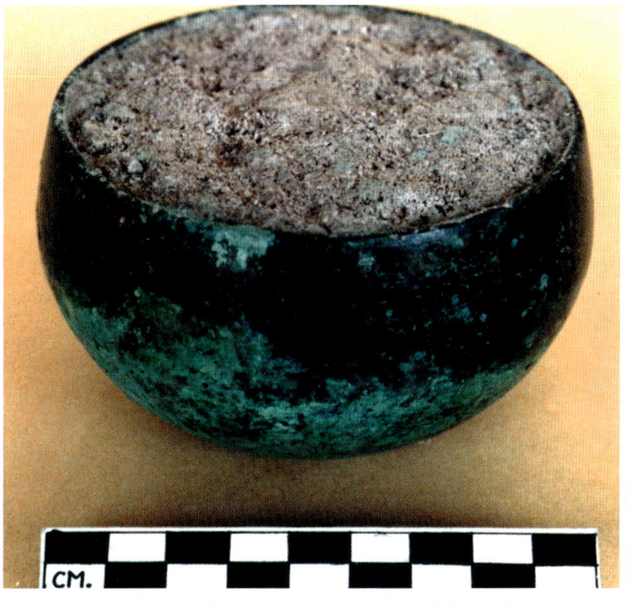

Figure 8. *Small bowl from context 79. Soil was left inside to preserve the bowl.*

fifty partial inhumations, recovering abundant pottery, iron tools and weapons, over 2,200 glass and stone beads, and over 150 bronzes. Table 2 shows the bronzes excavated in 1975–76 together with those excavated in subsequent seasons.

(a)

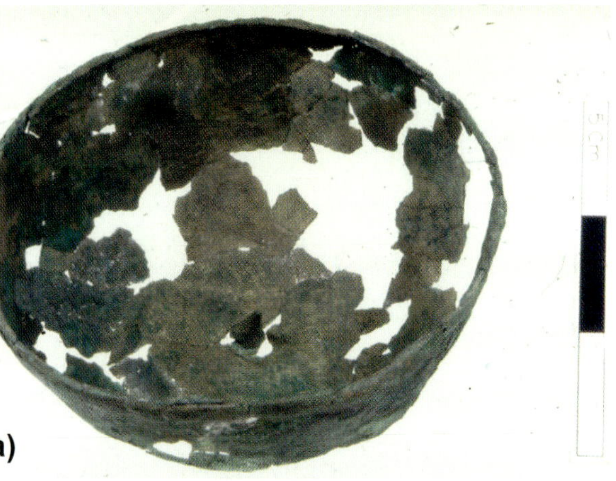

(b)

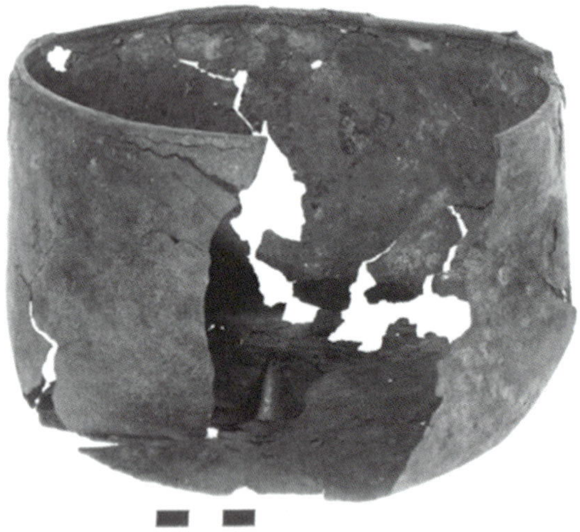

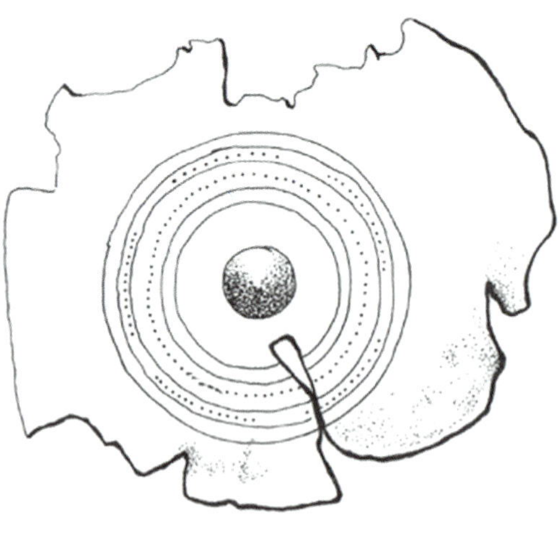

Figure 9. a. *High-tin knob base bowl from Ban Don Ta Phet after conservation. Photograph by B. Bellina.* ***b.*** *High-tin knob base canister from Ban Don Ta Phet after conservation. Photograph by the Conservation Department, Institute of Archaeology, UCL.*

The Bronze Vessels from BDTP

The most important technical study on Thai high-tin bronzes was undertaken by the Thai conservator Warangkhana Rajpitak, who undertook the analytical work while working as a research student at the Institute of Archaeology, UCL. Rajpitak worked on the finds from BDTP (Rajpitak 1979; Rapjitak and Seeley 1979) (tables 2 and 3), then later extended her research (Rajpitak 1983) to other sites in Thailand, examining in total over 277 bronze containers and ornaments, including forty high-tin bronze vessels and ornaments, and nearly one hundred fragments from more than sixteen sites.[3] As only a few of these objects, other than those from BDTP and the other sites mentioned in table 2, come from excavated, known, and dated contexts, the value of these later results is rather limited.

The walls of most bronze vessels from BDTP vary between 0.2 and 0.3 mm with rims that are thicker than the walls at around 2 to 3 mm. The surfaces show the fine, polished gray-green patina that generally characterizes copper alloys within the 22–28 wt.% tin range, although when freshly cast the metal has a yellowish golden color. The vessels include straight-sided, flat bottom cylindrical containers and ones with curved sides and bases (figs. 5–8). Diameters range from 5 to 17 cm and heights from 6 to 18 cm. Many were broken or slightly distorted (fig. 9). Some bronzes that were originally classified as lids are in reality the bases of unusual bowls, which have a central knob on the interior of the base surrounded by circles of incised lines and dots that have obvious parallels in bronze, stone, and ceramic from several parts of India (Glover 1990, 156) (fig. 10) and northern Vietnam (Janse 1962).

Rajpitak relied on x-ray fluorescence spectrometry as her main analytical procedure for her work on the BDTP bronzes[4] with some inductively coupled plasma (ICP) spectroscopy added in 1983. Thirty-three samples from eighteen objects from the 1975–76 season were analyzed, along with a further two small bells and a situla from the 1980–81 excavations. The analyses showed two main compositional groups. The first comprises high-tin bronzes with between 20 and 30 wt.% tin and traces of lead, arsenic, iron, silver nickel, bismuth, and cobalt. This group includes all the bowls and bracelets examined, including a distinctive type

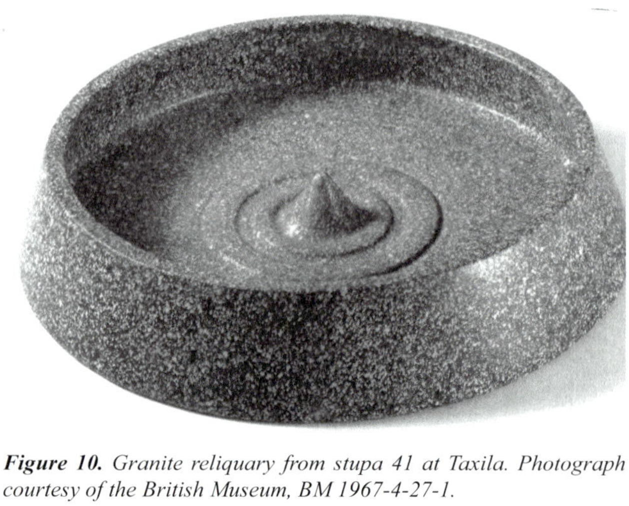

Figure 10. *Granite reliquary from stupa 41 at Taxila. Photograph courtesy of the British Museum, BM 1967-4-27-1.*

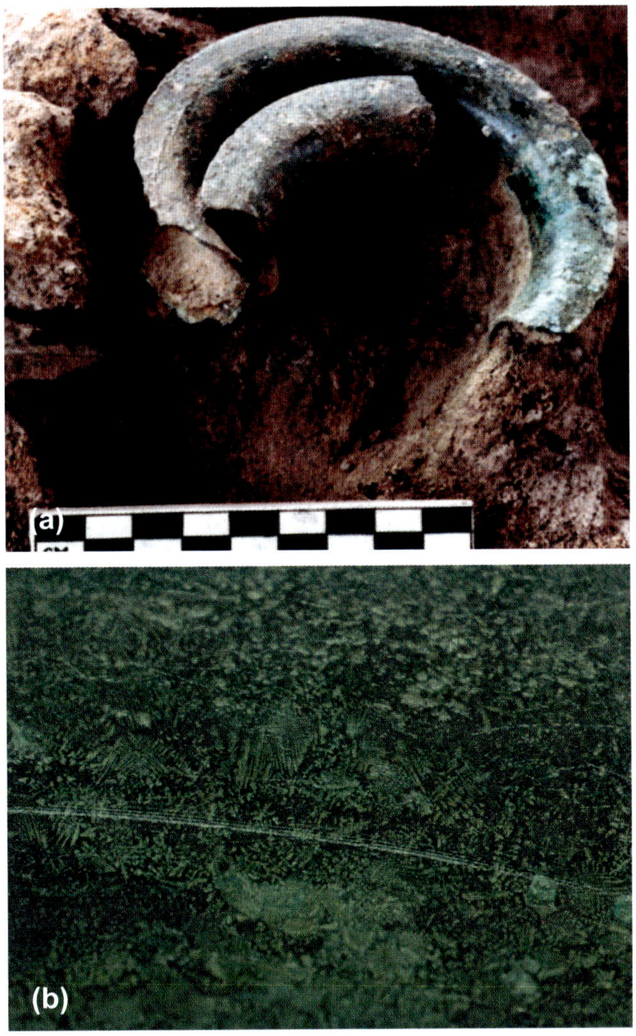

Figure 11. a. High-tin U-shaped anklet found during excavation at Ban Don Ta Phet, diam. 10.07 cm external and 7.05 cm internal. The object was as cast with no heat treatment and has a ***(b)*** *dendritic structure with alpha and delta eutectoids.*

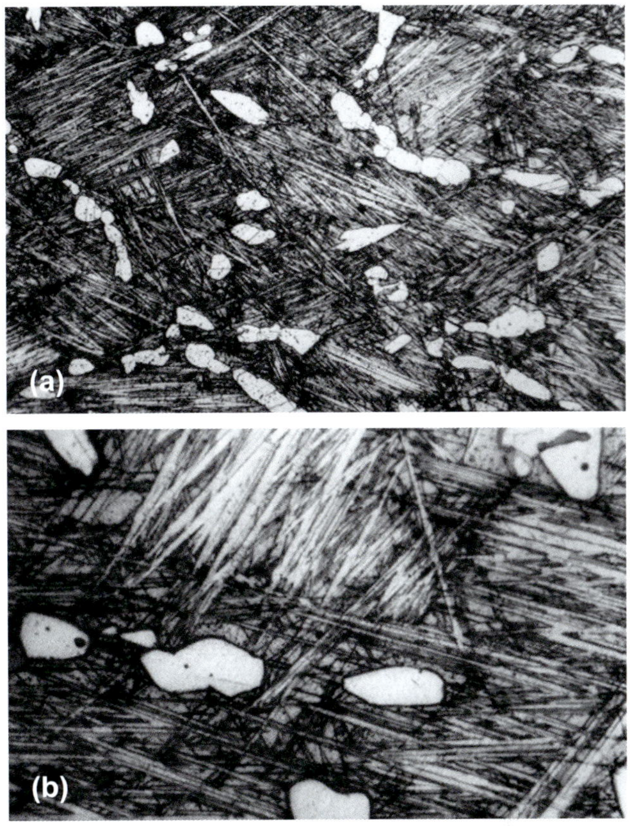

Figure 12. a and b. *Metallographic structure of a rim fragment of an undecorated high-tin cast bowl from BDTP, which has been heavily hot-worked and quenched from a temperature of above 650°C. Small twinned α-phase in β-phase needle structure. Composition determined by ICP: 73.63 wt.% Cu, 23.214 wt.% Sn, 0.011 wt.% As, 0.037 wt.% Sb, 0.039 wt.% Fe, 1.527 wt.% Ag, Total 98.469 wt.%. Vickers hardness typically > 220 HV. Images from Rajpitak (1979).*

of anklet with a broad U-shaped section (Glover 1990, fig. 8) (fig. 11), some small, thin, D-section bracelets that were usually found in multiple sets, one or two ladle pieces, and one finger ring. The second, much smaller group included two small bells and a situla. The latter objects are composed of ternary alloys with varying amounts of copper and tin and small quantities of lead. More recently, a few analyses have also been made of high-tin bronze finds and casting residue excavated at KSK.

Further excavations at BDTP in 1980–81 and 1984–85 were led by the Institute of Archaeology, UCL, in cooperation with the Thai FAD. Many of the artifacts, including some of the bronzes and glass and hard stone beads, suggested evidence for contacts with India (Glover 1990). Most bronze vessels had a high-tin composition, and microstructural examination and experiments on alloy manufacture have given us a clear picture of the techniques used to make them. The first stage of the process would have required the accurate formulation of the alloy, which may have been achieved by weighing the different constituents or by recognition of the correct color. The alloy would then have been cast into a blank, the shape of which might have roughly corresponded to the shape of the intended object, or been cast by the lost-wax process. Such an alloy is very fluid and is easy to cast, having a melting point of around 900°C. Although the casting of such thin-walled vessels would have required considerable skill, their manufacture would have been reasonably straightforward since the use of chaplets would not have been required because the mold would have been continuous around the rim as Reeves (1962) described in his work on casting in India (Rajpitak and Seeley 1979, 27).

Examination of polished and etched metallographic sections of fragments of some of the decorated and undecorated bowls confirmed that many of the objects from BDTP were of a very specific composition. Bronze alloys containing over 20 wt.% tin are composed of α-, β-, γ-, and δ-phases. The α-phase and δ-phases are stable at room temperatures, while the β-phase and γ-phase are stable at high temperature. High-tin bronzes are brittle at room temperature due to the appearance of the δ-phase, so cold-working of such bronzes is not possible and hot-working must be undertaken in the β- or γ- phase range of 586°C–798°C, during which the metal is extremely plastic. If the bronze is allowed to cool, the β-phase will be reabsorbed into the α- and γ-phases, resulting once again in a very brittle object. This brittleness can be alleviated, however, by quenching the bronze, giving rise to a martensitic structure of interlocking lamellae of phases. Although the object will still be somewhat brittle, the martensitic structure is stable and

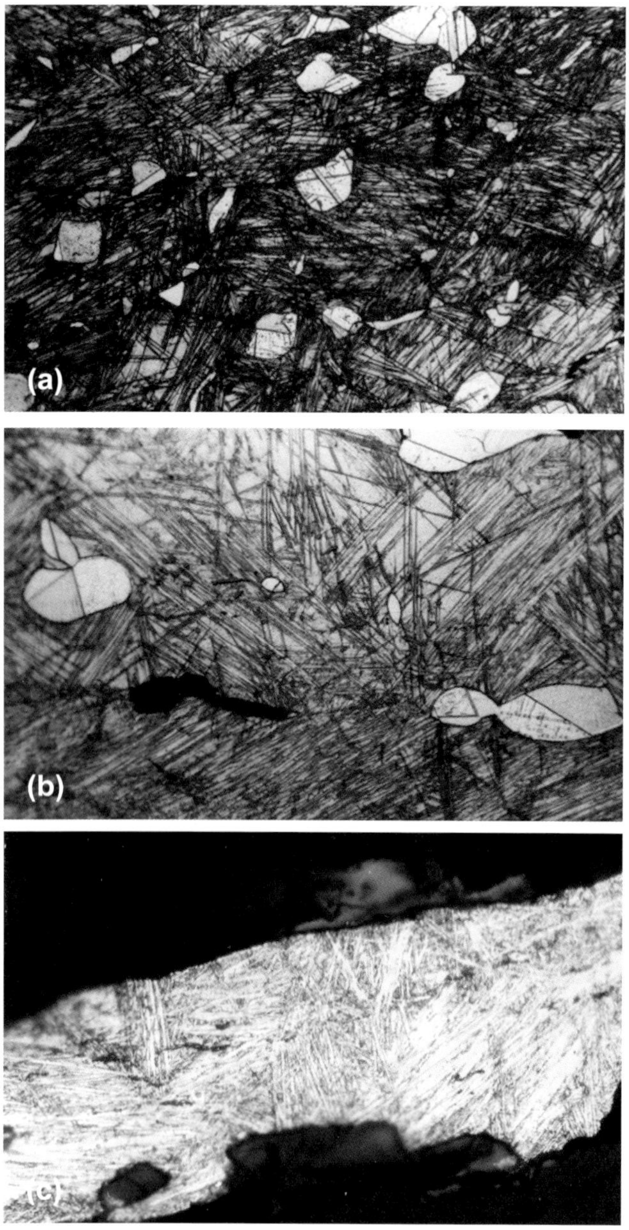

Figure 13. a and b. *Rim fragments of undecorated hemispherical bowls. Small amount of α-phase in the needle structure of β. Composition determined by XRF: 70.4 wt.% Cu, 29.5 wt.% Sn. Vickers hardness: 223.8 HV.* ***c.*** *Structure composed entirely of needles of β-phase martensite. Composition determined by ICP: 73.69–75.43 wt.% Cu, 20.21–21.42 wt.% Sn, 0.02–0.220 wt.% Pb, 0.010–1.58 wt.% S, 0.05–0.143 wt.% As, 0.08–0.24 wt.% Fe, 0.04–0.43 wt.% Sb, 0.01–0.08 wt.% Ni, 0.16–0.43 wt.% Ag, 0.09–0.44 wt.% Au, 0.08–0.19 wt.% Zn. Images from Rajpitak (1979).*

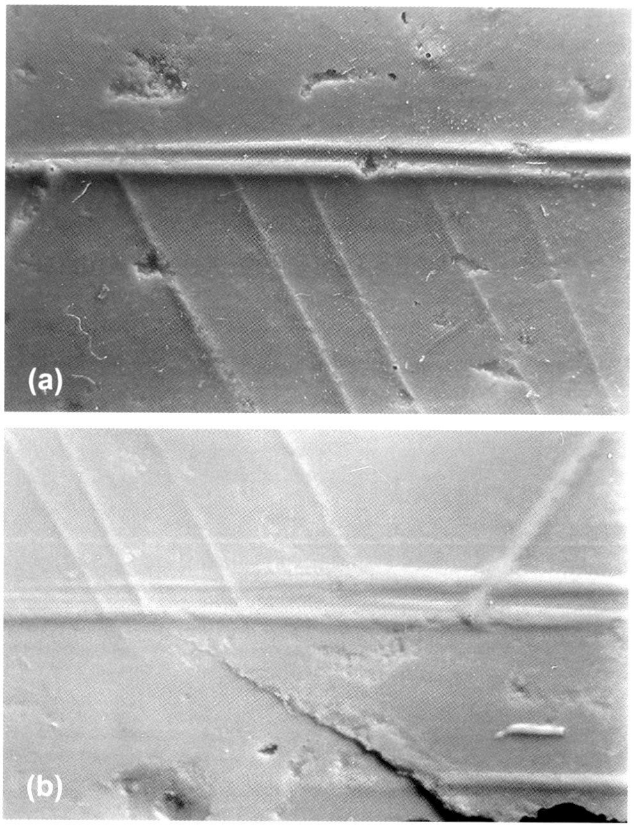

Figure 14. a and b. *Scanning electron microscopy (SEM) images of silicon rubber impressions of incised lines on bowls from Khao Jamook. These lines appear to have been cut with a cutting wheel.*

serviceable and can be mechanically finished using a lathe and decorated using a cutting wheel and a center-point drill, as seen on the BDTP bowls.

While metallography has been essential to our understanding of the methods of manufacture, it is worth mentioning that the structures observed were not necessarily those that would be predicted from the theoretical data. Electron microprobe analysis of individual phases has highlighted these discrepancies. Figure 12 shows microstructures of a rim fragment from an undecorated hemispherical bowl. The small twinned α-phase in a needle structure of the β-phase is entirely consistent with the elemental composition of 74 wt.% copper and 23 wt.% tin. Another rim fragment (fig. 13 a and b) shows a similar structure but has a composition in excess of 25 wt.% tin and should theoretically be composed entirely of needles of β-phase, whereas the fragment of the bowl in figure 13c is composed entirely of needles of martensite and has a composition of 74 wt.% copper and 21 wt.% tin. These examples clearly illustrate that the metal structures will vary not only with composition but also with the speed and the temperature from which they were quenched.

While a few of the artifacts were merely surface finished after casting, the great majority, and in particular the con-

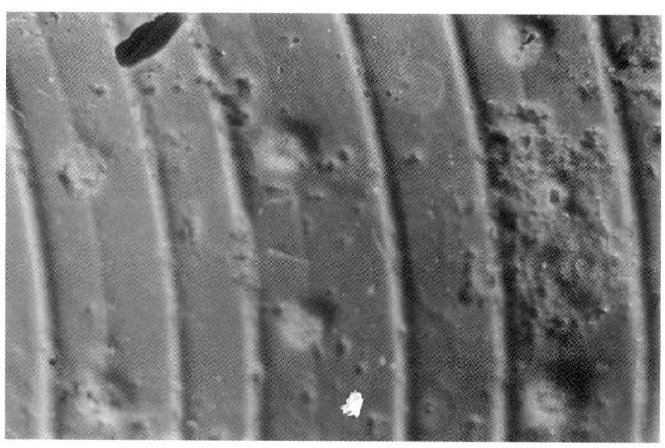

Figure 15. *SEM image of silicon rubber impressions of circular incised lines around the central knob of a bowl, which are sharp, deep, and even. It is clear that no metal has been thrown up on either side of the groove, which is characteristic of chasing. These lines appear to have been made with a cutting wheel.*

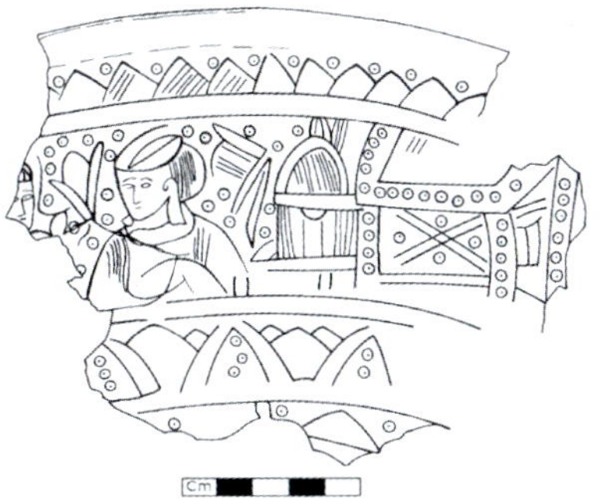

Figure 16. *Design on the largest surviving part of bowl A from the 1975 Thai excavations at Ban Don Ta Phet. Redrawn from a polyvinyl siloxene cast made by A. Bennett.*

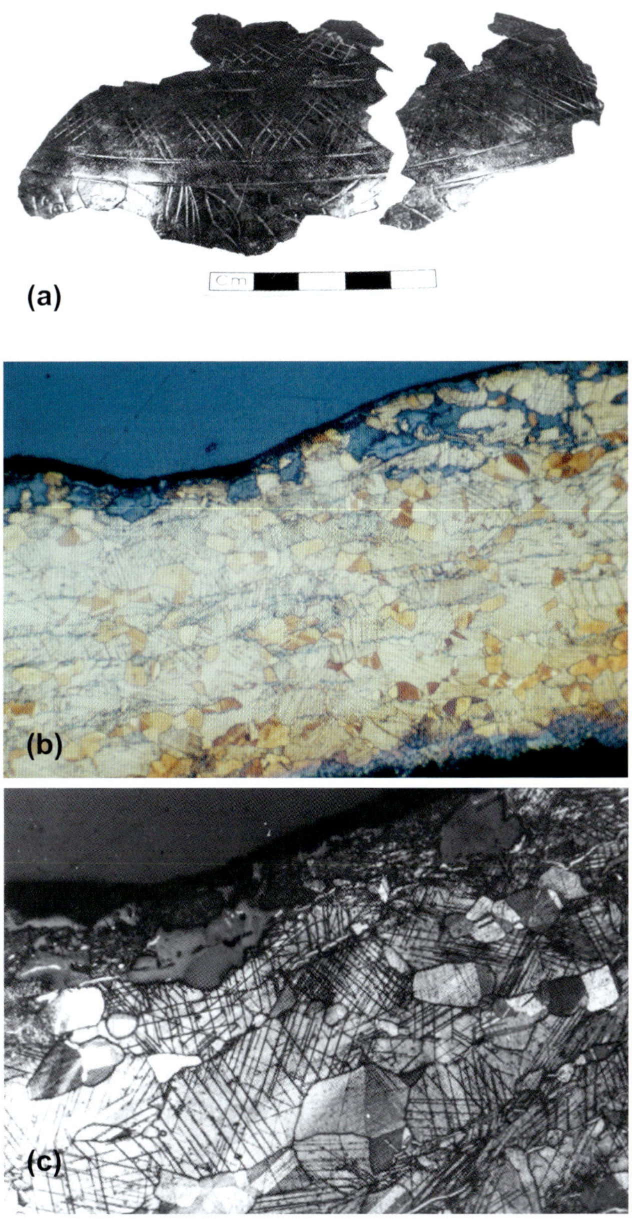

Figure 17. a. *Bowl fragment from Ban Don Ta Phet with incised lines. Metal thickness 0.7 mm.* ***b and c.*** *Twinned α-phase with areas of banded martensite that appears in the background of the large β-phase. Elongated cuprite inclusions can be seen running parallel to the surface, and heavy strain lines are visible. HV: 225.6 in the strained areas; Vickers hardness: 166.5 HV in the α-phase with background. Composition determined by XRF: 79.7 wt.% Cu, 20.3 wt.% Sn, trace elements Fe, Co, Ni, Bi, Ag. Image from Rajpitak (1979).*

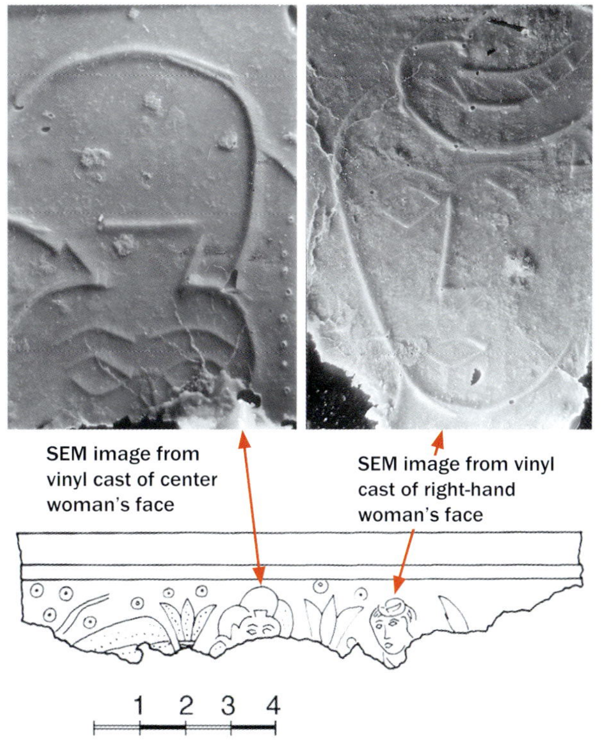

Figure 18. *Two SEM images from a polyvinyl siloxene cast of a fragment of a decorated high-tin bowl from Khao Jamook. The diagram at the bottom shows where the analyzed parts are located on the larger fragment. Short individual tool marks can be seen on the design of the hair. The tool was pointed at one end. Cast made by A. Bennett.*

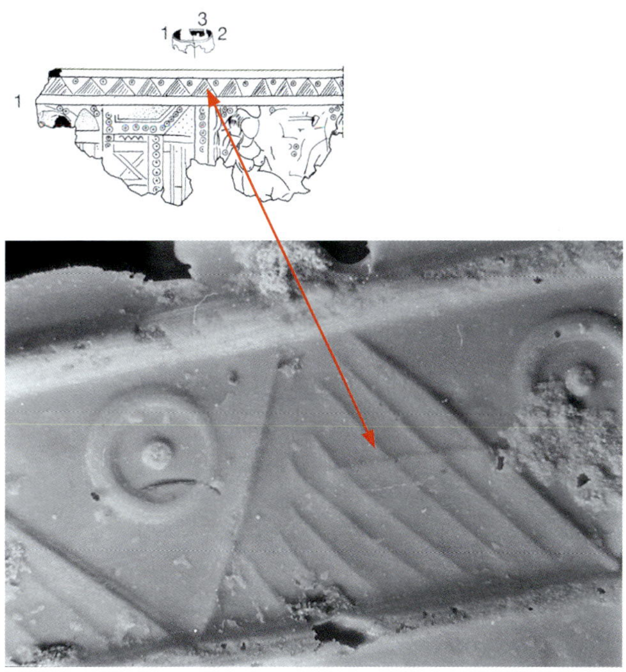

Figure 19. *SEM image from a polyvinyl siloxene cast of a fragment of a decorated high-tin bowl from Khao Jamook. The diagram at the top shows where the analyzed part is located on the larger fragment.*

tainers, were hot-worked in the 586°C – 798°C region from which they were then quenched. The combination of color and working properties of this alloy, the only one available at that time with a reasonable likeness to metallic gold prior to the introduction of brass, may have dictated the precise alloy composition to be used.

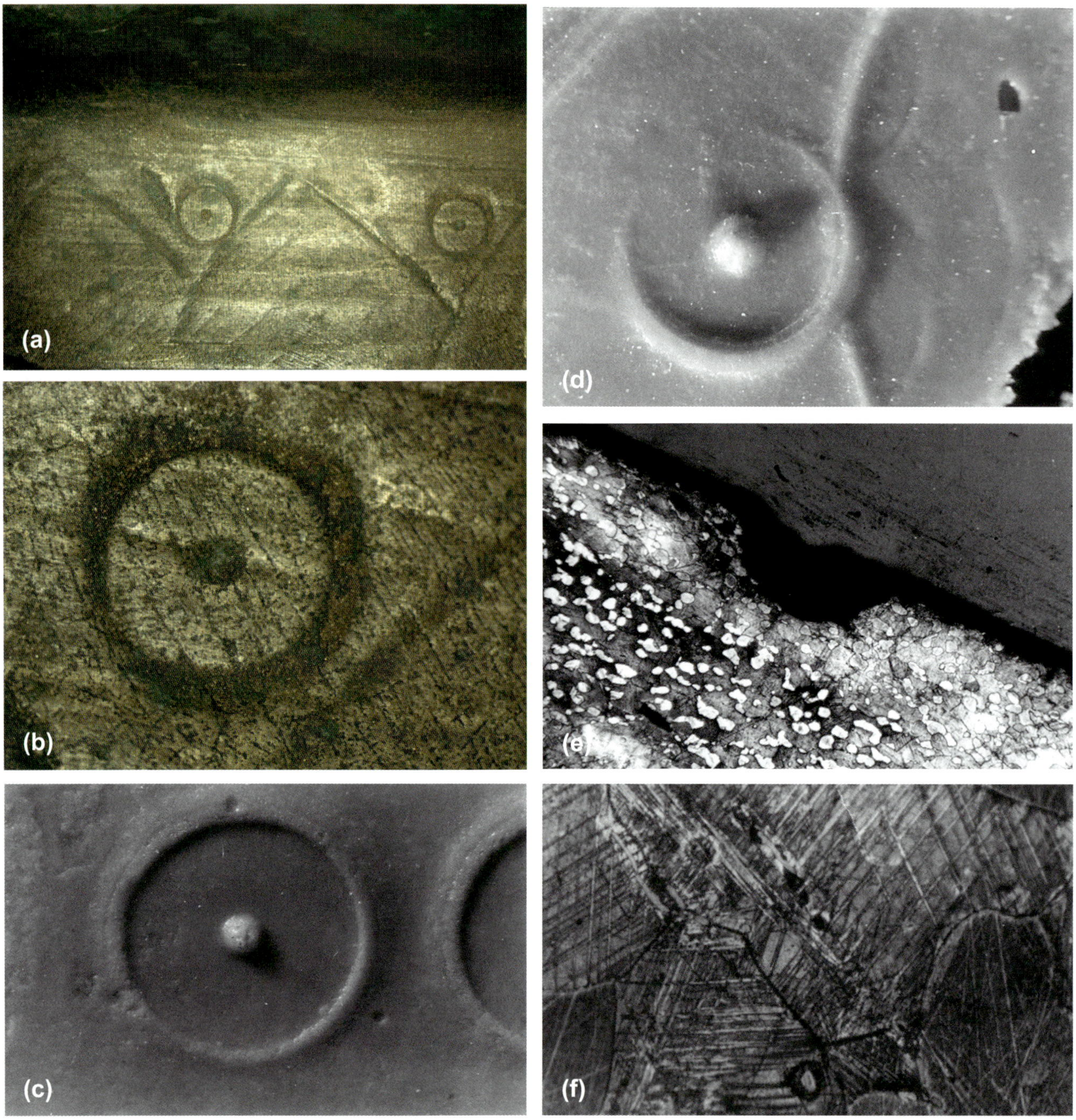

Figure 20. a and b. *Rim fragment of a decorated bowl from Ban Don Ta Phet showing the dot-and-circle design and extensive polishing.* ***c. and d.*** *Polyvinyl siloxene casts of the dot-and-circle decoration. The cut line around the inside of the circle in* ***(c)*** *is much sharper than the line outside. At the top left of the circle it is possible to see where the tool marks have overlapped. Such an overlap would be created if a center-point drill had been used.* ***e and f.*** *Microstructures cut through the dot-and-circle decoration, which show no evidence of any deformation.*

The extremely thin walls of the bowls and the surface striations suggest that there was a significant degree of finishing involved in their manufacture. This consisted of grinding and polishing using a simple lathe, with the addition of incised and drilled decoration where appropriate. Grinding and polishing were required not only to remove the oxide skin acquired during the heat treatments but also to remove any hammer marks resulting from hot-working and any other unevenness in the surface. Such irregularities can still be seen on present-day high-tin bronze vessels made in Kerala, southern India (Srinivasan 2010). The brittle nature of the alloy would have made the chasing of any decoration with a punch extremely difficult. The microstructure of samples has confirmed that no pressure was applied to the vessels. The lines cut just below the rim (fig. 14) as well as those used around the knobs on some bowls (see fig. 9b; fig. 15) and in the decorative scenes on the vessels (fig. 6) are sharp, deep, and even (fig. 17) with no signs of metal protrusion on either side of the groove—a characteristic of chasing. The short angular lines used to make the faces (fig. 18) and the hatched triangle panels were most probably made by handheld cutting tools (Bennett and Glover 1992, 200–204) (fig. 19). There was also no deformation of the metal in the areas of line and dot-and-circle decoration (figs. 20 e and f), indicating that metal must have been removed rather than compressed while making the designs. The even-

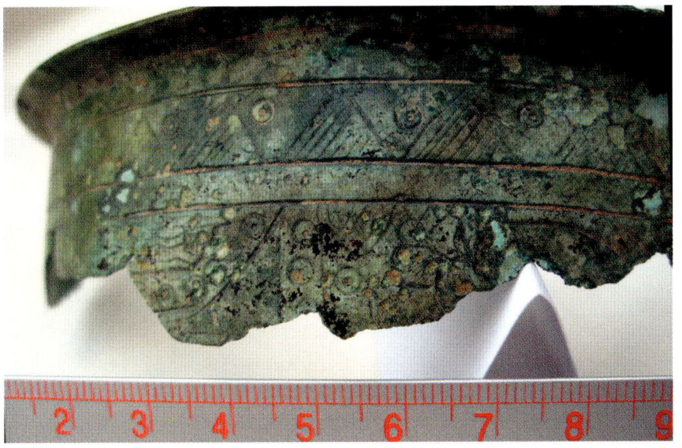

Figure 21. *A decorated bowl found by villagers in 2008 while digging for antiquities at Khao Sam Kaeo. It clearly belongs to the same group of high-tin bronzes as those from Ban Don Ta Phet and Khao Jamook. Photograph courtesy of B. Bellina and Dr. Bunchar Pongpanich.*

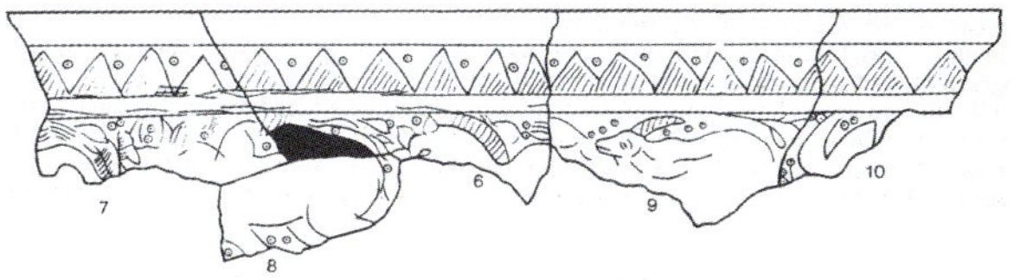

Figure 22. *Drawing of the rim of a decorated bowl from BDTP with animals in procession, including a horse, the earliest evidence of the horse in Thailand.*

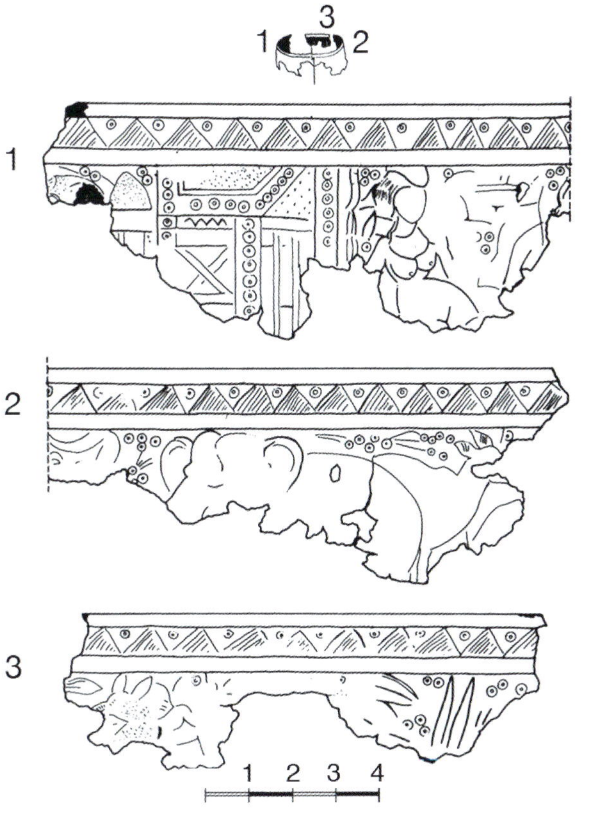

Figure 23. *Drawing of decorated high-tin bowl A from Khao Jamook with scene of women and animals in procession.*

ness and similar dimensions of the dot-and-circle decoration would have been impossible to achieve by hand engraving, and these motifs were probably made using a cutting wheel impregnated with abrasive powder, after which the bronze surfaces were extensively polished. Despite the hardness of this structure, which is typically in the region of 200–250 Hv, abrasives of garnet or corundum could have been used for this process.

Several of the bowls excavated at BDTP and others recovered from Khao Jamook (KJ) show scenes of people and animals, while the bases, where preserved, are decorated with open lotus flowers (Bennett and Glover 1992). More recently, a decorated bowl and the knob base of another one (fig. 21) were found by local villagers while working with a French–Thai team at Khao Sam Kaeo in peninsular Thailand. This team, despite looting, managed to recover an astonishing amount of information from this very extensive port-of-trade and manufacturing site, the earliest in Southeast Asia.

Decorated Bowls from Ban Don Ta Phet

The decorated high-tin bronze bowls from BDTP were first mentioned by Chin You-di (1976) and later studied by Bennett and Glover (1992, 191–206) together with those recovered from KJ. To the best of our knowledge, the sixteen surviving pieces of decorated bowls from BDTP belong to only two vessels. In an attempt to determine the types of tools used to produce the very sharp lines and dot-and-ring decorations, silicone rubber impressions of the surfaces were made and examined using reflex binocular and scanning electron microscopes.

BDTP Bowl A

BDTP bowl A (see fig. 16) is the largest and most splendid example of all the decorated bowls from Thailand, illustrating people and a house structure. The major surviving piece is, at the time of writing, displayed in the National Museum, Bangkok. About one quarter of the original vessel survives, and the golden, lustrous surface is remarkably uncorroded. Three small fragments can be joined to this vessel.

BDTP Bowl B

Bennett showed that the eight surviving pieces came from the same vessel, which had an estimated diameter of 22 cm with an everted rim. This vessel, BDTP bowl B (fig. 22), was decorated with cattle, a buffalo, a horse, and a zebu cow in procession from right to left. The image of the horse's head and part of its body is the earliest representation of a horse in Thailand. Other pieces depict cattle, buffalo, and the horns of another bovid. The wall thickness varies from 1.5 to 2 mm at the rim to 0.4 mm below the rim.

Decorated Bowls from Khao Jamook

The eleven fragments of bowls from KJ come from three different bowls; all appear to have been relatively shallow unrestricted forms with direct, slightly thickened rims. The outer surfaces of all the bowl fragments bear shallow incised scenes of animals, some seemingly grazing or in scenes with people and houses. One had an open lotus flower design on the base. A band of hatched triangles with a dot-and-circle motif below the rim was used elsewhere as an infill. The

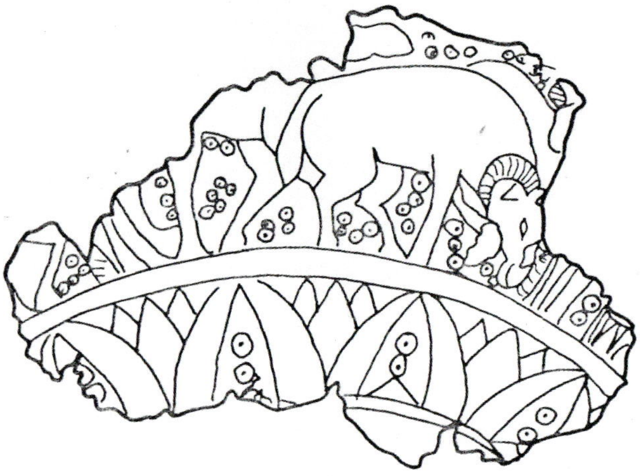

Figure 24. *Drawing of decorated high-tin bowl B from Khao Jamook showing a grazing animal above an open lotus flower at the base of the bowl.*

forms of the bowls, decorative techniques, and range of motifs are extraordinarily similar to the bowls found at BDTP and one from KSK. These may have been made in the same workshop and certainly belong to a single tradition. In addition, the KJ bowls are made of a similar high-tin bronze and were cast and polished down so that the body walls are about 0.5 mm thick. Semiquantitative analyses of the KJ bowls by Bennett indicate that they also had a very high tin content.

KJ Bowl A

KJ bowl A (fig. 23) was about 11 cm in diameter. The rim has a maximum thickness of 5.4 mm, and the surviving part of the body wall is only 0.5 mm thick. The decoration consists of a band of hatched triangles alternating with a dot-and-circle motif between double incised lines. Below this is a stylized building structure, perhaps a house, against which a naked woman with an elaborate headdress is leaning in languid fashion. To her right is the head and part of the upper right arm of another figure, possibly also a woman, and further to the right, separated by a few dots and circles is the substantial part of an elephant standing in front of a deer, which is facing in the opposite direction. This bowl reminds us very strongly of the largest fragment of bowl A from BDTP, on which a building structure is represented in a very similar manner with the upper part of two figures (at least one a woman) depicted in a way that suggests that they are looking over a balustrade or out of a window.

KJ Bowl B

KJ bowl B (see fig. 18) was about 14.2 cm in diameter and 1.15 mm thick at the lip, with the main body 0.5 mm thick. Less of the body of this bowl survives, and all pieces are severely corroded. This bowl lacks the upper band of hatched triangles and dot-and-circle design. The two women depicted on the bowl wear quite elaborate headdresses and/or hats. The size of the drill used to produce the dots and circles on KJ bowl B is identical to that used on bowl A, and the same tool may have been used to decorate them.

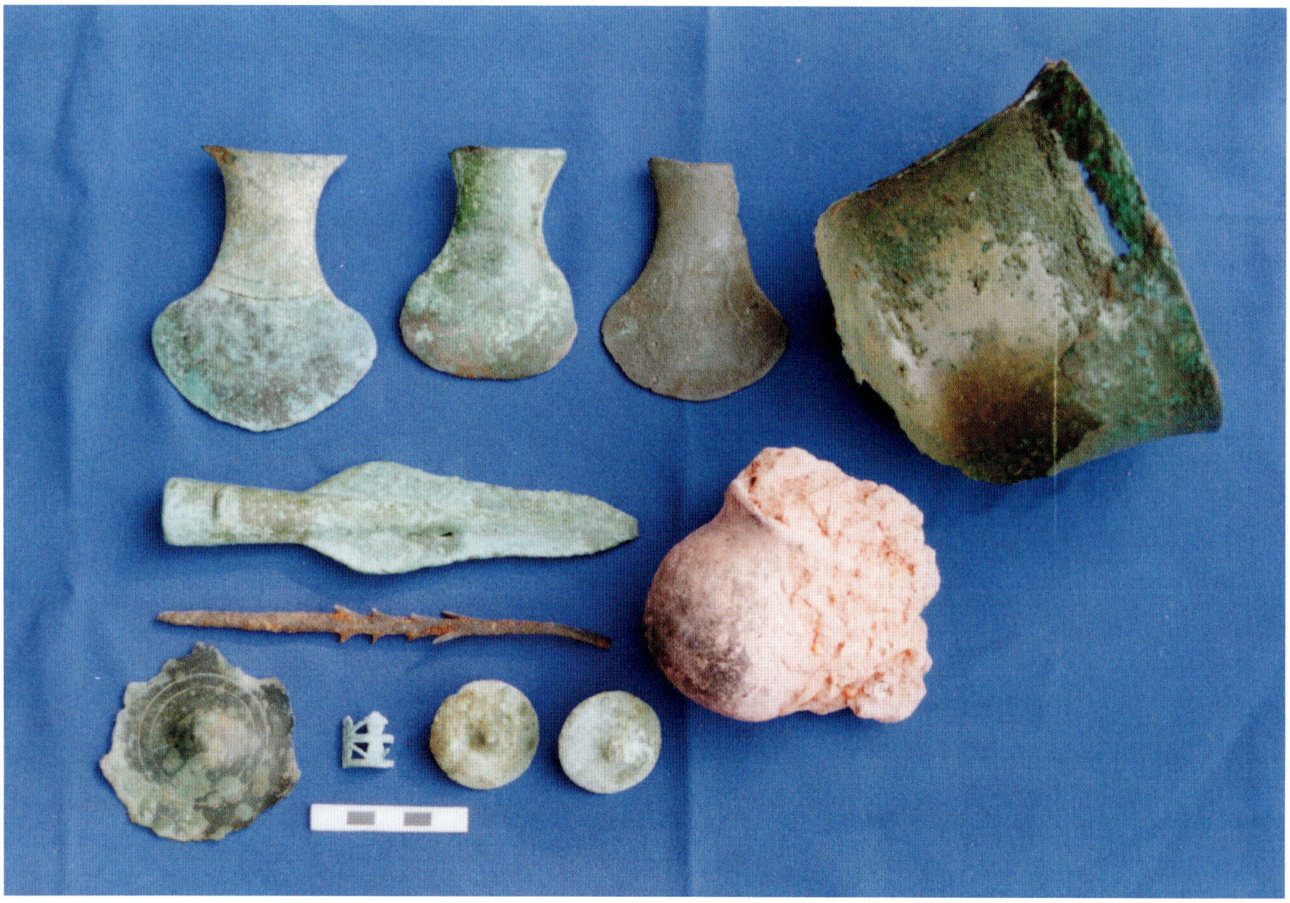

Figure 25. *Finds discovered during cleaning of a roadside ditch near Pak Beung, Chombung district, Ratchaburi province, Thailand. Lower left: Base fragment of a high-tin knob base bowl. Right: Small high-tin cup.*

KJ Bowl C

KJ bowl C (fig. 24) is represented by only a single substantial fragment from near the base, which has a stylized open lotus flower below grazing animals. As unlikely as it seems, one of these animals looks to be a grazing sheep or ram. Parts of other animals, perhaps cattle or horses, can also be seen. Bowl C is 1.8 mm thick above the sheep's back and thinner above the lotus flower. The tool used to make the dot-and-circle motif on bowl C was different from the one used on KJ bowls A and B.

Bronze Vessels from elsewhere in Thailand and Southeast Asia

Within Southeast Asia, high-tin bronze vessels with this composition and similar forms to those excavated at BDTP have been recovered from the tin gravels of western Malaysia (Batchelor 1978) and in Thailand at Ongbah cave on the Kwae Yai River (Sørensen 1973, fig. 22), at Pak Beung, Ratchaburi province (Bennett and Glover 1992, 188) (fig. 25), at Ban Pong Manao in Lopburi province (Natapintu pers. comm.), and at KSK in Chumphon province (Murillo-Barroso et al. 2010) (see figs. 21, 26). Similar vessels have also been found at Prohear, Prey Veng province in eastern Cambodia (Reinecke et al. 2009, 142–43); at Dong Son inhumations and Han-period (206 BCE–220 CE) brick tombs in northern Vietnam (Janse 1962); and from Tien Lanh, a late Sa Huynh burial in central Vietnam (Bui Chi Hoang 2008). Vessels decorated with very similar complex figurative designs have so far only been found at BDTP, KJ, and KSK.

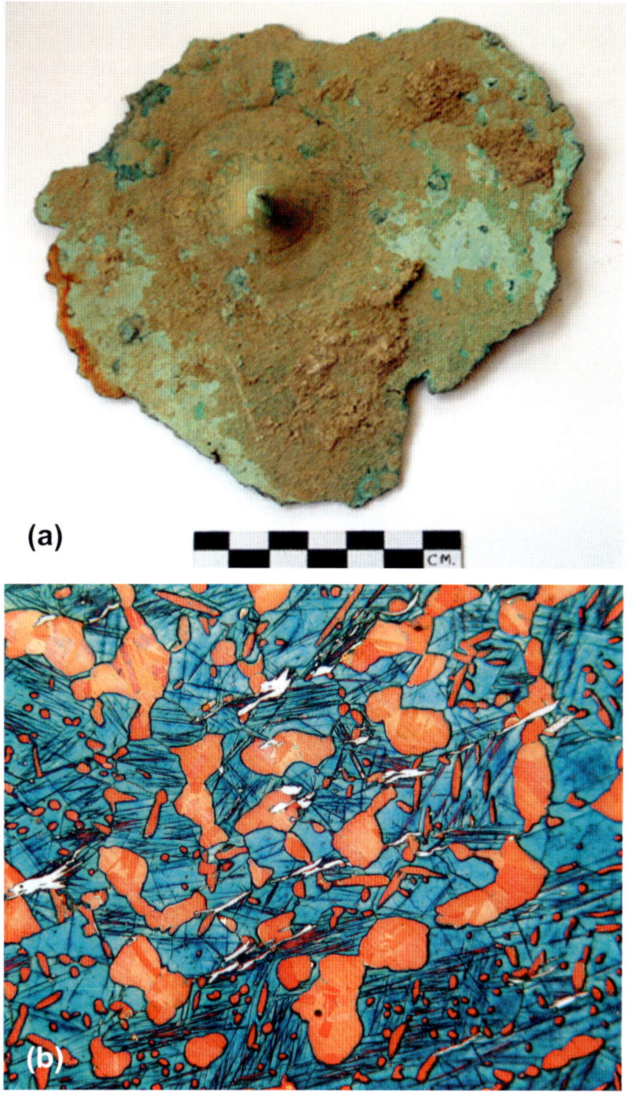

Figure 26. a. *Base fragment of a knob-base bowl (MA 73) found by villagers at KSK. Photograph courtesy of B. Bellina. Composition: 74.5–76.4 wt.% Cu, 22.3–24 wt.% Sn, less than 0.02 wt.% Pb (Murillo-Barroso et al. 2010, table 2).* ***b.*** *Metallographic structure of a knob-base bowl (MA 73) from KSK, etched, 200 Â, 0.5 mm width, plane-polarized light. Note the linear disposition of chalcosine and the equi-axed grains as evidence of hot-working, and the martensitic structure of the d compound characteristic of quenching. The orange phase is alpha bronze and the blue beta bronze, which may show signs of twinning. Courtesy of Murillo-Barroso et al. (2010).*

Vessels from South Asia

Rajpitak and Seeley (1979, 29–30) have already pointed to a small number of copper-alloy containers with similar compositions and forms found in southern India—from the Nilgiri Hills (fig. 27), from Adichanallur, and from Coimbatore in the peninsula—and in Taxila in Pakistan (table 4). Few of these are accurately dated, although the ones from the Bhir Mound at Taxila are roughly contemporary with those from BDTP and KSK. One bowl from Coimbatore was analyzed by Paramasivan (1941): it contained 23.6 wt.% Sn and was said to be typical of other vessels from the Nilgiri Hills (Breeks 1873) and Adichanallur (Rea 1915).[5] Craddock and Hook (2007) have published several analyses of other bowls from the Nilgiri Hills.

Since the early 1990s, however, metallurgical and ethnoarchaeological studies made by Indian scholars (Pillai et al. 1994; Srinivasan 2010) on South Asian bronzes have documented the existence of high-tin bronzes from a range of contexts from prehistoric to the present day. These include objects excavated from Indian Megalithic and Iron Age contexts, some going back into the first millennium BCE. Two high-tin knobbed bowls from Wari-Bateshar in Bangladesh have been published by Jahan (2006, 82). More recently,

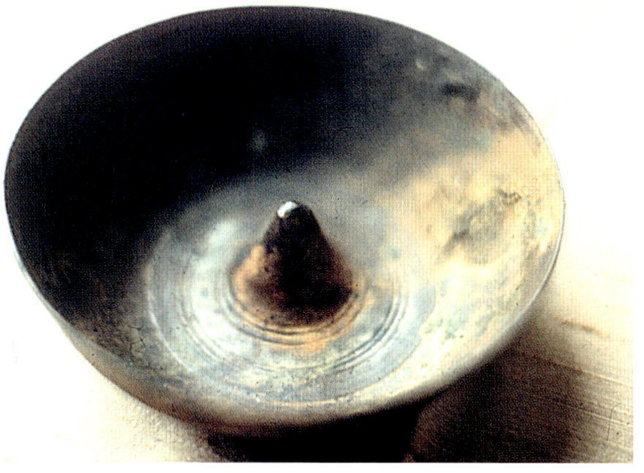

Figure 27. *Knob-base bowl from Taxilla, in the British Museum.*

Table 4. *Analyses of high-tin bronze vessels from India.*

Accession No.	Description (Part)	Provenance	Date	Cu	Sn	Pb	Zn	Ag	Au	Fe	Ni	As	Sb	Bi	Co	Cd	Mn
British Museum																	
1882-10-9,15	ring stand	Nilgiri Hills	2nd century BCE–5th century CE	69.9	10.2	8.72	10.5	0.054	0.002	0.32	0.078	0.32	0.17	0.02	0.005	0.0027	<0.001
1879-12-1,8	bowl	Nilgiri Hills	2nd century BCE–5th century CE	78.4	21.6	0.09	0.06	0.018	<0.005	0.56	0.092	0.43	0.03	0.01	0.020	<0.001	<0.001
1882-10-9,13	bowl	Nilgiri Hills	2nd century BCE–5th century CE	76.1	22.3	0.03	0.01	0.025	<0.005	0.40	0.548	0.13	0.03	0.01	0.018	<0.001	0.001
1882-10-9,2	bowl	Nilgiri Hills	2nd century BCE–5th century CE	78.0	22.7	0.09	0.02	0.010	<0.005	0.06	0.134	0.26	0.02	0.02	0.044	<0.001	0.001
1882-10-9,16	bowl	Nilgiri Hills	2nd century BCE–5th century CE	74.8	23.2	0.02	0.01	0.007	<0.005	1.92	0.084	0.02	0.02	0.02	0.168	<0.001	0.001
1904-7-21,1	bowl (handle)	Assam, Tezpur(?)	2nd century BCE–5th century CE	78.9	19.5	0.04	0.02	0.025	<0.005	0.08	0.431	0.99	0.02	0.02	0.104	<0.001	<0.001
1880-641	bowl (foot ring)	Nilgiri Hills	2nd century BCE–5th century CE	78.7	22.2	0.03	0.03	0.021	0.002	0.14	0.063	0.17	0.02	0.01	0.051	<0.001	0.001
	bowl* +	Kerala, Payangadi	18th century–20th century	80.5	22.9	0.30	0.37	0.080	na	0.04	0.060	0.04	0.18	na	0.001	na	na
	bowl +	Kerala, Irinjilakuda	18th century–20th century	81.3	20.3	nd	0.06	0.020	na	0.13	na	na	0.12	na	na	na	na
	bowl +	Tamil Nadu, Madurai	18th century–20th century	82.0	21.8	nd	0.14	na	na	0.17	na	na	0.16	na	na	na	na

All analyses were carried out by Michael Hughes using atomic absorption spectroscopy except the bowl marked with an asterisk, which was analyzed using inductively coupled plasma–atomic emission spectroscopy. All numbers given as weight percentages. Courtesy of Craddock and Hook 2007, table 3. nd = not detected; na = not analyzed. + See also Srinivasan and Glover (1995).

Datta et al. (2007) have published details of a furnace for the manufacture of high-tin bronze at Tilpi in western Bengal. These, together with studies revealing continuing traditions of making high-tin bronzes, especially the production of wrought and quenched high-tin bronze vessels and cymbals in Kerala, suggest the regular use of unleaded bronze from the very early stages of Indian prehistory (Srinivasan 2010).

Although there are similarities between the recent working processes and that which Rajpitak and Seeley (1979) inferred from their examination of the high-tin bronze containers from BDTP, there are also significant differences. The ingots for bowl-making in present-day Kerala are cast in open molds, not made by lost-wax casting in investment molds, and the Kerala craftsmen do not so extensively grind down and polish the inner and outer surfaces to leave them smooth. The evidence of hot-working, however, is clear to see in the finished product. Additionally, the complex decoration cut into the outer surfaces with a grinding wheel in Thailand has not been reported in India, except perhaps on the unique vase from Kundla in the Kulu valley, North India, which is reported to be of a ternary, rather than a binary, high-tin alloy.[6]

At present it still cannot be demonstrated for certain whether this distinctive alloy was developed independently in Thailand, where tin and copper sources are abundant, or was introduced from South Asia in the mid-first millennium BCE, but far more high-tin bronzes have been found and identified in Thailand and elsewhere in Southeast Asia than in South Asia. It clearly is time for us to reconsider the archaeological evidence in favor of a two-way exchange between South and Southeast Asia at the dawn of the historic period.

Acknowledgments

We gratefully acknowledge the support of the Freer and Sackler Galleries of the Smithsonian Institution and especially Paul Jett, Louise Cort, and Stacy Bowe, who greatly facilitated our attendance at the Fifth Forbes Symposium. We wish to dedicate this article to the late Professor Nigel Seeley who initiated research on the bronze from Ban Don Ta Phet and taught us much of what we know about the material.

Notes

1. From the account by Nearchus (fourth centry BCE), who was with Alexander's Macedonian army in the Indus valley, preserved in Strabo's *Geography* (15.1.67).
2. These were found during tin sluicing in hilly country near the Thai–Burmese border at 99° 11' E by 13° 41' N and were first published by Lualamai (1986). His figures were redrawn by Bennett, who pieced the fragments together and also noticed that some details had disappeared as a result of mechanical damage and corrosion since the first publication.
3. Some artifacts with sites listed by Rajpitak (1983) as "unknown" come from poorly documented museum collections. The artifacts belong to the Iron Age tradition known from Ban Chiang and come from northeast Thailand.
4. This is said to have an accuracy of between ± 2% and 5% and all elements of atomic number above manganese could be detected. Variability of ± 1% comes from the lack of homogeneity in the samples.
5. Leshnik (1974, 156) gives the composition of some South Indian bronze bowls as: Adichanallur—75 wt.% Cu, 23 wt.% Sn, 0.2 wt.% Pb, 0.4 wt.%. Fe; Maula Ali—79 wt.% Cu, 21 wt.% Sn. Rajpitak and Seeley (1979, 29) refer to the analysis of a bowl from Coimbatore, which had a tin content of 23.58 wt.%.
6. The Kundla vase, found in the nineteenth century in a ruined Buddhist temple and now in the British Museum, has not been studied exhaustively, but a surface analysis indicated a ternary composition (M. Hughes, letter of 3 February 1986, British Museum records, courtesy of M. Willis).

References

Batchelor, B. C. 1978. Post "Hoabinhian" coastal settlement indicated by finds in stanniferous Langat River alluvium near Dengkil, Selangor, peninsular Malaya. *Federated Malay States Museum Journal* 22:1–55.

Bennett, A., and I. C. Glover. 1992. Decorated high-tin bronzes from Thailand's prehistory. In *Southeast Asian Archaeology 1990: Proceedings of the Third Conference of the European Association of Southeast Asian Archaeologists,* ed. I. Glover, 187–208. Hull, UK: Centre for Southeast Asian Studies, University of Hull.

Breeks, J. W., ed. 1873. *An account of the primitive tribes and monuments of the Nilgiris.* London: India Museum.

Bui Chi Hoang. 2008. Di tich khao co hoc Tien Lanh [The Tien Lanh site]. In *Mot so van de khao co hoc o mien nam Vietnam* [Some archaeological achievements in southern Vietnam], ed. Thi Thu Nguyen, 245–77. Ho Chi Minh City: Social Sciences Publishing House.

Craddock, P., and D. Hook. 2007. The bronzes of the south of India: A continuing tradition? In *Scientific research on the sculptural arts of Asia: Proceedings of the Third Forbes Symposium at the Freer Gallery of Art,* ed. J. G. Douglas, P. Jett, and J. Winter, 75–89. London: Archetype Publications.

Datta, P. K., P. K. Chattopadhyay, and A. Ray. 2007. New evidence for high-tin bronze in ancient Bengal. *Society for Archaeological Sciences Bulletin* 30(2):13–16.

Dhepsuriyanon, Wasan 2001. *The archaeological site of Ban Don Ta Phet: Excavation results.* Bangkok: Fine Arts Department (in Thai, Saitsaya Saenpheung trans.).

Glover, I. C. 1990. Ban Don Ta Phet: The 1984–85 excavation. In *Southeast Asian archaeology 1986,* ed. I. C. Glover and E. Glover, 139–84. BAR International Series 561. Oxford: BAR.

Glover, I. C., B. A. P. Alvey, P. Charoenwongsa, and M. Kamnounket. 1983. The cemetery of Ban Don Ta Phet, Thailand: Results from the 1980–81 excavation season. In *South Asian archaeology 1981,* ed. B. Allchin, 319–30. Cambridge: Cambridge University Press.

Jahan, S. H. 2006. *Excavating waves and winds of (ex)change: A study of maritime trade in early Bengal.* BAR International Series 1533. Oxford: BAR.

Janse, O. R. T. 1962. Quelques réflections à propos d'un bol de type mégareén trouvé au Vietnam. *Artibus Asiae* XXV(4):280–92.

Leshnik, L.S. 1974. *South Indian "Megalithic" burials: The Pandukal complex*. Wiesbaden: Franz Steiner.

Lualamai, K. 1986. 2000 years old bronze bowls from Raj Buri. *Khan Chang*. Bangkok: Silpakorn University.

Murillo-Barroso, M., T. O. Pryce, B. Bellina, and M. Martinon-Torrés. 2010. Khao Sam Kaeo—An archaeometallurgical crossroads for trans-Asiatic technological traditions. *Journal of Archaeological Science* 37(7):1761–72.

Paramasivan, S. 1941. Investigations into ancient Indian metallurgy I: A prehistoric bronze bowl. *Proceedings of the Indian Academy of Science* 13:87–90.

Pillai, S. G. K., R. M. Pillai, A. D. Damodaran, and T. R. Ramachandran. 1994. The thermomechanical processing of high-tin bronzes: An old practice in a South Indian village. *Journal of Metals* 46(3):59–62.

Pryce, T. O., B. Bellina, and A. T. N. Bennett. 2006. The development of metal technologies in the upper Thai-Malay peninsula: Initial interpretation of the archaeometallurgical evidence from Khao Sam Kaeo. *Bulletin de l'Ecole française d'Extrême-Orient* 93:295–315.

Rajpitak, W. 1979. The technology of copper alloy artefacts from Ban Don Ta Phet, Thailand. BSc thesis, Institute of Archaeology, London.

———. 1983. The development of copper-alloy metallurgy in Thailand in the pre-Buddhist period with special reference to high-tin bronze. PhD diss., Institute of Archaeology, London.

Rajpitak, W., and N. Seeley 1979. The bronze bowls from Ban Don Ta Phet: An enigma of prehistoric metallurgy. *World Archaeology* 11(1):26–31.

Rea, A. 1915. *Catalogue of prehistoric antiquities from Adichanallur and Perumbair*. Madras Government Museum, India.

Reeves, R. 1962. *Cire perdue casting in India*. New Delhi: Crafts Museum.

Reinecke, A., V. Laychour, and S. Sonetra. 2009. *The first golden age of Cambodia: Excavation at Prohear*. Bonn: T. Müntzer.

Sørensen, P. 1973. Prehistoric iron implements from Thailand. *Asian Perspectives* 16(2): 134–73.

Srinivasan, S. 2010. Megalithic high-tin bronzes and peninsular India's "living prehistory." In *50 years of archaeology in Southeast Asia: Essays in honor of Ian Glover*, ed. B. Bellina, E. A. Bacus, and J. Wisseman Christie, 261–71. Bangkok: River Books.

Srinivasan, S. and I. Glover. 1995. Wrought and quenched, and cast high-tin bronzes in Kerala State, India. *Historical Metallurgy* 29(2): 69–88.

Thepsuriyanon, W. 2001. *The archaeological site of Ban Don Ta Phet: Excavation results*. Bangkok: Fine Arts Department.

You–di, C. 1976. *Ban Don Ta Phet: Preliminary excavation report*. Bangkok: Fine Arts Department.

A Flux that Binds? The Southeast Asian Lead Isotope Project

Thomas Oliver Pryce

ABSTRACT The continued lack of a widely agreed upon Southeast Asian prehistoric ceramic sequence means that archaeologists working in the region sometimes struggle to reliably identify and interpret medium- to long-range social interactions, including neighboring populations in East and South Asia. The Southeast Asian Lead Isotope Project (SEALIP) has been designed to contribute to our comprehension of regional connectedness for the late prehistoric Metal Age using copper-base and lead metal exchange networks as a social interaction proxy—a proxy that will of course omit social groups that did not for whatever reason develop or adopt metallurgical production or consumption behaviors. This paper outlines why archaeometallurgy is more than just the "history of metallurgy" and how such lead isotope research has the potential to furnish critical data for all scholars of Southeast Asian late prehistory. The current SEALIP database permits the two long-known prehistoric copper production centers, the Khao Wong Prachan valley in central Thailand and Phu Lon in northeast Thailand, to be reliably distinguished isotopically, fulfilling the Provenance Hypothesis and justifying the ongoing regional metal exchange program. The data also indicate that these two major copper producers were simultaneously importing copper-base metal, supporting previous suggestions that early Southeast Asian metal exchange was motivated more by gifting than commerce.

Archaeometallurgical Underperformance in the Far East

Lead is a metal that has been significantly overlooked by archaeologists working in Southeast Asia. Sanskrit names like *Suvarnabhumi* (Land of Gold) and *Suvarnadvipa* (Islands of Gold) (Wheatley 1961) have long assured an interest in the region's precious metal output (e.g., Miksic 1990; Richter 2000), just as the famous Southeast Asian Tin Belt (e.g., Schwartz et al. 1995) has garnered extensive speculation over this metal's long-term role in trans-Asiatic exchange (e.g., Bellwood 2007; Bronson 1977, 1992; Jacq-Hergoualc'h 2002)—a topic slowly approaching the realms of empiricism (e.g., Bennett and Glover 1992; Pryce, Murillo-Barroso, et al., in press). On copper-base metallurgy, debate has been raging on its "origins" in Southeast Asia since the late 1960s (e.g., Bayard 1972, 1979, 1981; Gorman and Charoenwongsa 1976; Higham 1975, 1996–97; Higham et al. 2011; Muhly 1981; Solheim 1968; Spriggs 1996–97; White 1986, 1988, 2008), though the current balance of evidence has resulted in a late second millennium BCE consensus for the inception of the Bronze Age, alongside the established mid-first millennium BCE accord for the Iron Age (e.g. Higham et al., in press; Higham and Higham 2009; Higham et al. 2011; Pigott and Ciarla 2007; cf. White and Hamilton 2009). Arguably more fruitful in terms of density of reliable data, the Thailand Archaeometallurgy Project's holistic approach to primary and secondary copper production in Thailand has been contributing invaluable information on prehistoric metallurgical behavior in the broadest sense since 1984 (Pigott and Natapintu 1988; e.g., Pigott et al. 1997; see also Smith 1973; Wheeler and Maddin 1976, for earlier work) and has recently been complemented by comparable investigations at Xepon in central Laos (Pryce et al. 2011; Sayavongkhamdy et al. 2009). Southeast Asian ferrous archaeometallurgy, despite an early start and enormous potential (e.g., Bennett 1982; Dizon 1988; Pigott and Marder 1984; Suchitta 1983), is in need of substantial development, but continued assaults (Pryce and Natapintu 2009) on the notion of ubiquitous lateritic iron sources will hopefully pay dividends in the form of regional studies on material and technological exchange (e.g., Biggs 2009).

In comparison, lead has scarcely been mentioned in the Southeast Asian literature, with only a few possible premodern mining sites noted during a 1984 survey by the Thailand Archaeometallurgy Project in Loei province, northern Thailand (Pigott 1986; see Sitthithaworn 1990). Lead is a major component of regional copper-base alloys, certainly from the Iron Age onward (e.g., Higham 1989, 240), but the only comments we generally hear are broad discussions of lead ore availability (e.g. Higham 1989, 136, 185) and the use of lead as a "flux" to "improve a copper-alloy's casting characteristics by reducing the melting point and viscosity, and increasing the workable casting range and rendering greater detail in the mould" (Murillo-Barroso et al. 2010, 1767). Statements like this are entirely correct but rather underplay the importance of lead's role in Southeast Asian metallurgical traditions. Although lead is relatively easy to smelt, its ores, like those of copper, tin, and high-quality iron, have a nonuniform distribution, and consumption of leaded copper or leaded bronze in one part of the region very probably involved the exchange of substantial quantities of lead from elsewhere. For example, the Iron Age leaded bronze Dong Son drums may each have a mass in the dozens of kilograms, which, with lead contents in excess of 10 wt.% (Murillo-Barroso et al. 2010, table 2), means that each artifact required several kilograms of lead metal. If we assume that alloys this rich in lead are due to the intentional addition of lead metal as opposed to the smelting of naturally lead-bearing copper ores, then Southeast Asia's leaded Dong Son bronzes alone require a technological and economic explanation for that high quantity of lead (see Pryce 2010). Where was it being made? When? By whom? And in what social context of production?

Lead's Potential Contribution to Southeast Asian Archaeology

Of the broad research questions presented above, the latter three require the surveying of Southeast Asia's known lead deposits (see, e.g., Pryce and Abrams 2010; Pryce, Brauns, et al. 2011, for copper), followed up with excavation of those mineralizations associated with premodern industrial activity (see, e.g., Pigott and Weisgerber 1998; Pigott et al. 1997, for copper), and technological analysis of the resulting assemblages (see, e.g., Pryce et al. 2010, for copper). This approach is promising but will take time. For the first question, however, lead offers some particular analytical affordances that have been dramatically underexploited in the Southeast Asian arena and that, as argued here, could result in rapid advances in our understanding of the region's early metal exchange networks.

It has been some decades since large international elemental analysis programs were instituted for European metal exchange research (e.g., Junghans et al. 1960, 1968, 1974), but this approach is now widely recognized to be problematic due to the differential partitioning of trace elements during metal production processes and the data conflation that all additive technologies suffer during alloying and recycling (e.g., Chippindale 2000; Pollard 2009; Pollard and Heron 2008)—though recent research suggests that these enormous databases can be exploited for information about past production processes, if not provenance as was originally intended (Bray 2009).

With extreme care, and taking into account elemental datasets, the provenance baton can be seen as having passed to lead isotope analysis (LIA). Lead, within widely accepted caveats, can be geochemically linked to its source mineralization via the analysis of its stable isotopes. This technique does not apply only to pure lead artifacts or to heavily leaded copper-base alloys; instead the detection limit of present-day instrumentation permits the determination of lead isotope ratios at trace element concentrations (parts per million or even parts per billion; see Niederschlag et al. 2003). As most copper ores contain trace amounts of lead, we are in a position to assess the exchange networks of both lead-base and copper-base alloys, although this dual usage does present some difficulties.

As per many parts of the world, Southeast Asia stands to benefit considerably from an anthropologically and geochemically well-founded diachronic reconstruction of its metals sector economy. However, our need is greater than that. With an area of ca. 5,000,000 sq. km and a present-day population of ca. six hundred million, the scope of the regional archaeological challenge could be compared against that for the European Union. A very well-attended Southeast Asian archaeological meeting nets in the region of five hundred scholars and students.[1] I am uncertain what the average attendance for a major European archaeology conference is, but I suspect that one of the significant differences will be that of "research density." Despite decades of commendable advances in Southeast Asian archaeology, we still lack a broadly agreed upon regional prehistoric ceramic sequence (White and Hamilton 2009, 358), which, combined with linguistic and publication barriers, presents a real problem for synthetic accounts of shared history between neighboring territories whose former populations we know through common sense were most probably in contact. I hope to demonstrate that archaeometallurgical methodologies tried and tested elsewhere can serve to ameliorate this situation, from the advent of the Bronze Age onward, by paying obeisance to geological rather than national boundaries. Lead, is thus, a flux that binds.

Lead Isotope Characterization and Provenance

Archaeological LIA is not new to Southeast Asia, having been carried out by Japanese researchers since 1993 (see Kakukawa et al. 2008 and references therein). However, the geographical and temporal extent of regional LIA sampling would benefit from a sharp expansion as well as an increased density of coverage to enable a systematic unveiling of premodern nonferrous regional metal exchange networks. This is the aim of the Southeast Asian Lead Isotope Project (SEALIP; Pryce, Pollard et al. 2011). To appreciate SEALIP's potential in regards to Southeast Asian archaeology, it is necessary to recognize LIA's own historic trajectory. The technique was originally developed in the 1930s as a geological dating method (Russell and Farquhar 1960, 14) but was co-opted for archaeological purposes in the late 1960s (Brill and Wampler 1967), its methodology being further modified in subsequent decades into a reliable and mature methodology for studying ancient metal exchange and provenance, though not without a significant period of negotiation in the late 1980s and early 1990s (reviewed in Pollard 2009).

The premise is that lead has four stable isotopes—^{204}Pb, ^{206}Pb, ^{207}Pb, and ^{208}Pb—of which ^{204}Pb is primeval lead and the others the end members of radioactive decay sequences. The ratios between the lead isotopes in rocks and minerals are related to the geological age of their formation and, as such, can be used to characterize those rocks and minerals from specific locations. Ore minerals also contain trace amounts of lead, or predominantly lead in the case of galena and cerussite, of which, crucially, the isotope ratios remain constant throughout metal production processes.[2] This is unlike trace element patterns, which, as already mentioned, shift significantly. Thus, in theory, the lead isotope ratios of archaeological copper-base and lead-base artifacts can be linked to the mineralization from which they were made. The reality of archaeological LIA requires the broaching of the following major difficulties.

1. Geographically separate mineralizations may have similar lead isotope signatures if they are of a comparable geological age.
2. Single mines may have a dispersed signature if comprised of multiple sequential mineralizations of different geological ages.
3. Most critically, metallurgy is an additive technology, meaning the same type of metal (e.g., copper) from multiple sources with different signatures can be blended, different types of metal (e.g., copper and tin) can be alloyed, and the resulting metal can be recycled many times, thoroughly mixing the original lead isotope (and trace element) signatures.
4. One must pay great attention to whether the alloy has been deliberately leaded. That is, if lead metal has been added to the alloy, it will totally overwhelm the lead isotope signature of the trace lead. The result is that we have two metal exchange systems, copper and lead, plotted on the same graphs with an uncertain boundary between them. Artifacts with 5–10 wt.% Pb are clearly intentionally leaded, but at 1–2 wt.% Pb interpretation is not so simple and context and discretion must be applied. Nevertheless, archaeological LIA is much the stronger for its disadvantages being well understood.

Another issue is a necessarily slavish obedience to the Provenance Hypothesis, which, among other measures, states that, "Inter-source variation must be greater than intra-source variation for successful source discrimination" (Wilson and Pollard 2001, 508). Variation in source here refers to isotopic variation in production signatures based upon analyses of smelting slag (Pryce, Brauns, et al. 2011; Pryce, Pollard et al. 2011). Unless one can distinguish between regional production signatures (e.g., central and northern Thailand), then one should simply give up, as the resultant lead isotope analyses of artifacts would be demonstrably meaningless. Even when sources can be distinguished, an artifact can only be considered "provenanced" when its lead isotope signature is "consistent" with a known source, and even then it could come from another source with an identical signature. If an artifact is not consistent with any of the known sources, then it is not provenanced but characterized, potentially awaiting provenance as more production systems are studied. Strict attention to context and alternate datasets (typological, technological, and elemental) is imperative to avoid disastrous misinterpretations.

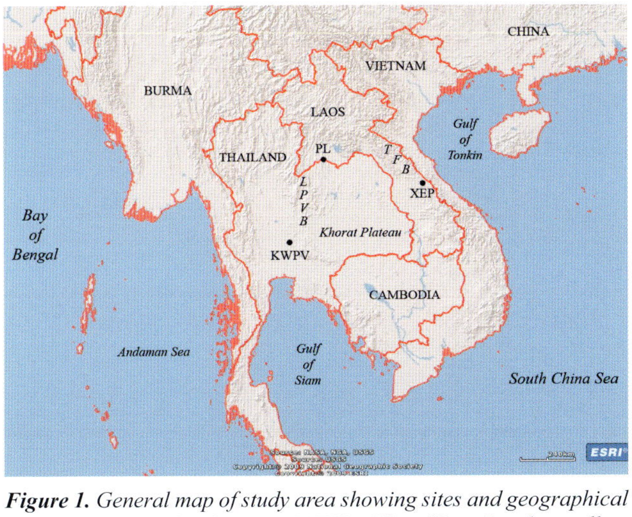

Figure 1. General map of study area showing sites and geographical features mentioned in the text. KWPV = Khao Wong Prachan valley; LPVB = Loei-Petchabun Volcanic Belt; PL = Phu Lon; TFB = Truongson Fold Belt; XEP = Xepon.

Primary Copper Source Differentiation in Thailand

Throughout the 1980s, 1990s, and most of the 2000s until Xepon arrived on the scene (Sayavongkhamdy et al. 2009; Pryce, Brauns et al. 2011), all that was known about prehistoric extractive metallurgy in Southeast Asia was derived from the efforts of the Thailand Archaeometallurgy Project team, working at either end of the Loei-Petchabun Volcanic Belt (see, e.g., Natapintu 1988) (fig. 1). In the northern Thai Loei province, the first-millennium-BCE copper mine and ore processing facility of Phu Lon on the banks of the Mekong River was investigated in two seasons in 1984 and 1985 (Pigott et al. 1992; Pigott and Weisgerber 1998). During 1986, 1990, and 1992, the team's focus shifted south to the Khao Wong Prachan valley in the central Thai Lopburi province, where several copper mineralizations were recorded in close association with two major copper production and settlement sites, Non Pa Wai and Nil Kham Haeng, along with other locations of interest in the vicinity (e.g., Ciarla 2005, 2007, 2008; Pigott et al. 1997). The Khao Wong Prachan area seems to transition to the Bronze Age in the terminal second millennium BCE, with copper-base founding and possibly copper smelting initiated at the same time or soon after and persisting with a trajectory of increasing intensity and technical competence until the early/mid-first millennium CE (Higham et al. 2011; Pryce et al. 2010; Pryce, Pollard et al. 2011). The long academic hegemony, unsought and well deserved, of the Thailand Archaeometallurgy Project has led to many assumptions about where Southeast Asia's copper was being produced (e.g., White and Pigott 1996, 158, for a well-reasoned example). The following datasets may provide the framework to put those assumptions to the test, but we must first establish that we can distinguish Phu Lon and Khao Wong Prachan copper, lest we fail the Provenance Hypothesis (Wilson and Pollard 2001). Though there may be sequential hydrothermal depositions of metal-bearing fluids at both these mineralizations, as they lie on the same fault system, the major formation of which is assumed to be a single event in geological time, this discrimination test is critical for the ongoing validity of SEALIP research.

Methodology

SEALIP's pilot study involved the LIA of three Metal Age (ca. 1000 BCE–ca. 500 CE) slag samples and one Bronze Age bronze ax from Non Pa Wai, and three Metal Age slags and three copper cordiform implements thought to be ingots from Nil Kham Haeng (Pryce, Pollard et al. 2011). This pilot study did indeed provide "Southeast Asia's first isotopically-defined prehistoric copper production system," with five of the six slag samples as well as the three ingots plotting closely together, and only the Bronze Age unleaded bronze ax proving inconsistent with the Khao Wong Prachan valley copper signature (Pryce, Pollard et al. 2011, fig. 4). Nevertheless, the outlying slag lead isotope determination undermined the integrity of that signature and, with more funding in hand, a further fourteen Non Pa Wai and Nil Kham Haeng slag samples from the Thailand Archaeometallurgy Project archive at the University of Pennsylvania Museum of Archaeology and Anthropology have been added to the SEALIP database. As per the pilot study, these samples were analyzed by multiple collector–inductively coupled plasma–mass spectrometry (MC-ICP-MS) at the Curt-Engelhorn-Zentrum Archäometrie gGmbH (CEZA) in Mannheim, Germany (see Niederschlag et al. 2003 for a detailed methodology). This central Thai material was accompanied to Germany by ten Phu Lon samples from the Penn Museum archive: seven copper mineral fragments, one slag fragment, one bronze casting "drip," and one bronze ax (fig. 2; table 1). The ax had previously been analyzed for major elements at the Museum Applied Science Center for Archaeology (Vincent C. Pigott, pers. comm.), but trace elements were included in the present study (table 2). Although slag is the preferred material for characterizing production systems, representing only those minerals actually used by ancient metalworkers and pooling the lead isotope signatures of other potential minor contributors like degrading ceramics and fuel ash (Pryce, Pollard et al. 2011), very little slag was recovered at Phu Lon. This was possibly due to recognition issues with fieldworkers inexperienced in metallurgical materials, but the result is that the Phu Lon production signature is primarily ore mineral-based at present.

Results

The lead isotope results of the thirteen additional slag samples from Non Pa Wai and Nil Kham Haeng have substantially firmed up the copper production signature of the Khao Wong Prachan valley, leaving sample SEALIP/NPW/1 as an apparent outlier (fig. 2, table 2). The Phu Lon copper production signature, based upon seven mineral and one slag samples, is far more dispersed than that for the Khao Wong Prachan valley, but most importantly, there is no overlap between those two signatures. Of Phu Lon's two copper-base samples, SEALIP/PL/9 and SEALIP/PL/10, the first appears consistent with the Phu Lon copper production signature whereas the second does not (see table 2).

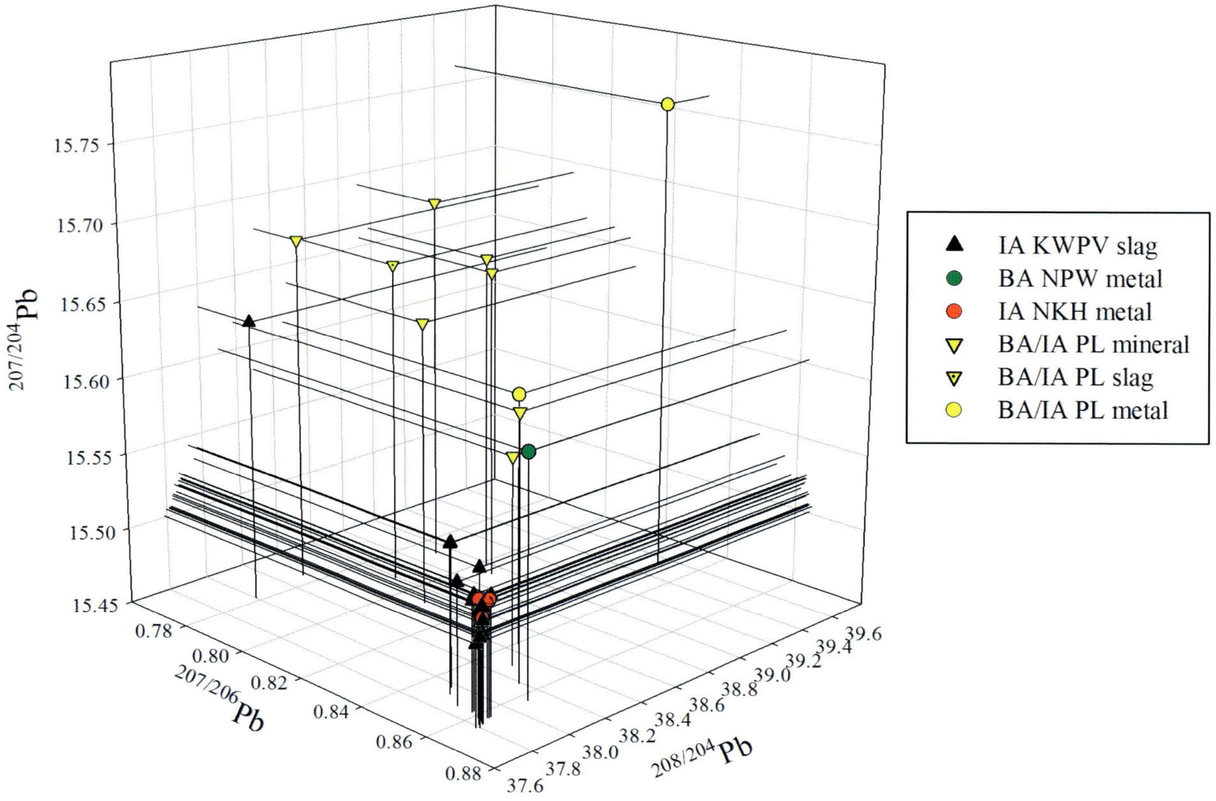

Figure 2. *Three-dimensional plot of lead data for artifacts. BA = Bronze Age; IA = Iron Age; KWPV = Khao Wong Prachan valley; NKH = Nil Kham Haeng; NPW = Non Pa Wai; PL = Phu Lon.*

Table 1. *Contextual information for samples used in the study.*

SEALIP ID	CEZA ID	Country	Site	Context	Catalogue	Date Found	Other ID	Reference	Artifact	Period (~)	Corrosion
SEALIP/NPW/1	MA-091895	Thailand	Non Pa Wai	7-1	9055	2/15/1992	NPWMS1	Pryce 2009	slag	500–300 BCE	
SEALIP/NPW/2	MA-091896	Thailand	Non Pa Wai	BC14	36809.2	1/1/1986	NPWMS13	Pryce 2009	slag	500–300 BCE	
SEALIP/NPW/3	MA-091897	Thailand	Non Pa Wai	8UNIT 12	11605	3/3/1992	NPWMS18	Pryce 2009	slag	500–300 BCE	
SEALIP/NPW/4	MA-093024	Thailand	Non Pa Wai	BA13	36807.1	3/31/1986	NPWMS5	Pryce 2009	slag	500–300 BCE	
SEALIP/NPW/5	MA-093025	Thailand	Non Pa Wai	MA7	36719	3/20/1986	NPWMS7	Pryce 2009	slag	500–300 BCE	
SEALIP/NPW/6	MA-093026	Thailand	Non Pa Wai	3A1	6807	1/27/1992	NPWMS9	Pryce 2009	slag	500–300 BCE	
SEALIP/NPW/7	MA-093027	Thailand	Non Pa Wai	AB4	36787	1/3/1986	NPWMS11	Pryce 2009	slag	500–300 BCE	
SEALIP/NPW/8	MA-093028	Thailand	Non Pa Wai	7SEQ1	15104	3/25/1992	NPWMS16	Pryce 2009	slag	500–300 BCE	
SEALIP/NPW/9	MA-093029	Thailand	Non Pa Wai	8UNIT 1 2	11485	3/2/1992	NPWMS17	Pryce 2009	slag	500–300 BCE	
SEALIP/NPW/10	MA-093030	Thailand	Non Pa Wai	8UNIT 1 3	14371	3/22/1992	NPWMS20	Pryce 2009	slag	500–300 BCE	
SEALIP/NPW/11	MA-092049	Thailand	Non Pa Wai		223	1986		Pigott et al. 1997	bronze ax	1300–1000 BCE	low
SEALIP/NKH/1	MA-091898	Thailand	Nil Kham Haeng	2A3-4	630	2/11/1990	NKHMS4	Pryce 2009	slag	300 BCE–300 CE	
SEALIP/NKH/2	MA-091899	Thailand	Nil Kham Haeng	1NEQ11	3790	—	NKHMS13	Pryce 2009	slag	301 BCE–300 CE	
SEALIP/NKH/3	MA-091900	Thailand	Nil Kham Haeng	2NWQ4A	2793	3/11/1990	NKHMS17	Pryce 2009	slag	302 BCE–300 CE	
SEALIP/NKH/4	MA-093031	Thailand	Nil Kham Haeng	2D5	4797	3/27/1990	NKHMS6	Pryce 2009	slag	303 BCE–300 CE	
SEALIP/NKH/5	MA-093032	Thailand	Nil Kham Haeng	4J4	2042	2/22/1990	NKHMS9	Pryce 2009	slag	304 BCE–300 CE	
SEALIP/NKH/6	MA-093033	Thailand	Nil Kham Haeng	3NEQ18	3659	3/16/1990	NKHMS10	Pryce 2009	slag	305 BCE–300 CE	
SEALIP/NKH/7	MA-093034	Thailand	Nil Kham Haeng	3NEQ6	515	2/15/1990	NKHMS12	Pryce 2009	slag	306 BCE–300 CE	
SEALIP/NKH/8	MA-093035	Thailand	Nil Kham Haeng	3NEQ18	4706	3/16/1990	NKHMS14	Pryce 2009	slag	307 BCE–300 CE	
SEALIP/NKH/9	MA-093036	Thailand	Nil Kham Haeng	3SEQ195	4294	3/22/1990	NKHMS19	Pryce 2009	slag	308 BCE–300 CE	
SEALIP/NKH/10	MA-093037	Thailand	Nil Kham Haeng	3SWQ18	3843	3/21/1990	NKHMS21	Pryce 2009	slag	309 BCE–300 CE	
SEALIP/NKH/11	MA-092047	Thailand	Nil Kham Haeng	Op.1	1	3/25/1986		Pigott et al. 1997	Cu ingot	310 BCE–300 CE	low
SEALIP/NKH/12	MA-092051	Thailand	Nil Kham Haeng	JCB cut	4891	1985		Pigott et al. 1997	Cu ingot	311 BCE–300 CE	low
SEALIP/NKH/13	MA-092046	Thailand	Nil Kham Haeng	JCB cut	1/85	1985		Pigott et al. 1997	Cu ingot	312 BCE–300 CE	low
SEALIP/PL/1	MA-100755	Thailand	Phu Lon	PL.A1A2.1	39635	3/18/1985		Pigott and Weisgerber 1998	malachite	1st mill. BCE	
SEALIP/PL/2	MA-100757	Thailand	Phu Lon	PL.B3.7	39626	3/8/1985		Pigott and Weisgerber 1998	malachite	1st mill. BCE	
SEALIP/PL/3	MA-100758	Thailand	Phu Lon	PL.B4.1	39631	3/12/1985		Pigott and Weisgerber 1998	malachite	1st mill. BCE	
SEALIP/PL/4	MA-100761	Thailand	Phu Lon	PL.B3.13	39627	3/8/1985		Pigott and Weisgerber 1998	ferruginous malachite	1st mill. BCE	
SEALIP/PL/5	MA-100764	Thailand	Phu Lon	PL.A2B2.4	39647	3/18/1985		Pigott and Weisgerber 1998	malachite	1st mill. BCE	
SEALIP/PL/6	MA-100760	Thailand	Phu Lon	PL.B3B4.2	39650	3/19/1985		Pigott and Weisgerber 1998	siliceous malachite	1st mill. BCE	
SEALIP/PL/7	MA-100750	Thailand	Phu Lon	PL.A1B1.4	39644	3/15/1985		Pigott and Weisgerber 1998	ferruginous malachite	1st mill. BCE	
SEALIP/PL/8	MA-100748	Thailand	Phu Lon	PLII.SS4.surface	39921	2/12/1985		Pigott and Weisgerber 1998	slag	1st mill. BCE	
SEALIP/PL/9	MA-100768	Thailand	Phu Lon	LF:PIN3.SM55.Level 4.30-4.50m		2/2/1985		Pigott and Weisgerber 1998	Cu/bronze drip	1st mill. BCE	low
SEALIP/PL/10	MA-092048	Thailand	Phu Lon	BN.π.1.F.4		3/13/1985		Pigott and Weisgerber 1998	bronze ax	1st mill. BCE	low

— = data missing; blank space = not relevant.

Table 2. *Raw lead isotope ratios for study samples plus trace elements for metal samples.*

SEALIP ID	Pb208toPb206	Pb207toPb206	Pb208toPb204	Pb207toPb204	Pb206toPb204	ppm Co	ppm Ni	ppm As	ppm Se	ppm Au	ppm Pb	ppm Bi	ppm Fe
SEALIP/NKH/1	2.1061±0.0002	0.86283±0.00002	37.895±0.001	15.525±0.001	17.993±0.002								
SEALIP/NKH/2	2.1082±0.0001	0.8648±0.00005	37.798±0.004	15.505±0.002	17.929±0.002								
SEALIP/NKH/3	2.1037±0.0001	0.86222±0.00001	37.864±0.008	15.519±0.003	17.999±0.003								
SEALIP/NKH/4	2.1014±0.0001	0.86132±0.00005	37.853±0.007	15.515±0.003	18.013±0.003								
SEALIP/NKH/5	2.1049±0.0001	0.86261±0.00003	37.878±0.009	15.523±0.003	17.995±0.003								
SEALIP/NKH/6	2.1066±0.0002	0.86421±0.00005	37.786±0.012	15.501±0.005	17.937±0.003								
SEALIP/NKH/7	2.1068±0.0001	0.86386±0.00005	37.815±0.007	15.505±0.002	17.949±0.001								
SEALIP/NKH/8	2.0681±0.0004	0.84648±0.00002	37.970±0.008	15.541±0.002	18.360±0.002								
SEALIP/NKH/9	2.0708±0.0001	0.84874±0.00002	37.925±0.008	15.544±0.004	18.314±0.004								
SEALIP/NKH/10	insufficient Pb												
SEALIP/NKH/11	2.1060±0.0001	0.86297±0.00003	37.883±0.003	15.523±0.001	17.988±0.001	878	1999	880	68.4	45.6	215	31.9	3748
SEALIP/NKH/12	2.1056±0.0001	0.86340±0.00003	37.836±0.002	15.514±0.001	17.969±0.001	1067	1824	721	71.8	55.7	168	27.7	10159
SEALIP/NKH/13	2.0996±0.0001	0.86066±0.00001	37.866±0.008	15.522±0.004	18.035±0.004	59	1644	274	143	20.3	75.5	4.4	44.6
SEALIP/NPW/1	1.8921±0.0001	0.77854±0.00006	37.999±0.001	15.635±0.001	20.083±0.001								
SEALIP/NPW/2	2.1036±0.0001	0.86266±0.00001	37.846±0.001	15.520±0.001	17.991±0.001								
SEALIP/NPW/3	2.1067±0.0001	0.86385±0.00003	37.832±0.002	15.513±0.001	17.958±0.001								
SEALIP/NPW/4	2.1059±0.0001	0.86354±0.00003	37.814±0.016	15.506±0.007	17.956±0.005								
SEALIP/NPW/5	2.0830±0.0002	0.85409±0.00008	37.869±0.009	15.527±0.004	18.180±0.002								
SEALIP/NPW/6	2.0949±0.0001	0.85831±0.00001	37.8840±0.01	15.522±0.005	18.084±0.005								
SEALIP/NPW/7	2.0950±0.0001	0.85848±0.00002	37.871±0.007	15.519±0.003	18.077±0.003								
SEALIP/NPW/8	2.0920±0.0001	0.85627±0.00001	37.951±0.008	15.534±0.003	18.141±0.002								
SEALIP/NPW/9	2.1050±0.0001	0.86299±0.00002	37.833±0.003	15.511±0.001	17.973±0.001								
SEALIP/NPW/10	2.1077±0.0001	0.86406±0.00001	37.825±0.003	15.506±0.001	17.946±0.001								
SEALIP/NPW/11	2.1064±0.0003	0.86261±0.00001	38.101±0.007	15.603±0.001	18.088±0.003	470	314	90.0	35.4	22.0	65.9	14.7	2968
SEALIP/PL/1	2.0056±0.0001	0.8067±0.00002	38.923±0.001	15.656±0.001	19.407±0.001								
SEALIP/PL/2	1.9534±0.0002	0.7877±0.00008	38.916±0.039	15.693±0.018	19.922±0.024								
SEALIP/PL/3	1.9898±0.0002	0.80853±0.00006	38.483±0.029	15.637±0.012	19.340±0.012								
SEALIP/PL/4	1.8949±0.0001	0.77549±0.00005	38.309±0.003	15.678±0.002	20.217±0.001								
SEALIP/PL/5	2.0917±0.0002	0.85557±0.00001	38.186±0.007	15.619±0.002	18.256±0.001								
SEALIP/PL/6	1.9955±0.0001	0.80163±0.00004	38.984±0.025	15.661±0.008	19.536±0.014								
SEALIP/PL/7	2.0834±0.0001	0.84786±0.00003	38.291±0.008	15.583±0.003	18.379±0.003								
SEALIP/PL/8	1.9526±0.0002	0.79256±0.00002	38.587±0.004	15.663±0.002	19.762±0.002								
SEALIP/PL/9	2.0826±0.0001	0.83043±0.00002	39.530±0.007	15.762±0.002	18.981±0.003								
SEALIP/PL/10	2.0711±0.0001	0.84066±0.00004	38.460±0.002	15.611±0.001	18.570±0.001	152	71.1	972	212	115	51.2	50.4	16930

Discussion and Conclusion

Phu Lon's dispersed signature is probably attributable to two potentially linked factors: first, the geological literature indicates that the Phu Lon mineralization was formed by multiple hydrothermal events (Kamvong and Zaw 2009), which could result in a wide distribution of lead isotope results; second, the Phu Lon lead isotope signature is based at present upon copper ore mineral samples, which, unlike slag, were not definitely selected by ancient metalworkers. At the time of writing, two further Phu Lon slag samples have been submitted for analysis to complement the existing sample, SEALIP/PL/8. If these three determinations form a discrete cluster within the Phu Lon mineral field, then it will be possible to suggest that these slag results are the core prehistoric Phu Lon copper production field. With respect to the dispersed Phu Lon mineral results, the Khao Wong Prachan valley slag outlier, SEALIP/NPW/1, could conceivably be the result of Khao Wong Prachan valley metalworkers having smelted Phu Lon ore minerals, but this is a very tentative interpretation of a single lead isotope determination.

Of the metal samples from Phu Lon, the bronze drip SEALIP/PL/9 is reasonably consistent with the Phu Lon copper mineral signature, which prompts a parsimonious interpretation of locally smelted copper being alloyed with imported tin in crucibles of the type previously studied by William Vernon (1996–97). The unleaded bronze ax SEALIP/PL/10 can only be interpreted as an import. That is, Phu Lon's former occupants were, at a major prehistoric copper production locale, also importing copper. This is strikingly similar to the situation evinced by the Bronze Age bronze ax sample SEALIP/NPW/11 from the Khao Wong Prachan valley (see Pryce, Pollard et al. 2011) (see fig. 2; table 2) and, as previously noted, is a "coals to Newcastle" predicament noted from elsewhere in the Asian Old World (Weeks 2007). As has been long suspected by scholars who prefer not to see artifacts in purely functional terms, it seems SEALIP has demonstrated that a bronze ax was not simply an ax in prehistoric Thailand.

The differentiation afforded between the lead isotope signatures of the two known prehistoric Thai primary copper production centers (see fig. 2) means that the Provenance Hypothesis (Wilson and Pollard 2001) is satisfied with respect to those centers, and thus the SEALIP program can, in theory, identify whether copper-base artifacts are consistent with the Khao Wong Prachan valley and Phu Lon copper production systems. There is a long way to go before we have a substantial regional reconstruction of the prehistoric copper-base and lead-base metals sector, but I hope the reader will agree that SEALIP appears to have legs.

Acknowledgments

My research was supported by a Leverhulme Trust Early Career Fellowship and a Junior Research Fellowship at St. Hugh's College, Oxford, in addition to a British Academy Small Research Award, a grant from John Fell Oxford University Press Fund and the Association of Southeast Asian Studies in the United Kingdom. I offer my sincere gratitude to Prof. Dr. Ernst Pernicka and Dr. Michael Brauns (Curt-Engelhorn-Zentrum Archäometrie) for their laboratory input, and also to Dr. Bérénice Bellina (CNRS), Prof. Charles Higham (University of Otago), and Prof. Vincent C. Pigott (UCL), who were kind enough to review this manuscript.

Notes

1. For example, the 19th Meeting of the Indo-Pacific Prehistory Association (IPPA) in Hanoi, December 2009, though by definition IPPA covers areas that fall outside of any definition of Southeast Asia.
2. Within instrumental error (Budd et al. 1995).

References

Bayard, D. T. 1972. Early Thai bronze: Analysis and new dates. *Science* 176:1411–12.

———. 1979. The chronology of prehistoric metallurgy in northeast Thailand: *Silabhumi* or *Samrddhabhumi*? In *Early South East Asia: Essays in archaeology, history and historical geography,* ed. R. B. Smith and W. Watson, 15–32. Oxford: Oxford University Press.

———. 1981. Temporal distribution and alloy variation in early bronzes from Non Nok Tha. *Current Anthropology* 22(6):697–99.

Bellwood, P. 2007. *Prehistory of the Indo-Malaysian archipelago.* Honolulu: University of Hawaii Press.

Bennett, A. 1982. Metallurgical analysis of iron artifacts from Ban Don Ta Phet, Thailand. PhD diss., Institute of Archaeology, London.

Bennett, A., and I. C. Glover. 1992. Decorated high-tin bronzes from Thailand's prehistory. In *Southeast Asian Archaeology 1990: Proceedings of the Third Conference of the European Association of Southeast Asian Archaeologists,* ed. I. Glover, 187–208. Hull, UK: Centre for Southeast Asian Studies, University of Hull.

Biggs, L. 2009. Iron technology in peninsular Thailand: Evidence from second century BC sites at Khao Sam Kaeo and Phu Khao Thong. MSc thesis, Institute of Archaeology, London.

Bray, P. 2009. Exploring the social basis of technology: Re-analysing regional archaeometric studies of the first copper and tin-bronze use in Britain and Ireland. PhD diss., School of Archaeology, University of Oxford.

Brill, R. H., and J. M. Wampler. 1967. Isotope studies of ancient lead. *American Journal of Archaeology* 71(1):63–77.

Bronson, B. 1977. Exchange at the upstream and downstream ends: Notes towards a functional model of the coastal state in Southeast Asia. In *Economic exchange and social interaction in Southeast Asia: Perspectives from prehistory, history and ethnography,* ed. K. E. Hutterer, 39–52. Ann Arbor: Center for South and Southeast Asian Studies, University of Michigan Press.

———. 1992. Patterns in the early Southeast Asian metals trade. In *Early metallurgy, trade and urban centres in Thailand and Southeast Asia,* ed. I. C. Glover, P. Suchitta, and J. Villiers, 63–114. Bangkok: White Lotus.

Budd, P., A. M. Pollard, B. Scaife, and R. G. Thomas. 1995. The possible fractionation of lead isotopes in ancient metallurgical processes. *Archaeometry* 37(1):143–50.

Chippindale, C. 2000. Capta and data: On the true nature of archaeological information. *American Antiquity* 65(4):605–12.

———. 2007. A preliminary report on LoRAP archaeological excavations at prehistoric Khao Sai on, Lopburi province, central Thailand. *East and West* 57:395–401.

———. 2008. The Thai–Italian "Lopburi Regional Archaeological Project" (LoRAP). Excavation at Khao Sai on-Noen Din 2008. Preliminary report. *East and West* 58:313–36.

Dizon, E. Z. 1988. An Iron Age in the Philippines? A critical examination. PhD diss., Department of Anthropology, University of Pennsylvania.

Gorman, C., and P. Charoenwongsa. 1976. Ban Chiang: A mosaic of impressions from the first two years. *Expedition* 18:14–26.

Higham, C. F. W. 1975. Aspects of economy and ritual in prehistoric northeast Thailand. *Journal of Archaeological Science* 2(4):245–88.

———. 1989. The later prehistory of mainland Southeast Asia. *Journal of World Prehistory* 3(3):235–82.

———. 1996–97. The social and chronological contexts of early bronze working in Southeast Asia. In *Ancient Chinese and Southeast Asian Bronze Age cultures,* ed. D. Bulbeck, 821–88. Taipei: SMC Publishing Inc.

Higham, C. F. W., and T. Higham. 2009. A new chronological framework for prehistoric Southeast Asia, based on a Bayesian model from Ban Non Wat. *Antiquity* 82:125–44.

Higham, C. F. W., T. Higham, R. Ciarla, K. Douka, A. Kijngam, and F. Rispoli. 2011. The origins of the Bronze Age of Southeast Asia. *Journal of World Prehistory* 24: 227–274.

Higham, C. F. W., T. F. G. Higham, and A. Kijngam. 2011. Cutting a Gordian Knot: The Bronze Age of Southeast Asia: Origins, timing and impact. *Antiquity* 85:583–98.

Jacq-Hergoualc'h, M. 2002. *The Malay peninsula: Crossroads of the maritime Silk Road (100 BC—1300 AD).* Leiden: Brill.

Junghans, S., E. Sangmeister, and M. Schröder. 1960. *Metallanalysen kupferzeitlicher und frühbronze-zeitlicher bondenfunde aus Europa. Studien zu den anfängen der metallurgie.* Vol. 1. Berlin: Mann Verlag.

———. 1968. *Kupfer und bronze in der frühen metallzeit Europas. Die materialgruppen beim stand von 12000 analysen. Studien zu den anfängen der metallurgie.* Vol. 2(1–3). Berlin: Mann Verlag.

———. 1974. *Kupfer und bronze in der frühen metallzeit Europas. Katalog der analysen nr. 10041-22000 (mit nachuntersuchungen der analysen nr. 1010040). Studien zu den anfängen der metallurgie.* Vol. 2(4). Berlin: Mann Verlag.

Kakukawa, S., S. Hieda, and Y. Hirao. 2008. Chemical analysis on bronze bracelets unearthed from the Phum Snay archaeological site in Cambodia and the identification of their production area. In *Preliminary report for the excavation in Phum Snay 2007,* ed. Y. Yasuda and C. Phoeurn, 60–65. Kyoto: Institute Research Center for Japanese Studies.

Kamvong, T., and K. Zaw. 2009. The origin and evolution of skarn forming fluids from the Phu Lon Deposit, Northern Loei Fold Belt, Thailand: Evidence from fluid inclusion and sulfur isotope studies. *Journal of Asian Earth Sciences* 34(2):624–33.

Miksic, J. 1990. *Old Javanese gold.* Singapore: Ideation.

Muhly, J. D. 1981. The origin of agriculture and technology—Aarhus, Denmark, November 21–25, 1978.2. Summary—The origin of agriculture and technology—West or East-Asia. *Technology and Culture* 22:125–48.

Murillo-Barroso, M., T. O. Pryce, B. Bellina, and M. Martinón-Torres. 2010. Khao Sam Kaeo: An archaeometallurgical crossroads for trans-Asiatic technological traditions. *Journal of Archaeological Science* 37(7):1761–72.

Natapintu, S. 1988. Current research on ancient copper-base metallurgy in Thailand. In *Prehistoric studies: The Stone and Metal Ages in Thailand,* ed. P. Charoenwongsa and B. Bronson, 107–24. Bangkok: The Thai Antiquity Working Group.

Niederschlag, E., E. Pernicka, T. Seifert, and M. Bartelheim. 2003. The determination of lead isotope ratios by multiple collector ICP-MS: A case study of Early Bronze Age artefacts and their possible relation with ore deposits of the Erzgebirge. *Archaeometry* 45(1):61–100.

Pigott, V. C. 1986. Pre-industrial mineral exploitation and metal production in Thailand. *MASCA Journal* 3:170–74.

Pigott, V. C., and R. Ciarla. 2007. On the origins of metallurgy in prehistoric Southeast Asia: The view from Thailand. In *Metals and mines: Studies in archaeometallurgy,* ed. S. La Niece, D. Hook, and P. Craddock, 76–88. London: Archetype Publications in association with the British Museum.

Pigott, V. C., and A. R. Marder. 1984. Prehistoric iron in Southeast Asia: New evidence from northeastern Thailand. In *Southeast Asian archaeology at the XV Pacific Science Congress,* ed. D. T. Bayard, 278–308. Dunedin, New Zealand: Otago University.

Pigott, V. C., and S. Natapintu. 1988. Archaeological investigations into prehistoric copper production: The Thailand Archaeometallurgy Project, 1984–86. In *The beginning of the use of metals and alloys,* ed. R. Maddin, 156–62. Cambridge, MA: MIT Press.

Pigott, V. C., S. Natapintu, and U. Theetiparivatra. 1992. The Thailand Archaeometallurgy Project 1984–85: Research in the development of prehistoric metal use in northeast Thailand. In *Early metallurgy, trade and urban centres in Thailand and Southeast Asia,* ed. I. C. Glover, P. Suchitta, and J. Villiers, 47–62. Bangkok: White Lotus.

Pigott, V. C., and G. Weisgerber. 1998. Mining archaeology in geological context: The prehistoric copper mining complex at Phu Lon, Nong Khai province, northeast Thailand. In *Metallurgica antiqua: In honour of Hans-Gert Bachmann and Robert Maddin,* ed. J. D. Muhly, 135–61. Bochum: Deutsches Bergbau-Museum.

Pigott, V. C., A. Weiss, and S. Natapintu. 1997. The archaeology of copper production: Excavations in the Khao Wong Prachan valley, central Thailand. In *South-East Asian archaeology 1992,* ed. R. Ciarla and F. Rispoli, 119–57. Proceedings of the Fourth International Conference of the European Association of South-East Asian Archaeologists. Rome, 28 September–4 October 1992. Serie Orientale Roma, Vol. 77. Rome: Istituto Italiano per l'Africa e l'Oriente.

Pollard, M. 2009. What a long strange trip it's been: Lead isotopes and archaeology. In *From mine to microscope: Advances in the study of ancient technology,* ed. A. Shortland, I. Freestone, and T. Rehren, 181–89. Oxford: Oxbow Books.

Pollard, M., and C. Heron. 2008. *Archaeological chemistry.* Cambridge: The Royal Society of Chemistry.

Pryce, T. O. 2010. Review of *The distribution of bronze drums in early Southeast Asia: Trade routes and cultural spheres* by Ambra Calò. British Archaeological Reports International Series 1913. *Antiquity* 84:905–6.

Pryce, T. O., and M. J. Abrams. 2010. Direct detection of Southeast Asian smelting sites by aster remote sensing imagery: Technical issues and future perspectives. *Journal of Archaeological Science* 37(12):3091–98.

Pryce, T. O., A. H. Bevan, R. Ciarla, F. Rispoli, J. L. Malakie, B. Hassett, and C. Castillo. In press. Intensive archaeological survey in Southeast Asia: Methodological and metallurgical insights from Khao Sai On, central Thailand. *Asian Perspectives.*

Pryce, T. O., M. Brauns, N. Chang, E. Pernika, A. M. Pollard, C. Ramsey, T. Rehren, V, Souksavatdy, and T. Sayavongkhamdy. 2011. Isotopic and technological variation in prehistoric Southeast Asian primary copper production. *Journal of Archaeological Science* 38(12): 3309–22.

Pryce, T. O., M. Murillo-Barroso, L. Biggs, M. Martinón-Torres, and B. Bellina. In press. The metallurgical industries. In *Khao Sam Saeo: A late prehistoric city of the upper Thai–Malay peninsula,* ed. B. Bellina. Paris: BEFEO.

Pryce, T. O. and S. Natapintu. 2009. Smelting iron from laterite: Technical possibility or ethnographic aberration? *Asian Perspectives* 48:249–64.

Pryce, T. O., V. C. Pigott, M. Martinón-Torres, and T. Rehren. 2010. Prehistoric copper production and technological reproduction in the Khao Wong Prachan valley of central Thailand. *Archaeological and Anthropological Sciences* 2(4):237–64.

Pryce, T. O., M. Pollard, M. Martinòn-Torres, V. C. Pigott, and E. Pernicka. 2011. Southeast Asia's first isotopically-defined prehistoric copper production system: When did extractive metallurgy begin in the Khao Wong Prachan valley of central Thailand? *Archaeometry* 53(1):146–63.

Richter, A. 2000. *Jewelry of Southeast Asia.* New York: Harry N. Abrams.

Russell, R. D., and R. M. Farquhar. 1960. *Lead isotopes in geology.* London and New York: Interscience.

Sayavongkhamdy, T., N. Chang, V. Souksavatady, and H. Cawte. 2009. The archaeology of Sepon, Lao PDR: Archaeometallurgy, unexploded bombs and collaborations. Paper presented at the 19th Congress of the Indo-Pacific Prehistory Association, 29 November –5 December 2009, Hanoi, Vietnam.

Schwartz, M. O., S. S. Rajah, A. K. Askury, P. Putthapiban, and S. Djaswadi. 1995. The Southeast Asian tin belt. *Earth-Science Reviews* 38:95–293.

Sitthithaworn, E. 1990. Metallogenic map of Thailand. Bangkok: Department of Mineral Resources.

Smith, C. S. 1973. Bronze technology in the East: A metallurgical study of early Thai bronzes, with some speculations on the cultural transmission of technology. In *Changing perspectives in the history of science, essays in honour of Joseph Needham,* ed. M. Teich and R. Young, 21–32. London: Heinemann.

Solheim, W. G. 1968. Early bronze in northeastern Thailand. *Current Anthropology* 9:59–62.

Spriggs, M. 1996–97. The dating of Non Nok Tha and the "Gakashuin Factor." In *Ancient Chinese and Southeast Asian bronze cultures,* ed. D. Bulbeck and N. Barnard, 941–48. Taipei: SMC Publishing.

Suchitta, P. 1983. The history and development of iron smelting technology in Thailand. PhD diss., Department of Anthropology, Brown University.

Vernon, W. W. 1996–97. The crucible in copper-bronze production at prehistoric Phu Lon, northeast Thailand: Analyses and interpretation. In *Ancient Chinese and Southeast Asian Bronze Age cultures,* ed. F. D. Bulbeck and N. Barnard, 809–20. Taipei: SMC Publishing.

Weeks, L. 2007. Coals to Newcastle, copper to Magan? Isotopic analyses and the Persian Gulf metals trade. In *Metals and mines: Studies in archaeometallurgy,* ed. S. La Niece, D. Hook, and P. Craddock, 89–96. London: Archetype Publications in association with the British Museum.

Wheatley, P. 1961. *The golden Khersonese.* Kuala Lumpur: Oxford University Press and Kuala Lumpur University Press.

Wheeler, T., and R. Maddin. 1976. The techniques of the early Thai metalsmith. *Expedition* 18(4):38–47.

White, J. C. 1986. A revision of the chronology at Ban Chiang and its implications for the prehistory of northeast Thailand. PhD diss., Department of Anthropology, University of Pennsylvania.

———. 1988. Early East Asian metallurgy: The southern tradition. In *The beginning of the use of metals and alloys,* ed. R. Maddin, 175–81. Cambridge, MA: MIT Press.

———. 2008. Dating early bronze at Ban Chiang, Thailand. In *Archaeology in Southeast Asia: From* Homo erectus *to the living traditions,* ed. J.-P. Pautreau, A. Coupey, V. Zeitoun, and E. Rambault, 91–104. Choice of Papers from the 11th EurASEAA Conference, Bougon, 25–29 September 2006. Chiang-Mai, Thailand: European Association of Southeast Asian Archaeologists.

White, J. C., and E. G. Hamilton. 2009. The transmission of early bronze technology to Thailand: New perspectives. *Journal of World Prehistory* 22(4):357–97.

White, J. C., and V. C. Pigott. 1996. From community craft to regional specialization: Intensification of copper production in prestate Thailand. In *Craft specialization and social evolution: In memory of V. Gordon Childe,* ed. B. Wailes, 151–75. University Museum Monograph 93. Philadelphia: The University Museum of Archaeology and Anthropology, University of Pennsylvania.

Wilson, L., and A. M. Pollard. 2001. The Provenance Hypothesis. In *Handbook of archaeological sciences,* ed. D. R. Brothwell and A. M. Pollard, 507–17. Chichester, UK: John Wiley & Sons.

Khmer Bronze Metallurgy during the Angkorian Period (Twelfth to Thirteenth Centuries): Technical Investigation of a New Selected Corpus of Artifacts from the National Museum of Cambodia, Phnom Penh

Brice Vincent, David Bourgarit, and Paul Jett

ABSTRACT Technical investigation was recently performed on bronze artifacts in the collection of the National Museum of Cambodia in Phnom Penh as part of an ongoing PhD research project on Khmer bronze metallurgy during the Angkorian period. Ninety-three drilled metal samples were removed from seventy-two objects and analyzed using inductively coupled plasma–atomic emission spectroscopy; the analytical effort was made possible by collaboration with the Centre de Recherche et de Restauration des Musées de France in Paris and the Smithsonian's Freer and Sackler Galleries in Washington, D.C. This work aimed to build on pioneering studies of Khmer bronzes undertaken in the late 1990s by the Walters Art Museum, Baltimore, and the Musée National des Arts Asiatiques–Guimet in collaboration with the Centre de Recherche et de Restauration des Musées de France, but with one of the main objectives being to focus on and provide referential data on bronze production in the twelfth and thirteenth centuries, which is the best represented period in Khmer bronze metallurgy.

Introduction

A new selected corpus of bronze artifacts from the National Museum of Cambodia in Phnom Penh was recently the subject of technical investigation as part of an ongoing PhD research project on Khmer bronze metallurgy during the Angkorian period (twelfth to thirteenth centuries). It was made possible by collaboration with the Centre de Recherche et de Restauration des Musées de France (C2RMF) in Paris and the Smithsonian's Freer and Sackler Galleries in Washington, D.C.

The Walters Art Museum, Baltimore, was one of the first institutions to apply archaeometallurgical research to Khmer "bronze" or copper-based metallurgy (Woodward 1997; Strahan 1988). Its conservation laboratory, as well as others, examined seventeen Khmer and Khmer-influenced bronze items from the Alexander B. Griswold collection, mainly statues dated from the eleventh century to the thirteenth century, using radiographic examinations as well as bulk-metal, surface, and clay-core analyses. In the late 1990s, in conjunction with an international exhibition, further analytical studies were launched in Paris by the Musée National des Arts Asiatiques–Guimet and the C2RMF (Bourgarit et al. 2003); researchers focused on seventy-five Khmer bronzes from the National Museum of Cambodia and Musée Guimet collections—not only statues but also ritual and utilitarian artifacts dated from the seventh century to the sixteenth century—and performed radiographic examinations along with combined bulk-metal and surface analyses.

Our technical investigation aimed to build on these pioneering studies and to employ most of their results and conclusions. However, one of the main objectives was to focus on and provide referential data on bronze production in the twelfth and thirteenth centuries, which is the best represented period in Khmer bronze metallurgy, so that important but still unclear technological issues might be addressed. Of particular interest was the issue of metal compositions, since the study of both alloying elements and impurity patterns may help to characterize the metallurgical practices of Khmer bronze craftsmen. In turn, such characterization may help to discriminate between different groups of production, be they defined by the period or the location within the Angkorian kingdom.

Corpus of Study

Seventy-two artifacts from the National Museum of Cambodia were first selected. To avoid methodological bias, the items were chosen to satisfy one or several of the following requirements.

First, all selected artifacts were dated to the twelfth and thirteenth centuries, specifically the early thirteenth century, on the basis of iconographical and stylistical features. Two main groups of artifacts were distinguished:

- those dated to the Angkor Wat style and period (between 1100 and 1175; twenty-eight items); and

- those dated to the Bayon style and period (between 1180 and 1230; forty-one items).

Style and chronology have to be clearly dissociated, since, as noted by Le Bonheur (1972, 129, n. 4) about Angkor Wat and Bayon-style bronzes, works in each of these styles can be dated long after the periods' developmental phases. The differences between the two stylistic traditions are thus not as clear as in stone sculpture, for which they were first defined, and this is especially the case with regard to bronzes. Hence, three of the objects could not be clearly assigned and were separated out (*ga* 195, *ga* 5685, and *ga* 5739).[1] It would have been possible to further refine these stylistic attributions and dating, but groups so defined would have been too small to permit statistical treatment.

Second, all selected artifacts were associated with an identified, or at least presumed, archaeological context. A wide variety of available archaeological documentation has been used to identify or confirm the find locations of the bronzes in Cambodia and neighboring countries, particularly the archives and pictures of the École française d'Extrême-Orient (EFEO) as well as the internal documentation and former catalogues of the National Museum of Cambodia. Only one statue without any provenance was kept in the selection because of its iconographical and stylistical features (*ga* 6861).

An attempted regional grouping was then possible and two main groups were distinguished:

- artifacts from the Angkor region (thirty-three items); and
- artifacts from other regions of the Angkorian kingdom (thirty-eight items): six from "East Angkor," two from "East," thirteen from "Southeast," seven from "South," and ten from "West." These regional subgroups are mostly defined from the provenance of the objects, which does not necessarily correspond to their production sites. They are working subgroups, which may help to address the issue of regional bronze workshops in ancient Cambodia (fig. 1).[2]

Third, the selected artifacts were representative of nearly all types of objects produced by Khmer bronze craftsmen during the twelfth and thirteenth centuries (fig. 2 a–h). Most are religious images of Buddhist and Brahmanical—Shaivaite or Vaishnavite—deities, and their sizes vary greatly, from small to large statuary. The corpus also includes a wide variety of objects, such as ritual paraphernalia, vehicle fittings, furniture parts, or statue attributes.

In order to further expand our analysis, all the objects securely dated to the twelfth and early thirteenth century that were analyzed in the late 1990s at the Walters Art Museum and the C2RMF were also included in the corpus of study. These are fourteen artifacts from the National Museum of Cambodia, twenty-eight from the Musée Guimet, and thirteen from the Walters Art Museum.[3]

We took the opportunity to refine the dating and provenance of several objects from the collections of the National Museum of Cambodia and the Musée Guimet. As for the bronzes from the Walters Art Museum, for which no archaeological context is known, they were selected because of their iconographical, stylistical, and technical features.[4] Finally, two more objects analyzed at the C2RMF after the year 2000 were also added: a Bayon lamp from the Musée Guimet (MA 6939; Bourgarit and Mille 2002); and an Angkor Wat statue of Prajñāpāramitā from the Musée des Arts Asiatiques, Nice (03IA25; Bourgarit and Borel 2003).

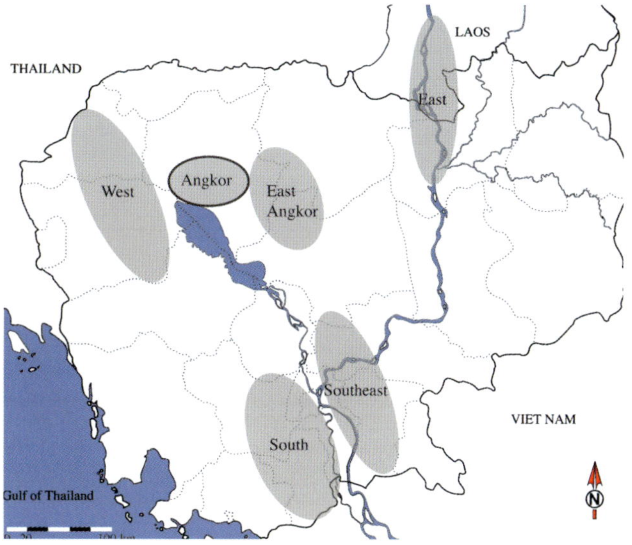

Figure 1. *Schematic location of the regional subgroups mentioned in the text. Photograph by Dominique Soutif, EFEO.*

The final corpus of study thus includes 129 artifacts, which correspond to 114 ensembles that divide into the proposed groupings noted below and in table 1.[5]

1. Forty-seven Angkor Wat-style objects (thirty-eight ensembles) and seventy-four Bayon-style works (sixty-eight ensembles). Eight other objects could not be clearly assigned to one of the stylistic traditions and were designated "undetermined."
2. Forty-six items from the Angkor region (thirty-nine ensembles); fifty-five from other regions (forty-seven ensembles): eight from "East Angkor," three from "East," fourteen from "Southeast," seven from "South," twelve from "West," ten from "Cambodia," and one from "Champa"; and twenty-eight without any provenance (twenty-eight ensembles).

Finally, some of the selected artifacts—mostly medium and large statues—have been classified as "major" objects, according to several features that may be relevant, such as their discovery location, size, or technical complexity, particularly for decoration (table 2). As attempted before on a larger scale but only regarding "aesthetic quality" (Bourgarit et al. 2003, 107), such a characterization may help us to understand the evolution of metallurgical practices during the twelfth and thirteenth centuries in relationship with patronage and commission. Such a socio-economical issue, however, is still poorly understood with respect to the Angkorian period.

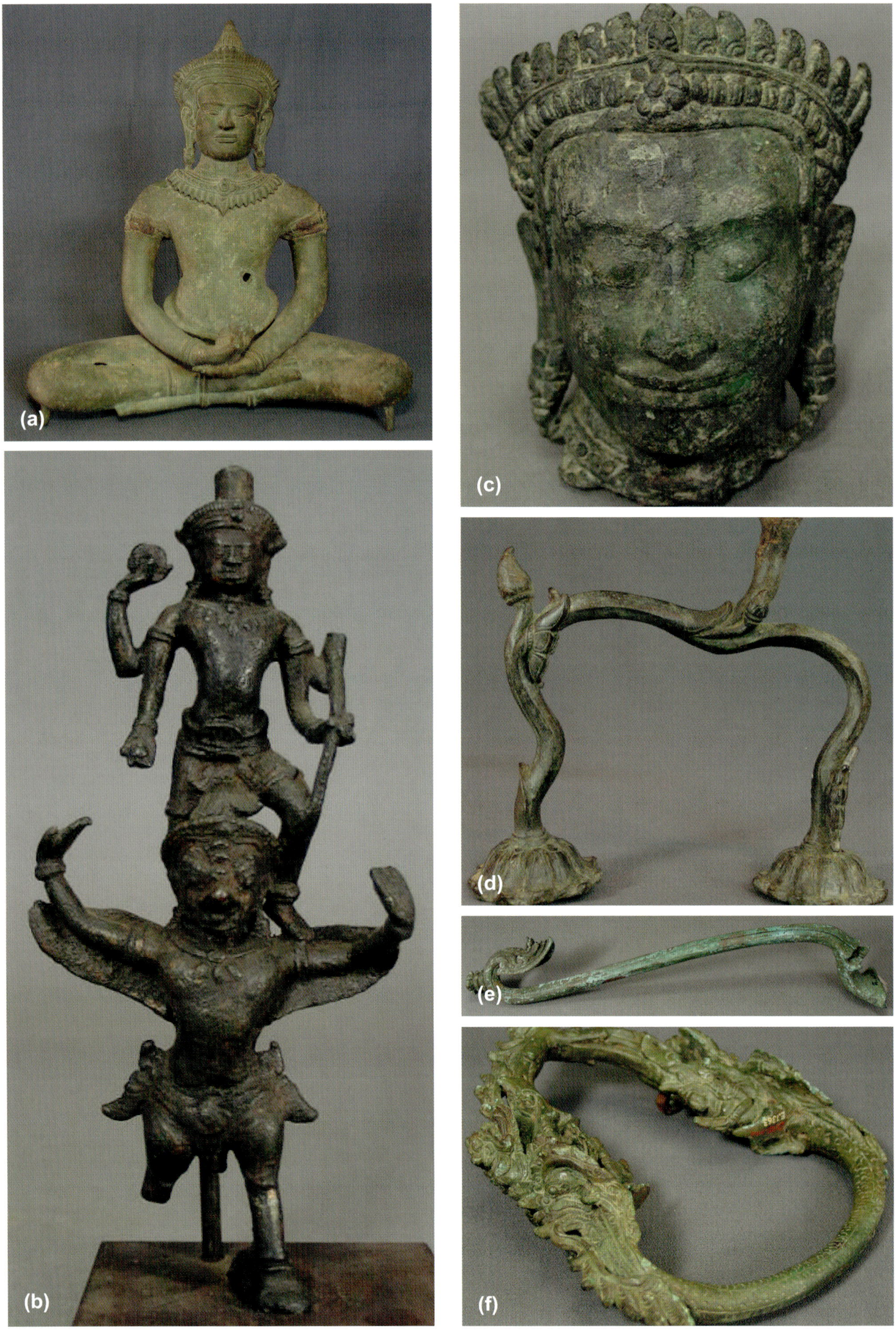
(a)
(b)
(c)
(d)
(e)
(f)

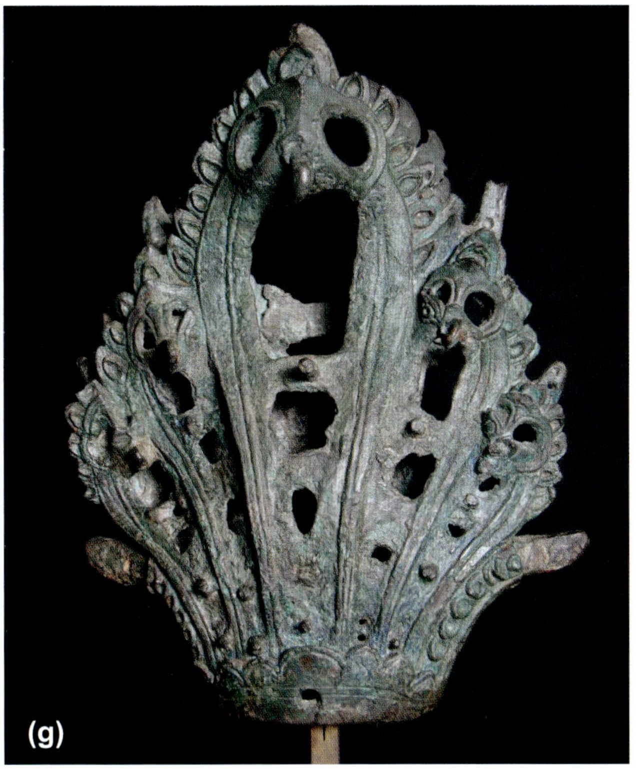

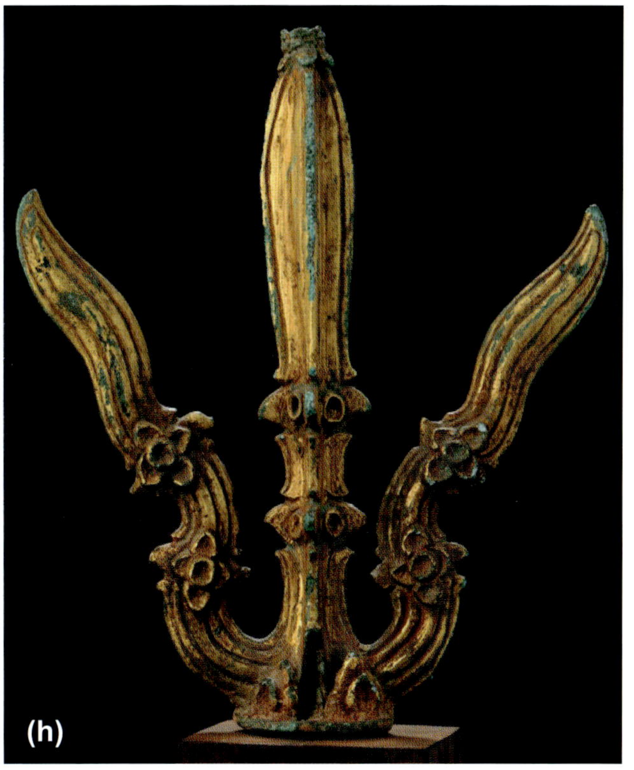

Figure 2.** Khmer artifacts from the new selected corpus.* ***a. *Adorned Buddha in* dhyānamudrā (nāga-*protected Buddha), Angkor Wat style, first half of the twelfth century, Preah Khan of Kampong Svay (Bakan), Preah Vihear province, h. 39.5 cm. NMC, ga 3378. Photograph by Brice Vincent.* ***b.*** *Visnu astride Garuda, Batheay, Angkor Wat style, first half of the twelfth century, Kampong Cham province, h. 16 cm. NMC,* ga *3773. Photograph by Brice Vincent.* ***c.*** *Head of male divinity, Bayon style, late twelfth–early thirteenth century, Royal Palace, Angkor Thom, Siem Reap province, h. 11.5 cm. NMC,* ga *4222. Photograph by Brice Vincent.* ***d.*** *Fragment of perfume censer or lamp(?), Bayon style, late twelfth–early thirteenth century, Tonle Sap banks, Angkor, Siem Reap province, h. 18 cm. NMC,* ga *3721. Photograph by Brice Vincent.* ***e.*** Nāga-*shaped foot, Angkor Wat style, twelfth century, Beng Mealea, Angkor, Siem Reap province, h. 33.5 cm. NMC,* ga *4254.2. Photograph by Brice Vincent.* ***f.*** *Palanquin ring, Bayon style, late twelfth–early thirteenth century, Wat Phu, Champasak province, Laos(?), h. 19 cm. NMC,* ga *4743. Photograph by National Museum of Cambodia.* ***g.*** Nāga-*shaped finial, Bayon style, late twelfth–early thirteenth century, Bayon, Angkor Thom, Siem Reap province(?), h. 46 cm. NMC,* ga *5165. Photograph by Brice Vincent.* ***h.*** *Trident, Bayon style, late twelfth–early thirteenth century, Preah Khan, Angkor, Siem Reap province, h. 25 cm. NMC,* ga *4266. Photograph by Brice Vincent.*

Experimental Procedures and Data Comparison

As already mentioned, the studied corpus is made up of results stemming from various analytical campaigns carried out by various research teams. The most recent analyses have been performed by the authors. Sampling was carried out in the Metal Conservation Laboratory of the National Museum of Cambodia. Samples weighing roughly 20 mg were removed by drilling from each object after eliminating corrosion by removing the surface layer. In the case of objects made of separately cast sections, most parts of interest were sampled, when possible. A total of ninety-three drilled metal samples was taken from the selected artifacts. The samples were examined using microscopy prior to analysis, and corrosion was removed in order to have clean metal for the procedure. Analysis by inductively coupled plasma–atomic emission spectroscopy (ICP-AES) was then performed on the drillings at the C2RMF, following a methodology specifically developed for ancient copper-alloy artifacts (Bourgarit and Mille 2003). See table A1 in the appendix for the analysis results.

The same methodology was used in the late 1990s study at the C2RMF for the analyses of objects from the Musée Guimet collection, and later for the Bayon lamp (MA 6939), to measure twenty-six elements. The analyses of objects from the National Museum of Cambodia collection performed at the same time were made using a direct current plasma spectrometer (DCP-AES), with only twelve elements characterized. In addition, one drilling of the Prajñāpāramitā from the Musée des Arts Asiatiques, Nice (03IA25), was analyzed using proton-induced x-ray emission (PIXE) at the C2RMF ion beam facility AGLAE (3 MeV, proton beam extracted in air with a spot size of 100 µm; see Calligaro et al. 1998). All three methods in use at the C2RMF have proven to give comparable results.

Finally, the objects of the Walters Art Museum collection were analyzed using ICP-AES at the Laboratory of Biomedical and Environmental Sciences, University of California at Los Angeles (Drayman-Weisser et al. 1997).

In order to optimize the weight of each ensemble within the statistical treatment, only one composition was used to represent the different parts of one ensemble in those cases where the results for each part of the ensemble were similar; in these instances, the mean values were calculated. In some cases, the results for different parts within an ensemble showed large discrepancies between each other, and in those cases, two or more compositions were used in order to properly represent the ensemble. Finally, 126 so-called "results" were considered.[6]

Table 1. The 129 Khmer artifacts studied.

	Designation	Museum	Inv. No.	Height (cm)	Type of Artifact	Style	Date	Provenance	Discovery Location	Metal Analysis	Laboratory & Date
Angkor Wat	adorned Buddha in *abhayamudrā*	NMC	*ga* 2081	79.5	large statue	Angkor Wat	2nd half of 12th C.	Angkor	Angkor Wat (Angkor)	DCP-AES	C2RMF (1996)
	Viṣṇu-Vāsudeva-Nārāyaṇa	NMC	*ga* 5291	40	medium statue	Angkor Wat	1st half of 12th C.	Angkor	Kapilapura (Angkor)	DCP-AES	C2RMF (1996)
	kneeling female figure (support for mirror)	NMC	*ga* 5476	35	medium statue	Angkor Wat	1st half of 12th C.	Angkor	Bayon (Angkor)	DCP-AES	C2RMF (1996)
	nāga-shaped finial	MG	MG 18889	34	vehicle fitting	Angkor Wat	12th C.	Angkor	Banteay Srei (Angkor)	DCP-AES	C2RMF (1999)
	nāga-shaped foot	NMC	*ga* 4254.1	33.5	furniture part	Angkor Wat	12th C.	Angkor	Beng Mealea (Angkor)	ICP-AES	C2RMF (2010)
	nāga-shaped foot	NMC	*ga* 4254.2	33.5	furniture part	Angkor Wat	12th C.	Angkor	Beng Mealea (Angkor)	ICP-AES	C2RMF (2010)
	nāga-shaped foot	NMC	*ga* 4730	21	furniture part	Angkor Wat	1st half of 12th C.	Angkor	Chau Say Tevoda (Angkor)	ICP-AES	C2RMF (2010)
	adorned *nāga*-protected Buddha	NMC	*ga* 5985	37	medium statue	Angkor Wat	12th C.?	Angkor	Srah Srang (Angkor)	ICP-AES	C2RMF (2010)
	adorned *nāga*-protected Buddha	NMC	*ga* 5986	71	large statue	Angkor Wat	2nd half of 12th C.	Angkor	Srah Srang (Angkor)	ICP-AES	C2RMF (2010)
	adorned Buddha in *dhyānamudrā* (*nāga*-protected Buddha)	NMC	*ga* 3378	39.5	large statue	Angkor Wat	1st half of 12th C.	East Angkor	Preah Khan of Kampong Svay / Bakan (Preah Vihear)	ICP-AES	C2RMF (2010)
	Vaishnavite triad astride Garuḍa (fragment of finial)	NMC	*ga* 3597	16.5	ritual paraphernalia	Angkor Wat	12th C.	East Angkor	"Christianville" (Kampong Thom)	ICP-AES	C2RMF (2010)
	Garuḍa (fragment of Buddhist finial)	NMC	*ga* 3605	25.5	ritual paraphernalia	Angkor Wat	12th C.	East Angkor	"Christianville" (Kampong Thom)	ICP-AES	C2RMF (2010)
	female figure (fragment of Buddhist finial)	NMC	*ga* 3611	16	ritual paraphernalia	Angkor Wat	12th C.	East Angkor	"Christianville" (Kampong Thom)	ICP-AES	C2RMF (2010)
	arcature (fragment of Buddhist finial)	NMC	*ga* 4081	16.5	ritual paraphernalia	Angkor Wat	12th C.	East Angkor	"Christianville" (Kampong Thom)	ICP-AES	C2RMF (2010)
	arcature (fragment of Buddhist finial)	NMC	*ga* 4082	15.5	ritual paraphernalia	Angkor Wat	12th C.	East Angkor	"Christianville" (Kampong Thom)	ICP-AES	C2RMF (2010)
	finial and stem with adorned *nāga*-protected Buddha	NMC	*ga* 2939	41	ritual paraphernalia	Angkor Wat	12th C.	Southeast	Tuol of Russei Pontea (Kampong Cham)	ICP-AES	C2RMF (2010)
	finial and stem with adorned *nāga*-protected Buddha	NMC	*ga* 4444	41	ritual paraphernalia	Angkor Wat	12th C.	Southeast	Tuol of Russei Pontea (Kampong Cham)	ICP-AES	C2RMF (2010)
	Viṣṇu astride Garuḍa	NMC	*ga* 3773	16	small statue	Angkor Wat	1st half of 12th C.	Southeast	Batheay (Kampong Cham)	ICP-AES	C2RMF (2010)
	Umā-Maheśvara on bull	NMC	*ga* 3778	14.5	small statue	Angkor Wat	early 12th C.	Southeast	Phnom Tuol (Kampong Cham)	ICP-AES	C2RMF (2010)
	female divinity	NMC	*ga* 4011	13.5	small statue	Angkor Wat	1st half of 12th C.	Southeast	Snay Pol (Prey Veng)	ICP-AES	C2RMF (2010)
	female divinity	NMC	*ga* 4090	13.5	small statue	Angkor Wat	1st half of 12th C.	Southeast	Snay Pol (Prey Veng)	ICP-AES	C2RMF (2010)
	arcature	NMC	*ga* 5348	46	medium statue	Angkor Wat	1st half of 12th C.	Southeast	Snay Pol (Prey Veng)	ICP-AES	C2RMF (2010)

adorned *nāga*-protected Buddha	NMC	*ga* 5356	20.5	medium statue	Angkor Wat	1st half of 12th C.	Southeast	Snay Pol (Prey Veng)	ICP-AES	C2RMF (2010)
finial with Vajrasattva	NMC	*ga* 5420	27	ritual paraphernalia	Angkor Wat	12th C.	Southeast	Vat Ha (Prey Veng)	ICP-AES	C2RMF (2010)
Dvārapāla	NMC	*ga* 5466	12.5	small statue	Angkor Wat	12th C.	Southeast	Vihear Thom (Kampong Cham)	ICP-AES	C2RMF (2010)
finial and stem with adorned *nāga*-protected Buddha	NMC	*ga* 5593	49.5	ritual paraphernalia	Angkor Wat	12th C.	Southeast	"Banteai Chha" (Kampong Cham)	DCP-AES	C2RMF (1996)
finial with Viṣṇu astride Garuḍa	NMC	*ga* 5676	32.5	ritual paraphernalia	Angkor Wat	12th C.	Southeast	Snay Pol (Prey Veng)	ICP-AES	C2RMF (2010)
Viṣṇu astride Garuḍa	NMC	*ga* 3596	13	small statue	Angkor Wat	1st half of 12th C.	South	Trapeang Kak (Takeo)	ICP-AES	C2RMF (2010)
nāga-shaped hook	NMC	*ga* 4729	18	vehicle fitting	Angkor Wat	12th C.	South	Vat Cheung Aek (Phnom Penh)	ICP-AES	C2RMF (2010)
adorned Buddha in *vitarkamudrā*	NMC	*ga* 5325	21	small statue	Angkor Wat	early 12th C.	South	"Phnom Eng Khmeng" (Kampong Speu)	ICP-AES	C2RMF (2010)
Lokeśvara	NMC	*ga* 5327	26	small statue	Angkor Wat	2nd half of 12th C.	South	Po (Takeo)	ICP-AES	C2RMF (2010)
adorned *nāga*-protected Buddha	NMC	*ga* 5357	41	medium statue	Angkor Wat	1st half of 12th C.	West	Anlong Sa (Banteay Meanchey)	ICP-AES	C2RMF (2010)
adorned *nāga*-protected Buddha	NMC	*ga* 3296	46	medium statue	Angkor Wat	late 12th–early 13th C.?	Champa	"Silver Towers" (Binh Dinh, Vietnam)	DCP-AES	C2RMF (1996)
dancing female figure	MG	C 6500	21	small statue	Angkor Wat	12th C.	Cambodia	?	ICP-AES	C2RMF (1999)
Lokeśvara	MG	MA 509	8	small statue	Angkor Wat	12th C.	Cambodia	?	ICP-AES	C2RMF (1999)
fragment of conch	WAM	54.2536	12	ritual paraphernalia	Angkor Wat	12th C.	?	?	ICP-AES	UCLA (1994)
Brahmanic triad (pedestal)	WAM	54.2715	8.5	small statue	Angkor Wat	1st half of 12th C.	?	?	ICP-AES	UCLA (1994)
Viṣṇu	WAM	54.2721	15.5	small statue	Angkor Wat	1st half of 12th C.	?	?	ICP-AES	UCLA (1994)
adorned Buddha in *vitarkamudrā*	WAM	54.2724	20	small statue	Angkor Wat	early 12th C.?	?	?	ICP-AES	UCLA (1994)
adorned Buddha in *vitarkamudrā* (body)	WAM	54.2727	57	large statue	Angkor Wat	2nd half of 12th C.	?	?	ICP-AES	UCLA (1994)
adorned Buddha in *abhayamudrā*	WAM	54.2728	18	small statue	Angkor Wat	early 12th C.	?	?	ICP-AES	UCLA (1994)
adorned *nāga*-protected Buddha	WAM	54.2730	20	small statue	Angkor Wat	2nd half of 12th C.	?	?	ICP-AES	UCLA (1994)
Prajñāpāramitā	NMC	*ga* 5333	15	small statue	Angkor Wat	2nd half of 12th C.	?	?	DCP-AES	C2RMF (1996)
Viṣṇu astride Garuḍa	MG	MA 539	14	small statue	Angkor Wat	1st half of 12th C.	?	?	ICP-AES	C2RMF (1999)
Lokeśvara	MG	MA 5960	63	large statue	Angkor Wat	2nd half of 12th C.?	?	?	ICP-AES	C2RMF (1999)
Prajñāpāramitā	MAA	03IA25	40	medium statue	Angkor Wat	1st half of 12th C.	?	?	PIXE	C2RMF (2003)
Lokeśvara	NMC	*ga* 6861	47	medium statue	Angkor Wat	2nd half of 12th C.	?	?	ICP-AES	C2RMF (2010)

cont'd

Table 1. *The 129 Khmer artifacts studied* cont'd.

	Designation	Museum	Inv. No.	Height (cm)	Type of Artifact	Style	Date	Provenance	Discovery Location	Metal Analysis	Laboratory & Date
Bayon	Buddhist triad (Lokeśvara, Buddha, Prajñāpāramitā)	NMC	*ga* 2424	49.5	medium statue	Bayon	late 12th–early 13th C.	Angkor	Prei Monti (Angkor)	DCP-AES	C2RMF (1996)
	Hevajramaṇḍala	NMC	*ga* 2494	20.5	small statue	Bayon	late 12th–early 13th C.	Angkor	Angkor region?	DCP-AES	C2RMF (1996)
	Hevajra	NMC	*ga* 4170	22.5	small statue	Bayon	late 12th–early 13th C.	Angkor	Royal Palace (Angkor Thom)	DCP-AES	C2RMF (1996)
	Lokeśvara	NMC	*ga* 5340	42	medium statue	Bayon	late 12th–early 13th C.	Angkor	North Khleang (Angkor Thom)	DCP-AES	C2RMF (1996)
	tripod	NMC	*ga* 5485	10	ritual paraphernalia	Bayon	late 12th–early 13th C.	Angkor	Tonle Sap banks (Angkor)	DCP-AES	C2RMF (1996)
	Buddhist triad (Lokeśvara, Buddha, Prajñāpāramitā)	MG	MG 18204	8	ritual paraphernalia	Bayon	late 12th–early 13th C.	Angkor	Angkor region	ICP-AES	C2RMF (1999)
	nāga-shaped finial	MG	MG 18888	29	vehicle fitting	Bayon	late 12th–early 13th C.	Angkor	West Mebon (Angkor)	ICP-AES	C2RMF (1999)
	fragments of *nāga*-shaped finial	NMC	*ga* 2989	37	vehicle fitting	Bayon	late 12th–early 13th C.	Angkor	Bayon (Angkor Thom)?	ICP-AES	C2RMF (2010)
	nāga-shaped finial	NMC	*ga* 5165	46	vehicle fitting	Bayon	late 12th–early 13th C.	Angkor	Bayon (Angkor Thom)?	ICP-AES	C2RMF (2010)
	Śiva	NMC	*ga* 3451	21	small statue	Bayon	late 12th–early 13th C.	Angkor	Royal Palace (Angkor Thom)	ICP-AES	C2RMF (2010)
	fragment of lamp	NMC	*ga* 3699	8.5	ritual paraphernalia	Bayon	late 12th–early 13th C.	Angkor	Bayon (Angkor Thom)?	ICP-AES	C2RMF (2010)
	fragment of lamp	NMC	*ga* 3846	22.5	ritual paraphernalia	Bayon	late 12th–early 13th C.	Angkor	Bayon (Angkor Thom)?	ICP-AES	C2RMF (2010)
	fragment of lamp	NMC	*ga* 4032	5	ritual paraphernalia	Bayon	late 12th–early 13th C.	Angkor	Bayon (Angkor Thom)?	ICP-AES	C2RMF (2010)
	fragment of lamp	NMC	*ga* 4152	11	ritual paraphernalia	Bayon	late 12th–early 13th C.	Angkor	Bayon (Angkor Thom)?	ICP-AES	C2RMF (2010)
	fragment of lamp	NMC	*ga* 5644	49	ritual paraphernalia	Bayon	late 12th–early 13th C.	Angkor	Bayon (Angkor Thom)?	ICP-AES	C2RMF (2010)
	fragment of perfume censer or lamp?	NMC	*ga* 3721	18	ritual paraphernalia	Bayon	late 12th–early 13th C.	Angkor	Tonle Sap banks (Angkor)	ICP-AES	C2RMF (2010)
	conch setting	NMC	*ga* 3788	17	ritual paraphernalia	Bayon	1196–1197	Angkor	Preah Khan (Angkor)	ICP-AES	C2RMF (2010)
	nāga-shaped finial	NMC	*ga* 4023	14.5	furniture part	Bayon	late 12th–early 13th C.	Angkor	Royal Palace (Angkor Thom)	ICP-AES	C2RMF (2010)
	female divinity	NMC	*ga* 4025	19.5	small statue	Bayon	late 12th–early 13th C.	Angkor	Royal Palace (Angkor Thom)	ICP-AES	C2RMF (2010)
	torso of adorned Buddha in *abhayamudrā*	NMC	*ga* 4047	15.5	small statue	Bayon	late 12th–early 13th C.	Angkor	Royal Palace (Angkor Thom)	ICP-AES	C2RMF (2010)

head of male divinity (Buddha?)	NMC	*ga* 4222	11.5	large statue	Bayon	late 12th–early 13th C.	Angkor	Royal Palace (Angkor Thom)	ICP-AES	C2RMF (2010)
Garuḍa-shaped finial	NMC	*ga* 4227	31	vehicle fitting	Bayon	early 13th C.	Angkor	Angkor Wat (Angkor)	ICP-AES	C2RMF (2010)
lion-shaped finial	NMC	*ga* 4228	28	vehicle fitting	Bayon	early 13th C.	Angkor	Royal Palace (Angkor Thom)	ICP-AES	C2RMF (2010)
coiffure of "radiating" Lokeśvara	NMC	*ga* 4250	9	large statue	Bayon	late 12th–early 13th C.	Angkor	Royal Palace (Angkor Thom)	ICP-AES	C2RMF (2010)
trident	NMC	*ga* 4266	25	statue attribute	Bayon	late 12th–early 13th C.	Angkor	Preah Khan (Angkor)	ICP-AES	C2RMF (2010)
Buddhist triad (Lokeśvara, female divinity, Prajñāpāramitā)	NMC	*ga* 4424	27	small statue	Bayon	late 12th–early 13th C.	Angkor	Srah Srang (Angkor)	ICP-AES	C2RMF (2010)
palanquin hook	NMC	*ga* 4443	18	vehicle fitting	Bayon	late 12th–early 13th C.?	Angkor	Angkor region?	ICP-AES	C2RMF (2010)
Shaivaite male divinity	NMC	*ga* 5292	15	small statue	Bayon	late 12th–early 13th C.	Angkor	Srah Srang (Angkor)	ICP-AES	C2RMF (2010)
Śiva	NMC	*ga* 5301	13.5	small statue	Bayon	late 12th–early 13th C.	Angkor	Srah Srang (Angkor)	ICP-AES	C2RMF (2010)
Lokeśvara	NMC	*ga* 5338	20	small statue	Bayon	late 12th–early 13th C.	Angkor	Banteay Kdei (Angkor)	ICP-AES	C2RMF (2010)
necklace of statue	NMC	*ga* 5627	7.5	statue attribute	Bayon	late 12th–early 13th C.	Angkor	Banteay Kdei (Angkor)	ICP-AES	C2RMF (2010)
miniature Buddhist shrine	NMC	*ga* 5630	19	small statue	Bayon	late 12th–early 13th C.	Angkor	Prei Monti (Angkor)	ICP-AES	C2RMF (2010)
Garuḍa-shaped finial	NMC	*ga* 5681	23.5	vehicle fitting	Bayon	early 13th C.	Angkor	Banteay Samre (Angkor)	ICP-AES	C2RMF (2010)
fragment of Garuḍa-shaped finial	NMC	*ga* 5683	16	vehicle fitting	Bayon	early 13th C.	Angkor	Royal Palace (Angkor Thom)	ICP-AES	C2RMF (2010)
palanquin ring	MG	MA 3563b	17	vehicle fitting	Bayon	late 12th–early 13th C.	East Angkor	Kampong Thom region	ICP-AES	C2RMF (1999)
mirror handle	MG	EO 801	21	ritual paraphernalia	Bayon	early 13th C.	East	Vat Phu (Champasak, Laos)	ICP-AES	C2RMF (1999)
palanquin hook	NMC	*ga* 4445	21	vehicle fitting	Bayon	late 12th–early 13th C.	East	"Phnom Dei Chral" (Stung Treng)	ICP-AES	C2RMF (2010)
palanquin ring	NMC	*ga* 4743	19	vehicle fitting	Bayon	late 12th–early 13th C.	East	Vat Phu (Champasak, Laos)?	ICP-AES	C2RMF (2010)
Lokeśvara	NMC	*ga* 2959	20	small statue	Bayon	late 12th?	Southeast	Kampong Cham region?	ICP-AES	C2RMF (2010)
palanquin hook	NMC	*ga* 3698	13.5	vehicle fitting	Bayon	late 12th–early 13th C.	South	Phnom Bayang (Takeo)	ICP-AES	C2RMF (2010)
female divinity	NMC	*ga* 4012	12.5	small statue	Bayon	late 12th–early 13th C.	South	Prey Sloek (Takeo)	ICP-AES	C2RMF (2010)
Garuḍa-shaped finial	NMC	*ga* 5680	31	vehicle fitting	Bayon	early 13th C.	South	"Phnom Sruoch" (Kampong Speu)	ICP-AES	C2RMF (2010)

cont'd

Table 1. *The 129 Khmer artifacts studied* cont'd.

Designation	Museum	Inv. No.	Height (cm)	Type of Artifact	Style	Date	Provenance	Discovery Location	Metal Analysis	Laboratory & Date
Vaishnavite triad (Śiva, Viṣṇu, Devī)	NMC	*ga* 2283	24.5	small statue	Bayon	late 12th–early 13th C.	West	Spean (Battambang)	DCP-AES	C2RMF (1996)
Prajñāpāramitā	NMC	*ga* 3540	25	small statue	Bayon	late 12th–early 13th C.	West	Battambang region?	ICP-AES	C2RMF (2010)
palanquin hook	NMC	*ga* 3783	20	vehicle fitting	Bayon	late 12th–early 13th C.	West	Mongkol Borei region (Banteay Meanchey)	ICP-AES	C2RMF (2010)
Buddhist triad (Lokeśvara, Buddha, Prajñāpāramitā)	NMC	*ga* 4417	18.5	small statue	Bayon	late 12th–early 13th C.	West	Dang Run (Banteay Meanchey)	ICP-AES	C2RMF (2010)
adorned Buddha in *vitarkamudrā*?	NMC	*ga* 5320	27	small statue	Bayon	late 12th–early 13th C.	West	Mong region (Battambang)?	ICP-AES	C2RMF (2010)
adorned Buddha in *vitarkamudrā*	NMC	*ga* 5410	14	small statue	Bayon	late 12th–early 13th C.	West	Paobun (Battambang)	ICP-AES	C2RMF (2010)
adorned Buddha in *vitarkamudrā*?	NMC	*ga* 5414	23.5	small statue	Bayon	late 12th–early 13th C.	West	Mong region (Battambang)?	ICP-AES	C2RMF (2010)
Lokeśvara	NMC	*ga* 5422	20	small statue	Bayon	late 12th–early 13th C.	West	Phnom Sampov (Battambang)	ICP-AES	C2RMF (2010)
nāga-shaped finial	NMC	*ga* 5679	21	vehicle fitting	Bayon	late 12th–early 13th C.	West	Battambang region	ICP-AES	C2RMF (2010)
Garuḍa-shaped finial	MG	MA 102	19	vehicle fitting	Bayon	late 12th–early 13th C.	Cambodia	?	ICP-AES	C2RMF (1999)
Prajñāpāramitā	MG	MA 508	12	small statue	Bayon	late 12th–early 13th C.	Cambodia	?	ICP-AES	C2RMF (1999)
tripod	MG	MA 541	12	ritual paraphernalia	Bayon	late 12th–early 13th C.	Cambodia	?	ICP-AES	C2RMF (1999)
Śiva	MG	MA 2110	18	small statue	Bayon	late 12th–early 13th C.	Cambodia	?	ICP-AES	C2RMF (1999)
conch	MG	MA 3564	25	ritual paraphernalia	Bayon	late 12th–early 13th C.	Cambodia	?	ICP-AES	C2RMF (1999)
tripod	MG	MA 3565	9	ritual paraphernalia	Bayon	late 12th–early 13th C.	Cambodia	?	ICP-AES	C2RMF (1999)
bell	MG	MA 18850	22	ritual paraphernalia	Bayon	late 12th–early 13th C.	Cambodia	?	ICP-AES	C2RMF (1999)
palanquin ring	MG	MG 18851b	19	vehicle fitting	Bayon	late 12th–early 13th C.	Cambodia	?	ICP-AES	C2RMF (1999)
adorned Buddha in *dhyānamudrā* (*nāga*-protected Buddha?)	WAM	54.2687	16	medium statue	Bayon	early 13th C.?	?	?	ICP-AES	UCLA (1994)
Lokeśvara	WAM	54.2689	16	small statue	Bayon	late 12th C.	?	?	ICP-AES	UCLA (1994)
Brahmanic triad (Śiva, Lakṣmī)	WAM	54.2715	12 (Śiva), 10.5 (Lakṣmī)	small statue	Bayon	late 12th–early 13th C.	?	?	ICP-AES	UCLA (1994)

“radiating” Lokeśvara	WAM	54.2726	53	large statue	Bayon	late 12th–early 13th C.	?	?	ICP-AES	UCLA (1994)
Brahmā	WAM	54.2734	29.5	large statue	Bayon	1st half of 13th C.	?	?	ICP-AES	UCLA (1994)
Śiva	WAM	54.2735	20	small statue	Bayon	early 13th C.	?	?	ICP-AES	UCLA (1994)
Yoginī	MG	MA 540	9	small statue	Bayon	late 12th–early 13th C.	?	?	ICP-AES	C2RMF (1999)
Hevajra	MG	MA 849	15	small statue	Bayon	late 12th–early 13th C.	?	?	ICP-AES	C2RMF (1999)
adorned Buddha in *varadamudrā*?	MG	MA 3788	15	small statue	Bayon	late 12th–early 13th C.	?	?	ICP-AES	C2RMF (1999)
female divinity	MG	MA 3789	21	small statue	Bayon	late 12th–early 13th C.	?	?	ICP-AES	C2RMF (1999)
“radiating” Lokeśvara	MG	MA 5940	39	medium statue	Bayon	late 12th–early 13th C.	?	?	ICP-AES	C2RMF (1999)
quadripod (support for lamp?)	MG	MA 5984	8	ritual paraphernalia	Bayon	late 12th–early 13th C.	?	?	ICP-AES	C2RMF (1999)
drum	MG	MA 5988	33	ritual paraphernalia	Bayon	late 12th–early 13th C.	?	?	ICP-AES	C2RMF (1999)
palanquin hook	MG	MA 5989	19	vehicle fitting	Bayon	late 12th–early 13th C.	?	?	ICP-AES	C2RMF (1999)
lamp	MG	MA 6939	57	ritual paraphernalia	Bayon	late 12th–early 13th C.	?	?	ICP-AES	C2RMF (2001)
Undetermined Vajra	MG	MA 1617	14	ritual paraphernalia	Angkor Wat / Bayon	12th–early 13th C.	Angkor	Angkor region	ICP-AES	C2RMF (1999)
back of statue	MG	MG 18175	20	medium statue	Angkor Wat / Bayon	12th–early 13th C.	Angkor	Angkor region	ICP-AES	C2RMF (1999)
bull of Śiva	NMC	*ga* 5739	41	large statue	Bayon / Post-Bayon	late 12th–13th C.	Angkor	West of Angkor Wat (Angkor)	ICP-AES	C2RMF (2010)
Vajra	NMC	*ga* 5483	12.5	ritual paraphernalia	Angkor Wat / Bayon	12th–early 13th C.	East Angkor	Preah Theat Khvav (Siem Reap)	DCP-AES	C2RMF (1996)
finial with adorned *nāga*-protected Buddha	NMC	*ga* 5685	16.5	furniture part / vehicle fitting?	Angkor Wat / Bayon	12th–early 13th C.	Southeast	Phdau (Kampong Cham)	ICP-AES	C2RMF (2010)
perfume censer or lamp?	NMC	*ga* 5467	16.5	ritual paraphernalia	Angkor Wat / Bayon	12th–early 13th C.	West	Dang Run (Banteay Meanchey)	DCP-AES	C2RMF (1996)
Haṃsa-shaped finial	NMC	*ga* 195	18.5	vehicle fitting	Angkor Wat / Bayon	12th–early 13th C.	West	Battambang region?	ICP-AES	C2RMF (2010)
perfume censer	MG	MA 5984	17	ritual paraphernalia	Angkor Wat / Bayon	12th–early 13th C.	?	?	ICP-AES	C2RMF (1999)

C2RMF = Centre de Recherche et de Restauration des musées de France; DCP-AES = direct current plasma–atomic emission spectroscopy; ICP-AES = inductively coupled plasma–atomic emission spectroscopy; MAA = Musée des Arts Asiatiques, Nice; MG = Musée Guimet, Paris; NMC = National Museum of Cambodia, Phnom Penh; PIXE = particle induced x-ray emission; UCLA = University of California, Los Angeles; WAM = Walters Art Museum, Baltimore.

Table 2. *Khmer artifacts of the studied corpus classified as "major" objects.*

	Designation	Inv. No.	Assembly	Decoration	Inscription	Other Examples (iconography, function, and/or size)
Angkor Wat	adorned Buddha in *abhayamudrā*	*ga* 2081	Y			Y
	Viṣṇu-Vāsudeva-Nārāyaṇa	*ga* 5291	Y			Y
	kneeling female figure (support for mirror)	*ga* 5476	?	gilding		N
	adorned *nāga*-protected Buddha	*ga* 5986	Y			Y
	adorned Buddha in *dhyānamudrā* (*nāga*-protected Buddha)	*ga* 3378	Y	gilding		Y
	pair of palanquins for Buddhist images	*ga* 3597, *ga* 3605, *ga* 3611, *ga* 4081, and *ga* 4082	Y	gilding		N
	Buddhist triad?	*ga* 4011, *ga* 4090, *ga* 5348, and *ga* 5356	Y			N
	adorned *nāga*-protected Buddha	*ga* 5357	Y			Y
	adorned *nāga*-protected Buddha	*ga* 3296	Y	fire gilding		Y
	Lokeśvara	MA 5960	Y			Y
	Prajñāpāramitā	03IA25	Y			N
	Lokeśvara	*ga* 6861	Y			Y
Bayon	Buddhist triad (Lokeśvara, Buddha, Prajñāpāramitā)	*ga* 2424	Y	fire gilding		N
	Hevajramaṇḍala	*ga* 2494	Y	gilding (fire gilding: Hevajra)		N
	Hevajra	*ga* 4170	Y	fire gilding, inlays		Y
	Lokeśvara	*ga* 5340	Y	gilding		Y
	pair of *nāga*-shaped finials	*ga* 2989 and *ga* 5165	Y	gilding (fire gilding?), inlays		N
	lamp	*ga* 3699, *ga* 3846, *ga* 4032, *ga* 4152, and *ga* 5644	Y	gilding (fire gilding?)		Y
	conch setting	*ga* 3788	N		K. 779	Y
	head of male divinity (Buddha?)	*ga* 4222	?			N
	Garuḍa-shaped finial	*ga* 4227	Y	inlays		Y
	coiffure of "radiating" Lokeśvara	*ga* 4250	?	gilding (fire gilding?), inlay		Y
	trident	*ga* 4266	N	gilding (fire gilding?), inlays		N
	necklace of statue	*ga* 5627	N	gilding (fire gilding?), inlays		Y
	miniature Buddhist shrine	*ga* 5630	N	gilding (fire gilding?)		N
	mirror handle	EO 801	N		K. 963	N
	Vaishnavite triad (Śiva, Viṣṇu, Devī)	*ga* 2283	Y	fire gilding		Y
	"radiating" Lokeśvara	54.2726	Y			Y
	Brahmā	54.2734	N			N
	"radiating" Lokeśvara	MA 5940	Y			Y
	lamp	MA 6939	Y			Y
Undetermined	bull of Śiva	*ga* 5739	Y	fire gilding	K. 1013	Y

The overall accuracy of the measurements was estimated by two means. On the one hand, four objects without any assembly were analyzed in several places (MA 541, MA 3565, MA 5984 [quadripod], and MA 5988). On the other hand, the different parts of the Bayon Buddhist triad (*ga* 2424) proved to be cast with the same metal batch (table 3): the variation

Table 3. *Metal variations: single assembled objects and pairs of finials.*

	Designation	Inv. No.	Number of Analyses	Parts Analyzed ("=" Indicates homogeneous parts; "/" indicates heterogeneous parts.)	Major Object
"Group A": Homogeneous ensembles	Vaishnavite triad	*ga* 2283	4	Śiva = Viṣṇu = Devī = pedestal	Y
	Buddhist triad	*ga* 2424	7	Lokeśvara = Buddha (figure = pedestal) = *nāga* (heads = body) = Prajñāpāramitā = pedestal	Y
	"radiating" Lokeśvara	MA 5940	3	body = arms	Y
	bell	MA 18850	2	bell = ball	
	"radiating" Lokeśvara	54.2726	2	body = PR arm	Y
	adorned *nāga*-protected Buddha	54.2730	2	Buddha (primary casting) = *nāga* heads (secondary casting)	
	adorned Buddha in dhyānamudrā (*nāga*-protected Buddha)	*ga* 3378	2	head = body	Y
	lamp	*ga* 3699, *ga* 3846, *ga* 4032, *ga* 4152, and *ga* 5644	6	fragments *ga* 3699 = *ga* 3846 = *ga* 4032 = *ga* 4152 = *ga* 5644	Y
	pair of *nāga*-shaped feet	*ga* 4254.1 and *ga* 4254.2	2	feet *ga* 4254.1 = *ga* 4254.2	
	finial with Vajrasattva	*ga* 5420	2	Vajrasattva = aureole	
	adorned *nāga*-protected Buddha	*ga* 5985	2	Buddha = *nāga* heads	
Group "B": Alloy variations	adorned Buddha in *abhayamudrā*	*ga* 2081	2	head / body	Y
	nāga-shaped finial	MG 18889	2	central head / side head	
	pair of palanquins for Buddhist images	*ga* 3597, *ga* 3605, *ga* 3611, *ga* 4081, and *ga* 4082	5	fragments *ga* 3597 = *ga* 3605 = *ga* 3611 / arcatures *ga* 4081 = *ga* 4082	Y
	Buddhist triad?	*ga* 4011, *ga* 4090, *ga* 5348, and *ga* 5356	5	Buddha *ga* 5356 = female divinity *ga* 4011 = female divinity *ga* 4090 / arcature *ga* 5348	Y
	Buddhist triad	*ga* 4417	4	Lokeśvara = Buddha = Prajñāpāramitā / pedestal	
"Group C": Impurities variations	Hevajramaṇḍala	*ga* 2494	3	Yoginī / miniature shrine (body = pedestal)	Y
	lamp	MA 6939	5	kneeling figure / main flower = main stem (?) / main stem (lower part) = support	Y
	Brahmanic triad	54.2715	2	Śiva / Lakṣmī	
	pair of *nāga*-shaped finials	*ga* 2989 and *ga* 5165	2	finial *ga* 2989 / finial *ga* 5165	Y
	Shaivaite triad?	*ga* 5292 and *ga* 5301	2	Shaivaite male divinity *ga* 5292 / Śiva *ga* 5301	
"Group D": Alloy and impurities variations	adorned *nāga*-protected Buddha	*ga* 3296	3	Buddha = *nāga* heads / *nāga* body	Y
	Lokeśvara	*ga* 5340	4	upper part (head = PR arm) / lower part (cloth / PL foot)	Y
	finial with adorned *nāga*-protected Buddha	*ga* 5593	3	Buddha / *nāga* central heads / stem	
	Lokeśvara	MA 5960	3	body / arms	Y
	pair of finials with adorned *nāga*-protected Buddha	*ga* 2939 and *ga* 4444	4	finial *ga* 2939 (*nāga* side heads) / finial *ga* 4444 (Buddha / *nāga* central heads / stem)	
	Buddhist triad	*ga* 4424	3	Lokeśvara / female divinity / Prajñāpāramitā	
	adorned *nāga*-protected Buddha	*ga* 5986	3	Buddha / *nāga* (heads = body)	Y

of composition of the seven analyzed parts was used as an indicator of accuracy as well. The results are reported in the appendix (table A2). A variation of up to 30% may be observed for most elements. Thus, a dispersion of less than 30% should be not considered as significant for any of the elements. In some cases, higher variations may occur, due to the proximity of the measured value with the detection limit. This is notably the case for gold. In such situations, larger variations may be accepted. Lead dispersion might be higher than 30% as well, due to the well-known heterogeneity of lead distribution in copper alloys.

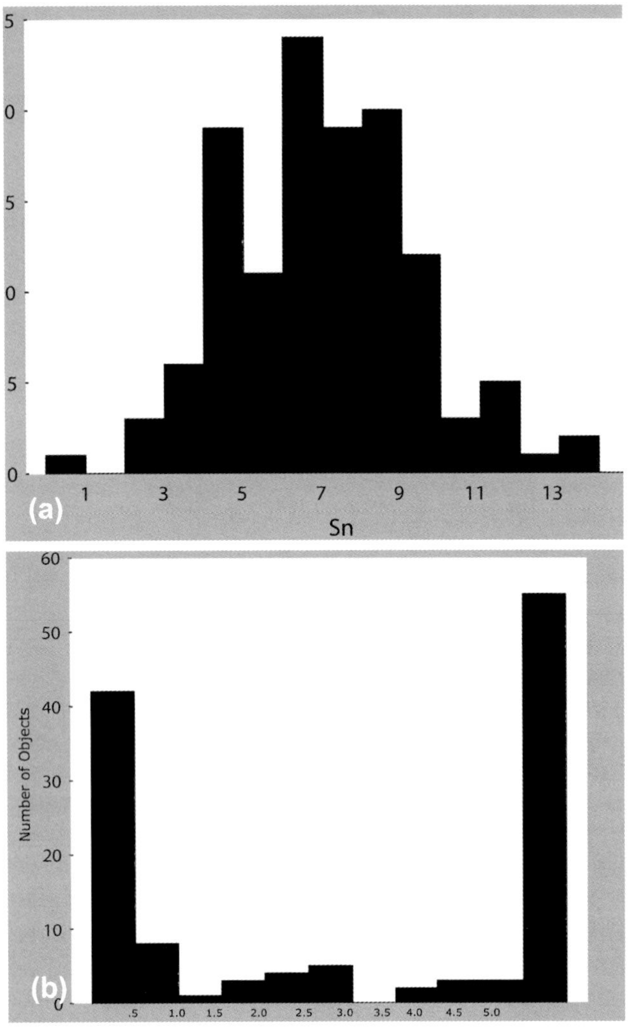

Figure 3. Distribution of the (**a**) tin (Sn) and (**b**) lead (Pb) contents within the 126 samples studied (wt.%)—that is, within the 113 ensembles dated from the early twelfth century to the early thirteenth century.

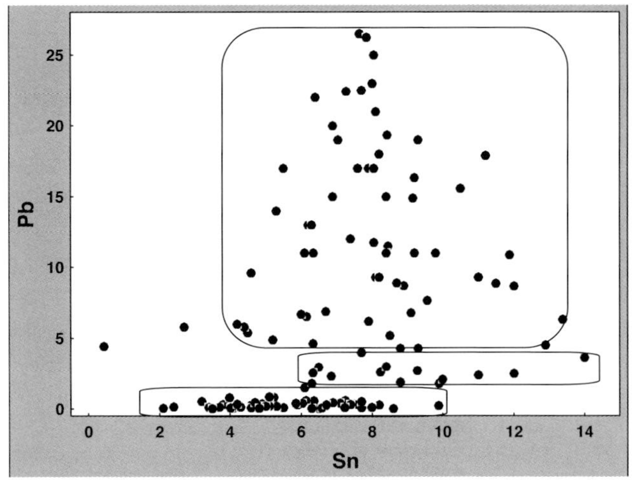

Figure 4. Tin (Sn) and lead (Pb) contents within the 126 samples studied (wt.%) showing three types of alloys.

The problems entailed in mixing the results of different analytical campaigns are well understood, but the types of analyses and results used in this study were believed to be reliably comparable.

Results

Khmer Copper-base Alloys (Twelfth–Thirteenth Centuries)

Bronzes and leaded bronzes

All the selected artifacts are composed of bronze or leaded bronze, that is, copper-tin alloys with or without lead, with one exception—an object found to be made of brass. Based on the statistical distribution of the tin and lead contents (fig. 3) and on the relationship between the tin and lead contents (fig. 4), three bronze alloys may be distinguished.

1. Bronze (fifty-one results)
   These objects are composed of copper alloys containing between 3 and 9 wt.% tin. Most of them bear less than 1 wt.% lead. A few extreme examples are observed, either low in tin (two results are down to 2 wt.%) or high in tin (one result shows up to 10 wt.%). Note that the Bayon Hevajramandala (*ga* 2494) is counted as a bronze, although its central figure of Hevajra is made of a remarkable silver-tin-copper alloy exhibiting 17 wt.% silver and 15 wt.% tin (Arc'Antique 1997, file no. 96408; Bourgarit et al. 2003, 111, 118, where it is mistakenly confused with a Hevajra statue found at the Royal Palace of Angkor Thom [*ga* 4170]).[7]
2. Leaded bronze (sixty-four results)
   These objects are composed of leaded bronze, that is alloys containing between 4 and 12 wt.% tin (sometimes up to 14 wt.%), and more than 3.5 wt.% lead, although the lead contents show large discrepancies, with values reaching up to 26 wt.%. No further sorting could be made on the basis of the lead content.
3. Low-lead bronze (ten results)
   An intermediate alloy composition based on lead levels as well may be sorted out, that is, bronzes containing between 1.5 and 3 wt.% lead and between 6 and 12 wt.% tin. Only ten results belong to this group, and more elemental analyses would be needed to characterize it.

In addition, one leaded brass composition has been identified: it is an adorned Buddha protected by the *nāga* associated to the Angkor Wat style and perhaps dated to the so-called "Angkor Wat period" as well (*ga* 5985). The statue exhibits 18 wt.% zinc and 4 wt.% lead (and 0.4 wt.% tin).

A chronological evolution of alloying practices

Contrary to what had been observed previously (Bourgarit et al. 2003, 111–12, 117–18), the sorting into three types of alloys may be partially related to chronological aspects. On the one hand, Bayon artifacts clearly show a high number of compositions with low lead content, that is, with less than 1 wt.% lead (figs. 5 and 6). It is especially true for the rare

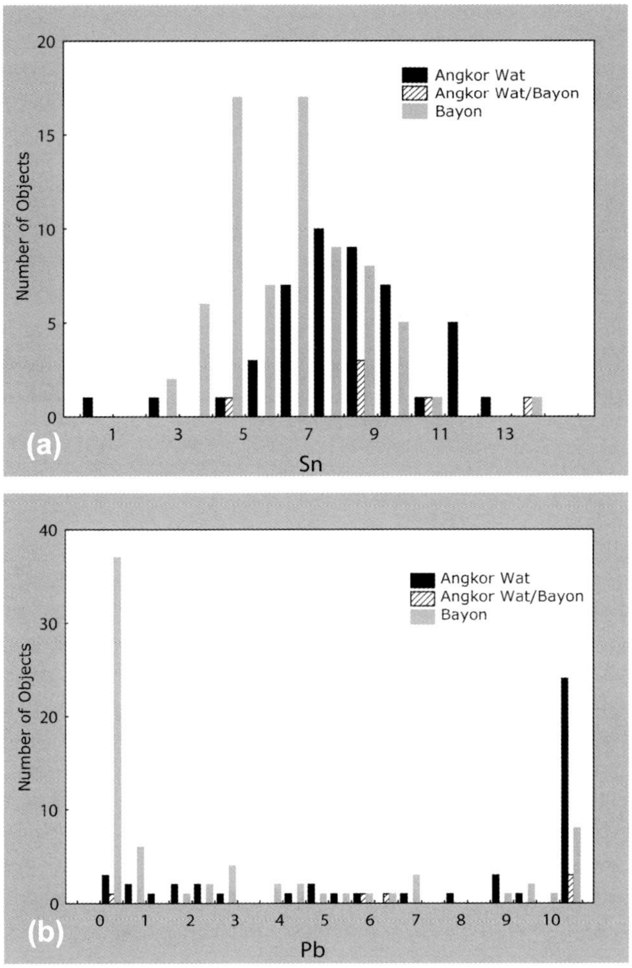

Figure 5. Distribution of the (a) tin (Sn) and (b) lead (Pb) contents within the 126 samples studied (wt.%), according to style.

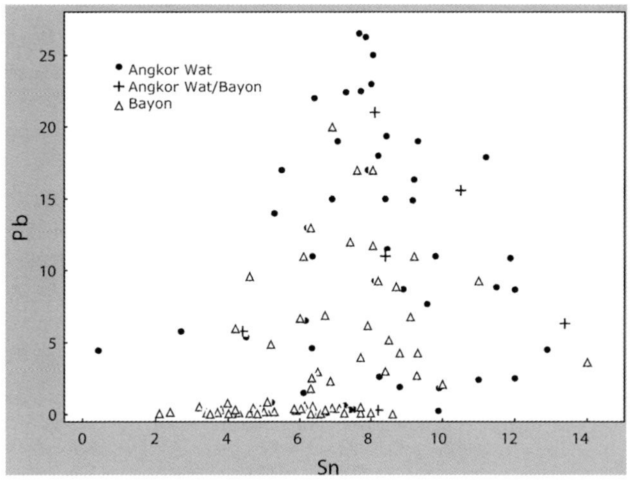

Figure 6. Tin (Sn) and lead (Pb) contents within the 126 samples studied (wt.%), according to style.

large statues of the corpus—more than 30 cm high—as well as for the finials. That said, lead may appear in some Bayon objects. Indeed, five results are low-lead bronze (between 2 and 3 wt.% lead), and another twenty-four results are leaded bronze. Note that small statues and, particularly, parts of triads are well represented in the leaded bronze group. Yet even for the latter artifacts, lead contents are relatively low when compared to the extremes of between 4 and 13 wt.% lead observed in Angkor Wat objects, with some contents as high as 26 wt.%. Only two atypical Bayon artifacts contain more than 13 wt.% lead (*ga* 2959 and MG 18204). Similarly, the lowest tin contents are seen within the Bayon bronzes (see figs. 5 and 6). More precisely, tin contents are distributed along two poles. The high-tin pole (forty-one results with 6 to 9 wt.%) is similar to what can be observed in the Angkor Wat objects. The low-tin pole (twenty-eight results with 2 to 5.3 wt.%) is specific to the Bayon corpus. Note that the low-tin group is marked by nearly unleaded bronzes, whereas the high-tin one yields both leaded and unleaded bronzes. Only three ritual artifacts contain 10 or more wt.% tin (*ga* 3788, EO 801, and MA 5988). Among them the conch setting (*ga* 3788) exhibits an atypical composition with relatively high levels of arsenic and antimony.

On the other hand, lead compositions are more widely distributed in the Angkor Wat artifacts (see fig. 5), and unleaded bronze compositions are rather rare (see fig. 6); only four items were identified as such, both statues and objects (*ga* 3296, *ga* 4254.1&2, MA 5960 [body], and MG 18889). On the contrary, leaded bronzes clearly represent a majority, with the highest lead contents of the corpus: twenty out of forty-six results contain 13 or more wt.% lead. Among them are several typical statues of the Angkor Wat style clearly dated to the first half of the twelfth century.[8] In addition, it is worth pointing out that the tin contents within the Angkor Wat group are generally higher than those encountered in the Bayon group. Hence, more than half of the items in the Angkor Wat group exhibit between 6 and 8.5 wt.% tin (twenty-four results). Only two atypical objects contained less than 5 wt.% (*ga* 5466 and *ga* 4729), whereas very high tin contents between 10 and 13 wt.% were also evidenced for some Buddha statues (*ga* 5325, 54.2715 [pedestal], 54.2724, and 54.2727 [body]) and a ritual conch (54.2536).[9] As mentioned above, another Buddha statue (54.2728) dated to the twelfth century, specifically the early twelfth century, has a very high tin content (22 wt.%), which is quite atypical.

A regional influence?

One may wonder whether the chronological evolution of alloying practices described above may actually pertain to the provenance of the artifacts rather than to their style. Almost half of the Bayon corpus comes from the Angkor region, whereas less than one quarter of the Angkor Wat corpus may be securely attributed to the "Angkor" group (fig. 7). Moreover, most Bayon objects from the Angkor region are unleaded bronzes. Note also that they stand for almost half of the low-tin bronzes. Interestingly, combined metal and clay-core analyses performed on the Khmer and Khmer-influenced objects from the Walters Art Museum collection led to the identification of a small group of statues low in tin—around 4 wt.%—with little, if any, lead (54.2689, 54.2715 [Śiva and Lakṣmī], 54.2726, and 54.2734, which are also part of the studied corpus; see Drayman-Weisser et al. 1997, 279); they have been dated to the Bayon period and

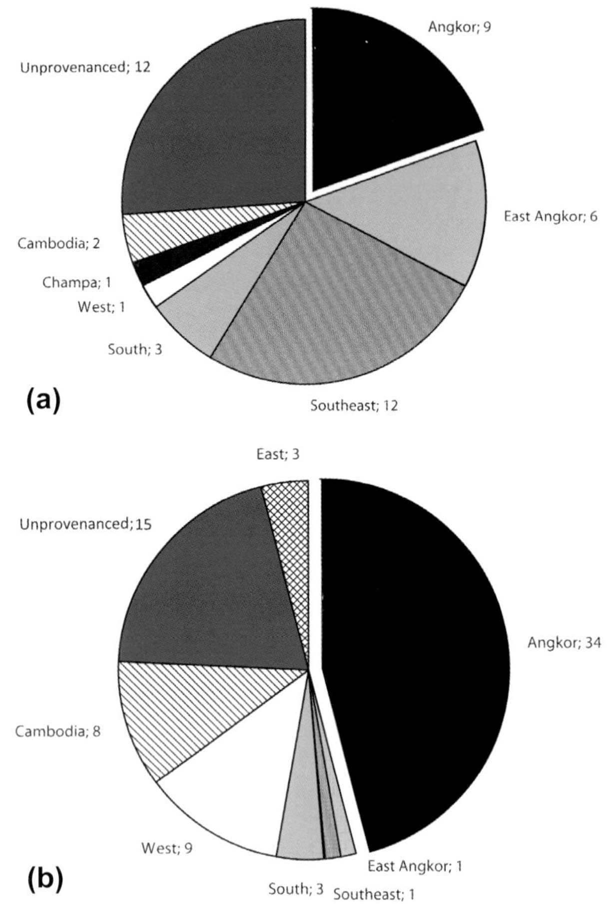

Figure 7. Style and provenance of the 129 Khmer artifacts studied. a. Angkor Wat style; b. Bayon style.

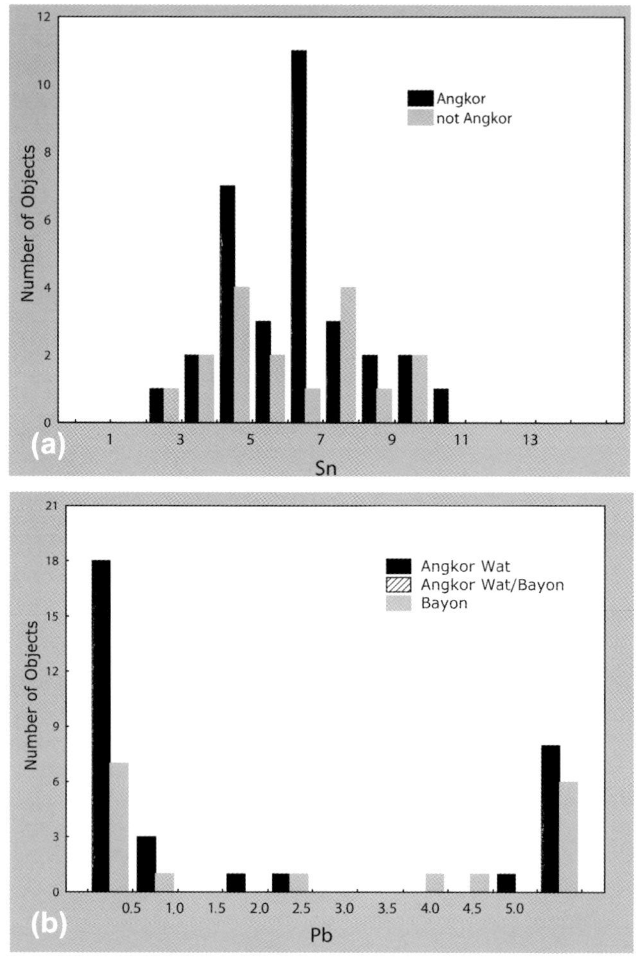

Figure 8. Distribution of the (a) tin (Sn) and (b) lead (Pb) contents within the seventy-four Bayon-style artifacts (wt.%), according to the provenance.

their production tentatively located in the Angkor region. The assumption of metallurgical practices specific to the Angkor region is thus quite interesting. Yet, the distribution of lead and tin contents in Bayon items found in the Angkor region is very similar to the one of the Bayon items found in other regions of the Angkorian kingdom (fig. 8): more analyses are needed to ascertain the geographical or the chronological hypothesis.

Another group of artifacts deserving some attention is the group of fourteen "Southeast" items, referring mostly to the Angkor Wat style. Among them, eight ensembles show similar alloy compositions high in both tin and lead, respectively 8–12 wt.% and 11–26.5 wt.%. Only one of them is unusual in exhibiting almost 1 wt.% zinc (*ga* 5676).[10] These objects may have been produced locally, since most of them have been found in one small region, which corresponds today to the western part of Kampong Cham province (Batheay and Prey Chhor districts) and the northern part of Prey Veng province (Sithor Kandal district). They are mostly Buddhist, with both statues (*ga* 2959, *ga* 4011, *ga* 4090, *ga* 5348, and *ga* 5356) and finials bearing Buddha or Vajrasattva images (*ga* 2939, *ga* 4444, *ga* 5420, *ga* 5593, and *ga* 5685). One statue of Visnu astride Garuda (*ga* 3773) and one Vaishnavite finial (*ga* 5676) are also part of the group.

In this frame, it would also be worth reconsidering the northeastern and central Thailand attributions proposed previously for some Khmer and Khmer-influenced objects from the Walters Art Museum collection on the basis of specific iconographical, stylistical, and technical features (Drayman-Weisser et al. 1997, 279; about the "Phimai tradition," see also Woodward 1997, 79; 2003, 211). Unfortunately, the supposed "Phimai" composition characterized by relatively high tin and lead contents (respectively 11–22 wt.% and 6–17 wt.%) relies only on the analysis of three Khmer bronzes (54.2536, 54.2715 [pedestal], and 54.2728, which are also part of the studied corpus) and possibly a Khmer-influenced work (54.2732; see Woodward 1997, cat. no. 39).[11]

Impurity Patterns

No correlation between major and minor elements could be found, except for a linear relationship between lead, antimony, and bismuth (fig. 9). Once antimony and bismuth are decorrelated from lead, most impurities present a similar pattern. Silver, arsenic, cobalt, nickel, and antimony contents show few variations, with similar mean contents for all elements of between 0.05 and 0.1 wt.%. However, some discrepancies may be observed.

First, the distribution of nickel content may be related to chronological aspects, with statistically higher values in the Bayon (between 0.2 and 0.3 wt.%) than in the Angkor Wat artifacts (between 0.05 and 0.1 wt.%) (fig. 10).

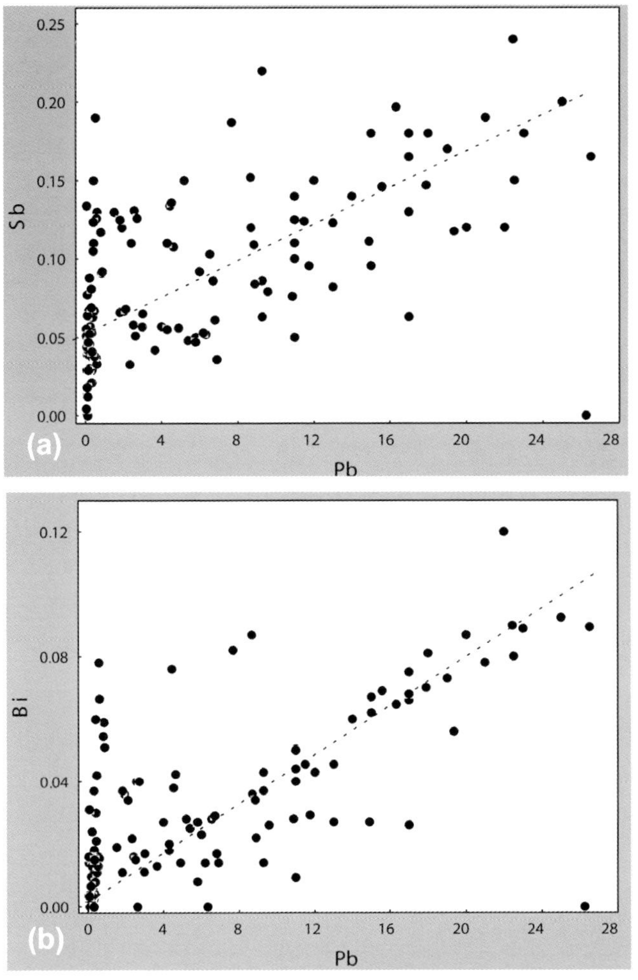

Figure 9. *Correlation between **(a)** lead (Pb) and antimony (Sb) and **(b)** lead (Pb) and bismuth (Bi) within the 126 samples studied (wt.%). A generally linear relationship may be represented by the dotted lines. Only the* Prajñāpāramitā *from the Musée des Arts Asiatiques, Nice (03IA25), does not follow this correlation, since the antimony and bismuth contents are below the detection limits of the particle induced x-ray emission (PIXE) analysis.*

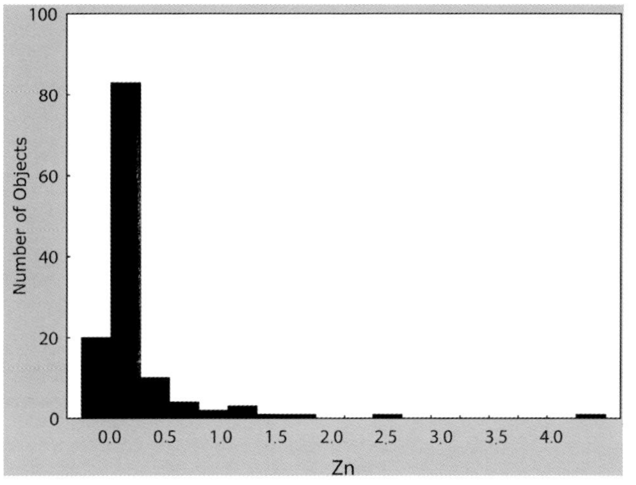

Figure 11. *Distribution of the zinc (Zn) contents within the 126 samples studied (wt.%).*

In addition, among the other impurities that show unusual patterns on isolated objects (see appendix, table A3), gold and zinc deserve some particular attention. The case of zinc has already been noted when discussing alloying practices. One can add to the unusual zinc-containing alloys twelve results of bronzes or leaded bronzes showing particularly high zinc contents ranging from 0.5 to 2.5 wt.% (fig. 11). Moreover, whereas most selected artifacts show less than 0.01 wt.% gold, some twenty gold-rich copper alloys were identified, all bearing more than 0.08 wt.% (fig. 12). In particular, four statues happen to contain up to between 0.43 and 0.5 wt.% gold (*ga* 3451, *ga* 4025, *ga* 5320, and *ga* 5422).

Metal Variations: Single Assembled Objects and Pairs of Finials

To follow up previous work carried out on metal variations within single objects composed of multiple parts (Bourgarit et al. 2003, 113–15, table 4), some more items have been looked at, but with an exclusive focus on the twelfth and early thirteenth century. Twenty-eight ensembles from this period, which also included pairs of finials, have thus been studied.[12]

This study led to four different situations (see table 3). Almost one-half of the studied ensembles (eleven out

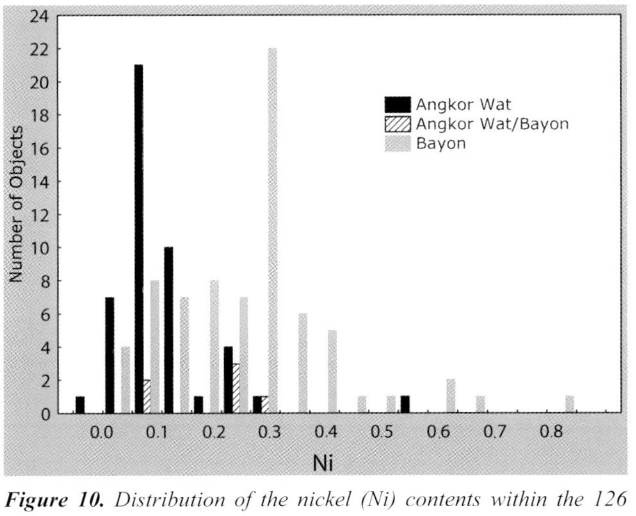

Figure 10. *Distribution of the nickel (Ni) contents within the 126 samples studied (wt.%), according to style.*

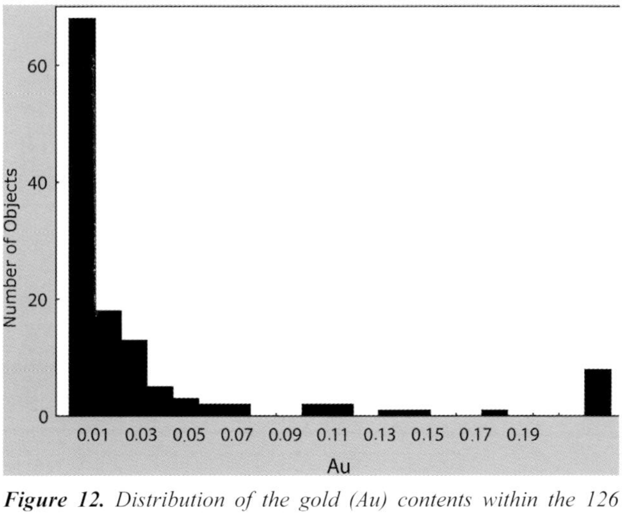

Figure 12. *Distribution of the gold (Au) contents within the 126 samples studied (wt.%).*

of twenty-eight) show a remarkable metal composition homogeneity, with similar contents in all analyzed parts in alloying elements as well as in impurities ("group A"). This is especially the case with five fragments, which have finally been identified as a single assembled lamp dated to the Bayon style (*ga* 3699, *ga* 3846, *ga* 4032, *ga* 4152, and *ga* 5644; see Groslier 1931, 114, 116, pl. 44-2 and 45-1, where this association was first proposed).[13] The other half of the ensembles show discrepancies either on the alloying elements ("group B": five ensembles), on the impurities ("group C": five ensembles), or on both ("group D": seven ensembles). Note that one more group has been added to the previous similar classification, namely "group C." No particular size, provenance, or quality of execution could be associated to one of these groups. Yet, of note is the fact that only Bayon artifacts are present in "group C."

Discussion

From Leaded Bronzes to Bronzes: Technical and/or Economical Constraints

Whereas tin and lead appear as the main alloying elements during the twelfth and early thirteenth century, a change in the choice of alloys seems to have occurred between the first half of the twelfth century and the late twelfth century. This change is marked by a move away from the use of leaded bronzes toward a greater employment of unleaded bronzes with slightly lower tin contents.

As far as technical constraints having had a role in this change, one can rule out the lead effect on the fluidity of liquid metal, in spite of the fact that this effect still has to be properly characterized.[14] There is as yet no indication of an increase of casting complexity within the Angkor Wat productions when compared to the Bayon ones, which might have justified variations of lead input in the melt. The gilding issue may, in turn, have much more influenced the change of alloying practices. Fire gilding seems to have become a common decorative technique from the late twelfth century onward, as attested by a series of fire-gilded bronze statues and objects clearly dated to the Bayon style and period as well as by other textual and archaeological evidences.[15] The spread of fire gilding may have promoted the use of unleaded copper alloys, as seen in the Roman and Tibetan "bronze" statuary. It has been observed that fire gilding was best carried out on unleaded copper alloys (Oddy et al. 1981, 92–94; Oddy 2000, 9–14). Lead can combine with mercury and can form white spots on the gilded surface.[16] Interestingly, the rule holds true with the few Khmer bronzes that were subjected to both elemental and surface analysis in the late 1990s. All seven statues from the collections of the National Museum of Cambodia and the Musée Guimet that were fire gilded were made of unleaded bronze (Arc'Antique 1997, file no. 96408 [*ga* 2494]; Bourgarit et al. 2003, 115–16, table 5). This group includes one bronze believed to date to the sixth century (MG 18318; Baptiste and Zéphir 2008, cat. no. 24); another associated with the Angkor Wat style but probably produced later than the Angkor Wat period (*ga* 3296; see note 15); four Bayon statues dated to the late twelfth or early thirteenth century (*ga* 2283, *ga* 2424, *ga* 4170, and MA 102);[17] and finally, another object of the studied corpus recently analyzed by x-ray fluorescence spectroscopy (XRF) at the Freer and Sackler Galleries, that being the large bull statue (*ga* 5739) roughly dated to the late twelfth century to thirteenth century (for which only the head was subject to elemental analysis; Cort and Jett 2010, 98–99, fig. 49). Because of these few examples, it thus seems that the importance of copper-alloy composition for fire gilding was known by Khmer bronze craftsmen from the earliest uses of the technique around the sixth/eighth centuries to at least the thirteenth century.

Economic constraints may have also promoted a change of alloying practices. In particular, it may be suggested that the wider use of lead during the Angkor Wat period was intended to replace copper, which seemed to be a more expensive or, at least, a more valuable metal. As seen through Khmer epigraphy, lead—and tin—was placed below copper in the hierarchy of metals (Pottier 1997, 193–94). Such an economical explanation would imply, in turn, that some copper supply problems would have entailed the use of cheaper lead in place of copper during the Angkor Wat period, before copper supply increased and copper value changed again from the beginning of the Bayon period. Whereas both supply and circulation of Angkorian copper remain poorly understood, the change in the copper impurity pattern we observed may indicate a change in copper sources. This will be further discussed in the following section.

New Data for a Study of Angkorian Copper Supply

Cambodia—as well as the Mekong delta—has often been described as a metal-poor area, and following Bennett Bronson (1992, 78, 97–98), a consensus emerged around the idea that most nonferrous metals (copper, tin, lead, zinc, and silver) were necessarily imported during the Angkorian period. Copper ores occasionally cited as possible sources for Angkorian copper are thus often located outside Cambodian territory. Two groups may be distinguished:

- "regional" copper ores located outside the Angkorian kingdom, in central Laos (Pottier 1997, 197) or the Malay peninsula (Baptiste and Zéphir 2004, 183); and
- "provincial" copper ores located in regions more or less closely linked to Angkor, in the central Thailand and Lopburi regions (Strahan 1997, 36) or in southern Laos, near Wat Phu temple (Hendrickson 2007, 230–31, fig. 10.6).

One may note however, that without any archaeological field surveys conducted in Cambodia to find traces of early mining (with the exception of that done for iron sources; see Hendrickson et al. 2010), the possibility of a "local" copper supply may not be completely rejected for the Angkorian period. Several small copper deposits do indeed exist, especially in the region of Preah Khan of

Kampong Svay (Bakan), located around 100 km east of Angkor (Hendrickson 2007, 230–31, fig. 10.6; see also Dottin 1971, 38–40).

Given a variety of possible ore sources, the variation of nickel contents found in bronzes dated between the first half of the twelfth century and the late twelfth century seems particularly interesting. Indeed, this change in the copper impurity pattern may indicate a change of copper source and thus the use of at least two different copper ores during the twelfth and early thirteenth century. This contradicts one of the conclusions of the analytical studies conducted by the C2RMF in the late 1990s (Bourgarit et al. 2003, 112–13, 117, fig. 12)—namely, that for almost one thousand years of Khmer copper-based metallurgy (from the sixth/seventh centuries to the sixteenth century), copper exhibits a uniform fingerprint indicating a single copper ore source.

Moreover, within the Bayon artifacts analyzed so far, it seems possible to distinguish two types of copper, as evidenced by the variation of lead, bismuth, and antimony contents. Indeed, a small group of seven objects (*ga* 2494; *ga* 2989 and *ga* 5165; *ga* 4023; *ga* 4025; MA 3565; and MG 18851b) bearing very high contents of antimony and bismuth do not clearly relate to the correlations already seen between the three elements. In this case, the bismuth and antimony contents would then have been the result of not only the lead source but also the copper source, and especially a high-bismuth-bearing copper ore (0.05 wt.%). Two atypical Angkor Wat artifacts also show similar bismuth and antimony contents (*ga* 3296 and *ga* 4254.1&2).[18]

In addition, the particularly high levels of gold in some items may testify to the use of particular copper ores naturally rich in gold by Khmer bronze craftsmen. Albeit limited in number, such copper ores are, indeed, evident in the Malay peninsula—one of the "foreign" sources for Angkorian copper cited earlier—and may have been imported into the Angkorian kingdom.[19]

One final comment to make with regard to the copper supply issue concerns one particular item that stands out because of its high arsenic content (1.5 wt.%); it is the Bayon conch setting (*ga* 3788) dated to 1196–1197 by an inscription, which also mentions that it was an offering of King Jayavarman VII (ca. 1181–1218) to one of the deities of the Preah Khan temple in Angkor (Soutif 2008). One wonders whether or not it is an arsenical bronze, which is an alloy known in the region by other more or less contemporary examples (Craddock 1981, 23–24; Jett 1991; Reedy 1997a; Woodward 1997, cat. nos. 15, 50, 81, 82, and 92; Hykin et al. 2007, 95–97). In this case, however, it appears that arsenic hardly had any alloying effect, since the leaded bronze with around 11 wt.% tin and 9 wt.% lead was already strongly alloyed. Therefore, the arsenic should rather be considered as a copper impurity, as should the antimony, which is also high (0.22 wt.%). In addition, nine other Bayon artifacts corresponding to four ensembles, which are all low-tin bronzes, also bear high arsenic contents of around 0.9 wt.%: a lamp (*ga* 3699, *ga* 3846, *ga* 4032, *ga* 4152, and *ga* 5644), a quadripod (MA 5984), a pair of finials (*ga* 2989 and *ga* 5165), and a palanquin hook (*ga* 4445). Similarly, all these objects—the conch setting included—may raise the question of a deliberate selection of a high-arsenic-bearing copper ore.

Finally, note that contrary to the copper source, the lead source seems not to have changed during the twelfth and early thirteenth century, as seen by the continuity of its impurity pattern marked by antimony and bismuth. However, nothing may be said concerning tin sources.

Brass: An Exception in Khmer Copper-Based Metallurgy

The present study strongly confirms what had been already stated regarding brass and the production of Khmer copper-alloy statues and ritual artifacts (Bourgarit et al. 2003, 111–12, 116–17): brass does definitely not pertain to any Khmer alloying practice—at least during the pre-Angkorian and Angkorian periods. As mentioned earlier, only one item in the studied corpus happened to be made of a brass alloy—the Angkor Wat adorned Buddha protected by the *nāga* (*ga* 5985).

The earliest example of brass known in the Angkorian period is a fragmentary Visnu astride Garuda from the Musée Guimet, presumably dated to the eleventh to twelfth centuries (MA 2592; Bourgarit et al. 2003, appendix 2; Baptiste and Zéphir 2008, cat. no. 167). A Buddha in *bhūmiśparsamudrā* from the same collection, first attributed to the Bayon style but which may be slightly later (thirteenth–fourteenth centuries), contains large amounts of zinc as well (MG 9807; Bourgarit et al. 2003, 117, appendix 2).[20] However, contrary to the Angkor Wat *nāga*-protected Buddha, which is a leaded brass (4.5 wt.% lead), both statues are tin brasses (respectively 6 and 4 wt.% tin) and may reflect another metallurgical tradition. A pre-Angkorian example of brass was also recently identified in the collection of the National Museum of Cambodia (*ga* 6940; Bourgarit 2009; Jett 2010a, 82–83, 86, fig. 36; Jett 2010b, 50–51, fig. 4): it is a Khmer Maitreya dated to the second half of the seventh century and made of a quaternary alloy or leaded tin brass (9.5 wt.% zinc, 4.5 wt.% lead, and 2 wt.% tin). It would be the earliest example of brass known in Southeast Asia. Closest in age to the Maitreya figure would be a leaded brass Buddha from Thailand, attributed to the Dvāravatī period and dated to the ninth century (Woodward 1997, cat. no. 14).

According to the current state of knowledge, only four out of the around 170 copper-based Khmer statues and ritual artifacts (from the sixth/seventh centuries to the sixteenth century) that have been subject to elemental analysis are made of brass. Moreover, the associated alloys show significant variations, being of three different types (leaded brass, red brass [a copper, tin, and zinc alloy], and leaded red brass). Thus, it clearly appears that in ancient Cambodia zinc was not a common alloying element for copper—unlike tin and lead—and no continuous tradition of brass production may be identified. Based on the example of neighboring Thailand, it seems, in fact, that the introduction of brass in the region took place quite late, probably around the fifteenth century through the influence of Chinese metallurgy (Woodward 1997, 186–87).[21]

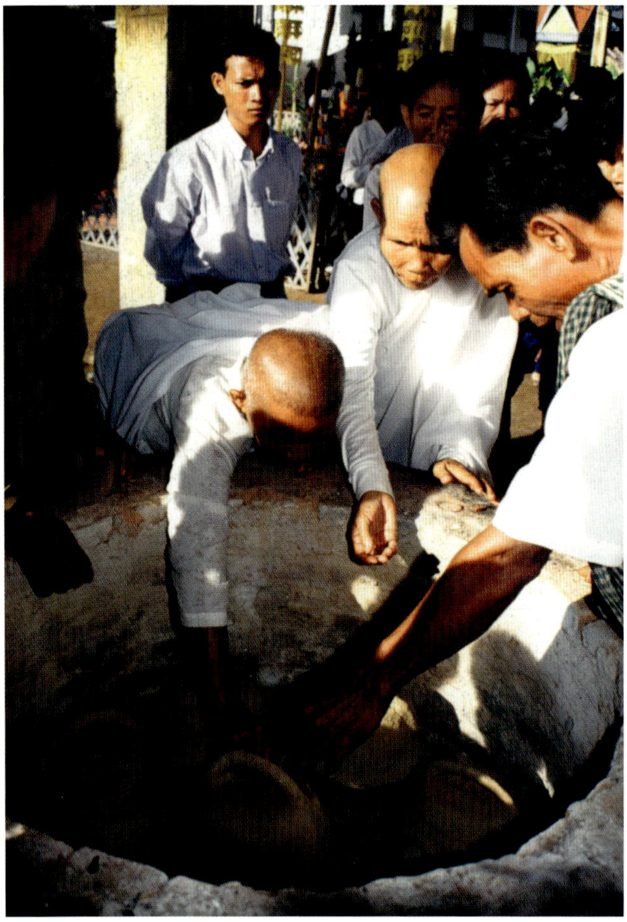

Figure 13. *Devotional offerings of jewelry during the casting of a Buddha image, Wat Prek Phneou, Kandal province, Cambodia, 1994. Photograph by Ang Choulean and Ashley Thompson.*

It is beyond the scope of this paper to discuss the reasons for these occurrences of brass in Khmer metallurgy. That said, it may be interesting to have a look at the items bearing high levels of zinc impurities, namely between 0.5 and 2.5 wt.%. The import of high-zinc-bearing copper or lead ores may not be discarded, although this may not explain zinc contents of more than 1 wt.% (Craddock 1981, 22). One should also consider the possibility of remelting earlier zinc-containing bronzes. Unfortunately the studied corpus is still too small—and still not thoroughly enough documented—to discuss this possibility. One last possible hypothesis is that small amounts of zinc were added deliberately. Whereas the three explanations just noted have been previously put forward and discussed (Bourgarit et al. 2003, 117),[22] the present study will focus briefly on the third idea—namely, the possibility that zinc was added deliberately due to particular foundry practices.

Foundry Practices: Recycling and Devotional Offerings

Zinc contents of between 1 and 4 wt.% may improve the casting properties of copper alloys, as noticed, for example, in the work of modern founders in Patan, Nepal (Craddock 1981, 22), as well as in Western countries (J.M. Welter, Société française de métallurgie et de matériaux, pers. comm., 2011; see also Brunhuber 1986).[23] At least one statue of the studied corpus may be a good example of such a foundry practice: it is a large statue of an adorned Buddha in *abhayamudrā* (*ga* 2081), whose head, body, and hands were cast separately. Two samples from the head and the body were analyzed: all impurities present a similar pattern except zinc, which is surprisingly higher in the head (1.6 wt.%) than in the body (0.2 wt.%). As for the alloying elements, the tin content is also higher in the head (11.5 wt.%) than in the body (6 wt.%). That said, as already discussed regarding the effects of lead, the benefit of such additions to enhance the casting properties is not straightforward, since the head does not seem, *a priori*, more difficult to cast than the body. So for this particular case, the theory that there was a deliberate addition of zinc for ritual reasons or during devotional offerings may not be completely rejected, especially due to the importance of the head in a Buddha image. Such ritual practices are, at least, attested to within a Theravāda context by modern castings of Buddha images in today's Cambodia. Various metal artifacts are commonly offered by Buddhist devotees, who may also add gold and silver jewelry directly to the crucibles used for the casting of the Buddha's head (fig. 13). This hypothesis of a ritual and deliberate addition would, in turn, give to zinc a particular status, which is not incompatible with its rarity in Khmer copper-base metallurgy.

Such devotional offerings, especially of jewelry, may also be invoked to explain the relatively high gold contents observed. When speaking of gold in Khmer copper alloys, one cannot avoid mentioning the famous Khmer *samrit,* described in both written and oral tradition as the ultimate and noblest alloy, particularly because of the large quantities of gold—and silver—in its composition.[24] Although large quantities of gold have never been observed in a Khmer bronze object, the relatively high gold contents measured in the studied corpus deserve some discussion. In addition to the geochemical explanation previously proposed, the remelting of gilded artifacts may be a possible explanation, at least for the lowest values. But a more tentative suggestion, proposed previously as well (Bourgarit et al. 2003, 112–13, 118, fig. 11) and already discussed for zinc, would be the ritual and deliberate addition of gold in the melt. The Angkor Wat adorned Buddha protected by the *nāga* (*ga* 3296) may represent such devotional offerings, since the Buddha is much richer in gold than the *nāga*'s body (respectively 0.16 wt.% and 0.01 wt.%, all other impurities contents being similar; see "group B" in table 3). Nevertheless, the fact that the *nāga*'s heads are also rich in gold (0.18 wt.%) is a bit striking.

With the current state of knowledge, it is difficult to say how widespread such ritual practices might have been. Did they belong to a broad Khmer metallurgical tradition or to specific workshop habits? One cannot help imagining that some norms, royal and/or religious, may have controlled such practices. In this frame, it should be remembered that five artifacts among those bearing the highest gold contents were discovered during the 1950s excavations of the Royal Palace of Angkor Thom and dated to the last occupation of the site, which was during the Bayon period (*ga* 3451, *ga* 4023, *ga* 4025, *ga* 4047, and *ga* 4250). However, it is not true for all Bayon items coming from that same archaeological site—nine artifacts in total. The five studied items may nevertheless be linked to a few others unearthed in contemporary Buddhist sanctuaries founded by Jayavarman VII—that is, the Bayon (*ga* 2989 and *ga* 5165), Preah Khan (*ga* 4266), and Banteay Kdei temples (*ga* 5627). All these gold-rich bronzes, both statues and objects, may thus be associated with a single workshop, certainly located within the Angkorian capital and in close relationship with the royal authority.

As far as foundry practices are concerned, it has previously been proposed (Bourgarit et al. 2003, 113–15) that the differences of metal composition homogeneity within single assembled items, whether statues or objects, may reflect several operating modes and thus specific workshop habits as well. In the ensembles of "group A" (see table 3), all parts were cast from a unique metal batch, whereas in the ensembles of the other groups separate pours were used to cast the multiple parts. In these latter situations, different methods of production may be evident. For example, one may propose for several ensembles belonging to "group B" that separate pours were done within the same workshop, but on different days, as part of a serial production mode. It may also be proposed that, still for "group B," specific alloy recipes were selected by Khmer bronze craftsmen depending on the type of assembled part to be cast. This may particularly have been the case for ensembles, including arches, finials, or other decorative ornamentation with statues, as was encountered twice within the studied corpus. The first example is an adorned Buddha protected by the *nāga* with its arch (*ga* 5348 and *ga* 5356), which, with two other female divinities (*ga* 4011 and *ga* 4090), may have constituted a single Buddhist triad. All parts show a remarkable compositional homogeneity, except in the case of the lead contents. The three statues contain 16–17 wt.% lead, whereas the two different parts of the arch exhibit a higher lead content of 24–26 wt.%. Such differences are clearly not due to analytical bias. The second example of such an ensemble is a series of fragments that was previously part of two palanquins used for Buddhist images (*ga* 3597, *ga* 3605, *ga* 3611, *ga* 4081, and *ga* 4082; see Estève and Vincent 2010). Once again, all parts are made with almost the same metal. Yet the metal composition of the arches differs in that lead contents are lower (around 10 wt.% versus more than 14 wt.% in the other parts) and zinc impurities are higher (between 1 and 1.4 wt.% versus less than 0.5 wt.%). While arches entail a particular casting technique—namely, solid casting as opposed to hollow casting for the other parts—technological reasons may not explain the variations of lead and/or zinc contents. If that were the case, less lead and zinc would have been found in the arches. The ritual reason is even less convincing. Perhaps these variations simply testify to the absence of precise alloy control and an intensive use of scrap metal.

The use of scrap metal also probably explains the alloys found in the "group D" ensembles, whose multiple parts are clearly made of different metals. Yet extreme cases such as the Angkor Wat Lokeśvara (MA 5960) may also indicate modern additions. The body and arms of the statue show large discrepancies in both alloying elements and impurities. Two hypotheses may thus be proposed: either the arms have been restored in a later period, or they have only recently been assembled. Finally, the case of relatively large impurities variations ("group C"), which had not been observed in the previous campaign (Bourgarit et al. 2003, 113–15, table 4), is difficult to interpret. One may actually wonder whether the metal composition homogeneity in alloying elements is not due to chance. If so, this small group of five ensembles would be similar to "group D."

As noticed in the previous study (Bourgarit et al. 2003, 115), the use of one or the other method of production could not be related to any criterion, including the quality of execution. Unfortunately, again, this may be due to the still-too-small corpus studied.

Conclusion

Promisingly, Khmer bronze metallurgy of the twelfth and early thirteenth century starts to be characterized by more referential data. Hence, several features specific to Khmer metallurgy have been reassessed, such as the absence of zinc as one of the typical alloying elements used and the remarkable high gold contents in some objects. Moreover, while previous studies pointed toward a uniform picture of metallurgical practices during the whole Angkorian period, the particularly large and well-documented corpus under study has allowed us to distinguish groups that were previously only defined by stylistical and chronological features, such as Angkor Wat and Bayon bronze productions. Alloying practices seem to have evolved from lead-containing to unleaded bronzes, possibly as a consequence of a change in copper sources, a theory further supported by observed changes in the copper impurity pattern. As a direct consequence, besides iconographical and stylistical features, technical features also can be considered to help greatly in attributing objects to one or the other of these metallurgical traditions.

Yet, a number of issues have still been scarcely studied, such as the identification of regional bronze workshops or the chronological development of the trends observed for the twelfth and early thirteenth century. An overall technical investigation of an enlarged corpus, including elemental analysis of both metals and clay cores and the characterization of casting techniques, shall greatly enhance our understanding of the evolution of Khmer bronze metallurgy.

Appendix. Additional Tables

Table A1. *The analytical results for seventy-three Khmer artifacts (ninety-one results)*

	Designation	Inv. No.	Section	Sample	Sn	Pb	Zn	Ag	As	Au	Ba	Bi	Cd	Co
Angkor Wat														
Angkor	*nāga*-shaped foot	*ga* 4254.1		inside hole near top	7.4	0.62	<0.0010	0.12	0.16	0.0063	<0.00002	0.0677	<0.00001	<0.00002
	nāga-shaped foot	*ga* 4254.2		inside hole near top	7.1	0.58	<0.00080	0.12	0.15	0.0058	<0.00002	0.0652	<0.00001	<0.00003
	nāga-shaped foot	*ga* 4730		bottom edge	5.5	17	<0.00089	0.10	0.25	0.002	<0.00002	0.075	<0.00001	0.0066
	adorned *nāga*-protected Buddha	*ga* 5985	Buddha	bottom edge, PR side	0.53	4.8	20	0.13	0.14	0.0029	<0.00002	0.0919	<0.00001	<0.00001
			nāga heads	tang, PL side	0.33	4.1	17	0.10	0.11	0.0016	<0.00002	0.0597	<0.00001	<0.00002
	adorned *nāga*-protected Buddha	*ga* 5986	Buddha	bottom of tang, back	5.9	7.5	<0.0010	0.092	0.23	0.0048	<0.00002	0.0556	<0.00001	0.012
			nāga heads	bottom edge, PR side	5.4	3.0	0.32	0.16	0.28	0.0229	<0.00002	0.038	<0.00001	0.0153
			nāga heads	bottom of tang, PL side	7.7	3.4	0.33	0.13	0.29	0.0182	<0.00002	0.0325	<0.00001	0.036
East Angkor	adorned Buddha in *dhyānamudrā* (*nāga*-protected Buddha)	*ga* 3378	head	bottom edge of necklace, back and PL side	7.6	22	0.021	0.087	0.31	0.003	<0.00002	0.0816	<0.00001	0.0225
			body	bottom of tang, PR side	7.8	23	0.031	0.085	0.28	0.0014	<0.00002	0.0777	<0.00001	0.0122
	Vaishnavite triad astride Garuḍa (fragment of finial)	*ga* 3597		PL arm (Garuḍa)	5.3	14	0.021	0.073	0.42	0.0101	<0.00002	0.0604	<0.00001	0.018
	female figure (fragment of Buddhist finial)	*ga* 3605		PL foot	6.9	20	0.14	0.096	0.34	0.0048	<0.00002	0.0758	<0.00001	0.0211
	Garuḍa (fragment of Buddhist finial)	*ga* 3611		PL wing	7.2	18	0.46	0.100	0.33	0.0078	<0.00002	0.0702	<0.00001	0.0239
	arcature (fragment of Buddhist finial)	*ga* 4081		bottom edge near tang, PR side	6.1	10	1.0	0.15	0.31	0.0105	<0.00002	0.0471	<0.00001	0.017
	arcature (fragment of Buddhist finial)	*ga* 4082		bottom edge near tang, PL side	6.6	12	1.4	0.11	0.28	0.0084	<0.00002	0.0544	<0.00001	0.0166
Southeast	finial and stem with adorned *nāga*-protected Buddha	*ga* 2939	*nāga* side heads	bottom edge, PL side	9.3	19	0.18	0.11	0.32	0.0065	<0.00002	0.0726	<0.00001	0.0187
	finial and stem with adorned *nāga*-protected Buddha	*ga* 4444	Buddha	tang, PR side	12	2.5	0.18	0.16	0.27	0.0275	<0.00002	0.0152	<0.00001	0.0445
			nāga central heads	side head, PR side	9.6	12	0.051	0.21	0.37	0.0074	<0.00002	0.0466	<0.00001	0.0277
			stem	bottom edge, PL side	7.3	11	0.75	0.14	0.34	0.0121	<0.00002	0.0437	<0.00001	0.0252
	Viṣṇu astride Garuḍa	*ga* 3773		under pedestal, PL foot (Garuḍa)	7.9	17	0.046	0.13	0.31	0.0074	<0.00002	0.0655	<0.00001	0.0167
	Umā-Maheśvara on bull	*ga* 3778		PR foot (Śiva)	6.1	1.5	0.10	1.3	0.54	0.24	<0.00002	0.0187	<0.00001	0.0922
	female divinity	*ga* 4011		under pedestal, PL foot	9.2	17	0.021	0.081	0.35	0.004	<0.00002	0.0658	<0.00001	0.0259
	female divinity	*ga* 4090		under pedestal, PL foot	9.0	16	0.015	0.080	0.34	0.0025	<0.00002	0.0619	<0.00001	0.0226
	arcature	*ga* 5348	finial	bottom of floral pattern, front	7.5	26	0.025	0.085	0.30	0.0017	<0.00002	0.093	<0.00001	0.0128
			arcature	bottom of tang, PL side (two drillings)	8.6	24	0.017	0.082	0.33	0.0029	<0.00002	0.0921	<0.00001	0.0245
	adorned *nāga*-protected Buddha	*ga* 5356	Buddha	bottom of band near tang, PR side	9.4	16	0.021	0.089	0.35	0.0041	<0.00002	0.0659	<0.00001	0.0264
	finial with Vajrasattva	*ga* 5420	Vajrasattva	bottom edge, PL side (two drillings)	7.2	30	0.024	0.089	0.27	0.0006°	<0.00002	0.0968	<0.00001	0.0111
			aureole	bottom edge near tang, PR side (two drillings)	8.1	23	0.025	0.090	0.30	0.00097°	<0.00002	0.0818	<0.00001	0.0127
	Dvārapāla	*ga* 5466		under pedestal, back PR corner	2.7	5.8	<0.00091	0.30	0.13	0.0159	<0.00002	0.0271	<0.00001	<0.00002
	finial with Viṣṇu astride Garuḍa	*ga* 5676		PL wing, back (Garuḍa)	9.8	11	0.90	0.28	0.27	0.0243	<0.00002	0.0401	<0.00001	0.0273
South	Viṣṇu astride Garuḍa	*ga* 3596		under pedestal, PR foot (Garuḍa)	8.8	1.9	0.0029°	0.16	0.22	0.0204	<0.00002	0.0364	<0.00001	0.0055
	nāga-shaped hook	*ga* 4729		top of tang	4.5	5.4	1.4	0.17	0.23	0.0089	<0.00002	0.0246	<0.00001	0.0163
	adorned Buddha in *vitarkamudrā*	*ga* 5325		bottom of tang	11	2.4	0.27	0.085	0.67	0.0269	<0.00002	0.0162	<0.00001	0.0165
	Lokeśvara	*ga* 5327		bottom of tang	8.9	8.7	0.29	0.15	0.35	0.062	<0.00002	0.036	<0.00001	0.043
West	adorned *nāga*-protected Buddha	*ga* 5357	*nāga* body	inner edge, back and PR side	8.0	23	0.14	0.10	0.34	0.0099	<0.00002	0.089	<0.00001	0.0222
Cambodia	Lokeśvara	*ga* 6861		under pedestal, back PL corner	8.4	15	0.0089	0.10	0.36	0.0056	<0.00002	0.0673	<0.00001	0.0217
?	Prajñāpāramitā	03IA25			7.8	26.2	<0.09	<0.11	<0.37	<0.05	nm	<0.2	< 0.07	< 0.06
Bayon														
Angkor	fragments of *nāga*-shaped finial	*ga* 2989		broken edge, back and PR side	6.2	0.11	<0.0011	0.96	1.1	0.0889	<0.00002	0.0028°	<0.00001	0.0322
	nāga-shaped finial	*ga* 5165		bottom edge, PL side	4.0	0.30	0.37	0.57	0.79	0.0259	<0.00002	0.0038	<0.00001	0.0583
	Śiva	*ga* 3451		PR leg	4.2	0.17	0.0034	0.14	0.30	0.4385	<0.00002	0.0096	<0.00001	0.0481
	lamp	*ga* 3699	lotus fruit	top surface	6.3	0.073	<0.00096	0.060	0.81	0.0031	<0.00002	0.0051	<0.00001	0.0123
		ga 3846	stem with lotus bud	under lotus bud (two drillings)	5.4	0.094	<0.00093	0.068	0.89	0.133	<0.00002	0.0020°	<0.00001	0.0315
		ga 4032	lotus fruit	top surface	5.9	0.073	<0.00093	0.056	0.76	0.0036	<0.00002	0.0032	<0.00001	0.0116

All values given in wt%. PL = proper left; PR = proper right; nm = not measured; x° = value under the limit of quantification (shall not be considered as a quantitative value).

Cr	Fe	Ge	Hg	In	Mg	Mn	Mo	Ni	P	S	Sb	Se	Te	Ti	U	V	W
<0.00008	0.0062	<0.0002	0.00077	0.0012	0.0016	0.0011	<0.0001	0.052	0.0185	<0.0011	0.13	0.00085	0.0048	<0.00003	0.0020°	<0.00005	0.0032
<0.00008	0.0024	<0.0002	0.0007	0.0013	0.00007°	0.00092	<0.0001	0.051	0.0122	<0.00086	0.13	0.00077	0.0056	<0.00003	0.0017°	<0.00005	0.0022
<0.00008	0.073	<0.0002	0.00062	0.0029	<0.00004	<0.00001	<0.0001	0.048	0.0014°	0.20	0.13	0.00052°	0.0055	<0.00003	0.0011°	<0.00005	0.0027
<0.00008	0.042	<0.0002	0.00062	0.0013	<0.00004	<0.00001	<0.0001	0.034	0.00091°	0.067	0.17	0.00074	0.0038	<0.00003	0.0014°	<0.00005	0.0019
<0.00008	0.025	<0.0002	0.00058	0.0016	<0.00004	<0.00001	<0.0001	0.035	0.0018°	0.064	0.098	0.00023°	0.0033	<0.00003	0.0015°	<0.00005	0.0024
<0.00008	0.097	<0.0002	0.00059	0.0027	<0.00004	<0.00001	<0.0001	0.080	0.0016°	0.16	0.13	0.0015	0.006	<0.00003	0.0024°	<0.00005	0.0035
<0.00008	0.12	<0.0002	0.00056	0.0019	<0.00004	<0.00001	<0.0001	0.17	0.0018°	0.14	0.093	0.0028	0.0051	<0.00003	0.0025°	<0.00005	0.0035
<0.00008	0.26	<0.0002	0.0005	0.0023	<0.00004	<0.00001	<0.0001	0.18	<0.0010	0.10	0.10	0.0018	0.0034	<0.00003	0.0027°	<0.00005	0.0024
<0.00008	0.28	<0.0002	0.00043	0.0036	<0.00004	<0.00001	<0.0001	0.056	0.0054	0.41	0.15	0.0012	0.0048	<0.00003	<0.0011	<0.00005	0.0111
<0.00008	0.23	<0.0002	0.00042	0.0045	<0.00004	<0.00001	<0.0001	0.045	0.0032°	0.40	0.15	0.0014	0.0046	<0.00003	<0.00092	<0.00005	0.0025
<0.00008	0.30	<0.0002	0.00045	0.0029	<0.00004	<0.00001	<0.0001	0.13	0.0016°	0.35	0.14	0.00037°	0.0038	<0.00003	<0.00096	<0.00005	0.0025
<0.00008	0.23	<0.0002	0.00046	0.0031	<0.00004	<0.00001	<0.0001	0.088	0.0012°	0.35	0.16	0.001	0.0049	<0.00003	0.00096°	<0.00005	0.0021
<0.00008	0.27	<0.0002	0.00045	0.0027	<0.00004	<0.00001	<0.0001	0.078	0.0012°	0.30	0.18	0.00072	0.0036	<0.00003	0.0017°	<0.00005	0.0025
<0.00008	0.28	<0.0002	0.00047	0.0021	<0.00004	<0.00001	<0.0001	0.096	0.0046	0.20	0.12	0.00074°	0.0048	<0.00003	<0.0011	<0.00005	0.0028
<0.00008	0.27	<0.0002	0.00053	0.0023	<0.00004	<0.00001	<0.0001	0.092	0.0013°	0.21	0.13	0.0012	0.004	<0.00003	<0.0010	<0.00005	0.0029
<0.00008	0.19	<0.0002	0.00041	0.0029	<0.00004	<0.00001	<0.0001	0.079	0.0010°	0.32	0.17	0.0015	0.0048	<0.00003	<0.00086	<0.00005	0.0023
<0.00008	0.60	<0.0002	0.00054	0.0022	<0.00004	<0.00001	<0.0001	0.25	0.0071	0.18	0.058	0.00048°	0.0036	<0.00003	<0.00099	<0.00005	<0.00027
<0.00008	0.20	<0.0002	0.00045	0.0032	<0.00004	<0.00001	<0.0001	0.17	0.0055	0.27	0.15	0.0025	0.0065	<0.00003	<0.00091	<0.00005	0.0024
<0.00008	0.25	<0.0002	0.00055	0.0028	<0.00004	<0.00001	<0.0001	0.14	0.0023°	0.23	0.098	0.0012	0.0043	<0.00003	<0.00082	<0.00005	0.0018
<0.00008	0.23	<0.0002	0.00043	0.0022	<0.00004	<0.00001	<0.0001	0.077	0.0028°	0.32	0.18	0.00089	0.0032	<0.00003	<0.00095	<0.00005	0.0027
<0.00008	0.35	<0.0002	0.00043	0.0021	0.00022	<0.00001	<0.0001	0.51	0.0027°	0.33	0.13	0.0013	0.0033	<0.00003	<0.0010	<0.00005	0.0025
<0.00008	0.28	<0.0002	0.00048	0.0043	<0.00004	<0.00001	<0.0001	0.069	0.0013°	0.36	0.20	0.0011	0.004	<0.00003	0.0014°	<0.00005	0.0027
<0.00008	0.22	<0.0002	0.00041	0.004	<0.00004	<0.00001	<0.0001	0.062	<0.0011	0.29	0.19	<0.00022	0.0039	<0.00003	<0.0010	<0.00005	0.0027
<0.00008	0.22	<0.0002	0.00041	0.0036	<0.00004	<0.00001	<0.0001	0.047	0.0026°	0.44	0.22	0.00094	0.0049	<0.00003	0.00093°	<0.00005	0.0021
<0.00008	0.25	<0.0002	0.00048	0.0042	<0.00004	<0.00001	<0.0001	0.068	0.0014°	0.38	0.18	0.0017	0.0048	<0.00003	0.0019°	<0.00005	0.0026
<0.00008	0.28	<0.0002	0.00043	0.0041	<0.00004	<0.00001	<0.0001	0.071	0.0027°	0.37	0.20	0.00069°	0.004	<0.00003	0.0013°	<0.00005	0.0028
<0.00008	0.25	<0.0002	0.00033	0.0055	<0.00004	<0.00001	<0.0001	0.048	0.0026°	0.54	0.20	0.0021	0.0058	<0.00003	<0.00087	<0.00005	0.002
<0.00008	0.20	<0.0002	0.00054	0.0042	<0.00004	<0.00001	<0.0001	0.046	0.0017°	0.44	0.13	0.0012	0.0054	<0.00003	<0.0011	<0.00005	0.0028
<0.00008	0.033	<0.0002	0.00077	0.0013	0.0008	<0.00001	<0.0001	0.11	0.0061	<0.00098	0.050	0.0018	0.0036	<0.00003	0.0011°	<0.00005	0.0027
<0.00008	0.79	<0.0002	0.00054	0.0025	<0.00004	<0.00001	<0.0001	0.14	0.0034	0.17	0.11	0.0008	0.003	<0.00003	0.0014°	<0.00005	<0.00027
<0.00008	0.043	<0.0002	0.0004	0.0016	<0.00004	<0.00001	<0.0001	0.091	0.0028°	<0.0010	0.12	0.0017	0.0045	<0.00003	0.0017°	<0.00005	0.0029
<0.00008	0.25	<0.0002	0.00068	0.0016	<0.00004	<0.00001	<0.0001	0.11	0.0034	0.086	0.048	0.0038	0.0039	<0.00003	0.0011°	<0.00005	0.002
<0.00008	0.50	<0.0002	0.00058	0.0023	<0.00004	<0.00001	<0.0001	0.12	0.0018°	0.10	0.11	0.0049	0.003	<0.00003	0.0021°	<0.00005	0.0020°
<0.00008	0.23	<0.0002	0.00053	0.0026	<0.00004	<0.00001	<0.0001	0.23	0.0029	0.19	0.12	0.001	0.0044	<0.00003	<0.0011	<0.00005	0.0018
<0.00008	0.59	<0.0002	0.00048	0.0031	<0.00004	0.00008	<0.0001	0.080	0.0041	0.41	0.18	0.0011	0.0038	<0.00003	0.0010°	<0.00005	0.0018
<0.00008	0.29	<0.0002	0.00054	0.0035	<0.00004	<0.00001	<0.0001	0.074	0.0064	0.31	0.18	0.00051°	0.0037	<0.00003	<0.0011	<0.00005	0.0037
nm	0.35	nm	nm	nm	nm	nm	nm	< 0.03	nm	nm	< 0.08	nm	nm	nm	nm	nm	nm
<0.00008	0.047	<0.0002	0.0185	0.0028	0.0019	<0.00001	<0.0001	0.89	0.0179	0.18	0.10	0.0026	0.0059	<0.00003	0.0025°	<0.00005	0.0032
<0.00008	0.61	<0.0002	0.00066	0.0028	<0.00004	0.00033	<0.0001	0.63	0.0028°	0.29	0.076	0.0034	0.0071	<0.00003	<0.0011	<0.00005	<0.00027
<0.00008	0.83	<0.0002	0.00066	0.0027	<0.00004	<0.00001	<0.0001	0.27	0.0064	0.13	0.067	0.0018	0.0094	<0.00003	0.0010°	<0.00005	<0.00027
<0.00008	0.26	<0.0002	0.00078	0.0018	<0.00004	0.00084	<0.0001	0.60	0.0028°	0.12	0.079	0.0057	0.0075	<0.00003	0.0013°	<0.00005	0.0025
<0.00008	0.23	<0.0002	0.0151	0.0018	<0.00004	<0.00001	0.00026°	0.59	0.0138	0.21	0.080	0.0056	0.0067	<0.00003	0.0011°	<0.00005	0.003
<0.00008	0.24	<0.0002	0.001	0.0019	<0.00004	0.00088	<0.0001	0.56	0.0030°	0.13	0.072	0.0049	0.0064	<0.00003	<0.00097	<0.00005	0.0024

cont'd

Table A1. *The analytical results for seventy-three Khmer artifacts (ninety-one results)* cont'd.

	Designation	Inv. No.	Section	Sample	Sn	Pb	Zn	Ag	As	Au	Ba	Bi	Cd	Co
		ga 4152	lotus fruit with horse	back PL leg (existing hole)	7.0	0.19	<0.0013	0.045	0.87	0.0011°	<0.00002	0.0021°	<0.00001	0.0554
		ga 5644	stem with lotus leaf	bottom edge	6.9	0.040	<0.00078	0.036	0.91	0.0041	<0.00002	0.00089°	<0.00001	0.0323
			stem with lotus bud	bottom edge	8.0	0.025	<0.00100	0.029	0.71	0.0024	<0.00002	0.0024°	<0.00001	0.0294
	fragment of perfume censer or lamp?	*ga* 3721		bottom of foot	8.5	5.2	0.35	0.10	0.34	0.221	<0.00002	0.0275	<0.00001	0.0289
	conch setting	*ga* 3788 (K. 779)		inner groove	11	9.3	0.20	0.15	1.5	0.0168	<0.00002	0.0373	<0.00001	0.0254
	nāga-shaped finial	*ga* 4023		side edge, back and PL side	7.1	0.42	0.034	0.13	0.18	0.1111	<0.00002	0.021	<0.00001	0.0045
	female divinity	*ga* 4025		under pedestal, PR foot	4.7	0.42	<0.00097	0.16	0.30	0.4436	<0.00002	0.0135	<0.00001	0.0442
	torso of adorned Buddha in *abhayamudrā*	*ga* 4047		bottom of cloth, broken edge	4.2	6.0	0.23	0.045	0.55	0.0598	0.00082	0.0225	<0.00001	0.0449
	head of male divinity (Buddha?)	*ga* 4222		broken neck, PL side	4.1	0.069	<0.0010	0.021	0.49	0.0024	<0.00002	0.0014°	<0.00001	0.0366
	Garuḍa-shaped finial	*ga* 4227		bottom of tang, PR side	6.9	0.45	<0.00098	0.095	0.078	0.0109	<0.00002	0.0418	<0.00001	0.0005
	lion-shaped finial	*ga* 4228		PL foot	7.7	0.097	<0.00096	0.17	0.35	0.0056	0.00069	0.0036	<0.00001	0.0278
	coiffure of "radiating" Lokeśvara	*ga* 4250		broken edge	4.8	0.20	<0.00097	0.032	0.34	0.1062	<0.00002	0.0027°	<0.00001	0.0407
	trident	*ga* 4266		bottom edge near tang	5.3	0.20	0.20	0.23	0.22	0.0422	<0.00002	0.0134	<0.00001	0.0014
	Buddhist triad (Lokeśvara, female divinity, Prajñāpāramitā)	*ga* 4424	Lokeśvara	under pedestal, PL foot	6.3	1.8	0.078	0.050	0.53	0.0043	<0.00002	0.0113	<0.00001	0.13
			female divinity	under pedestal, front PL corner (two drillings)	9.2	11	0.40	0.15	0.30	0.0173	<0.00002	0.0502	<0.00001	0.022
			Prajñāpāramitā	under pedestal, PR foot	5.2	4.9	0.17	0.083	0.38	0.0112	<0.00002	0.0138	<0.00001	0.0388
	palanquin hook	*ga* 4443		inside ring	9.1	6.8	0.073	0.069	0.38	0.0662	<0.00002	0.0165	<0.00001	0.0332
	Shaivaite male divinity	*ga* 5292		bottom of tang	6.3	13	0.28	0.063	0.39	0.0248	<0.00002	0.0269	<0.00001	0.0909
	Śiva	*ga* 5301		under pedestal, PL foot	8.7	8.9	0.25	0.21	0.34	0.0219	<0.00002	0.0221	<0.00001	0.0402
	Lokeśvara	*ga* 5338		under pedestal, front PR corner	6.5	0.048	<0.00080	0.13	0.080	0.0127	<0.00002	0.0144	<0.00001	0.0836
	necklace of statue	*ga* 5627		bottom edge	4.6	0.095	<0.00075	0.24	0.48	0.0936	<0.00002	0.004	<0.00001	0.0097
	miniature Buddhist shrine	*ga* 5630		bottom of foot	6.7	0.24	<0.00092	0.048	0.30	0.0036	<0.00002	0.0036	<0.00001	0.0244
	Garuḍa-shaped finial	*ga* 5681		inside tube	3.2	0.55	2.5	0.12	0.28	0.0209	<0.00002	0.0159	<0.00001	0.0461
	fragment of Garuḍa-shaped finial	*ga* 5683		broken edge, PR side	3.8	0.33	0.032	0.050	0.41	0.0173	<0.00002	0.0042	<0.00001	0.0498
East	palanquin hook	*ga* 4445		broken edge of ring	3.7	0.14	0.0031°	0.015	0.89	0.0224	<0.00002	0.0015°	<0.00001	0.0865
	palanquin ring	*ga* 4743		back of *nāga* head	4.6	9.6	0.27	0.078	0.58	0.0064	<0.00002	0.0258	<0.00001	0.0262
Southeast	Lokeśvara	*ga* 2959		under pedestal, PL foot	7.6	17	0.19	0.29	0.33	0.1414	0.0004	0.0257	<0.00001	0.0283
South	palanquin hook	*ga* 3698		inside ring	3.4	0.12	0.062	0.018	0.63	0.0048	0.00054	0.0019°	<0.00001	0.0547
	female divinity	*ga* 4012		bottom of tang	6.1	11	0.017	0.020	0.36	0.0013°	<0.00002	0.0093	<0.00001	0.0317
	Garuḍa-shaped finial	*ga* 5680		inside tube	7.7	4.0	0.019	0.083	0.17	0.0052	0.0012	0.0272	<0.00001	0.0253
West	Prajñāpāramitā	*ga* 3540		under pedestal, PR foot	7.4	12	0.69	0.12	0.41	0.0177	<0.00002	0.0433	<0.00001	0.0312
	palanquin hook	*ga* 3783		inside ring	9.3	4.3	0.041	0.15	0.42	0.0416	0.00073	0.0182	<0.00001	0.03
	Buddhist triad (Lokeśvara, Buddha, Prajñāpāramitā)	*ga* 4417	Lokeśvara	under pedestal, PR foot	7.7	12	0.51	0.19	0.37	0.0192	<0.00002	0.0274	<0.00001	0.0338
			Buddha	bottom of tang, PL side	8.1	13	0.50	0.13	0.36	0.0241	<0.00002	0.0311	<0.00001	0.0376
			Prajñāpāramitā	under pedestal, PR foot	7.9	10	0.52	0.18	0.37	0.0189	<0.00002	0.0307	<0.00001	0.0349
			pedestal	bottom edge, back PR corner	8.5	12	1.1	0.13	0.39	0.0285	<0.00002	0.0276	<0.00001	0.0459
	adorned Buddha in *vitarkamudrā*?	*ga* 5320		under pedestal, PR foot	4.6	0.28	0.011	0.11	0.35	0.5094	<0.00002	0.0075	<0.00001	0.0474
	adorned Buddha in *vitarkamudrā*?	*ga* 5414		under pedestal, PR foot	4.8	0.32	0.0023°	0.15	0.28	0.2317	<0.00002	0.0177	<0.00001	0.069
	adorned Buddha in *vitarkamudrā*	*ga* 5410		under pedestal, PR foot	2.4	0.16	0.076	0.018	0.44	0.0016	<0.00002	0.0013°	<0.00001	0.0538
	Lokeśvara	*ga* 5422		bottom of tang	4.9	0.37	0.051	0.15	0.44	0.4262	<0.00002	0.0079	<0.00001	0.0404
	nāga-shaped finial	*ga* 5679		bottom of *nāga* side head, PR side	6.0	6.7	0.75	0.087	0.30	0.1304	<0.00002	0.0291	<0.00001	0.0253
?	lamp	MA 6939	kneeling figure	under pedestal	6.6	1.3	0.049	0.049	0.68	0.0035	<0.00002	0.0069	<0.00002	0.02
			main flower	lower cercle	6.8	2.7	0.065	0.084	0.55	0.0065	<0.00002	0.011	<0.00002	0.022
			main stem	upper part	5.4	2.4	0.11	0.064	0.56	0.0058	<0.00002	0.0099	<0.00002	0.015
			main stem	lower part	7.3	4.6	0.054	0.13	0.34	0.015	<0.00002	0.014	<0.00002	0.019
			support	bottom edge	6.4	3.8	0.084	0.109	0.35	0.0081	0.00033	0.014	<0.00002	0.027
Undetermined														
Angkor	bull of Śiva	*ga* 5739 (K. 1013)		tang, PL side	5.5	0.10	<0.00100	0.019	0.27	0.002	<0.00002	0.0017°	<0.00001	0.0008
Southeast	finial with adorned *nāga*-protected Buddha	*ga* 5685		side edge, PR side	8.1	21	0.060	0.10	0.38	0.011	<0.00002	0.0777	<0.00001	0.0161
	Haṃsa-shaped finial	*ga* 195		bottom edge, PL side	8.2	0.29	<0.0010	0.083	0.27	0.022	<0.00002	0.0024°	<0.00001	0.0147

Cr	Fe	Ge	Hg	In	Mg	Mn	Mo	Ni	P	S	Sb	Se	Te	Ti	U	V	W
<0.00010	0.35	<0.00025	0.0041	0.0021	<0.00004	<0.00001	<0.0001	0.43	0.0071	0.64	0.11	0.0171	0.0118	<0.00003	0.0019°	<0.00005	0.0035
0.018	0.58	<0.0002	0.00066	0.0017	<0.00004	0.0011	0.023	0.70	0.0028°	0.25	0.076	0.0058	0.0062	<0.00003	0.00098°	0.013	0.0238
<0.00008	0.089	<0.0002	0.00058	0.0016	<0.00004	<0.00001	<0.0001	0.60	<0.0011	0.27	0.046	0.0034	0.0043	<0.00003	0.0014°	<0.00005	0.0033
<0.00008	0.21	<0.0002	0.00069	0.0015	<0.00004	<0.00001	<0.0001	0.29	0.0024°	0.14	0.15	0.0022	0.0043	<0.00003	<0.0010	<0.00005	0.0028
<0.00008	0.26	<0.0002	0.0005	0.0017	<0.00004	<0.00001	<0.0001	0.17	0.0023°	0.18	0.22	0.00092	0.0039	<0.00003	0.0016°	<0.00005	0.0029
<0.00008	0.085	<0.0002	0.0154	0.0014	<0.00004	0.00017	<0.0001	0.16	0.0217	<0.00080	0.11	0.0061	0.0035	<0.00003	0.0011°	<0.00005	0.0018
<0.00008	1.0	<0.0002	0.00083	0.0024	<0.00004	<0.00001	<0.0001	0.29	0.0042	0.13	0.15	0.0027	0.0112	<0.00003	0.0012°	<0.00005	<0.00027
<0.00008	3.3	<0.0002	0.00061	0.004	<0.00004	0.0007	<0.0001	0.39	0.0041	0.43	0.092	<0.00019	0.0031	<0.00003	<0.00091	<0.00005	<0.00027
<0.00008	0.18	<0.0002	0.00067	0.0022	<0.00004	<0.00001	<0.0001	0.40	0.0019°	0.25	0.029	0.0011	0.0036	<0.00003	0.0012°	<0.00005	0.0029
<0.00008	0.043	<0.0002	0.00074	0.0013	<0.00004	0.00078	<0.0001	0.17	0.0105	<0.0010	0.067	0.00056°	0.0042	<0.00003	0.0013°	<0.00005	0.0033
<0.00008	0.38	<0.0002	0.0009	0.0017	0.00036	0.0015	<0.0001	0.16	0.0665	0.19	0.039	0.0025	0.0073	<0.00003	<0.0010	<0.00005	0.0022
<0.00008	0.27	<0.0002	0.00069	0.0021	<0.00004	<0.00001	<0.0001	0.35	0.0033°	0.34	0.030	0.0027	0.0027	<0.00003	0.0013°	<0.00005	0.0026
<0.00008	0.061	<0.0002	0.0057	0.0014	<0.00004	<0.00001	<0.0001	0.27	0.0057	<0.0010	0.044	0.00039°	0.0012	<0.00003	0.0014°	<0.00005	0.0029
<0.00008	0.69	0.00043°	0.00044	0.0028	<0.00004	<0.00001	<0.0001	0.58	0.0022°	0.42	0.066	0.003	0.0039	<0.00003	<0.00095	<0.00005	<0.00027
<0.00008	0.19	<0.0002	0.00053	0.0022	<0.00004	<0.00001	<0.0001	0.12	0.0022°	0.22	0.14	0.0013	0.0038	<0.00003	<0.00074	<0.00005	0.0016
<0.00008	0.84	<0.0002	0.00057	0.0023	<0.00004	<0.00001	<0.0001	0.30	0.0019°	0.30	0.056	0.0011	0.0035	<0.00003	<0.0011	<0.00005	<0.00027
<0.00008	0.38	<0.0002	0.00062	0.0014	0.00025	0.00034	<0.0001	0.26	0.006	0.13	0.061	0.0027	0.0041	<0.00003	<0.0011	<0.00005	0.0027
<0.00008	0.87	<0.0002	0.00041	0.0027	<0.00004	<0.00001	<0.0001	0.41	0.0041	0.31	0.082	0.0013	0.0037	<0.00003	0.0010°	<0.00005	<0.00027
0.0068	0.66	<0.0002	0.00055	0.0023	<0.00004	0.00063	0.0099	0.26	0.0057	0.16	0.084	0.0029	0.0048	<0.00003	0.0013°	0.0043	0.0094
<0.00008	0.95	<0.0002	0.00055	0.0037	<0.00004	<0.00001	<0.0001	0.17	0.00095°	0.059	0.0023	0.0014	0.0157	<0.00003	<0.00083	<0.00005	<0.00027
<0.00008	0.20	<0.0002	0.0209	0.00089	<0.00004	<0.00001	<0.0001	0.21	0.00099°	0.051	0.064	0.0031	0.0034	<0.00003	<0.00079	<0.00005	0.0021
0.00011°	0.077	<0.0002	0.0013	0.0017	<0.00004	<0.00001	0.00078	0.12	0.0055	<0.00099	0.065	0.0047	0.0063	<0.00003	0.0020°	<0.00005	0.0039
<0.00008	1.0	<0.0002	0.00071	0.0029	<0.00004	0.0009	0.00067	0.31	0.0202	0.16	0.037	0.0018	0.0078	<0.00003	<0.0011	<0.00005	<0.00027
<0.00008	0.34	<0.0002	0.00075	0.0024	<0.00004	<0.00001	<0.0001	0.25	0.0041	0.30	0.029	0.0017	0.0047	<0.00003	<0.00075	<0.00005	0.0011
<0.00008	1.4	<0.0002	0.0203	0.0019	<0.00004	<0.00001	<0.0001	0.30	0.004	0.38	0.029	<0.00022	0.0019	<0.00003	<0.0011	<0.00005	<0.00027
<0.00008	0.27	<0.0002	0.00057	0.0023	<0.00004	<0.00001	<0.0001	0.38	<0.0011	0.20	0.079	0.003	0.0051	<0.00003	0.0016°	<0.00005	0.0026
<0.00008	0.46	<0.0002	0.00056	0.0019	0.00099	0.00045	0.00012°	0.29	0.0082	0.14	0.063	0.00079°	0.0032	<0.00003	0.0017°	<0.00005	0.003
<0.00008	4.2	<0.0002	0.0008	0.0039	0.0012	0.00091	0.00039	0.30	0.0304	0.39	0.042	<0.00023	0.0021	<0.00003	<0.0011	<0.00005	<0.00027
<0.00008	0.37	<0.0002	0.00045	0.0018	<0.00004	<0.00001	<0.0001	0.30	0.0019°	0.40	0.050	0.0019	0.0035	<0.00003	<0.0011	<0.00005	0.0025°
<0.00008	0.17	<0.0002	0.0004	0.0014	0.0066	0.00007	<0.0001	0.088	0.0272	0.074	0.057	0.00082	0.0034	<0.00003	<0.00096	<0.00005	0.0029
<0.00008	1.1	<0.0002	0.00057	0.0032	<0.00004	<0.00001	<0.0001	0.19	0.0015°	0.24	0.15	0.0011	0.0077	<0.00003	<0.00087	<0.00005	<0.00027
<0.00008	0.56	<0.0002	0.00069	0.002	0.0025	0.00028	0.00061	0.13	0.0462	0.11	0.11	0.0017	0.0049	<0.00003	0.0015°	<0.00005	0.002
<0.00008	0.47	<0.0002	0.00039	0.0026	0.00021	<0.00001	<0.0001	0.26	0.0023°	0.20	0.097	0.002	0.0046	<0.00003	0.0011°	<0.00005	0.0016°
<0.00008	0.45	<0.0002	0.00046	0.0024	<0.00004	<0.00001	<0.0001	0.25	0.0021°	0.23	0.098	0.0023	0.0031	<0.00003	0.0011°	<0.00005	0.0021
<0.00008	0.49	<0.0002	0.00049	0.0022	0.00017	<0.00001	<0.0001	0.26	0.0032°	0.19	0.094	0.0022	0.005	<0.00003	0.0021°	<0.00005	0.0019°
<0.00008	0.39	<0.0002	0.00046	0.0025	<0.00004	<0.00001	<0.0001	0.32	0.0024°	0.22	0.093	0.0022	0.0048	<0.00003	0.0015°	<0.00005	0.0022
<0.00008	1.1	<0.0002	0.00064	0.0022	<0.00004	<0.00001	<0.0001	0.33	0.0017°	0.17	0.065	0.002	0.0092	<0.00003	<0.0011	<0.00005	<0.00027
<0.00008	1.1	<0.0002	0.00064	0.003	<0.00004	<0.00001	<0.0001	0.27	0.0016°	0.23	0.053	0.0048	0.0146	<0.00003	<0.0011	<0.00005	<0.00027
<0.00008	1.9	<0.0002	0.00067	0.0031	0.00034	0.00012	0.00016°	0.35	0.005	0.46	0.039	0.0012	0.0041	<0.00003	<0.00091	<0.00005	<0.00027
<0.00008	0.85	<0.0002	0.0006	0.002	<0.00004	<0.00001	<0.0001	0.34	0.0041	0.20	0.063	0.0023	0.0079	<0.00003	<0.00091	<0.00005	<0.00027
<0.00008	1.2	<0.0002	0.00056	0.0026	<0.00004	<0.00001	<0.0001	0.27	0.0033°	0.18	0.086	0.00057°	0.004	<0.00003	<0.00096	<0.00005	<0.00027
<0.00008	0.036	<0.00028	0.0019	0.0013	0.00037	<0.00001	<0.0003	0.33	<0.016	0.076	0.053	0.0013°	0.0034°	0.00013°	<0.00052	<0.00005	<0.00027
<0.00008	0.094	<0.0004	0.0019	0.0014	0.00076	<0.00001	<0.0003	0.30	<0.019	0.12	0.058	0.0018°	0.0034°	0.00013°	<0.0006	<0.00005	<0.00027
0.00007°	0.072	<0.0004	0.0015	0.0015	0.00036	<0.00001	<0.0003	0.29	<0.019	0.077	0.064	0.0025	0.0032°	0.00009°	<0.0006	<0.00005	<0.00027
<0.00008	0.070	<0.0004	0.0021	0.0014	0.00026	<0.00001	<0.0003	0.25	<0.019	0.12	0.054	0.0015°	0.0037	0.00006°	<0.0006	<0.00005	<0.00027
<0.00008	0.14	<0.0004	0.0018	0.0017	0.0003	<0.00001	<0.0003	0.27	<0.019	0.14	0.055	0.0015°	0.0040°	0.00008°	<0.0006	<0.00005	<0.00027
<0.00008	0.0026	<0.0002	0.00072	0.0019	<0.00004	<0.00001	<0.0001	0.29	0.0049	<0.0011	0.014	0.00076	0.0034	<0.00003		<0.00005	0.0031
<0.00008	0.25	<0.0002	0.00043	0.0035	<0.00004	<0.00001	<0.0001	0.064	0.0036	0.41	0.19	0.00095	0.0047	<0.00003		<0.00005	0.0024
<0.00008	0.071	<0.0002	0.0021	0.0015	<0.00004	<0.00001	<0.0001	0.30	0.0053	<0.0011	0.038	0.0017	0.0103	<0.00003		<0.00005	0.003

Table A2. *Estimation of global error measurements.*

Designation	Inv. No.	Analyses and Variation	Sn	Pb	Ag	As	Au	Bi	Co	Fe	Ni	Sb	Zn
tripod	MA 541	analysis 1 (wt.%)	6	0.38	0.08	0.09	0.004	0.05	nd	0.08	0.03	0.07	0.009
		analysis 2 (wt.%)	1.4	0.25	0.02	0.02	0.002	0.04	nd	0.05	0.002	0.02	0.007
		variation	24	66	25	17	65	74	—	63	7	28	69
tripod	MA 3565	analysis 1 (wt.%)	6.9	0.42	0.19	0.09	0.002	0.06	nd	0.05	0.04	0.09	0.005
		analysis 2 (wt.%)	1.4	0.16	0.13	0.01	0.0002	0.03	nd	0.02	0.004	0.03	0.007
		variation	20	39	71	11	10	52	—	42	11	32	141
perfume censer (quadripod)	MA 5984	analysis 1 (wt.%)	3.5	0.03	0.13	0.84	0.0004	0.002	0.05	0.35	0.39	0.06	0.01
		analysis 2 (wt.%)	0.14	0.001	0.14	0.03	0.0006	0	0.001	0.01	0.01	0.002	0.001
		variation	4	4	112	3	141	0	3	4	2	4	12
drum	MA 5988	analysis 1 (wt.%)	14	3.7	0.05	0.29	0.01	0.01	0.06	0.28	0.26	0.04	0.6
		analysis 2 (wt.%)	0	0.07	0	0.007	0.0001	0	0	0.04	0	0.001	0.04
		variation	0	2	0	2	1	0	0	13	0	3	6
Buddhist triad	*ga* 2424	analysis 1 (wt.%)	7.2	0.1	0.14	0.06	0.02	nd	0.04	0.05	0.14	nd	nd
		analysis 2 (wt.%)	0.7	0.04	0.02	0.007	0.01	nd	0.006	0.01	0.01	nd	nd
		variation	10	37	11	13	46	—	13	31	8	—	—

nd = not detected

Table A3. *Atypical metal compositions (thirty-nine artifacts).*

	Designation	Inv. No.	Ag (<0.3)	Au (<0.25)	As (<0.9)	Zn (<0.4)	Fe (<2)	Co (<0.07)	Sb+Bi >Pb	Sb+Bi < Pb	Bi corr Pb (<LD)	Sb corr Pb (0.06)	Sn (>5 for Angkor Wat and <10 for Bayon)	Pb (>1 for Angkor Wat and < 13 for Bayon)	Ni (<0.2 for Angkor Wat and > 0.2 for Bayon)
Angkor Wat	kneeling female figure (support for mirror)	*ga* 5476							0.27 / 8						
	nāga-shaped finial	MG 18889												0.3	
	nāga-shaped feet	*ga* 4254.1 and *ga* 4254.2							0.2 / 0.6					0.6	
	Umā-Maheśvara on bull	*ga* 3778	1.3					0.09							0.61
	finial with Viṣṇu astride Garuḍa	*ga* 5676				0.9									
	nāga-shaped hook	*ga* 4729				1.4							4.5		
	adorned *nāga*-protected Buddha	*ga* 3296							0.15 / 0.8					0.2–0.8	
	Brahmanic triad (pedestal)	54.2715							0.24 / 9						
	Viṣṇu	54.2721													0.25
	adorned Buddha in *vitarkamudrā*	54.2724						0.11							
	adorned Buddha in *abhayamudrā*	54.2728											22		
	Prajñāpāramitā	*ga* 5333						0.5							
	Lokeśvara (body)	MA 5960												0.3	0.25
Bayon	Hevajramaṇḍala	*ga* 2494							0.2 / 0.6		0.02–0.08	0.04–0.13			
	Buddhist triad	MG 18204												20	0.03
	pair of *nāga*-shaped finials	*ga* 2989 and *ga* 5165	0.8						0.09 / 0.2						0.8
	Śiva	*ga* 3451		0.44											
	conch setting	*ga* 3788			1.5				0.26 / 9		0.04	0.22	11		0.51
	nāga-shaped finial	*ga* 4023							0.13 / 0.4		0.02	0.11			
	female divinity	*ga* 4025		0.44					0.2 / 0.4		0.01	0.15			
	torso of adorned Buddha in abhayamudrā	*ga* 4047					3.3						4.2		0.39
	Garuḍa-shaped finial	*ga* 5681				2.5									
	Lokeśvara	*ga* 2959								0.09 / 17				17	
	palanquin hook	*ga* 3698					4.2								
	female divinity	*ga* 4012								0.05 / 11					
	Prajñāpāramitā	*ga* 3540				0.69									
	Buddhist triad	*ga* 4417				0.66									

cont'd

Table A3. *Atypical metal compositions (thirty-nine artifacts).* cont'd

	Designation	Inv. No.	Ag (<0.3)	Au (<0.25)	As (<0.9)	Zn (<0.4)	Fe (<2)	Co (<0.07)	Sb+Bi >Pb	Sb+Bi < Pb	Bi corr Pb (<LD)	Sb corr Pb (0.06)	Sn (>5 for Angkor Wat and <10 for Bayon)	Pb (>1 for Angkor Wat and < 13 for Bayon)	Ni (<0.2 for Angkor Wat and > 0.2 for Bayon)
	adorned Buddha in *vitarkamudrā*?	*ga* 5320		0.5											
	Lokeśvara	*ga* 5422		0.43											
	nāga-shaped finial	*ga* 5679				0.75									
	tripod	MA 541													0.03
	tripod	MA 3565							0.11 / 0.3		0.04	0.07			0.04
	palanquin ring	MG 18851b							0.2 / 0.5		0.01	0.2			
	Lokeśvara	54.2689													0.04
	drum	MA 5988											14		
Undetermined	perfume censer or lamp?	*ga* 5467											13		
	perfume censer	MA 5984						0.09						17	
	All values given in wt.%														

Notes

1. The accession numbers of all the Khmer bronzes from Cambodian and foreign collections cited in the article are given systematically.
2. The "East Angkor" group corresponds to objects discovered in the provinces of Kampong Thom, Preah Vihear, and Siem Reap (Chikreng district); the "East" group to objects from the provinces of Stung Treng in Cambodia and Champasak in neighboring Laos; the "Southeast" group to objects from the provinces of Kampong Cham and Prey Veng; the "South" group to objects from the provinces of Kampong Speu and Takeo as well as from the municipality of Phnom Penh; and the "West" group to objects from the provinces of Banteay Meanchey and Battambang.
3. The Brahmanic triad from the Griswold collection (54.2715) was counted twice, since it is made of an Angkor Wat-style pedestal and two statues of an incomplete Bayon-style triad (Woodward 1997, cat. no. 27). The same happened for the perfume censer from the Musée Guimet (MA 5984), since the quadripod and the receptacle with its lid are clearly not associated and may have been assembled only recently (Bourgarit et al. 2003, 113). Hence, the quadripod was counted as part of the Bayon corpus, and the receptacle with its lid as part of the "undetermined" objects dated from the twelfth century to the early thirteenth century.
4. Five statues or parts of statues out of seventeen objects were, however, excluded, because they fell outside the ranges of the twelfth and early thirteenth century and were either too ancient (Woodward 1997, cat. no. 17) or too recent (cat. nos. 22, 27 [Visnu], 39, 40, and 41); or because of unusual technical features (cat. no. 31 [head]).
5. The term "ensemble" is used to describe single objects or groups of objects—that is, pairs of finials, parts of a single assembled object (statues, triads, parts of statuary, finials, and furniture parts), and fragments of a single object.
6. Only one Buddha from the Walters Art Museum collection (54.2728) was not included in the statistical treatment because of its unusually high tin content (22 wt.%). The statue will nevertheless not be excluded from discussion, since apart from its high tin content, it constitutes one of the rare known bronzes supposed to have been produced in the Phimai region in the early twelfth century (Woodward 1997, cat. no. 18).
7. Another example of a silver-tin-copper alloy is identified in Khmer bronze metallurgy: a Koh Ker-style female divinity, which is dated to the mid-tenth century and is now in the Musée Guimet (MA 2239; Baptiste and Zéphir 2008, cat. no. 44). It also shows a very high silver content (19 wt.%), but with only 5 wt.% tin (Bourgarit et al. 2003, 111, 118, appendix 2).
8. Recent analysis by x-ray fluorescence spectroscopy (XRF) performed on a Khmer enthroned Buddha in *dhyānamudrā* from the Freer Gallery of Art (F1937.29a–d), which may also be dated to the first half of the twelfth century, confirmed this result: it is a leaded bronze containing around 20 wt.% lead.
9. Another Angkor Wat-style artifact may be added to the list: a Visnu-Vāsudeva-Nārāyana from the Museum of Indian Art, Berlin (MIK II 1139), which may be dated to the first half of the twelfth century. Elemental analysis of the statue shows that it is a leaded bronze exhibiting 10.2 wt.% tin and 5.1 wt.% lead (Riederer 1988).
10. Two other "Southeast" statues may not be included within this group: a *dvārapāla* figure (*ga* 5466), with 2.7 wt.% tin and 5.8 wt.% lead, and a Umā-Maheśvara on bull (*ga* 3778) that shows an atypical composition with 1.3 wt.% silver and 0.24 wt.% gold, plus 6.1 wt.% tin and 1.5 wt.% lead (see table 3).
11. The unknown or at least presumed provenance of the objects seems also quite problematic, and petrographic analysis may not always support it; indeed, the thin-sections of only two of these "Phimai" bronzes may have been studied (54.2715 [pedestal], and 54.2732; see Reedy 1997b, table 3).
12. Among the ensembles analyzed in the late 1990s, six were finally excluded:

   - the Angkor Wat Visnu-Vāsudeva-Nārāyana (*ga* 5291), since the analytical data regarding its arms may not have been used for statistical treatment;
   - the Bayon tripods (MA 541 and MA 3565) and drum (MA 5988), since no assembly may have been evidenced;
   - the Bayon Hevajra (MA 849), since its red brass pedestal is known to be a later addition (Bourgarit et al. 2003, 114; see also Baptiste and Zéphir 2008, cat. no. 92); and
   - the Angkor Wat/Bayon perfume censer (MA 5984), for the reasons mentioned previously (see note 4).

   Note also that the Bayon Lokeśvara (MA 5940) was added to the twenty-eight ensembles studied, although its different parts have not been analyzed by ICP-AES. We have analytical data for both the body and arms of the statue, which were subject to elemental analysis by ultraviolet emission spectrometry in the early 1990s (Hurtel 1991).
13. Only one fragment of the lamp (*ga* 3846) exhibits different gold and mercury contents (respectively 0.13 wt.% and 0.015 wt.%, whereas the four other fragments show 0.003 wt.% gold and 0.001 wt.% mercury). These differences may be due to a contamination during the sampling of the object, which appears, through visual examination, to have been fire-gilded.
14. The only systematic study of lead effect on the fluidity of copper-based alloys known by the authors so far is the experiments carried out in the 1950s on British Middle and Late Bronze Age materials (Brown and Blin-Stoyle 1959, cited by Craddock 1981, 23). It has been proven that the addition of lead, up to a maximum of 2 wt.%, enhances the fluidity of the molten metal; lead additions also decrease the melting temperature of the alloy.
15. Bunker 2008, 300, 302; Bunker and Latchford 2008, 130. See also Gerschheimer and Vincent (2010) for a reappraisal of the beginnings of the use of fire gilding in ancient Cambodia (sixth/eighth to eleventh centuries).
16. The foundry practices were, however, different in ancient China. Evidence shows that fire gilding was used continuously during nearly two-and-a-half millennia in spite of changes occurring over time in the alloy composition of the metals used as substrates, including high-leaded bronzes and brasses (Jett and Chase 2000, 150–51).
17. Other items from the same collections, which may be identified only through visual examination as fire-gilded, are also unleaded bronzes: a Hevajra of the second half of the eleventh century (*ga* 5331; Bunker and Latchford, 2008, 56) as well as nine Bayon-style artifacts corresponding to seven ensembles from the new selected corpus (*ga* 2989 and *ga* 5165; *ga* 3846 and *ga* 5644; *ga* 4250; *ga* 4266; *ga* 5422; *ga* 5627; and *ga* 5630).
18. As noticed earlier, the Angkor Wat-related Buddha (*ga* 3296) shares several technical features with a series of artifacts of the Bayon corpus (unleaded bronze composition, fire gilding, high bismuth and antimony contents). It would thus be a good example of an Angkor Wat-style statue produced later than the Angkor Wat period, probably in the late twelfth or even early thirteenth century, that is, the Bayon period. This late dating is also attested to by its discovery location: the statue was found with other Khmer sculptures in the Vietnamese province of Binh Dinh, within the Bahn It sanctuaries (the "Silver Towers" of the French archaeological documentation), which were part of the Cham capital Vijaya, modern Quy Nho'n (Schweyer 2007, 59, 63, pl. 2 and 3). Hence, the Buddha image was certainly imported from Angkor for ritual purpose during the

Khmer rule of the region (1203–1220) and produced during this period or slightly before. Moreover, because of the same technical features, except fire gilding, the Angkor Wat-related *nāga*-shaped feet (*ga* 4254.1 and *ga* 4254.2) may also have been produced during the Bayon period.

19. The raw copper extracted from these ores may contain more than 500 wt. ppm gold (A. Lossin, Norddeutsche Affinerie, pers. comm., 2004).
20. A Ganeśa from the same collection was also thought to be a brass (inv. MA 1829; Bourgarit et al. 2003, 117), but, following the new alloy terminology proposed by David Bourgarit and Nicolas Thomas (to be submitted), would be a red brass (8.5 wt.% zinc and 7 wt.% tin). Later than the Bayon style as well, it should be dated to the second half of the thirteenth century.
21. This strongly contrasts Khmer metallurgy with that of eastern India, Tibet, and the Himalayan region, for instance, where zinc is used largely in the alloys for metal statues (see, among others, Werner 1972; Craddock 1981; Reedy 1997a, 281–90).
22. Yet the lead–zinc linear correlation as proposed in figure 14, in order to sustain the hypothesis of zinc coming from ores located at the Thai/Burmese border, is highly questionable.
23. Zinc amounts of more than 1 wt.% help with the removal of oxygen from copper and have a beneficial alloying effect (Craddock 1981, 22), whereas higher amounts lower the solification temperature (Schmidt et al. 1992).
24. Formulas for *samrit* are known by at least two Siamese manuscripts from the former Vajiranana National Library first published by George Cœdès, who identified these recipes as possibly inherited from Khmer metallurgical tradition (Cœdès 1923, 15). Indeed, the four formulas cited in the manuscripts give the composition of ideal alloys, which are all characterized by very high gold contents and may thus correspond to the Khmer beliefs associated with *samrit*. One of these formulas is, for instance, an alloy of nine metals, or *navaloha,* composed of nine parts gold, eight of silver, seven of copper, six of zinc, five of mercury, four of tin, three of iron, two of bismuth, and one of lead. Such nine-metal compositions may also be a reflection of Khmer metallurgical tradition, since alloys of nine metals, also called *samrit pram buon krasat,* were still practiced by Khmer bronze craftsmen in the early twentieth century for both statues and bells, but different metals and quantities were used than those mentioned in the Siamese manuscripts (Groslier 1921–23, 414–16).

References

Arc'Antique. 1997. Angkor et dix siècles d'art khmer: Restauration des œuvres métalliques du musée national de Phnom Penh. Arc'Antique, Nantes, France.

Baptiste, P., and T. Zéphir. 2004. Review of *Adoration and glory: The golden age of Khmer art* by E. C. Bunker and D. A. J. Latchford. *Arts Asiatiques* 59(1):182–84.

———. 2008. *L'Art khmer dans les collections du musée Guimet.* Paris: Réunion des Musées Nationaux.

Bourgarit, D. 2009. Étude de sept statues bouddhiques, 7ème–9ème siècle, Royaume khmer et Chine, Musée national du Cambodge. Compte-rendu d'étude C2RMF R18765.

Bourgarit, D., and T. Borel. 2003. Étude d'une statue de Prajnaparamita supposée khmère, XIIe–XIIIe siècles, Cambodge. Rapport interne C2RMF 3162. C2RMF, Paris.

Bourgarit, D., and B. Mille. 2002. Étude d'une lampe en bronze, Cambodge, fin XIIe–début XIIIe siècle (Khmer). Rapport interne C2RMF 2956. C2RMF, Paris.

———. 2003. The elemental analysis of ancient copper-based artifacts by inductively coupled plasma–atomic emission spectrometry (ICP-AES): An optimized methodology reveals some secrets of the Vix crater. *Measurement Science and Technology* 14(9):1538–55.

Bourgarit, D., B. Mille, T. Borel, P. Baptiste, and T. Zéphir. 2003. A millennium of Khmer bronze metallurgy: Analytical studies of bronze artifacts from the Musée Guimet and the Phnom Penh National Museum. In *Scientific research in the field of Asian Art: Proceedings of the First Forbes Symposium at the Freer Gallery of Art,* ed. P. Jett with J. G. Douglas, B. McCarthy and J. Winter, 103–26. London: Archetype Publications.

Bronson, B. 1992. Patterns in the early Southeast-Asian metals trade. In *Early metallurgy, trade, and urban centres in Thailand and Southeast Asia,* ed. I. Glover, P. Suchitta, and J. Villiers, 63–114. Bangkok: White Lotus.

Brown, M. A., and A. E. Blin-Stoyle. 1959. A sample analysis of British Middle and Late Bronze Age materials using optical spectrometry. *Proceedings of the Prehistoric Society* 25:188–209.

Brunhuber, E. 1986. *Guss aus kupferlegierungen.* Berlin: Schiele & Schön.

Bunker, E. C. 2008. Amalgam gilding in Khmer culture. In *Interpreting Southeast Asia's past: Monument, image and text,* ed. E. A. Bacus, I. C. Glover, and P. D. Sharrock, 296–305. Singapore: NUS Press.

Bunker E. C., and D. A. J. Latchford. 2008. *Khmer gold: Gifts for the gods.* Chicago: Media Art Resources.

Calligaro, T., J.-C. Dran, H. Hamon, B. Moignard, and J. Salomon. 1998. An external milli-beam for archaeometric applications on the AGLAEIBA facility of the Louvre Museum. *Nuclear Instruments and Methods B* 136(1):339–43.

Cœdès, G. 1923. *Bronzes khmèrs.* Étude basée sur des documents recueillis par M. P. Lefèvre-Pontalis dans les collections publiques et privées de Bangkok et sur les pièces conservées au Palais Royal de Phnom Penh, au Musée du Cambodge et au Musée de l'EFEO. Ars Asiatica 5. Paris and Brussels: G. Van Oest.

Cort, L. A., and P. Jett. 2010. *Gods of Angkor: Bronzes from the National Museum of Cambodia.* Washington, DC: Smithsonian Institution.

Craddock, P. T. 1981. The copper alloys of Tibet and their background. In *Aspects of Tibetan metallurgy,* ed. W. A. Oddy and W. Zwalf, 1–31. British Museum Occasional Paper 15. London: British Museum.

Dottin, O. 1971. *Géologie et mines du Cambodge.* Report 70 RME. Bureau de Recherches Géologiques et Miniéres, Paris.

Drayman-Weisser, T., J. A. Lauffenburger, and D. K. Strahan. 1997. Appendix C: Metal analyses. In *The sacred sculpture of Thailand: The Alexander B. Griswold Collection,* ed. H. W. J. Woodward, 278–87. Baltimore: Walters Arts Gallery.

Estève, J., and B. Vincent. 2010. L'about inscrit du Musée national du Cambodge (K. 943): Nouveaux éléments sur le bouddhisme tantrique à l'époque angkorienne. *Arts Asiatiques* 65:109–20.

Gerschheimer, G., and B. Vincent. 2010. L'épée inscrite de Boston (K. 1048, 1040–41 de notre ère). *Arts Asiatiques* 65:133–58.

Groslier, G. 1921–23. L'art du bronze au Cambodge. *Art et Archéologie Khmers* 1(4):413–23.

———. 1931. *Les Collections khmères du musée Albert-Sarraut à Phnom Penh.* Ars Asiatica 16. Paris and Brussels: G. Van Oest.

Hendrickson, M. 2007. Arteries of empire: An operational study of transport and communication in Angkorian Southeast Asia (9th to 15th centuries CE). PhD diss., Department of Archaeology, University of Sydney.

Hendrickson, M., D. Penny, T. O. Pryce, Phon Kaseka, and So Malay. 2010. Industries of Angkor Project: Preah Khan of Kompong Svay (Bakan). December 2009 Field Campaign Report. University of Sydney, Sydney, Australia.

Hurtel, L. P. 1991. Étude d'une statuette d'Avalokiteçvara irradiant. Rapport interne C2RMF 1834. C2RMF, Paris.

Hykin, A., R. Newman, and J. Cummins. 2007. Casting compassion: The technical study of a large bronze from Tibet. In *Scientific research on the sculptural arts of Asia: Proceedings of the Third Forbes Symposium at the Freer Gallery of Art,* ed. J. G. Douglas, P. Jett, and J. Winter, 90–100. London: Archetype Publications.

Jett, P. 1991. Technologische Studie zu den vergoldeten Guanyin-figuren aus dem Dali-Königreich. In *Der goldschatz der drei pagoden: Buddhistische kunst des Nanzhao-Königreichs in Yunnan, China,* ed. A. Lutz, 68–74. Zürich: Museum Rietberg.

———. 2010a. A technical study of the Kampong Cham figure group. In *Gods of Angkor: Bronzes from the National Museum of Cambodia,* ed. L. A. Cort and P. Jett, 78–87. Washington, DC: Smithsonian Institution.

———. 2010b. Buddhist bronzes in Cambodia: A newly discovered cache. *Orientations* 41(5):48–52.

Jett, P., and W. T. Chase. 2000. The gilding of metals in China. In *Gilded metals: History, technology and conservation,* ed. T. Drayman-Weisser, 145–55. London: Archetype Publications.

Le Bonheur, A. 1972. Un bronze d'époque préangkorienne représentant Maitreya. *Arts Asiatiques* 25:129–54.

Oddy, W. A. 2000. A history of gilding with particular reference to statuary. In *Gilded metals: History, technology and conservation,* ed. T. Drayman-Weisser, 1–19. London: Archetype Publications.

Oddy, W. A., M. Bimson, and S. La Niece. 1981. Gilding Himalayan images: History, tradition and modern techniques. In *Aspects of Tibetan metallurgy,* ed. W. A. Oddy and W. Zwalf, 1–31. British Museum Occasional Paper no. 15. London: British Museum.

Pottier, C. 1997. Nouvelles données sur les couvertures en plomb à Angkor. *Bulletin de l'École française d›Extrême-Orient* 84:183–220.

Reedy, C. L. 1997a. *Himalayan bronzes: Technology, style, and choices*. Newark and London: University of Delaware Press.

———. 1997b. Appendix B: Analyses of the clay cores. In *The sacred sculpture of Thailand: The Alexander B. Griswold Collection,* ed. H. W. J. Woodward, 266–77. Baltimore: Walters Arts Gallery.

Riederer, J. 1988. MIK II 1139. Rathgen-Forschungslabor, Staatliche Museen zu Berlin, Germany.

Schmidt, R. F., D. G. Schmidt, and M. Sahoo. 1992. Copper and copper alloys. In *ASM metal handbook,* Vol. 15, *Casting,* ed. D. M. Stefanescu, 1697–734. Metals Park, OH: ASM International.

Schweyer, A.-V. 2007. The confrontation of the Khmers and the Chams in the Bayon period. In *Bayon: New perspectives,* ed. J. Clark, 50–71. Bangkok: River Books.

Soutif, D. 2008. La monture de conque inscrite du musée national de Phnom Penh. Relecture de l'inscription angkorienne K. 779. *Aséanie* 22:47–61.

Strahan, D. K. 1988. A technical study of 11–12th c. Khmer bronze sculpture. In *The beginning of the use of metals and alloys*, ed. R. Maddin, 97–102. Cambridge, MA: MIT Press.

———. 1997. Bronze casting in Thailand. In *The sacred sculpture of Thailand: The Alexander B. Griswold Collection,* ed. H. W. J. Woodward, 27–41. Baltimore: Walters Art Museum.

Werner, O. 1972. *Spektralanalytische und metallurgische untersuchungen um Indischen bronzen*. Leiden: Brill.

Woodward, H. W. Jr. 1997. *The sacred sculpture of Thailand: The Alexander B. Griswold Collection*. Baltimore: Walters Arts Gallery.

———. 2003. *The art and architecture of Thailand.* Leiden & Boston: E. J. Brill.

Casting of Ancient Drums in North Vietnam

Trinh Sinh

ABSTRACT The Metal Age in North Vietnam occurred about 4,000 years BP to 2,000 years BP and has been divided into four archaeological cultures: Phung Nguyen, Dong Dau, Go Mun, and Dong Son. The earliest traces of metalworking in North Vietnam occur in the Phung Nguyen culture, though this culture produced only a few bronze objects. The Dong Dau culture produced a remarkable quantity of bronze objects, mostly from copper-tin alloys. A later culture, the Go Mun culture used similar alloys to those of the Dong Dau people. Archaeologists have found some fragments of stone and ceramic for casting bronze artifacts and some traces of kilns. The Dong Son culture (seventh century BCE to first/second century CE) produced a great quantity of bronze artifacts, including large drums. Dong Son alloys include leaded bronze, often containing high percentages of lead. Evidence also suggests the Dong Son people used ceramic molds to cast their large drums. Hundreds of Heger type I drums have been discovered in North Vietnam. These large drums are sometimes referred to as Dong Son drums because they are considered the most important artifacts left by the Dong Son culture. These drums are known to have been excellent percussion instruments and great symbols of power. In this study, eighteen bronze Heger type I drums have been studied for alloy composition by spectrographic analysis (Trinh 1988) and for their casting methods.

The Metal Age in North Vietnam

The Metal Age in North Vietnam occurred from about 4,000 years BP to 2,000 years BP, and has been divided into four archaeological cultures: Phung Nguyen, Dong Dau, Go Mun, and Dong Son.

The Phung Nguyen Culture

The Phung Nguyen culture has been dated to approximately 4000–3500 BP. The ^{14}C method was used to date some sites from this culture, yielding the following results: Trang Kenh site, Hai Phong province, dated to 3405 ± 100 BP (^{14}C sample number: Bln 891; see Kohl and Quitta 1978 regarding samples with "Bln" designation); Dong Cho site, Ha Tay province, dated to 3800 ± 60 BP (Bln 3081); and Ma Dong site, Ha Tay province, dated to 4145 ± 60 BP (Bln 1277). The last site does not belong to the Phung Nguyen culture but is of equivalent date.

Phung Nguyen culture sites contain the earliest traces of metalworking in North Vietnam, with small fragments of corroded bronze found in cultural layers in Go Bong, Xom Ren, Dong Xau, and Dong Vong. Chemical analysis of these small pieces demonstrated that they are composed of copper-tin alloys rather than pure copper metal. In the Phung Nguyen culture, bronze artifacts are rare; instead these people mainly used stone for making productive tools, weapons, and ornaments.

The Dong Dau Culture

The Dong Dau culture has been dated to approximately 3500–3100 BP, as shown by ^{14}C dates that yielded the following results: Vuon Chuoi site, Ha Tay province, dated to 3070 ± 100 BP (Bln 894); Dong Dau site, Vinh Phuc province, dated to 2960 ± 150 BP (Bln 3810); and Thanh Den site (layer 5), Vinh Phuc province, dated to 3390 ± 70 BP (^{14}C sample number R-9755/3).

The Dong Dau culture produced a remarkable quantity of bronze objects, including axes, chisels, fish hooks, brushes, point heads, stitches, spearheads, and arrows. The Dong Dau people continued the tradition of using stone tools, but they understood metallurgy well. Spectrographic analyses of some artifacts excavated from Dong Dau culture sites (Sinh 1990, 49–58) demonstrate that these people used mostly copper-tin alloys. We analyzed a total of thirty-three artifacts, with results showing that twenty-seven specimens were copper-tin (CuSn) bronze, four specimens were copper-antimony (CuSb) bronze, and two specimens were copper-arsenic (CuAs) bronze (approximately 6 wt.% As). Alloys of CuSb and CuAs were used mainly for arrows, and these alloys contain from 2.9 to 6.5 wt.% Sb and from 2.3 to 5.4 wt.% As.

The Go Mun Culture

The Go Mun culture has been dated to about 3100–2700 BP as shown by ^{14}C dates that yielded the following results:

Vinh Quang site, Ha Tay province, dated to 3045 ± 120 BP (Bln 829); and Thanh Den site (later layers), Vinh Phuc province, dated to 2940 ± 60 BP (Bln 2955) and 2860 ± 60 BP (Bln 2981).

In the Go Mun culture, the quantity of bronze artifacts increases and includes axes, chisels, fish hooks, spearheads, javelins, arrows, sickles, and statues. Bronze tools and weapons had an important role in agricultural production and war (Trinh 1992, 19–26). The Go Mun people did not have the technology to cast drums and other large bronze objects, but they made many small beautiful bronzes. They did not understand the role and use of lead for metallurgy. The alloys they used were similar to those of the Dong Dau people. We used spectrographic methods to analyze nine artifacts excavated from Go Mun culture sites and one artifact from the Dong Son site (lower level, dated equivalent to the Go Mun culture) (Trinh 1990, 49–58). We can see that these ancient people also mostly used CuSn bronzes. A total of ten specimens was analyzed, revealing that eight specimens were CuSn bronze, one specimen was CuSb bronze, and one specimen was a copper-tin-lead (CuSnPb) bronze (and belonged to a mixed cultural level).

Pre-Dong Son Cultures in Summary

The three above cultures are called pre-Dong Son cultures (i.e., before 2600 BP). About 4000–3500 BP the people of the Phung Nguyen culture were the first to know about the manufacture of bronze objects, and they produced CuSn bronzes. At that early time, bronzes were very rare, and only a few types were produced. To date, archaeologists have found only a few bronze pieces, and no mold or kiln for metal production has been found. We hope that in the future more samples of these bronzes can be found and studied.

About 3500–2700 BP, Dong Dau and Go Mun cultures used highly skilled metalworking techniques. They cast bronze objects *in situ* in their small villages. Archaeologists have found seven fragments of stone molds for casting both arrows and axes, as well as some ceramic molds for casting point heads (Le Xuan Diem and Hoang Xuan Chinh 1983, 40). At the Go Mun site, some molds were found (Ha Van Phung and Nguyen Duy Ty 1982). The negative shapes inside the molds are similar to the bronze artifacts dating to the same time. At the Dong Dau site, we found some traces of kilns for casting bronzes, although these kilns are small, perhaps only used within an agricultural village.

Dong Dau and Go Mun people knew how to create and use alloys. Those most common are CuSn alloys for making bronze objects and CuSb and CuAs alloys for making arrows. The elements antimony and arsenic were present only in small proportions in the bronzes, but these elements were very important when these people created their first alloys. Some scientists say that these CuAs alloys, sometimes called arsenical copper, appeared very early elsewhere in the world, but "[a]ccording to the analytical evidence available at present, arsenical copper was not used in Southeast Asia or in East Asia" (Muhly 1988, 12). Our analyses demonstrate that in North Vietnam arsenical copper alloys were indeed present about 3,500–3,100 years ago. Wang Dadao suggests that a lower limit of 2 wt.% is necessary for the determination of an intentionally added element in an alloy (Wang 1986, 59). In the Dong Dau culture, the contents of elemental arsenic are between approximately 2.3 and 5.4 wt.%. This arsenic content in the CuAs bronzes of the Dong Dau culture suggests an intentional CuAs alloy. Also, five samples from the Dong Dau and Go Mun cultures contain antimony in the alloys. Perhaps antimony had a significant role in tin alloys in North Vietnam. Research has shown that antimony was used in some of the earliest bronze alloys in the world (Rapp 1988, 21–27).

The Dong Son Culture

The Dong Son culture has been dated from about 2600 BP to 2000 BP. Many bronze artifacts have been discovered, including drums, vessels, spears, javelins, arrows, and daggers. A total of 555 samples has been analyzed by spectrographic methods and found to be composed of elemental copper and eleven copper-alloy types (table 1).

Table 1: *Alloy compositions of bronze artifacts from the Dong Son culture.*

Alloy Composition	No. of Samples	Percent of Samples
Cu	65	11.7
Cu As	10	1.8
Cu Sn	111	20.2
Cu Pb	16	2.9
Cu Sn As	3	0.5
Cu Sn Pb	192	34.6
Cu Pb Sn	141	25.4
Cu As Pb	4	0.7
Cu Sn Pb As	5	0.9
Cu Sn As Pb	2	0.4
Cu Pb Sn As	5	0.9
Cu As Pb Sn	1	0.2
Total	555	100.0

Source: Trinh (1989).

A total of 66% of the artifacts was found to be composed of alloys containing unusually high amounts of lead. In most of these (62.4%), the alloys contain both lead and tin.

Generally, alloys of pre-Dong Son cultures (Phung Nguyen, Dong Dau, and Go Mun) were different from those of the Dong Son culture. In pre-Dong Son cultures, bronze alloys contained no lead, while in the Dong Son culture bronze alloys contained a high lead content.

Metal Alloys Used for Casting

What were the sources of raw materials for early metallurgy in North Vietnam? This question continues to present a challenge to archaeologists. North Vietnam has many mines of copper and tin. Copper mines are located in Lao Cai

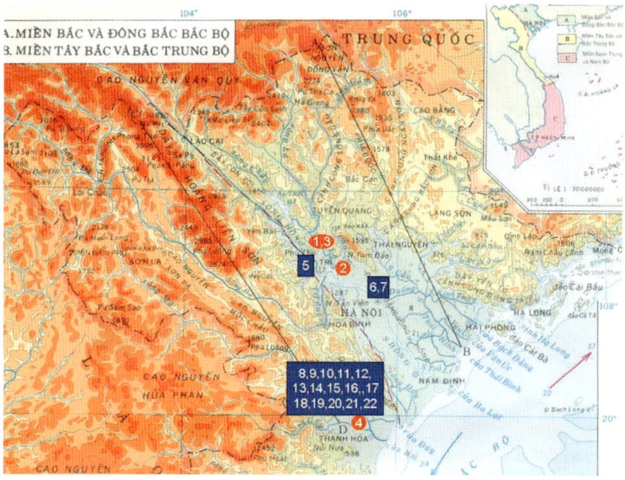

Figure 1. Map showing areas discussed in the text. Red circles indicate archaeological sites and their cultures: 1 = Phung Nguyen; 2 = Dong Dau; 3 = Go Mun; 4 = Dong Son. Blue rectangles indicate sites where Heger type I bronze drums cited in the paper were unearthed: 5 = Thuong Nong; 6 = Co Loa I; 7 = Ha Noi II; 8 = Hoang Son; 9 = Dong Son IV; 10 = Xuan Lap I; 11 = Thanh Van; 12 = Quang Thang II; 13 = Dong Hoa I; 14 = Thiet Cuong; 15 = Cam Thuy; 16 = Dinh Cong V; 17 = Quan Hoa; 18 = Dinh Cong I; 19 = Quang Thang I; 20 = Vinh Ninh; 21 = Quang Thang (miniature drum); 22 = Nui Nap (miniature drum). Numbers 1, 3, and 5 belong to Phu Tho province. Number 2 belongs to Vinh Phuc province. Numbers 4 and 8–22 belong to Thanh Hoa province. Numbers 6 and 7 belong to Hanoi capital.

province, Bac Giang province, Son La province, Cao Bang province, and along the Da River (Son La and Hoa Binh provinces), and tin mines are located in Cao Bang province, Nghe An province, and on Tam Dao Mountain (Thai Nguyen and Vinh Phuc provinces).

Did people living more than 2,700 years ago exploit copper and tin mines for casting bronze? To date archaeologists have not found ancient mines, slag, or tools for mining in North Vietnam.

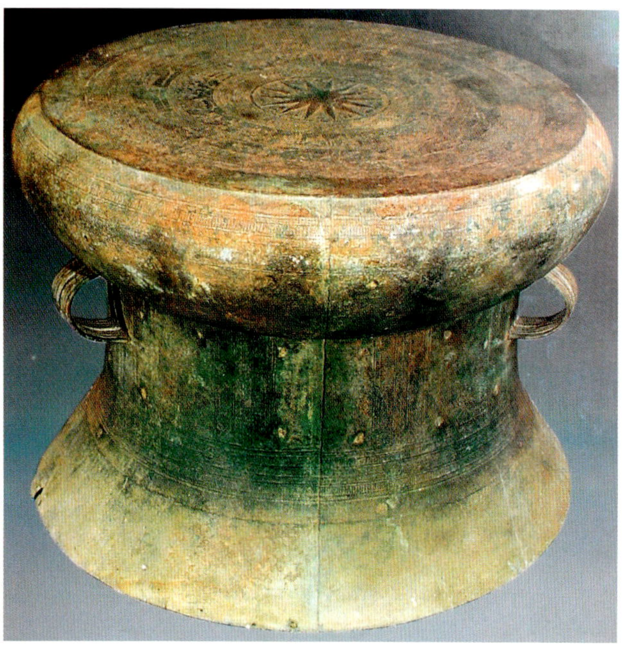

Figure 2. Co Loa I drum. Source: Ha Thuc Can and Van Huyen Nguyen (1989).

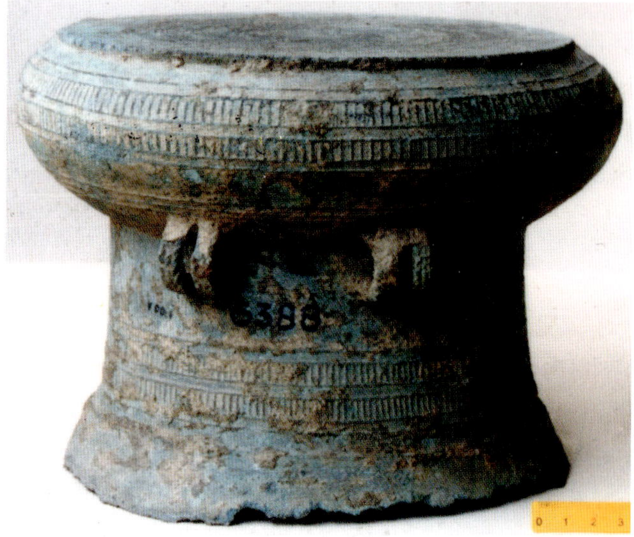

Figure 3. Ha Noi II drum.

An indication that bronze casting may have occurred in North Vietnam in ancient times is the prevalence of copper, tin, and lead mines that could have been the source of raw materials. However, archaeologists need to conduct more research to find evidence of mining during the Metal Age.

Molds and Other Items Related to Bronze Casting

Many stone and ceramic molds for casting have been found. These molds appear in the cultural layers of Dong Dau, Go Mun, and Dong Son cultures, at such sites as the Go Mun site, Lang Ca site (Trinh and Hong 1980), and Lang Vac site. The stone molds are typically bifacial and composed of sandstone. Only a few molds for bronze casting of large objects have been found. Perhaps ceramic molds were used for these large castings, and these molds were immediately destroyed after use.

In addition to the discovery of sandstone and ceramic molds, many other items related to bronze casting have

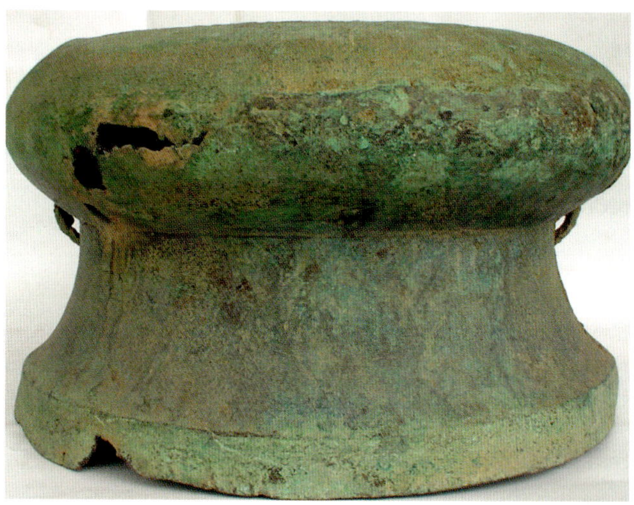

Figure 4. Thuong Nong drum.

Table 2. *Eighteen Heger type I drums from North Vietnam analyzed by spectrographic methods.*

Group	Drum Type, Province of Origin	Sample Location	Specimen
A	Co Loa I, Ha Noi	LS	39613
B	Hoang Son, Thah Hoa	LS	39297
		US	39303
		T	39305
		P	39327
	Dong Son IV, Thanh Hoa	LS	39299
		P	39312
		T	39321
		US	39440
	Xuan Lap I, Thanh Hoa	P	39309
		LS	39374
		C	39386
		US	39389
		T	39390
	Quang Thang, II, Thanh Hoa	P	39318
		US	39417
		T	39428
	Thiet Cuong, Thanh Hoa	LS	39333
		T	39379
		US	39394
		P	39430
	Cam Thuy, Thanh Hoa	US	39339
		P	39354
		T	39382
		C	39387
		LS	39410
	Dinh Cong V, Thanh Hoa	T	39340
	Dinh Cong I, Thanh Hoa	T	39349
	Quang Thang I, Thanh Hoa	LS	39400
		US	39413
		P	39421
		T	39441
	Ha Noi II, Ha Noi	US	39595
	Vinh Ninh, Thanh Hoa	P	39645
		LS	39646
		C	39647
		T	39648
		US	39649
C	Thanh Van, Thanh Hoa	T	39311
		US	39385
	Dong Hoa I, Thanh Hoa	US	39319
		P	39324
		LS	39353
		T	39369
D	Thuong Nong, Phu Tho	US	39445
		LS	39469
E	Quan Hoa, Thanh Hoa	LS	39344
		Frog figure on the tympan	39345
		T	39352
		P	39362
		US	39443
		C	39424
Miniature Drums	Quang Thang (miniature drum), Thanh Hoa	LS	39185
	Nui Nap (miniature drum), Thanh Hoa)	LS	39300

All drums are stored in the provincial museums of their origin.
C = chaplet (spacer); LS = lower section; P = pair, T = tympan; US = upper section.

been found. Vietnamese archaeologists have discovered slag, vestiges of foundries, and ceramic crucibles. In the Lang Ca, Ru Tran, and Dong Son sites, Vietnamese scholars have found large ceramic crucibles for casting. One of them, discovered in Lang Ca, would have been large enough to cast up to 12 kg of bronze (Vu Thi Ngoc Thu and Nguyen Duy Ty 1978, 39).

Dong Son Drums

Hundreds of Heger type I drums have been discovered in North Vietnam (Huyen et al. 1987). Vietnamese archaeologists call these Dong Son drums because they are the most important artifacts left by the Dong Son culture. They were known to be excellent percussion instruments and great symbols of power. Eighteen bronze Heger type I drums were studied by spectrographic analysis. As shown in the map in figure 1, these Heger type I drums were discovered in three provinces: Phu Tho, Ha Noi (Red River valley), and Thanh Hoa (Ma River valley). We classify these drums based on their shape and decoration, and their names are based on the location of excavation (Huyen et al. 1987).

- Group A: Co Loa I (1 drum) (fig. 2)
- Group B: Hoang Son, Dong Son IV, Xuan Lap I, Quang Thang II, Thiet Cuong, Cam Thuy, Dinh Cong V, Dinh Cong I, Quang Thang I, Ha Noi II (fig. 3), and Vinh Ninh (11 drums)
- Group C: Thanh Van and Dong Hoa I (2 drums)
- Group D: Thuong Nong (1 drum) (fig. 4)
- Group E: Quan Hoa (called Đ group in Vietnamese archaeological documents) (1 drum)
- Miniature drum group: Quang Thang and Nui Nap (2 drums)

Vietnamese archaeologists believe drums belonging to group A were the earliest and those belonging to group E were the latest. The miniature drums, which tend to be of later date, probably served as votive objects at funeral ceremonies.

Fifty-four samples were taken for analysis from different positions on eighteen Heger Type I drums: tympana,

Table 3. *Elemental composition and alloy type of the eighteen Heger type I drums from North Vietnam.*

Specimen No.	Cu	Sn	Pb	Zn	Bi	Ag	Sb	As	Fe	Ni	Co	Au	Alloy Type
39297	base	18	15	0.006	0.047	0.1	0.17	0.61	0.36	0.022	0.016	0.0052	Cu-Sn-Pb
39303	base	25	22	0.0075	0.047	0.06	0.19	1	0.33	0.0098	0.0074	0.0035	Cu-Sn-Pb-As
39305	base	19	7.9	—	0.072	0.1	0.13	0.42	0.36	0.032	0.018	0.0005	Cu-Sn-Pb
39327	base	10	12	0.006	0.026	0.06	0.095	0.37	1.1	0.014	0.0047	0.0006	Cu-Pb-Sn
39299	base	5.1	22	0.055	0.04	0.0015	0.045	0.37	0.24	0.0043	0.002	—	Cu-Pb-Sn
39312	base	5.1	22	0.099	0.052	0.0046	0.031	0.12	0.086	0.0052	0.0013	—	Cu-Pb-Sn
39321	base	6.2	24	0.064	0.043	0.0046	0.037	0.18	0.1	0.0098	0.0013	—	Cu-Pb-Sn
39440	base	5.1	25	0.017	0.055	0.0024	0.028	0.1	0.15	0.0047	0.0017	—	Cu-Pb-Sn
39309	base	7.5	25	0.0075	0.066	0.0024	0.031	0.42	0.056	0.014	0.002	—	Cu-Pb-Sn
39374	base	7.5	19	0.0036	0.056	0.0046	0.037	0.79	0.002	0.022	0.0027	—	Cu-Pb-Sn
39386	base	9.1	19	0.0093	0.13	0.034	0.21	2.7	0.018	0.02	0.0039	—	Cu-Pb-Sn-As
39389	base	12	27	0.0075	0.051	0.019	0.049	1.2	0.018	0.032	0.0033	—	Cu-Pb-Sn-As
39390	base	13	25	0.0093	0.047	0.06	0.054	1.6	0.0027	0.032	0.0033	—	Cu-Pb-Sn-As
39311	base	8.2	13	0.0075	0.051	0.061	0.5	0.42	0.075	0.035	0.023	0.0007	Cu-Pb-Sn
39385	base	7.5	15	0.0036	0.047	0.16	0.46	0.69	0.024	0.11	0.021	0.0081	Cu-Pb-Sn
39318	base	5.6	15	0.086	0.15	0.069	0.13	2	0.1	0.0068	0.003	0.0001	Cu-Pb-Sn-As
39417	base	4.6	21	0.014	0.079	0.061	0.11	1.9	0.0046	0.027	0.011	0.0003	Cu-Pb-Sn-As
39428	base	6.2	19	0.0093	0.11	0.16	0.11	2.5	0.065	0.0068	0.0047	0.0003	Cu-Pb-Sn-As
39319	base	12	12	0.04	0.19	0.1	0.37	0.33	0.6	0.017	0.002	—	Cu-Pb-Sn
39324	base	13	22	0.0093	0.15	0.06	0.37	0.54	0.44	0.024	0.0033	—	Cu-Pb-Sn
39353	base	5.1	12	0.006	0.19	0.06	0.17	0.26	0.4	0.022	0.0018	—	Cu-Pb-Sn
39369	base	7.5	18	0.002	0.25	0.06	0.21	0.21	0.15	0.017	0.0017	—	Cu-Pb-Sn
39333	base	1.5	4.9	0.099	0.04	0.06	0.087	1.4	0.36	0.032	0.011	0.0009	Cu-Pb-Sn-As
39379	base	1.2	8.8	0.011	0.072	0.06	0.13	1.2	0.037	0.032	0.028	0.0018	Cu-Pb-Sn-As
39394	base	3.5	9.7	0.0093	0.055	0.16	0.67	3.8	0.056	0.18	0.047	0.0052	Cu-Pb-Sn-As
39430	base	1.1	3.8	0.086	0.028	0.019	0.19	0.69	0.15	0.035	0.03	0.0014	Cu-Pb-Sn
39339	base	10	25	0.029	0.056	0.06	0.19	0.69	1.2	0.012	0.0081	—	Cu-Pb-Sn
39354	base	4.2	11	0.006	0.028	0.0046	0.049	0.26	0.4	0.0039	0.0027	—	Cu-Pb-Sn

39382	base	8.2	22	0.02	0.043	0.034	0.095	0.42	0.7	0.011	0.0057	—	Cu-Pb-Sn
39387	base	0.029	0.2	0.04	—	0.0069	0.049	0.29	5.7	0.027	0.018	—	Cu
39410	base	4.6	25	0.0047	0.043	0.06	0.065	0.18	0.51	0.0068	0.0057	—	Cu-Pb-Sn
39340	base	8.2	27	0.017	0.38	0.1	0.59	1.6	0.065	0.06	0.0012	—	Cu-Pb-Sn-As
39344	base	9.1	18	0.0093	0.0093	0.011	1.3	2	0.33	0.02	0.0047	—	Cu-Pb-Sn-As-Sb
39345	base	10	17	0.047	0.007	0.011	1.1	2.2	0.33	0.022	0.0033	—	Cu-Pb-Sn-As-Sb
39352	base	3.8	7.9	0.0047	0.0025	0.0019	0.0074	1.3	0.33	0.0075	0.0012	—	Cu-Pb-Sn-As
39362	base	7.5	12	0.0027	0.0057	0.1	0.015	1.9	0.037	0.027	0.0022	—	Cu-Pb-Sn-As
39443	base	4.6	18	0.0036	0.0028	0.0019	0.5	2.2	0.56	0.0062	0.0013	—	Cu-Pb-Sn-As
39424	base	4.6	28	0.17	0.007	0.0015	1.9	5.8	2.1	0.014	0.0039	—	Cu-Pb-Sn-As-Sb
39349	base	7.5	9.7	0.011	0.75	0.06	0.095	0.69	0.007	0.013	0.002	—	Cu-Pb-Sn
39400	base	6.8	22	0.024	0.062	0.06	0.059	1.2	1.3	0.0098	0.0052	—	Cu-Pb-Sn-As
39413	base	5.1	18	0.0093	0.04	0.0046	0.028	0.21	0.81	0.0057	0.0033	—	Cu-Pb-Sn
39421	base	5.1	19	0.0047	0.051	0.0019	0.024	0.23	0.99	0.0062	0.0033	—	Cu-Pb-Sn
39441	base	6.2	22	0.0075	0.037	0.019	0.037	0.42	1.1	0.0075	0.0052	—	Cu-Pb-Sn
39445	base	0.68	0.025	—	—	0.06	0.0062	0.033	0.0046	0.0003	0.0011	—	Cu
39469	base	0.57	0.013	—	0.0002	0.06	0.0036	0.02	0.021	0.0001	—	—	Cu
39595	base	10	32	—	0.015	0.019	0.0088	3.1	0.056	0.0082	0.003	—	Cu-Pb-Sn-As
39645	base	1.5	7.9	0.0014	0.01	0.0002	0.16	0.1	0.32	0.0068	0.0027	0.0003	Cu-Pb-Sn
39646	base	3.5	13	0.0036	0.028	0.0008	0.0016	0.12	0.12	0.013	0.0052	0.0007	Cu-Pb-Sn
39647	base	0.43	22	0.002	0.04	0.0019	0.034	1.2	0.0005	0.032	0.0011	—	Cu-Pb-As
39648	base	4.2	19	0.024	0.023	0.0015	0.0025	0.16	0.44	0.013	0.0057	0.0003	Cu-Pb-Sn
39649	base	4.6	18	0.002	0.04	0.0046	0.0033	0.29	0.056	0.027	0.012	0.002	Cu-Pb-Sn
39613	base	8.2	7.9	0.006	0.015	0.034	0.045	0.16	0.38	0.009	0.0057	—	Cu-Sn-Pb
39185	base	9.1	17	0.002	0.011	0.011	0.095	0.37	0.13	0.013	0.0043	—	Cu-Pb-Sn
39300	base	0.11	1.6	0.029	0.079	0.13	0.034	2.5	0.24	0.0082	0.0033	—	Cu-Pb-As

All values given are wt.%. All specimen numbers and analytical machine made by the Laboratory in the Institute of Archaeology, Academy of Science, Moscow, Russia.

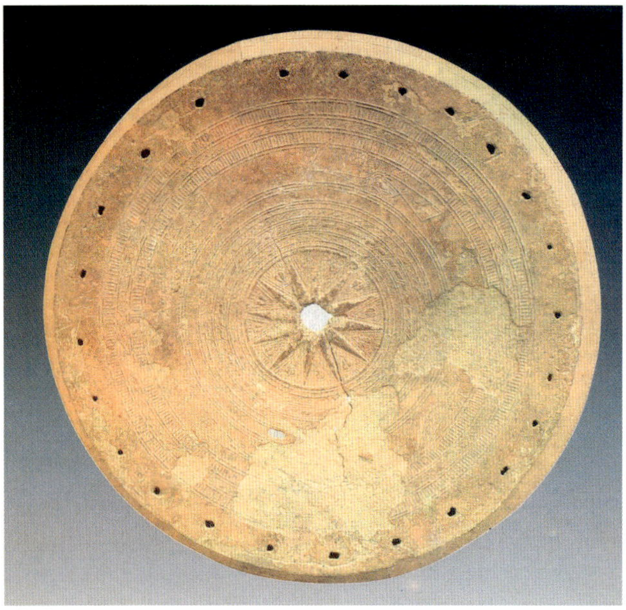

Figure 5. *Chaplets on Phu Duy drum. Source: Ha Thuc Can and Van Huyen Nguyen (1989).*

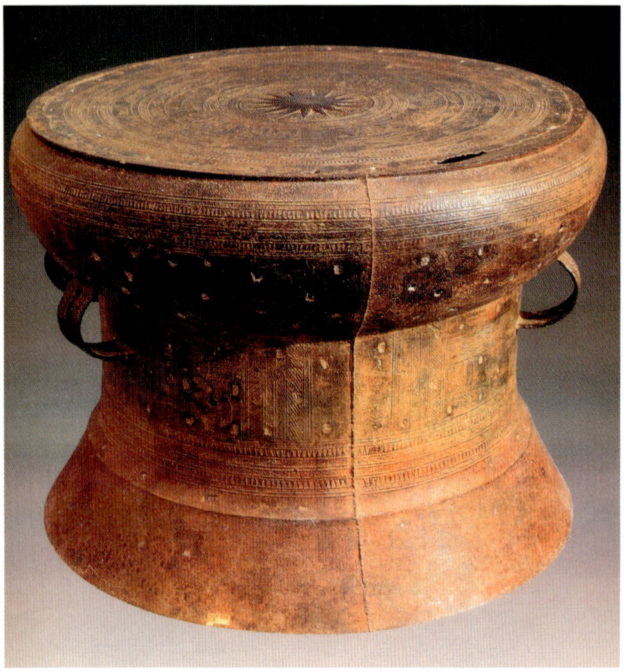

Figure 6. *Chaplets on Hoang Ha drum. Source: Ha Thuc Can and Van Huyen Nguyen (1989).*

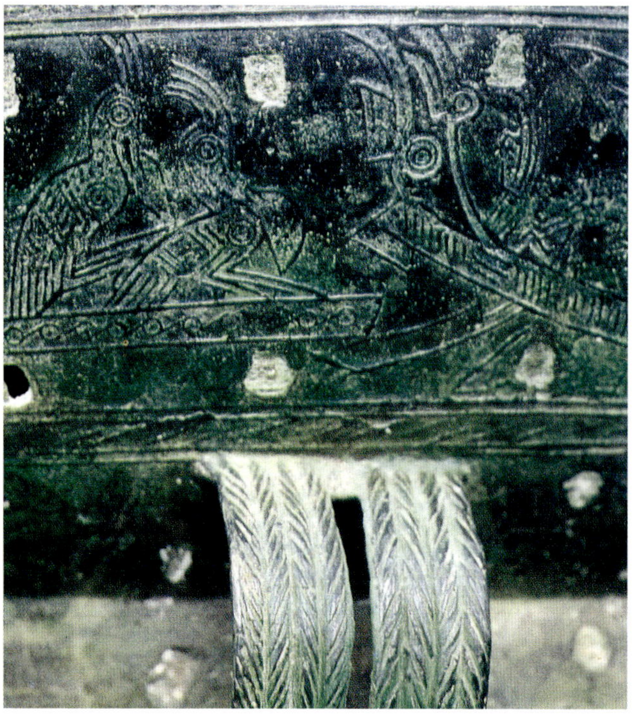

Figure 7. *Chaplets on Mieu Mon drum. Source: Ha Thuc Can and Van Huyen Nguyen (1989).*

upper section, handle pairs, lower section, chaplets (spacers), and frog decoration on the tympana. The position of the samples and origins of these drums are listed in table 2. Table 3 shows the results of spectrographic analysis on samples from the eighteen drums. Although some researchers believe that at least 2 wt.% of an element must be present to indicate it was added intentionally to an alloy (Wang 1986, 59), here a lower limit of 1 wt.% is taken to indicate intentional addition. For this reason, arsenic and antimony are considered to be alloying elements in Heger type I drums of North Vietnam (see table 3).

Iron is present in some samples from the drum collection: in the samples from Hoang Son, Cam Thuy, Quan Hoa, and Quang Thang I, for instance, as much as 5.7 wt.% iron was present, suggesting that iron was present in the copper ore.

The results of the spectrographic analysis of Heger type I drums lead us to some conclusions.

1. Some variations in the elemental contents occur throughout the drums. These variations are not large, however.
2. The use of chaplets (spacers) in drum casting requires further study. One chaplet sample from the Cam Thuy drum was found to be composed of copper, which is quite different from the alloy of the various sections (Cu-Pb-Sn). Chaplets often differ in composition from surrounding bronze, suggesting they were added as spacers between the walls of molds. Traces of chaplets can be found on drums' tympana, for example, as seen on the Phu Duy bronze drum (fig. 5) or on the body of the Hoang Ha bronze drum (fig. 6). Chaplets are usually square in shape, as on the Mieu Mon drum body (fig. 7). Interestingly, we can count about 250–300 chaplets on one Vinh Ninh drum (Sinh 1996, 50).
3. The chemical composition of the Thuong Nong drum is pure copper. Samples from the upper section and lower section were analyzed. This drum is pot-shaped, similar to the drum group in Wanjiaba, Yunnan province, Southwest China (Murowchick 1989). Some archaeologists believe that the Thuong Nong drum has a later date, although others believe it is very early, possibly belonging to the system "pre-Heger drums."
4. The two miniature drums (Quang Thang and Nui Nap drums) have chemical compositions similar to many bronze drums. These miniature drums were cast for use in funeral ceremonies.

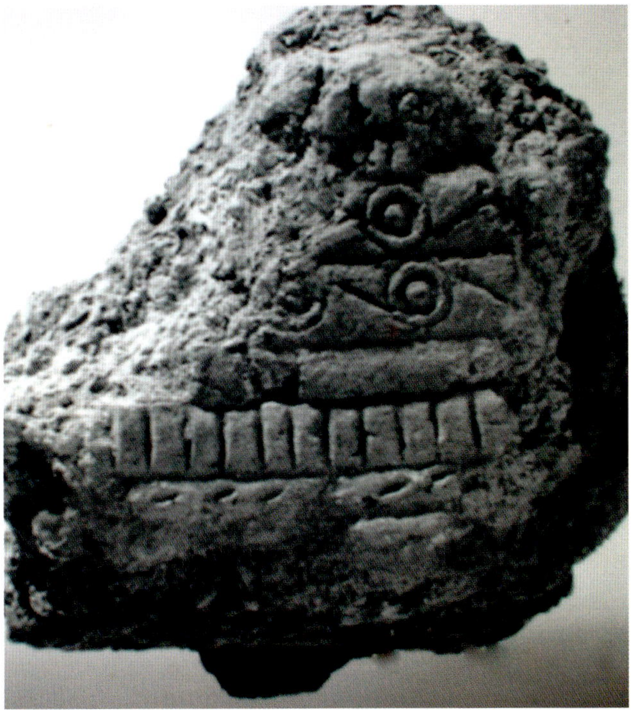

Figure 8. *Fragment of a ceramic mold for casting a bronze drum found at the Luy Lau site. Photo courtesy of the Institute of Archaeology in Vietnam.*

5. Except for the Thuong Nong drums, all others are composed of Cu-Pb-Sn alloys, and some have an addition of arsenic and antimony in small amounts.
6. The lead content of some drums is very high, up to approximately 32 wt.% lead in the Cu-Pb-Sn alloy of one Ha Noi II drum. This drum was probably not used as a percussion instrument because the high lead content would produce a very unpleasing sound.
7. The overall average content of lead and tin in our drum collection is 16.88 wt.% and 7.35 wt.%, respectively (excluding Thuong Nong copper drum and samples from chaplets). The lead content is very high and twice the proportion of tin.
8. Dong Son people used a high lead alloy for bronze casting. In pre-Dong Son cultures, people in North Vietnam had no knowledge of lead for use in bronze alloys. The use of lead considerably changed bronze casting in the Dong Son culture. Because of the chemical properties of lead, the Dong Son people could cast large objects such as drums.

Casting Methods for Bronze Drums

Molds for Casting

Only one piece of a ceramic mold has been discovered, found in Luy Lau, Bac Ninh province, in 1998 by M. Nishimura, a Japanese scholar. He announced this artifact at the Institute of Archaeology in Vietnam in 1998 (fig. 8). Ceramic molds were suitable for the casting of large objects with sophisticated decorations. The molds to cast bronze

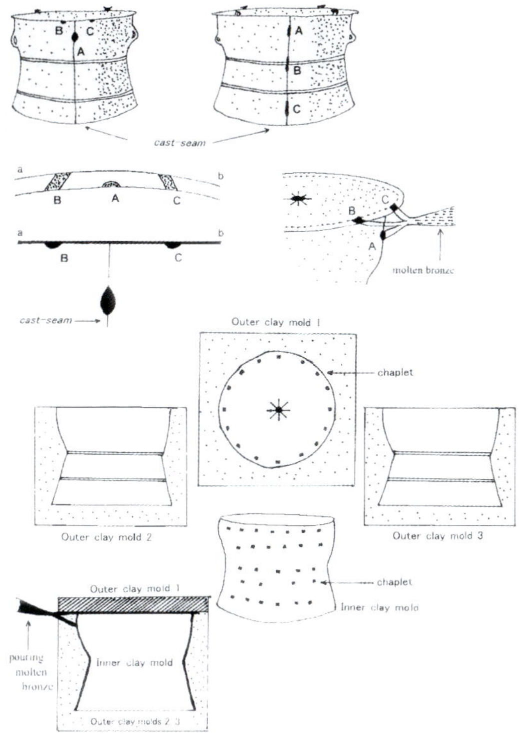

Figure 9. *Diagram showing some ideas about the methods used for casting Heger type II bronze drums.* ***top.*** *Two types of casting-seams on drums: type 1 (left) and type 2 (right). On each, three points labeled A, B, and C indicate the pouring locations for the molten bronze.* ***middle.*** *Detail of traces A, B, and C points of type 1 viewed from the bottom and the side.* ***bottom.*** *Mold system including three outer clay molds and an inner clay mold. Chaplets are placed between the inner and outer mold sections.*

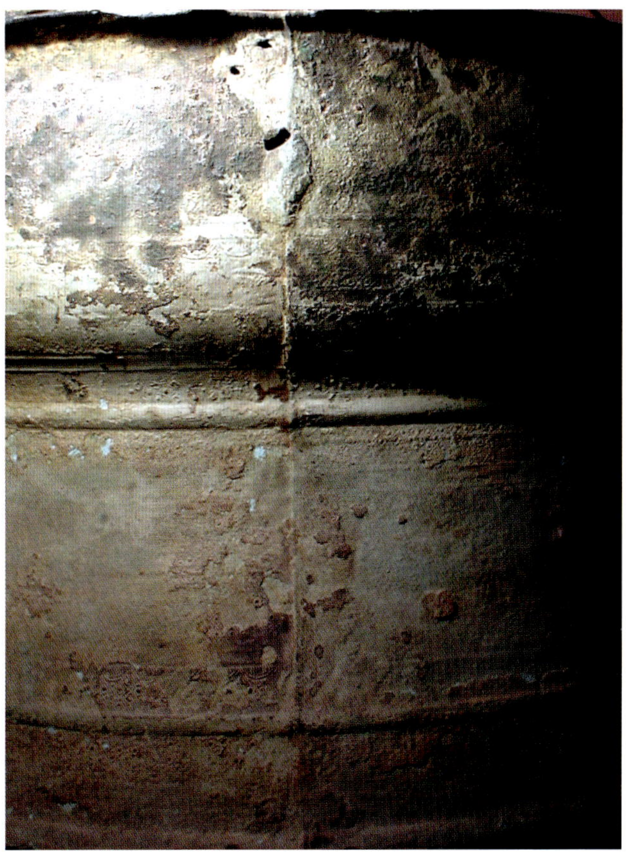

Figure 10. *Trace of points A, B, and C where metal was poured in at the edge of the drum's tympanum. Heger type II drum, HN 6660, Hanoi Museum.*

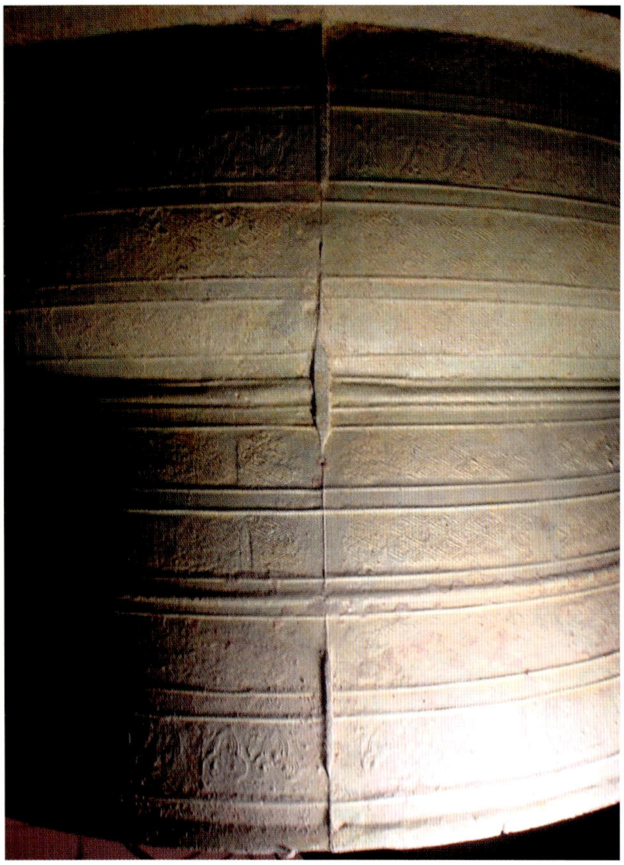

Figure 11. Trace of points A, B, and C where metal was poured into the side of the drum. Heger type II, HN 10496, Hanoi Museum.

drums were made from a loose mixture of alluvial soil, rice husk, and wood ash and then fired to make a ceramic (Trinh 1999). To make drums, trivalve mold systems were used with spacers placed between the inner and outer mold pieces to keep them separated during casting. The spacers were distributed in equidistant positions on the tympanum and upper, middle, and lower sections of the drum. Carved designs such as humans, birds, deer, and houses were incorporated into the soil mold before firing.

We do not understand exactly how the molten bronze was poured into the molds to create Dong Son drums of the Heger I type; however, traces of the pouring procedure are preserved in the metal of Heger II drums dating to 2000–2200 BP (called Muong culture drums in Vietnam). These traces suggest some variation in the method of pouring molten bronze into a mold (fig. 9). Sometimes the metal was poured in at the edge of the drum's tympanum, other times into the side of the drum. Traces of two methods of pouring were found on many Heger II bronze drums (figs. 10 and 11).

In North Vietnam, some modern villages are home to people who still cast bronze objects using traditional methods with ceramic molds (fig. 12). They are also casting experimental bronze drums for scientific study. In Thanh Hoa province, a well-known bronze casting site, Tra Dong in Thieu Hoa district casts bronze drums for comparative study.

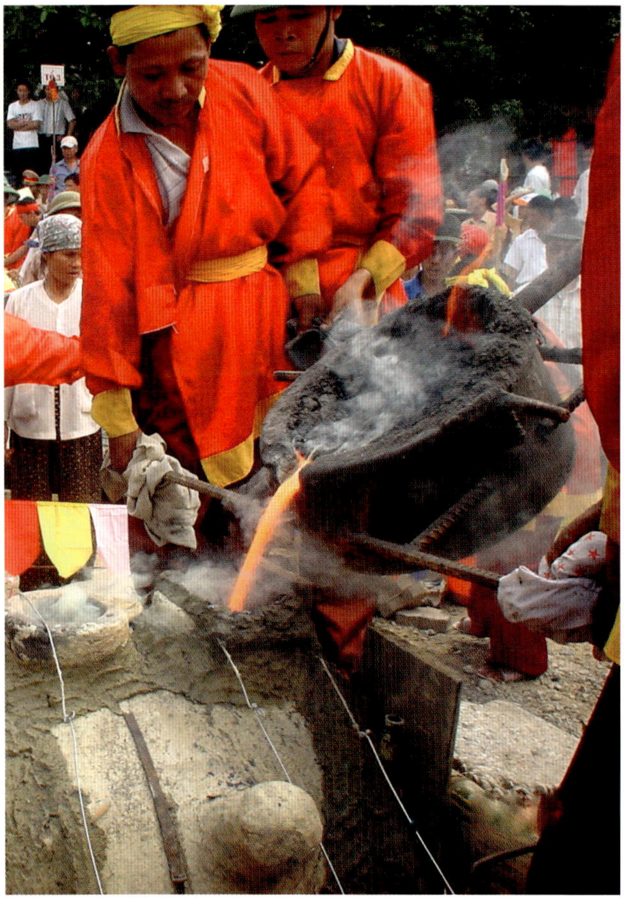

Figure 12. Modern bronze casting using traditional methods with ceramic molds.

Acknowledgments

I thank the Freer Gallery of Art and Arthur M. Sackler Gallery, Smithsonian Institution, with particular thanks to Paul Jett and Janet G. Douglas for their help with this paper. I would also like to thank the International Center for East Asian Archaeology and Cultural History, Boston University, where I was a visiting scholar in 2007–08, with special thanks to Prof. Dr. Robert E. Murowchick. The Institute of Archaeology in Vietnam, where I have been working for about forty years, has provided me with many archaeological documents related to the metal finds in Vietnam.

References

Ha Thuc Can and Nguyen Van Huyen. 1989. *The bronze Dong Son drums: A collective work of archaeologists*. Hanoi: Ha Thuc Can.

Ha Van Phung and Nguyen Duy Ty. 1982. Di chi khao co hoc go mun [Archaeological site in Go Mun]. Hanoi: *Nha xuat ban khoa hoc xa hoi.*

Huyen Pham Minh, Nguyen Van Huyen, and Trinh Sinh . 1987. *Trong Dong Son* [Dong Son drums]. Hanoi: *Nha xuat ban khoa hoc xa hoi.*

Kohl, G., and H. Quitta. 1978. Berlin radiocarbon dates V. *Radiocarbon* 20(3):386–97.

Le Xuan Diem and Hoang Xuan Chinh. 1983. Di chi khao co hoc Dong Dau [Archaeological site in Dong Dau]. Hanoi: *Nha xuat ban khoa hoc xa hoi.*

Muhly, J. D. 1988. The beginnings of metallurgy in the Old World. In *The beginning of the use of metals and alloys*, ed. R. Maddin, 2–20. Cambridge, MA: MIT Press.

Murowchick, R. E. 1989. The ancient bronze metallurgy of Yunnan and its environs: Development and implications. PhD diss., Harvard University.

Rapp, G. Jr. 1988. On the origins of copper and bronze alloying. In *The beginning of the use of metals and alloys*, ed. R. Maddin, 21–27. Cambridge, MA: MIT Press.

Trinh Sinh. 1988. Metallurgy and making metal in the millennium II–I BC in North Vietnam. PhD diss., Institute of Archaeology, Academy of Sciences in Russia.

———. 1989. Lead bronze alloys: The Hung King and Dong Son cultures. *Khao co hoc* [Archaeology] 2:43–50.

———. 1990. Phan tich quang pho hien vat dong o van hoa Dong Dau va Go Mun [Spectrographic analysis of bronze artifacts in Dong Dau, Go Mun cultures]. *Khao co hoc* [Archaeology] 1990(4):49–58.

———. 1992. Nhung tac dong kinh te xa hoi cua nghe luyen kim [The socio-economic effect of metallurgy]. *Khao co hoc* [Archaeology] 1992(4):19–26.

———. 1996. Qua nhung lan thuc nghiem duc trong dong [From experiments of casting of bronze drums]. *Khao co hoc* [Archaeology] 1996(2):42–52.

———. 1999. Ky thuat tao khuon qua manh khuon duc trong o Luy Lau [A piece of ceramic mold for casting bronze drums in Luy Lau site]. *Nhung phat hien moi ve Khao co hoc nam* [New Archaeological Discoveries] 1999:178–79.

Trinh Sinh and Ngo Si Hong. 1980. Vai Nhan xet ve lang ca (Vinh Phu) [Some remarks about Lang Ca site (Vinh Phu)]. *Khao co hoc* [Archaeology] 4:19–30.

Vu Thi Ngoc Thu and Nguyen Duy Ty. 1978. Bo Dung cu duc dong o Lang Ca (Vinh Phu) [Collection of artifacts for bronze casting in Lang Ca site, Vinh Phu province]. *Khao co hoc* [Archaeology] 2:36–9.

Wang Dadao. 1986. *Yunnan tonggu* [The bronze drums of Yunnan]. Kunming, China: Yunnan jiaoyu chubanshe.

West Asian Copper Alloys

Ten Millennia of Metallurgy in Western Asia

Paul T. Craddock

ABSTRACT This paper explores the subject of innovation, specifically relating to the metals and metallurgy of Western Asia and the evidence and likelihood of diffusion to the rest of Eurasia. The whole subject of the diffusion of culture has been one of the central themes of Old World archaeology for about a century, with Egypt and Western Asia at its center. The perception of the spread of metallurgy has undergone considerable change through the twentieth century.

Introduction

The earliest copper-based metallurgy is found in that broad region of precocious technology stretching from Anatolia to North India and dates from about the eighth millennium BCE.[1] In the old days of *ex oriente lux* it could be believed that metallurgy diffused from these primary areas around the Old World, as formulated on a proper archaeological basis by Childe (1958). Although to some degree the dating evidence supports the concept of a spread out from a center, more detailed local studies across Eurasia suggest numerous independent inventions of metallurgy (a suggestion championed by Renfrew from the 1960s in numerous publications; see Renfrew 1969, 1973, for example), which were possibly inspired by the appearance of metal artifacts from neighboring regions (Craddock 1999). As with all archaeological paradigms, this scenario is in a state of flux,[2] and theories supporting diffusion from Western Asia are currently back in favor, both for the very inception of metallurgy (Roberts et al. 2009) as well as for the later spread of furnace smelting (Amzallag 2009).[3] Evidence from the Balkans, however, rebuts this (Radivojević et al. 2010), unless, of course, the Balkans are included within the core area of technological development at this time. This changing situation applies not just to Western Asia. For example, in the 1970s at the height of the rejection of diffusion, the independent discovery of metallurgy in Southeast Asia was much vaunted (Bayard 1980, for example), but more recent discussion centers on whether metallurgy was introduced into the region from Central Asia, China, or possibly South Asia (Pigott and Ciarla 2007).

The diffusionist position was probably best expressed by that arch-diffusionist Joseph Needham. He believed that simple technologies were likely to have developed independently in a large number of areas but that more complex technologies were likely to have only developed in a few sophisticated societies, from which they spread to the more primitive cultures (Needham 1954, 228–29). This paradigm can be applied to the spread of metallurgy across the ancient world (Craddock 1999). The initial smelting processes were not technologically challenging, based as they were on the reduction of copper ores in hearths not very different from those used for domestic purposes (Craddock 2000). It may have been necessary to introduce the more complex processes, such as slag smelting and lost-wax casting, to Eurasia, although these advanced technologies may not always have been appropriate to the needs of the simpler communities.

The Inception of Metallurgy

Sites such as Tell Maghzaliyeh, Iraq (Rydina and Yakhontova 1985); Ali Kosh, Iran (Pigott 1999; Thornton 2009); and Mehrgarh, Pakistan (Moulherat et al. 2002) attest to the first metal artifacts, dating from the seventh millennium BCE onward, while artifacts dating to the late ninth millennium BCE have been found at Cayönü Tepesi in Anatolia (Maddin et al. 1991). All of these artifacts are made of native copper and seemingly are included in assemblages of small, prized items, usually nonfunctional and often made of exotic materials, including semiprecious stones and minerals, including the various copper ores found among the early Neolithic communities. These must often have originated considerable distances from the communities that obtained them, although there is no evidence of a reciprocal interest in the contemporary hunter-gatherer cultures of Europe, Africa, or southern India.

Evidence for the first smelted metal both in Western Asia and in the Balkans comes from sometime around the sixth–fifth millennia BCE (Radivojević et al. 2010). At Catal Hüyük in Anatolia, for example, fragments of a crucible that contained remains of burnt and partially vitrified copper ore, in which copper had presumably been smelted, were found (Craddock 1995, 125), and at Tal-i-Iblis in Iran, similar evidence has been found (Thornton 2009; Frame, in this vol.). The progression through native, annealed, melted, and finally, smelted copper is easy to postulate (Craddock 1995, 122). The only question is why this development took so long; it is presumably a reflection on the low ranking of metals as materials of economic significance for tools or weapons.

The smelting of oxidized copper ores is not technically difficult, and there is no reason why the technology should not have developed independently (contra Roberts et al. 2009).[4] The only novel parameter would have been attaining a temperature of about 1100°C, sufficient to melt the forming copper. It is improbable that any community would think to heat fragments of green rock at high temperature for extended duration, thereby using considerable quantities of charcoal (especially by volume), which, although potentially available, was probably little used and not deliberately prepared before the advent of metallurgy.[5] The inspiration could have been provided by the appearance of metal artifacts from other cultures and the recognition that these were of the same material as the small quantities of native metals found in the more local deposits. It is interesting to note that the metals of the first Eurasian metal age outside of Western Asia are copper and gold, which both occur in the native state and could be recognized, but that lead, silver, and tin appeared generally much later, quite simply because there were no native white metals to inspire further investigation of their ore deposits (Craddock 1999; Roberts et al. 2009).

The simple nonslagging hearth or crucible processes enabled metal to be produced quickly and relatively easily by nonspecialists. Because requirements were small in scale, with individual communities periodically visiting the ore deposits as the need arose and carrying out the smelting in the vicinity of the mine (or less commonly carrying the ore back to the main settlements for smelting), the efficiency of the process was of relatively minor importance. The ore and fuel were sufficiently abundant not to be a problem. However, once societies became more complex and centralized, based in large settlements, the demand for metal would have grown rapidly and the inefficiency of the smelting process would have become a major constraint. Back in the 1970s, Renfrew (1972) and others emphasized the role that metals played in the evolution of complex societies; perhaps this was a two-way process, with the demands of the evolving societies stimulating the technology of extractive metallurgy. Processes of longer duration to enable greater quantities to be produced in a single smelt, thus saving fuel, as well as higher temperatures and more reducing conditions to enable more metal to be extracted from the ore were required.

Central to this was the development of slagging processes contained within a furnace, whereby the predominant waste material associated with the mineral in the ore, usually varying amounts of silica and iron oxides, was separated from the forming metal by the formation of a molten slag. This gave better separation of metal and waste and enabled a much greater proportion of the metal to be extracted from the ore. Molten slag occupied a smaller volume than the semivitrified solid waste, with the added potential of being tapped from the smelting unit. This allowed repeated recharging of the smelting unit until a considerable volume of metal had accumulated, which was much more economic in fuel usage than repeated small short-run smelts, no small consideration in many of the ore locations in Western Asia where wood was scarce.

Surviving scatters or heaps of slag provide evidence for the slagging process. These slag smelting operations seem to have commenced around 3000 BCE and have been identified at a number of locations in the eastern Mediterranean and in Western Asia, at sites such as Chrysokamino on Crete (Betancourt 2006), Arslantepe in Anatolia (Palmieri et al. 1999), Wadi Dara on the Red Sea (Castel et al. 1998), Feinan in Jordan (Hauptmann 2000) and Shahr-I Sokata (Hauptmann et al. 2003) and Tepe Hissar in Iran (Thornton 2009). Further evidence for the slagging process in Western Asia is provided by the metal itself, with the iron content of the copper rising rapidly after about 3000 BCE throughout Western Asia (Craddock and Meeks 1987), suggesting that slagging was rapidly adopted by the suppliers to the major cultures.[6] Thus, for example, while the mines or smelters that supplied copper to the Harappan cities of the Indus valley have yet to be identified, the high iron content of their copper indicates that the slagging process was already in use.

Although the introduction of slagging was extremely important, it must not be seen as the only change. The earliest slag-smelting units seem to have been deliberately sited to take advantage of the prevailing winds, and only later were bellows adopted, as evidenced by survival of the tuyeres, the clay bellows tubes.[7] Thus slagging was one of a series of major changes through the Bronze Age of Western Asia, during which furnaces became larger and the process hotter and more reducing and thereby more efficient (as measured by the copper content of the slags). The processes carried out at sites such as Timna in Israel (Rothenberg 1990), Feinan (Hauptmann 2000), and El Maysar in Oman (Hauptmann 1985; Craddock et al. 2003) in the later part of the second millennium BCE at the end of the Bronze Age were very different from those carried out almost 2,000 years earlier at the inception of the slagging process; the final development was the smelting of iron, which developed out of the smelting of copper.

The furnace-slagging processes seem to have been confined to supplying the major Bronze Age civilizations of Western Asia, and later China from the early second millennium BCE. Outside of these areas, in Europe, southern South Asia, Southeast Asia, and sub-Saharan Africa, it seems that the simpler processes continued, as evidenced by the absence of slag heaps at mines or tuyeres and the general low iron content of the copper.[8] Furnace slagging technology only seems to have fully spread through Eurasia and Africa with iron smelting, for which it was essential, from the late second millennium BCE onward.

Alloying

The story of early alloying practice is complicated. Early copper artifacts often contain small amounts of arsenic (Thornton 2010), which can greatly improve the working properties of the metal (Budd 1992).[9] Arsenic can also impart a beautiful silvery sheen to the surface of the copper, although this can come about quite naturally.[10] Copper deposits are, however, often intermixed with those of arsenic. Thus the presence of arsenic in the copper could be entirely accidental and unappreciated by the smith,[11] or it could be a deliberate addition for a specific reason.

At Shahr-I Sokata in Iran in the third millennium BCE, the copper prills contained within the slags and the copper ingots have low arsenic contents, but the contemporary artifacts are of arsenical copper, suggesting alloying had taken place (Thornton 2009). An alloy could be prepared by adding arsenic minerals to molten copper in a closed crucible. The common arsenic sulfide minerals realgar and orpiment are often found with copper deposits and may well have already have been sought as pigments and thus should have been available to the early smiths. Arsenic was not prepared as a pure metal until the postmedieval period. Alternatively, arsenic-rich copper ore could have been selected: the presence of arsenic in a copper ore would have been immediately recognized by the distinctive garlic odor emitted when the ore was roasted or smelted.

Trying to understand the reason for selecting an arsenical copper adds a complication. It might be thought that where there was a selection the hard, strong arsenical copper would be reserved for tools and weapons and the pure copper for decorative items. However, in the most famous group of arsenical metalwork, the hoard from Nahal Mishmar belonging to the Chalcolithic Ghassulian culture and dated to the mid-fourth millennium BCE, the "elite" or "prestige" maceheads, crowns, and standards are made of an antimony-arsenical copper but the utilitarian flat axes are made of copper (Tadmor et al. 1995). The reason for the apparent reversal is immediately obvious once the entire hoard is viewed. The extraordinary dry and stable environment of the cave in which the hoard lay ensured that the metal surfaces retained their original appearance, and the beautiful steely gray sheen on the antimony-arsenical copper maceheads almost exactly matches that of the maceheads of polished hematite.

Further demonstration of the awareness and appreciation of the surface potential of arsenical copper is provided by the well-known small copper figurines from Horoztepe in Turkey (Smith 1973). In these, the underlying copper is not arsenical, and arsenic-rich plating is broken up into striking patterns, showing that it must have been deliberately applied. The evidence for the deliberate selection of arsenical copper and for arsenic plating outside Western Asia is less certain.[12]

Only in the third millennium BCE did tin bronze become established, as evidenced by the bronzes at Troy and Ur, although for many centuries tin bronze was used alongside arsenical copper rather than in place of it in the Mediterranean and in Western Asia. This is in stark contrast with what occurred in Western Europe, where tin bronze very rapidly replaced copper in the late third millennium BCE. The relative scarcity of tin through the Mediterranean and Western Asia must be the reason for this. Kestel-Göltepe in Turkey is the only known source of tin that was being worked in the region in antiquity (Yener et al. 2003). Kestel was a hard rock deposit, which was probably not very rich and seems to have been very thoroughly worked out by the second millennium BCE. Thereafter Western Asia is likely to have had to rely on imports from distant sources.

As with the spread of the concept of copper smelting, no real compelling evidence for the diffusion of bronze usage exists; instead tin and tin bronze were discovered afresh wherever major deposits of tin were encountered. Thus, in the British Isles there were several centuries of copper and gold usage before, very suddenly and with no real change in artifact typology, copper and arsenical copper were comprehensively replaced by bronze containing about 10 wt.% tin (Craddock 1979). Presumably the potential of the vast tin deposits in southwest England had been discovered, possibly while the tin-bearing gravels were being sifted for gold.

Finally, there is brass, final in the sense that it was the last of the major alloys of antiquity and also in the sense that it was probably the last major metallurgical innovation to come from within Western Asia. As with tin bronze, its discovery and usage was, to a degree, dependent on the availability and type of ores. Major deposits of smithsonite—zinc carbonate—occur abundantly in Western Europe and in China. That mineral when ground up and added to molten copper in a closed crucible with charcoal would have produced an acceptable brass. However, both Europe and China had excellent tin deposits, and thus there was no need for a substitute for bronze. In contrast, Western Asia had only major deposits of sphalerite—zinc sulfide—from which the sulfur had to be almost totally removed before a satisfactory brass could be produced. But, as already noted, there were few major deposits of tin in Western Asia, and thus a substitute would be very viable.

Sporadic occurrences of copper or bronze artifacts containing several percent of zinc and dating from almost the inception of metallurgy have been found throughout much of the world (Craddock 1998; Thornton 2007). In any group, the artifacts' zinc contents are variable but typically tend to be just a few percent, with only a very few over 5 wt.%. Furthermore, they occur as very minor components of otherwise similar artifacts of copper or bronze. Thus they would seem most likely to be accidental productions in which the copper ore happened to contain zinc (contra Thornton 2007).

Zinc does seem to occur more frequently in copper or bronze artifacts from the late second millennium BCE in Western Asia. The copper alloys from Nuzi in Mesopotamia dating from the later part of the second millennium BCE include a significant percentage of copper alloys in which the zinc content lies between 10 and 20 wt.% often unaccompanied by other alloying elements (Eremin, in this vol.), strongly suggesting that these brasses, for so they must be termed, were made deliberately, although currently the methods of production are not known. These alloys are possibly to be associated with contemporary literary references to a prized "copper of the mountain," which was adopted by the Greeks as *oreichalkos* and latterly certainly was brass (Halleux 1973).[13] The earliest regular series of brass artifacts so far encountered date from the first century BCE and include the brass coins from Bithynia and Phyrgia in Anatolia (Burnett et al. 1982) and the Roman *Alesia* type brooches (Istenič and Šmit 2007). Dating from around the same time are descriptions of the preparation of the zinc oxide, necessary for brass production, from the zinc sulfide ores by a complex process that would continue in Western Asia through the Middle Ages.[14] Although brass eventually became widely adopted over much of Eurasia, this technique of preparing zinc oxide seems never to have penetrated much beyond Western Asia, quite simply because of the availability of oxidized zinc deposits elsewhere.[15]

Fabrication

Having dealt with the smelting and alloying of metals in ancient Western Asia and their likely impact on surrounding cultures, it is necessary to investigate some of the principal fabrication methods. Here, even more than with metal production, the requirements of the communities making the artifacts are of great importance in determining which fabrication methods were adopted. Perhaps the most distinctive fabrication methods are the various methods of casting. Single molds were soon joined by two-piece molds that enabled the easy production of reasonably simple three-dimensional shapes, and such molds have remained in use everywhere ever since. There is no need to invoke diffusion for their spread, and it is noticeable that in Western Europe, for example, the first copper artifacts were cast in single molds for centuries, strongly suggesting that the first metallurgists there had no knowledge of the more advanced technologies of Western Asia. Once again this is indicative of independent local development, though advances were also possibly inspired by the appearance of a few artifacts cast in two-piece molds from more advanced metalworking centers.

The situation with the technically more advanced lost-wax casting method is more complex (Davey 2009). It clearly developed early in Western Asia, with the earliest known examples dating from the late fifth millennium BCE and coming from Mehrgarh, Baluchistan (Mille et al. 2004), and the method developed rapidly as exemplified by the complex castings in the Nahal Mishmar hoard (Tadmor et al. 1995). The crowns and scepters are complex shapes that could only have been cast in one piece by using the lost-wax technique, and the maceheads and scepters were hollow cast around clay cores.

If it is difficult to demonstrate the diffusion of two-piece-mold technology from Western Asia, it is even more difficult with lost-wax technology. In the whole of European prehistory, the technique never became popular beyond the Aegean. Recent studies on Late Bronze Age goldwork have shown that the technique was known (Armbruster 1995; Meeks et al. 2008), but for the bulk copper-alloy metalwork, two-piece molds sufficed. Complex pieces were usually built up from assemblages of cast and hammered components. Elsewhere the situation is diverse. In West Africa, lost-wax castings have a long history and may well have been the main casting method employed from the inception of copper-alloy metallurgy there. However, detailed study of the earliest complex castings, admittedly from as late as the ninth–tenth century CE, suggests that this was an indigenous technology (Craddock et al. 1997). South Asia seems to have used lost-wax casting for complex pieces, such as figurines, from at least Harappan times without a break until the present. In total contrast, East Asia developed the technology of piece-molding, and lost-wax casting seems to be a relatively late development, utilized beginning only about two thousand years ago. Whether it was introduced along with brass by Buddhist monks from South Asia or developed indigenously is still debated.[16] Thus overall it is not possible to demonstrate an early spread of lost-wax casting out of Western Asia. The appearance of the technique at a number of widely separated centers around the Old World probably represents independent developments.

The occurrence of hollow lost-wax casting is even more selective. Once again as with lost-wax casting itself, the idea of minimizing the amount of metal used by running it around a core is not necessarily a late development; to return yet again to the Nahal Mishmar hoard, many of the maceheads are true hollow lost-wax castings, cast around a completely enclosed clay core. However, most large-scale works in copper alloy were of sheet metal hammered around a wooden or bitumen former, with just a few massive thick hollow castings, mainly in copper, from Mesopotamia, which seem to have been isolated prestige productions (Moorey 1994, 254–65). Only from the early first millennium BCE do large hollow lost-wax castings of some complexity, in which the metal is no more than few millimeters in thickness, appear. The earliest examples come from Egypt (Taylor et al. 1998), but the process rapidly spread across Western Asia (if it had not developed there already), down into southern Arabia (Mille, in this vol.), and also into the Mediterranean and South Asia. Both direct and indirect hollow lost-wax processes were used. In the former process the wax original was molded onto a clay core and lost during the casting process. In the latter process a negative plaster piece mold was taken of the original, which was thereby preserved, and a wax positive of the required thickness taken from the assembled plaster mold. A core was then inserted into the hollow wax and the process carried forward much as before with the direct process. Both techniques were adopted and greatly refined by the Greeks.[17] Thereafter the indirect process became established in Europe, but elsewhere only the direct process is attested.

The Overall Influence of Western Asia Copper-Based Metal Technologies on the Old World

Most, if not all, copper-based metal technologies used in the ancient Old World seem to have developed first in Western Asia. However, it is difficult to establish their diffusion to the rest of Eurasia. The technologies of smelting, alloying, and casting seem to have developed indigenously just about everywhere, although development was possibly influenced by the stimulus of metal artifacts deriving ultimately from Western Asia.

The first and probably the major technology that is likely to have diffused from Western Asia across the Old World was furnace-based slag smelting. However, through most of Eurasia with the notable exception of China, slag smelting was not adopted in the Bronze Age but was spread by iron workers, although the process was then rapidly adopted for nonferrous smelting.

The other technology likely to have diffused from Western Asia was the use of brass. As discussed, it seems that the vast distances separating the major tin deposits utilized in prehistory mean that tin bronze was likely to have been an independent discovery at each of these locations. In contrast, the production of brass by direct methods of reacting carefully oxidized zinc sulfide minerals with molten copper certainly developed in Western Asia through the late second and first millennnium BCE, and the usage of the alloy was disseminated across Europe and the Mediterranean by the Romans. At much the same time, the first brasses were reaching China from the northwest of India, although production of brass took much longer to become established in East Asia. In both Western Europe and in China, major deposits of oxidized zinc ores occur, and there was never any need to adopt the complicated oxidation/sublimation techniques that were necessary to treat the sulfidic zinc ores of Western Asia.

New development in metal technology within Western Asia largely ceased at the end of the first millennium BCE. Big advances continued to be made but no longer in the "cradle of civilization." Crucible steel was already being produced in Central and South Asia, and the manipulation of cast iron was being mastered in East Asia. In the first millennium CE, the production of zinc metal began in South Asia, followed soon afterward in East Asia. In the medieval period, extractive metallurgy in Western Asia already seems to have been in general decline, with many metals, notably copper, being imported into the region (La Niece et al., in this vol.).

Whatever the underlying socio-economic reasons for this decline in Western Asia, the basic technical problem over the past two millennia was the failure to utilize water power and fossil fuels in metallurgy. Water power was utilized in both Europe and East Asia to drive the bellows, enabling furnace volumes to greatly increase, as well as to power the hammers used to forge the metals. The endemic problems of limited charcoal supplies began to be addressed by the use of fossil fuel through the development of both the reverbatory furnace and coking techniques, which occurred in East Asia in the first millennium CE and almost a thousand years later, apparently independently, in Europe. There is no evidence of this activity within Western Asia, which thus placed its traditional metal industries at an increasingly severe disadvantage.

Notes

1. The technical and cultural homogeneity of the various precocious metallurgical groups within the area defined as West Asia is, however, much debated. See Yener (2000) or Thornton (2009), for example.
2. In a prescient note back in 1986, James Muhly, writing about the move away from diffusion to independent discovery, stated that "opinion has swung from one extreme to the other and we are now at the far side of the swing of the pendulum."
3. See Thornton et al. (2010) for a rebuttal of this paper, although Amzallag's basic concept of the diffusion of metallurgy from West Asia is not challenged.
4. Several years of experience of conducting smelting experiments with adult students over five-day periods has shown that this is not a problem (Timberlake 2007). The copper ore, malachite or azurite, mixed with charcoal, is contained within a crucible. The crucible is of the most basic materials and construction, rough clay mixed with some sand, dung, and straw, shaped as one would form a pastry pie base and lightly baked in an open fire (no potter would call it a ceramic, much less a refractory). The smelting unit is a simple hearth in which the lidded crucible sits surrounded by charcoal. With air supplied, either through a simple bag bellows or by blowpipes, the copper is easily smelted after approximately an hour. The results are usually blobs of copper weighing a few hundred grams (typically representing about 50% of the theoretical maximum recovery). The debris is predominantly the charcoal ash, together with some semivitrified clinker contained (or not) in a now barely recognizable crucible, but no true durable slag.
5. Every wood fire produces some charcoal, and at its simplest charcoal can be produced by rapidly extinguishing a well-established log fire. This method is extremely wasteful but was used in India by the Agaria iron smelters (Elwin 1942, 121–22) for example, where the requirements were small and the wood resources seemingly inexhaustible.
6. The adoption of a slagging process did not invariably lead to an increase in the iron content of the copper. If the furnace conditions were not reducing enough to produce wustite (FeO) or metallic iron, then iron could not enter the copper. Thus, for example, the Bronze Age metalwork from the Aegean has low iron values even though slag heaps from the Early Bronze Age are abundant. A detailed analysis of some of these slags (Georgikopolou et al. 2011) has shown that the iron is principally in the form of magnetite (Fe_3O_4), which could not have been absorbed by the forming copper draining through the slag.

   Conversely, a number of early smelting sites in Europe have produced tiny amounts of slag with entrapped prills of copper. Analyses of the latter have shown that they sometimes contain substantial traces of iron, leading to claims that these are representative of the freshly smelted copper before it was refined. An alternative explanation is to regard the iron-rich copper prills as having experienced very different local conditions—that is surrounded by a slag—and thus they are different from the bulk of the smelted copper that did not.
7. At Chrysokamino there are the remains of a strange vessel claimed as a pot bellows, but if this is so, it apparently operated without tuyeres to protect the nozzle (Betancourt 2006).
8. For example, the iron contents of the indigenous Late Bronze Age bronzes of the Iberian peninsula are very much lower than those of the bronzes from the contemporary Phoenician settlements in the south of Iberia (Craddock and Meeks 1987).
9. Arsenic contents of 2–3 wt.% lower the melting point and make the molten metal more fluid and thus easier to cast. If correctly worked, the resulting metal can be much harder and stronger than pure copper.
10. The sheen is produced by the formation of the delta phase as the volatile arsenic concentrates on the surface as the molten alloy solidifies. Furthermore, this surface enrichment can survive hammering, and thus an enriched surface on a wrought artifact could still be a survival of the original casting (Meeks 1993).
11. Metallographic examination of Copper Age artifacts of arsenical copper from Central Europe have shown that the metal was often unworked or in an annealed state, and thus the potential of the alloy was not realized (Budd 1992).
12. For example, the very detailed metallographic study of an arsenic-enriched surface of a Copper Age Millarian dagger from the southeast of Spain led Meeks (1993) to finally conclude that it was "probably deliberately plated." Other Bronze Age daggers of arsenical copper have been perhaps over-confidently described as deliberately plated on the basis of their appearance alone.
13. These early alloys have quite low zinc contents, below 15 wt.%, and thus would appear no different from bronze. By the end of the first millennium BCE, the alloy regularly contained zinc in the range of 15 to 25 wt.%, which would have appeared more golden, and it was perhaps then that the name changed its meaning from "mountain" to "golden" in the Latin *aurichalcum*.

14. The zinc sulfide was roasted to produce very impure zinc oxide, which was then smelted under reducing conditions to produce a vapor of metallic zinc. This was extremely reactive and promptly reoxidized to give dense white clouds of pure zinc oxide, which in turn could be collected in cool chambers above the furnace flues onto baffles or directly onto iron or clay bars suspended in the flues (Craddock and Eckstein 2003).
15. The process was in use in the last centuries of the first millennium BCE at Zawar in Rajasthan, northwest India, on the periphery of Western Asia (Craddock and Eckstein 2003).
16. Hua (1983) published some extremely intricate bronze vessels of the mid-first millennium BCE as being lost-wax castings, but this has been refuted by Zhou et al. (2009), who demonstrated that they are made up of many piece-molded components welded together. They go on to suggest that lost-wax casting was introduced approximately a thousand years later, initially to cast Buddhist statuettes. Both Hua and Zhou et al. assume that the technology was imported.
17. This is the conclusion of a very recent joint study by the author at the British Museum and Benoît Mille at the Musée de Louvre.

References

Amzallag, N. 2009. From metallurgy to Bronze Age civilisations: The synthetic theory. *American Journal of Archaeology* 113(4):497–519.

Armbruster, B. R. 1995. Rotary motion-lathe and drill. In *Prehistoric gold in Europe,* ed. G. Morteani and J. P. Northover, 399–423. Dordrecht: Kluwer Academic.

Bayard, D. T. 1980. An early indigenous bronze technology in southeast Thailand. In *The diffusion of material culture,* ed. H. H. E. Loofs-Wissowa, 191–214. Manoa: University of Hawaii Press.

Betancourt, P. P. 2006. *The Chryskamino metallurgy workshop and its territory*. Hesperia Supplement 36. Princeton, NJ: American School of Classical Studies at Athens.

Budd, P. 1992. Alloying and metalworking in the Copper Age of Central Europe. *Bulletin of the Metals Museum of the Japan Institute of Metals* 17:3–14.

Burnett, A., P. T. Craddock, and K. Preston. 1982. New light on the origins of *orichalcum*. In *Proceedings of the 9th International Congress of Numismatists,* ed. T. Hackens and R. Weiller, 263–68. Publication 7. Brussels: Association Internationale des Numismatists Professionels.

Castel, G., B. Mathieu, G. Pouit, M. El Hawari, G. Shaaban, H. Hellal, T. Abdallah, and A. Ossama. 1998. Wadi Dara copper mines. In *Proceedings of the First International Conference on Ancient Egyptian Mining and Metallurgy and Conservation of Metallic Artifacts,* ed. F.A. Esmael, 15–31. Cairo: Ministry of Culture.

Childe, V. G. 1958. *The prehistory of European society*. Pelican: London.

Craddock, P. T. 1979. Deliberate alloying in the Atlantic Bronze Age. In *Proceedings of the 5th Atlantic Colloquium,* ed. M. Ryan, 369–85 and microfiche. Dublin: Stationery Office.

———. 1995. *Early metal mining and production*. Edinburgh: Edinburgh University Press.

———. 1998. Zinc in classical antiquity. In *2000 years of zinc and brass,* ed. P. T. Craddock, 1–6. British Museum Occasional Paper 50. London: British Museum Press.

———. 1999. Paradigms of metallurgical innovation in prehistoric Europe. In *The beginnings of metallurgy,* ed. A. Hauptmann, E. Pernicka, T. Rehren, and Ü. Yalçin, 175–92. Der Anschnitt, Beiheft 9. Bochum: Deutsches Bergbau-Museum.

———. 2000. From hearth to furnace. *Paléorient* 26(2):151–65.

Craddock, P. T., J. Ambers, D. R. Hook, R. M. Farquhar, V. E. Chikwendu, A. C. Umeji, and T. Shaw. 1997. Metal sources and the bronzes from Igbo-Ukwu, Nigeria. *Journal of Field Archaeology* 24:405–29.

Craddock, P. T., and K. Eckstein. 2003. Production of brass in antiquity by direct reduction. In *Mining and metal production through the ages,* ed. P. T. Craddock and J. Lang, 216–30. London: British Museum Press.

Craddock, P. T., S. La Niece, and D. R. Hook. 2003. Evidences for the production, trading and refining of copper in the Gulf of Oman during the third millennium BC. In *Man and mining: Studies in honour of Gerd Weisgerber on the occasion of his 65th birthday,* ed. T. Stöllner, T. G. Koerlin, G. Steffens, and J. Cierny, 103–12. Der Anschnitt, Beiheft 16. Bochum: Deutsches Bergbau-Museum.

Craddock, P. T. and N. D. Meeks. 1987. Iron in ancient copper. *Archaeometry* 29(2):187–204.

Davey, C. J. 2009. The early history of lost wax casting. In *Metallurgy and civilisation: Eurasia and beyond,* ed. J. Mei and T. Rehren, 147–54. London: Archetype Publications.

Elwin, V. 1942. *The Agaria*. Oxford: Oxford University Press.

Georgikopolou, M., Y. Bassiaskos, and O. Philanoitou. 2011. Seriphos surfaces: A study of copper slag heaps and copper sources in the context of Early Bronze Age Aegean metal production. *Archaeometry* 53(1):123–45.

Halleux, R. 1973. L'orichalque et le laiton. *Antique Classique* 42:64–81.

Hauptmann, A. 1985. *5000 jahre kupfer in Oman: Die entwicklung der kupfermetallurgie vom 3. jahrtausend bis zur neuzeit*. Der Anschnitt, Beiheft 4. Bochum: Deutsches Bergbau-Museum.

———. 2000. *Zur frühen metallurgie des kupfers in Fenan/Jordanien*. Der Anschnitt, Beiheft 11. Bochum: Deutsches Bergbau-Museum.

Hauptmann, A., T. Rehren, and S. Schmitt-Strecker. 2003. Early Bronze Age copper metallurgy at Shahr-I Sokhta (Iran) reconsidered. In *Man and mining: Studies in honour of Gerd Weisgerber on the occasion of his 65th birthday,* ed. T. Stöllner, T. G. Koerlin, G. Steffens, and J. Cierny, 197–213. Der Anschnitt, Beiheft 16. Bochum: Deutsches Bergbau-Museum.

Hua, J. 1983. Marquis Zeng Yu's wine vessel and tray and the origins of the lost-wax method. *Ziran Kexueshi Yanjiu* [Studies in the History of Natural Sciences] 4:352–59.

Istenič, J., and Ž. Šmit. 2007. The beginning of the use of brass in Europe with particular reference to the south Alpine region. In *Metals and mines: Studies in archaeometallurgy,* ed. S. La Niece, D. R. Hook, and P. T. Craddock, 140–47. London: Archetype Publications in association with the British Museum.

Maddin, R., T. Stech, and J. D. Muhly. 1991. Cayönü Tepesi. In *Découverte du* métal, ed. C. Eluère and J. P. Mohen. Paris: Picard.

Meeks, N. D. 1993. Surface characterization of tinned bronzes, high tin bronze, tinned iron and arsenical copper. In *Metal plating and patination,* ed. S. La Niece and P. T. Craddock, 247–75. Oxford: Butterworth.

Meeks, N. D., P. T. Craddock, and S. P. Needham. 2008. Bronze Age penannular gold rings from the British Isles: Technology and composition. *Jewellery Studies* 11:13–30.

Mille, B., R. Bessenval, and D. Bourgarit. 2004. Early "lost-wax casting" in Balochistan (Pakistan): The "Leopards Weight" from Shahi-Tump. In *Persiens antike pracht, bergbau-handwerk-archäologie,* ed. T. Stöllner, R. Slotta, and A. Vatandoust. Der Anschnitt, Beiheft 12, 274–80. Bochum: Deutsches Bergbau-Museum.

Moorey, P. R. S. 1994. *Ancient Mesopotamian materials and industries: The archaeological evidence*. Oxford: Clarendon Press.

Moulherat, C., M. Tengberg, J. F. Haquet, and B. Mille. 2002. First evidence of cotton at Neolithic Mehrgarh, Pakistan. *Journal of Archaeological Science* 29(12):1393–401.

Muhly, J. D. 1986. Prehistoric background leading to the first use of metals in Asia. *Bulletin of the Metals Museum of the Japan Institute of Metals* 11:21–42.

Needham, J. 1954. *Science and civilisation in China*, Vol. 1, *Introductory orientations*. Cambridge: Cambridge University Press.

Palmieri, A. M., M. Frangipane, A. Hauptmann, and K. Hess. 1999. Early metallurgy at Arslantepe during the late Chalcolithic and Early Bronze Age 1A–1B periods. In *The beginnings of metallurgy,* ed. A. Hauptmann, E. Pernicka, T. Rehren, and Ü. Yalçin, 141–48. Der Anschnitt, Beiheft 9. Bochum: Deutsches Bergbau-Museum.

Pigott, V. C. 1999. The development of metal production on the Iranian plateau: An archaeological perspective. In *The archaeometallurgy of the Asian Old World,* ed. V. C. Pigott, 73–106. Philadelphia: University of Pennsylvania Museum of Archaeology and Anthropology.

Pigott, V. C., and R. Ciarla. 2007. On the origins of metallurgy in prehistoric Southeast Asia. In *Metals and mines: Studies in archaeometallurgy,* ed. S. La Niece, D. R. Hook, and P. T. Craddock, 76–88. London: Archetype Publications in association with the British Museum.

Radivojević, M., T. Rehren, E. Pernicka, D. Šljivar, M. Brauns, and D. Borić. 2010. On the origins of extractive metallurgy: New evidence from Europe. *Journal of Archaeological Science* 37(11):2775–87.

Renfrew, C. 1969. The autonomy of the south-east European Copper Age. *Proceedings of the Prehistoric Society* 35:12–47.

———. 1972. *The emergence of civilisation*. London: Methuen.

———. 1973. Sitagroi and the independent invention of metallurgy in Europe. In *Actes du VIIIe Congrès international des sciences préhistoriques et protohistoriques, Rapports et corapports, Beograd 1971,* Vol. 2, ed. G. Novak, 473–81. Belgrade: University of Belgrade.

Roberts, B. W., C. P. Thornton, and V. C. Pigott. 2009. Development of metallurgy in Eurasia. *Antiquity* 83:1012–22.

Rothenberg, B. 1990. *The ancient metallurgy of copper*. London: Institute for Archaeo-Metallurgical Studies.

Rydina, N. V., and L. K. Yakhontova. 1985. The earliest copper artifact from Mesopotamia. *Sovetskaia Archeologija* 1985–2:155–66.

Smith, C. S. 1973. An examination of an arsenic plated bronze bull from Horoztepe. In *Application of science in examination of works of art,* ed. W. J. Young, 96–102. Boston: Museum of Fine Arts.

Tadmor, M., D. Kedem, F. Bergemann, A. Hauptmann, E. Pernicka, and S. Schmitt-Strecker. 1995. The Nahal Mishmar hoard from the Judean desert: Technology, composition and provenance. *Atiqot* 27:95–148.

Taylor, J., P. T. Craddock, and F. Sherman. 1998. Egyptian hollow-cast bronze statues of the early first millennium BC: The development of a new technology. *Apollo* 148(437):9–14.

Thornton, C. P. 2007. Of brass and bronze in prehistoric Southwest Asia. In *Metals and mines: Studies in archaeometallurgy,* ed. S. La Niece, D. R. Hook, and P. T. Craddock, 123–35. London: Archetype Publications in association with the British Museum.

———. 2009. Iranian plateau: Escaping the Levantine paradigm. *Journal of World Prehistory* 22:301–27.

———. 2010. The rise of arsenical copper in southeastern Iran. *Iranica Antiqua* 45:31–50.

Thornton, C. P., J. M. Golden, D. J. Killick, V. C. Pigott, T. Rehren, and B. W. Roberts. 2010. A Chalcolithic error. *American Journal of Archaeology* 114(2):305–15.

Timberlake, S. 2007. The use of experimental archaeology/archaeometallurgy for the understanding and reconstruction of Early Bronze Age mining and smelting technologies. In *Metals and mines: Studies in archaeometallurgy,* ed. S. La Niece, D. R. Hook, and P. T. Craddock, 27–36. London: Archetype Publications in association with the British Museum.

Yener, K. A. 2000. *The domestication of metals: The rise of complex metal industries in Anatolia*. Leiden: Brill.

Yener, K. A., A. Adriaens, B. Earl, and H. Özbal. 2003. Analyses of metalliferous residues, crucible fragments, experimental smelts, and ores from Kestel tin mine and the tin processing site of Göltepe, Turkey. In *Mining and metal production through the ages,* ed. P. T. Craddock and J. Lang, 181–97. London: British Museum Press.

Zhou, W., Y. Dong, Q. Wan, and C. Wang. 2009. New research on lost-wax casting in ancient China. In *Metallurgy and civilisation: Eurasia and beyond,* ed. J. Mei and T. Rehren, 73–89. London: Archetype Publications.

Copper-Zinc Alloys at Nuzi: Dilemmas from an Early Excavation

Katherine Eremin, Susanna Kirk, Andrew Shortland, and Megan Richters

ABSTRACT The metal assemblage from the Late Bronze Age city of Nuzi in Mesopotamia was examined as part of a multidisciplinary study of the materials assemblage from this site. Analysis of the copper alloys indicates significant zinc levels in several artifacts. Since Bronze Age copper-zinc alloys remain rare, the presence of such alloys at Nuzi is important for understanding their role and origin. However, detailed study of the metalwork assemblage shows that some caution is required in interpreting the data, the majority of which comes from nondestructive surface analysis. Some artifacts originally attributed to the Late Bronze Age have been shown to be intrusive or of questionable date. In other cases, zinc is restricted to the surface and attributed to previous conservation treatments rather than the original alloy. Given the frequent absence of detailed treatment records for artifacts from early twentieth-century excavations, such treatment effects can confuse interpretations based on surface analysis alone. Careful consideration of the copper-alloy assemblage from Nuzi suggests that copper-zinc alloys were fairly rare and were deliberately used for small decorative items.

Introduction

The role of copper-zinc alloys in antiquity has been much debated, often controversially, with many early finds being discredited or questioned (for example, Craddock 1978a, 1978b, 1980, 1988, 1998). More convincing are two early brass artifacts from Nuzi, northern Mesopotamia (Bedore and Dixon 1998), which were also discussed and related to other early brass in ancient Near Eastern assemblages, particularly three early brass artifacts from Tepe Yahya, southeast Iran, by Thornton and Ehlers (2003). A possible link between brass and bronze in prehistoric Southwest Asia was explored by Thornton (2007). These studies clearly showed that brass occurred in the third to first millennium BCE but suggested it was fairly rare and possibly reserved for small decorative items due to its distinctive golden color. Other studies lack such early Near Eastern brass (Moorey 1994; Stech 1999; Pigott 1999; De Ryck et al. 2005; Frame 2010). Frame (2010, 1710) notes that although there are early examples of copper-zinc alloys from the third to the early second millennium BCE, "the archaeological contexts for many of these remain questionable and the presence of copper-zinc artifacts during the third to early second millennia BCE remains rare." Only nine artifacts were analyzed in the initial studies of Nuzi metalwork (Bedore and Dixon 1998), which raised the questions of whether these two copper-zinc artifacts are anomalous within the assemblage from this specific site and what proportion of such alloys truly date to the mid-second millennium BCE. We hence initiated a program to examine all the metalwork from Nuzi to determine the prevalence and chronology of copper-zinc alloys at this site. In the course of this study, it became apparent both that this was an extremely rich collection of Mesopotamian metalwork and that the stratigraphy of the site, and hence the dating of many items, was less well defined than initially thought. This is particularly true for many items of personal ornament, which lack the distinctive typologies found in other artifact types, such as vessels, arrowheads, or armor. This paper discusses some of these issues and the wider implications for archaeometallurgical studies of collections gathered in excavations during the late nineteenth to the early twentieth century.

The Site

The Late Bronze Age (LBA; ca. 1550–1200 BCE) city of Nuzi, located at the modern site of Yorghan Tepe, about 11 km southwest of Kirkuk in modern Iraq, was excavated between 1925 and 1931 under the auspices of the American School of Oriental Research (ASOR), the new Baghdad Museum, and Harvard University. The initial seasons were directed by Edward Chiera; his assistant, Richard Starr, an employee of the Fogg Art Museum, directed the final two seasons and published the excavation (Starr 1937, 1939). The excavations uncovered the bulk of the LBA city, recorded as Nuzi on cuneiform tablets, of which over five thousand were found at the site. The site covered around 200 square meters, including temple and palace complexes and residential areas. The city was a regional capital of the Mitanni empire and was sacked and destroyed around 1340

BCE, the destruction layer being termed stratum II. Nuzi was never extensively resettled, which resulted in limited disruption of the stratigraphy of the site. The excavations concentrated on stratum II but also uncovered artifacts from the prehistoric (fifth–fourth millennia BCE), Ga.Sur (late third millennium BCE) and Parthian (second century BCE–third century CE) levels and from later Sasanian and Islamic graves. Recent examination of the finds reveals more mixing of periods than initially thought, as discussed later. Although the site was well excavated for the time, not all finds were noted in the field notebooks and many objects were not marked with field numbers. This caused some problems with interpretation of the finds and their significance, as discussed under results.

The Materials Assemblage

The finds were divided between the excavators and the Baghdad Museum, as the legislation of the time demanded. The bulk of the objects that came to the United States are held in the Semitic Museum, Harvard University, with a smaller selection at the Harvard Art Museums, the Peabody Museum of Archaeology and Ethnology, Harvard University, and the University of Pennsylvania Museum of Archaeology and Anthropology. The assemblage is unique for the Near Eastern LBA in terms of the number and range of objects, with many thousands of beads (glass, faience, Egyptian blue, stone, shell), glass vessels, ceramics, and metals. As discussed in the excavation report (Starr 1937, 1939), the metals in the Nuzi (LBA) levels include iron, silver, gold, lead, copper, and bronze, and cover a range of typologies, such as weapons, armor, tools, jewelry, ornaments, and personal items. According to Starr (1939, 470), "Copper and bronze were the two most common metals and were found in a variety of forms and in considerable quantity." Tests of specimen pieces suggested slightly more copper than bronze; most were extremely corroded and developed bronze disease on exposure to air (Starr 1939). Although Starr does not describe the analytical technique used, spectrographic and microchemical analyses were performed on some pieces (Gettens 1933, 120, 135). In addition to the LBA metals from the destruction layer, there were earlier metal artifacts from the Ga.Sur levels, including rare gold and silver, some lead, and, most commonly, copper (Starr 1939).

Analysis

All the metalwork was analyzed by nondestructive x-ray fluorescence spectroscopy (XRF) in the storeroom and display of the Semitic Museum, Harvard University. This analysis contrasts with that done in the 1930s when Starr noted, "[S]ince it was impossible to test each piece [of copper alloy], all have been referred to herein under the single name of copper" (Starr 1939, 470). The XRF system used was a Bruker Tracer with a rhodium tube, operated at 40 kV and 1.4 mA with a Ti-Al filter. The objects were placed on the XRF without being cleaned, and most were analyzed in two or more areas. However, not all pieces of armor were analyzed in cases in which multiple pieces were in a single bag with one find spot and object number (103 individual pieces in 1930.76.37). Following XRF analysis, a subset was selected for sampling and analysis by scanning electron microscopy with energy-dispersive spectrometry (SEM-EDS). Mounted and polished samples were examined at the Museum of Fine Arts, Boston, with a JEOL JSM-6460 LV scanning electron microscope at 20 kV with an Inca X-sight Oxford Instruments energy dispersive detector.

Results

The XRF analysis confirmed that the metal assemblage was dominated by copper alloys, as shown in table 1. Defining exact numbers of objects is complicated by the fact that many items are fragmentary, of uncertain original function, or form components of one or more larger items (for example, individual plates of armor). The presence of iron at Nuzi and any relationship to rare early copper-zinc alloys would be interesting given the rarity of iron in the Near East in the LBA (Moorey 1994, 287–88). However, Starr considered all but two artifacts to be late—namely, an iron grip in the handle of a copper dagger (Starr 1937, plate 125 KK; 1939, 470) and a bead from temple A, neither of which is at Harvard. The field notebooks show most iron artifacts either lack recorded field numbers or come from late graves, late pavements, the surface, or "dirt."

Table 1. *Division of Nuzi assemblage into metal types.*

Metal Type	No. of Objects/Pieces Analyzed
Copper alloys	~ 930
Iron	~ 75
Lead	~ 40
Silver	~ 10
Gold	3

Numbers are approximate due to difficulties in defining individual objects.

The copper alloys include a variety of tools, weapons and armor, jewelry and other personal items, decorative pieces such as figures, utilitarian items such as nails and wires, and a large group of unknown items (fragmentary, damaged, or lacking any obvious typology). The most common pieces are chisels and sickles, pins, needles and nails, and armor for men (specific armor for chariots was not identified and identifiable armor for horses was rare). In Starr's opinion, some items, such as spearheads, adzes, and swords or daggers, were rare due to looting rather than an original lack of such items, noting that "the quantity of metal remaining on the site is no true indication of what was originally there" (Starr 1939, 470). This is confirmed by translated texts describing Nuzi as a regional store of weapons and armor (Kendall 1974). The current numbers are a fraction of those described, suggesting most were removed from the site by either the fleeing population or the looting invaders. Jewelry and small personal items make up only about 8% of the assemblage, again attributed to removal of such small valuable items from the site. Hence, any attempt to assess the proportions of different alloys at the site must take into account that not only has much of the metalwork been removed in antiquity but also the proportions of different typologies now present probably do not reflect that at the site during use. If specific alloys were preferred for different object types, this discrepancy will affect any discussion of proportions of different alloys. There is no evidence of primary production of any materials at Nuzi and surprisingly little evidence of secondary working, although a few molds and rare crucible fragments indicate some metalworking and texts mention working of metals to produce arms and armor to store in the city (Kendall 1974).

Table 2. *Division of copper alloys into distinct alloy types based on surface XRF.*

Alloy from Surface XRF	Pieces Analyzed
Copper	407
Bronze (Cu-Sn)	320
Copper + Tin/Bronze	87
Brass (Cu-Zn)	24
Mixed Cu-Zn-Sn	13

The total of analyzed pieces is lower than the estimated total of copper alloy pieces as not every individual piece of armor plate was analyzed.

The alloy attributions based on the XRF data are summarized in table 2. It should be remembered that a significant quantity of metalwork was retained by the Baghdad Museum. Distinguishing copper from corroded low-tin bronze by surface analysis alone is difficult (as discussed later), and hence there are some artifacts that could originally have been either low-tin bronze or dirty copper (trace amounts of tin). The proportion of bronze is probably underestimated, and further work is needed to refine these groups. The current data agree with the greater proportion of copper noted by Starr (1939, 470). The results show that copper-zinc alloys account for 4% of the copper-alloy finds held in the Semitic Museum, Harvard University, and the Harvard Art Museums, a small but significant proportion. Corrosion and surface accretions make it impossible to quantify the composition without sampling, but XRF analysis suggests some variability with different mixtures of copper, zinc, lead, and/or tin. As suspected from earlier studies (Shortland et al. 2008), the copper-zinc alloys appear to have been used mainly for jewelry, including finger rings, earrings, and bracelets, small personal items such as pins, and items of unknown function (including fragmentary pieces). As noted before, no weapons, armor, or large tools contain significant zinc, with these instead divided between copper and bronze (both having varying lead content). Since jewelry accounts for only about 8% of the total copper alloy assemblage and may have been selectively removed from the site, zinc-containing alloys may have been more prevalent during the LBA. The results thus suggest that copper-zinc alloys were a small but important component of the alloy assemblage. However, any discussion of their significance requires confidence in both the chronology of the assemblage and the alloy attributions from surface XRF analysis.

Chronology

Although most material dates to the LBA, examination of the excavation reports and field notebooks indicates that metalwork was found at all levels as well as on the surface. Some pieces are recorded as coming from late graves or surface deposits and thus are clearly late. Others can be confidently assigned to the LBA based on typology, such as the armor and some weapons. Many pieces, however, lack either a definitive typology or a recorded context, and thus their dating remains problematic. Unfortunately, many of the zinc-containing alloys occur in pieces with a wide typological range and/or lack of context. Even where the context is recorded and believed to correspond to a stratum II level, some care is needed as there are intrusive materials in several LBA contexts. This is clearest for glass beads and vessels, which can often be confidently assigned to later periods based on their technology and composition (Kirk 2009; K. Eremin, S. Kirk, A. Shortland, M. Walton, and P. Degryse, in preparation). Kirk (2009) discusses the disruption of the site in later periods and the effect of this on the interpretation of the glass assemblage in some detail and concludes that a large area of the site, particularly the northwestern residential area (which includes temple A), has suffered from some disturbance and could contain later items.

Careful consideration allows us to divide the context of the copper-zinc pieces into a variety of categories, as shown in table 3. However, even items from areas of secure stratum II context can be shown typologically to postdate the LBA. A good example of this is provided by a long-handled "spoon," Semitic Museum number 1930.71.70, found in room H2 in temple A. Starr published this as a spoon and notes, "The mode of design is not seen elsewhere in Nuzi, and is remarkable for its restraint and its pleasing geometrical effect" (Starr 1937, plate 126 U; 1939, 485). However, this artifact and its decoration find close parallels among published Roman cosmetic and surgical tools for collecting and dispensing liquids, semiliquids, and powders, such as the ligulae, scoops, cyathiscomeles, and spatulae discussed by Milne (1907, 61–62, 77–79), Waldbaum (1983, 105–8), and Bliquez (1994, 48–50), and is undoubtedly Roman to Byzantine/Sasanian in date. Since temple A contains abundant LBA material and this piece was accepted as LBA by Starr, this reduces confidence in any attribution that does not have a good typological basis. Unfortunately, this may be the case for much of the simple jewelry. As yet, the chronology of the metalwork has not been clearly defined, but it appears that none of the copper pieces from the third-millennium-BCE levels contain any significant zinc. Thus it seems that copper-zinc alloys are a feature of the mid-second millennium BCE and later, but not earlier, and that some care is needed in ascribing specific items to the LBA.

Table 3. *Context of copper-zinc alloy artifacts.*

Object No.	Typology	Field No.	Provenance	Cu	Zn	Pb	Sn	Disturbance
1930.72.28	anklet	30-2-250	H18	x	x	nd	nd	none recorded
1930.71.70	spoon	30-1-83	H2, temple A	x	x	x	tr	none recorded
1930.71.425B	flat object	none	L4 wall	x	x	x	nd	none recorded
1930.71.425A	flat object	none	L4 wall	x	x	tr	x	none recorded
1930.71.426	flat object	none	L4 wall	x	x	tr	x	none recorded
1930.77.44	flake	1299	L4	x	x	tr	nd	none recorded
1930.71.492	wire	29-12-294	F27	x	x	x	x	minor
1930.72.9	finger ring	29-1-127	P334	x	x	x	tr	minor
1930.72.13	bracelet	29-11-58	G8	x	x	x	nd	minor
1930.72.17	finger ring	29-12-36	G18	x	x	x	tr	minor
1930.72.31	brooch	30-2-389	G57	x	tr	x	x	minor
1930.72.37	anklet	30-12-110	H5	x	x	x	x	minor
1930.72.50	filigree	1129	palace	x	x	x	tr	variable
1930.77.28	unidentified	30-2-208	G29	x	x	x	nd	variable
1930.71.470	unidentified	none	none	x	x	x	x	unknown
1930.71.480	pin	none	none	x	tr	x	x	unknown
1930.71.508	unidentified	none	none	x	x	x	x	unknown
1930.71.532	nail	none	none	x	x	x	x	unknown
1930.72.54	earring	none	none	x	x	x	x	unknown
1930.72.56	bracelet	none	none	x	x	x	x	unknown
1930.72.57	bracelet	none	none	x	x	x	x	unknown
1930.72.61	bracelet	none	none	x	x	x	nd	unknown
1930.72.62	bracelet	none	none	x	x	x	x	unknown
1930.71.252	pin	30-2-371	surface	x	x	x	tr(?)	surface
1930.72.15	bracelet	29-11-209	F6-8, above P.1.	x	x	x	tr	late context
1930.72.24	pendant	29-12-305	F19	x	x	x	nd	considerable
1930.72.30	finger ring	30-2-384	G50	x	x	x	nd	variable
1930.72.45	unidentified	27-28:1016	shil 7	x	x	x	x	considerable
1930.72.58A	earring	none	GR. 36	x	x	tr	x	late grave
1930.72.58B	earring	none	GR. 36	x	x	x	x	late grave
1930.72.59	earring	29-12-33B	GR. 17	x	x	x	x	late grave
1930.72.60	ring	30-2-384	G50	x	x	x	nd	variable
1930.77.39	tweezer	1007A	shil 7 (dirt)	x	x	nd	nd	unstratified
1930.77.40	handle	1007B	shil 7 (dirt)	x	x	nd	nd	unstratified

Alloy types based on nondestructive XRF. nd = not detected; tr = trace; x = present.

As shown in table 4, SEM-EDS analysis of the small number of pieces sampled shows a variety of compositions. In most cases, the SEM-EDS and XRF analyses show fairly close agreement despite the surface corrosion. The compositions are consistent with those reported by Thornton and Ehlers (2003) and Thornton (2007). A variety of microstructures was observed in the Nuzi samples, including worked examples with strongly aligned inclusions and more granular (probably annealed) examples. All of these contained inclusions of zinc-sulfide and/or zinc-(copper)-sulfide and lead, shown in figure 1. Similar inclusions were seen in the samples examined by Thornton and Ehlers (2003) and Thornton (2007). The SEM-EDS results in this study do not show a clear distinction between the compositions of pieces considered to be definitely early (such as 1930.72.17 and 1930.72.37) and those known to be later (such as 1930.72.58 from grave 36 and 1930.72.59 from grave 17). These definitely later pieces are, however, distinct in containing inclusions of copper oxide rather than sulfides. Hence, while the compositions may not allow us to clearly differentiate between the different chronological groups, the phases present may prove more diagnostic. For the artifacts listed in tables 3 and 4, currently available data from consideration of the typology, field records, and analyses support the association of nine artifacts with the LBA: items 1930.72.13, .15, .17, .37, .56, .57, .60, .61, and .62. Of the others, ten are believed on the same criteria to postdate the LBA: items 1930.71.70, .508 and .532; 1930.72.45, .54,

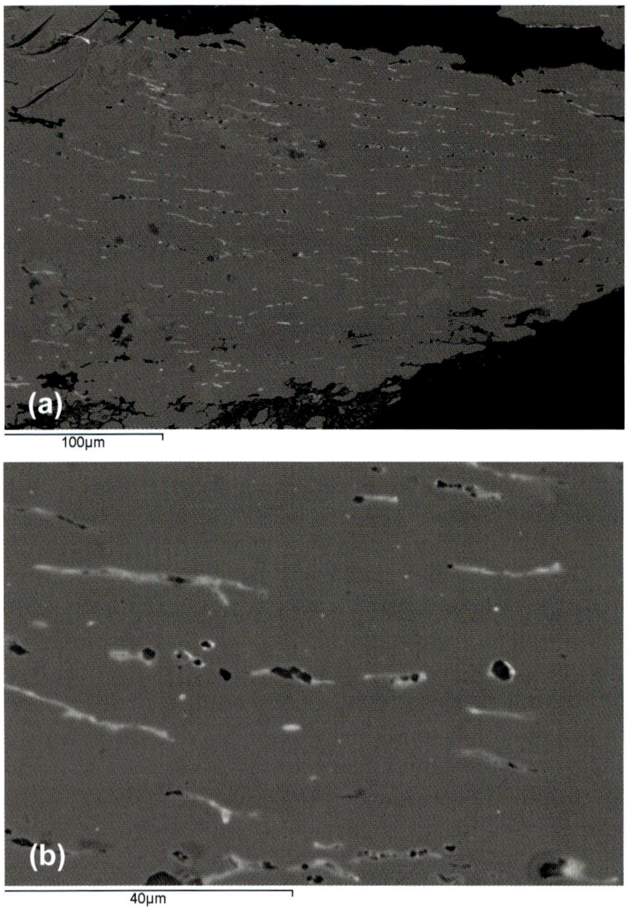

Figure 1. ***a.*** *Backscattered electron (BSE) image of polished sample of finger ring 1930.72.17, Semitic Museum, Harvard University, showing the aligned inclusions of lead (white) and zinc-(copper)-sulfide (dark).* ***b.*** *BSE image showing detail of inclusions in finger ring 1930.72.17, lead (white) and zinc-(copper)-sulfide (dark).*

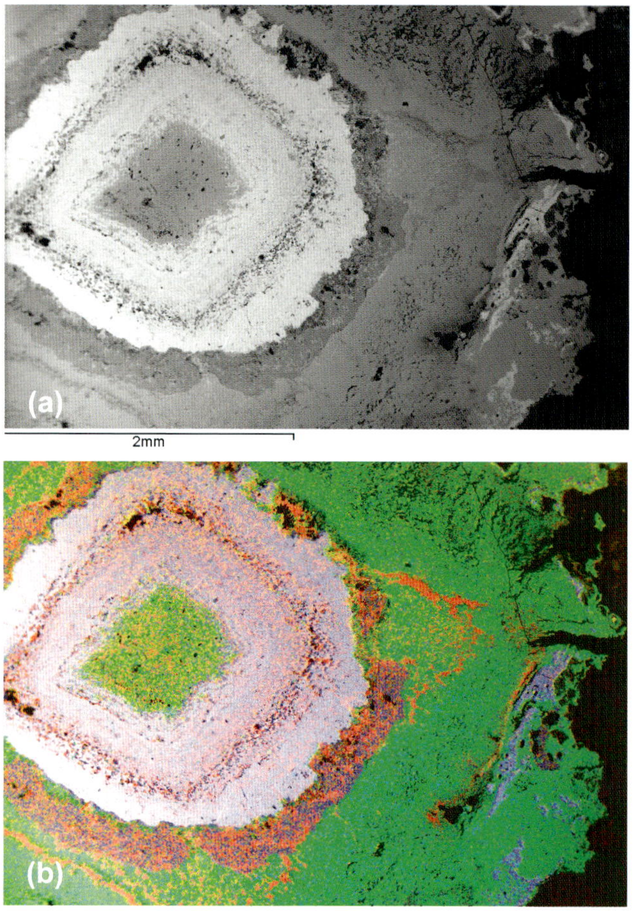

Figure 2. ***a.*** *BSE image of a section through the core and surface layers of nail 1930.71.306.* ***b.*** *Overlay of elemental maps and BSE image for nail 1930.71.306, Semitic Museum, Harvard University, showing the tin-poor surface. Copper: blue; chlorine: green; tin: red.*

.58A, .58B, and .59; and 1930.77.39 and .40. Further investigation is required to determine the likely date of the other fifteen pieces and for the present they remain problematic: items 1930.71.252, .425A, .425B, .426, .470, .480, and .492; 1930.72.9, .24, .28, .30, .31, .50; and 1930.77.28 and .44.

Conservation

The majority of the analyses undertaken to date are nondestructive XRF, hence they will be heavily influenced by any chemical alteration of the surface due to surface accretions from burial, corrosion, and/or conservation. Many artifacts have visible burial accretions, often with significant levels of iron and calcium. Although these affect the analyses and would limit quantification, they are unlikely to affect alloy attribution. However, repeat analyses on some artifacts show considerable variation in the copper to tin ratio from different areas of a single object. This is particularly worrying in cases where the surface shows very low to no detectable tin, whereas the broken cross-section (a closer guide to original alloy) has high levels of tin. Hence, analysis of the surface leads to an attribution of copper, whereas analysis of the interior results in bronze. The explanation is clear from SEM-EDS of mounted cross-sections of such pieces. An example of this is wall nail 1930.71.306. SEM shows a thick surface layer of copper chloride corrosion with tin present only in isolated lamellae around a bronze core (fig. 2). In this example, XRF of the interior as represented by the cross-section of the broken shaft indicated bronze with significant tin, whereas XRF of the outer surface of the shaft lacked detectable tin and suggested copper. Although this is probably an extreme case, it shows that care is needed in distinguishing bronze from copper in this burial environment based purely on surface XRF analyses. No clear examples of dezincification were seen in the comparison of surface XRF and SEM data, but only a limited number of pieces have been sampled and analyzed by SEM to date. Reduction of surface zinc to copper ratios due to corrosion is relatively common and could lead to brass or a mixed zinc-containing alloy being incorrectly assigned. However, comparison of some XRF data and data from drilled samples actually indicates the reverse problem, as discussed in the following paragraphs.

Although the metalwork has not been conserved recently, some pieces were conserved soon after excavation. It is therefore important to assess whether conservation treatments at the time affected the surface composition. This is suggested by a unique censer, Harvard Art Museums 1928.183, shown in figure 3, one of a number of pieces studied by Rutherford John Gettens, the first chemist in

Table 4. *Comparison of SEM-EDS and XRF data for some copper-zinc alloys.*

Object No.	Typology	Provenance	Cu	Zn	Pb	Sn	XRF Data	Disturbance
1930.72.17	ring	G18	81.2	16.5	2.2	0.2	Cu, Zn, Pb, tr Sn	minor
1930.72.37	anklet	H5	82.3	16.1	1.6	0.1	Cu, Zn, Pb, Sn	none recorded
1930.72.50	filigree	palace	85.2	13.5	1.0	0.3	Cu, Zn, Pb, tr Sn	variable
1930.72.57	bracelet	none	75.2	20.1	4.4	0.2	Cu, Zn, Pb, Sn	unknown
1930.72.62	bracelet	none	84.8	11.7	1.2	2.2	Cu, Zn, Pb, Sn	unknown
1930.71.480	pin	none	85.0	3.1	3.5	7.8	Cu, Zn, Pb, Sn	unknown
1939.72.58A	earring	GR. 36	80.3	18.1	0.2	1.4	Cu, Zn, tr Pb, Sn	late grave
1938.72.59	earring	GR. 17	79.2	19.7	0.3	0.7	Cu, Zn, Pb, Sn	late grave
1930.72.60	ring	G50	84.8	10.3	2.5	2.1	Cu, Zn, Pb	variable
1930.71.252	pin	surface	85.6	12.2	0.2	2.1	Cu, Zn, Pb, tr Sn	surface

SEM-EDS results are given in wt.%. tr = trace.

an art museum in the United States and a key member of the conservation laboratory in the new Fogg Art Museum, now part of the Harvard Art Museums. (For a discussion of Rutherford J. Gettens and his role in conservation, see Bewer 2001, 2010, and in this vol.). Surface XRF analysis indicated the presence of copper, zinc, lead, tin, and arsenic, with considerable variation in zinc and lead from point to point. However, inductively coupled plasma–mass spectroscopy (ICP-MS) analysis undertaken in a study of ancient bronze artifacts in the Harvard Art Museums indicated a bulk composition with below 0.1 wt.% zinc (unpublished ICP-MS analysis Josef Riederer, Rathgen-Forschungslabor). There is thus a discrepancy between the surface and bulk analyses. The censer was lacquered in Iraq but developed green efflorescence, bronze disease, about six months after entering the Fogg Art Museum (Gettens 1931). The relatively new technique of electrochemical reduction was used to remove the chlorides that caused bronze disease as well as encrustations of soil (Gettens 1931). The treatment consisted of electrolysis with a 2% solution of caustic soda for two months to remove the disfiguring corrosion. Object files in the Straus Center for Conservation (SCC), Harvard Art Museums, still contain Gettens's original notes showing the proposed treatment for the censer, as illustrated in figure 4. Electrochemical reduction left grayish deposits of tin oxide on the surface, which were removed by wrapping the censer in strips of zinc sheet and immersing it overnight in a 10% solution of caustic soda. Following this, the gray deposit was mostly absent and could be brushed off as needed (Gettens 1931). The object files also contain the original pre- and postconservation photos taken by Gettens, which show the strong visual improvement (fig. 5), although visual inspection today shows a grayish surface remaining in some areas.

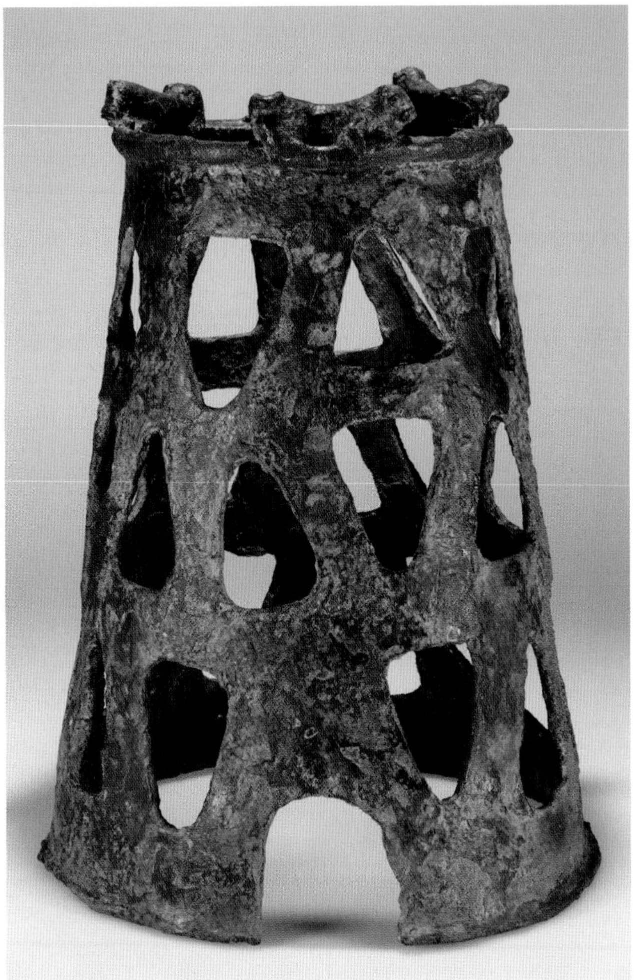

Figure 3. *Censer 1928.183, Harvard Art Museums, showing current appearance. Copyright Harvard Art Museums.*

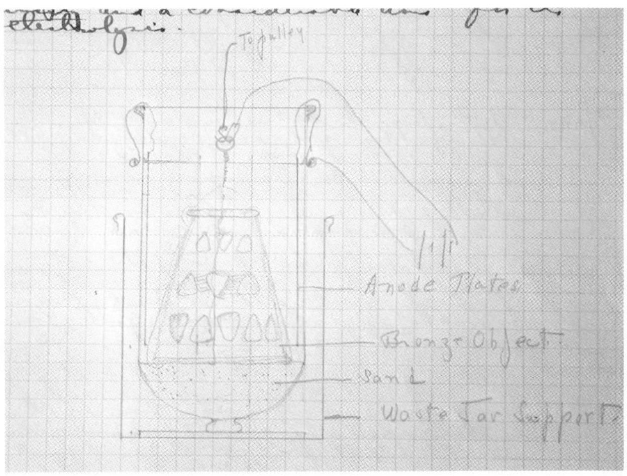

Figure 4. *Gettens's notes showing the proposed electrochemical reduction of censer 1928.183, Harvard Art Museums.*

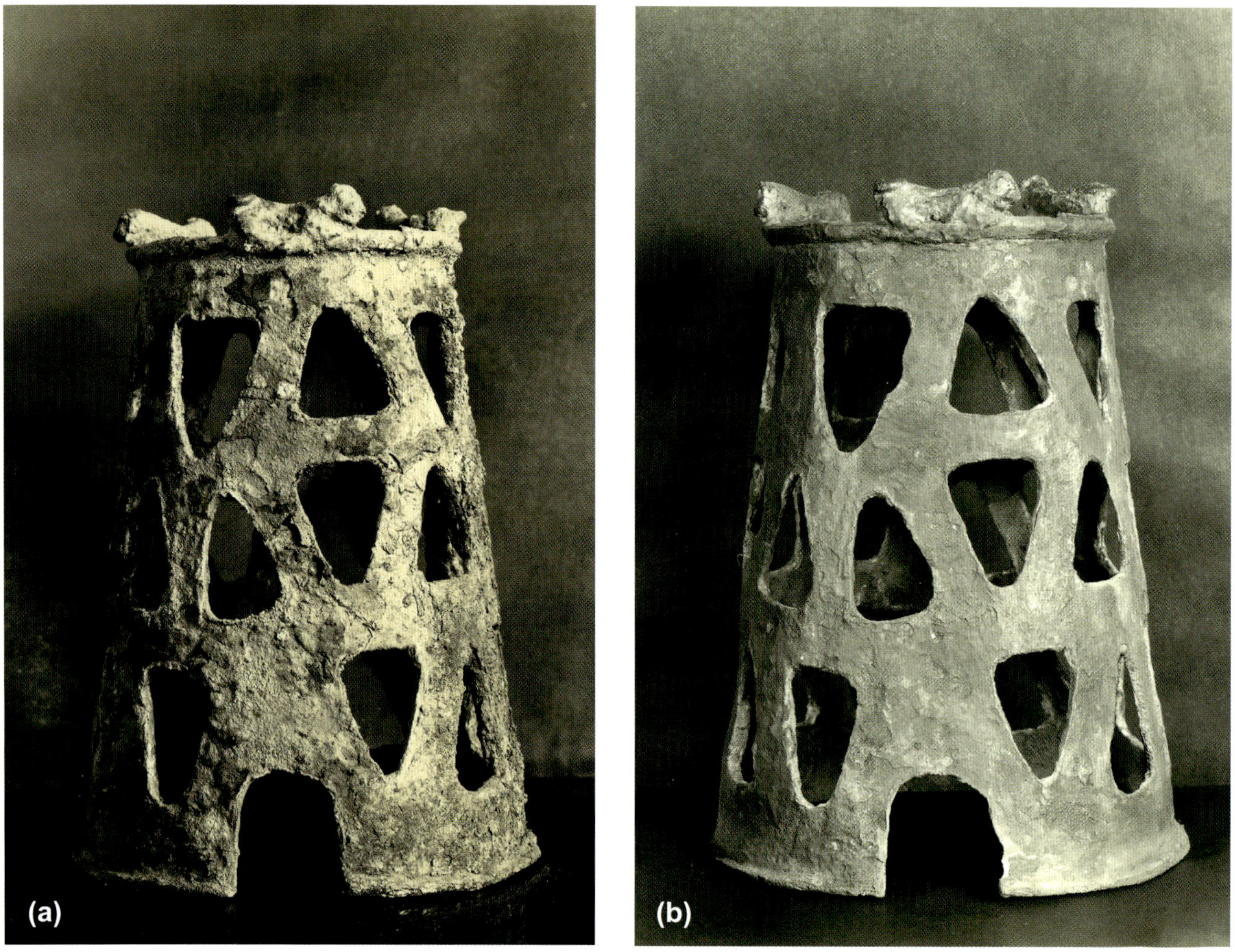

Figure 5. *Gettens's photos showing the censer* ***(a)*** *before treatment and* ***(b)*** *immediately after treatment.*

A number of other copper alloy objects were treated similarly (Gettens 1931, 1933). Some were lacquered after treatment, and one sickle was patinated with glacial acetic acid and ammonia, followed by exposure to carbon dioxide (Gettens 1931, 283). This sickle can be identified as Semitic Museum 1930.71.355 by its distinctive bright green surface, which is very different from that of any other piece. Gettens's original notes on the objects to be treated are shown in figure 6a and the current appearance of the sickle in figure 6 b and c. This sickle is also distinctive in showing some zinc in the XRF spectra.

In some instances it is surprisingly hard to identify exactly which pieces were treated, even with the descriptions and images in Gettens's publications, as the objects lacked Semitic Museum numbers at the time and few are unique artifacts. For example, although the Straus Center archives contain Gettens's original notes and figures, we cannot identify the wire and flower-headed nail also depicted in figure 6a, although the nail is also illustrated, presumably after treatment, in Starr's excavation report (Starr 1937, pl. 125 L1 and L2). Examination of a unique inscribed copper plaque from room F24, the principal room in a high-status house from the northwestern residential area (Starr 1937, pl. 27 A; 1939, 214, 217), shows that it has been conserved, although we cannot find any record of this. Comparison of the plate with the current condition of the piece again indicates conservation prior to publication. XRF of this plaque did not detect surface zinc, suggesting a different treatment or surface response. Gettens's notes in the SCC reveal that he used a variety of treatments on corroded copper nails from Nuzi between 1930 and 1935. Some of these notes postdate Gettens (1933), suggesting experimentation continued after publication. For example, notes dated June 1935 describe the treatment of nine copper nails as: No. 1, reduction by zinc and sodium hydroxide; No. 2, electrolysis in sodium sulfate; No. 3, electrolysis in potassium cyanide; No. 4, treatment with citric acid; No. 5, electrolysis in distilled water; and No. 6, treatment with sodium sesquicarbonate. In contrast, Gettens (1933) describes six copper nails treated by the electrochemical method of Fink and Eldridge (1925) and then mechanically cleaned. A number of the nails were destroyed by the treatment (Gettens 1933, 133, and SCC notes) and there is no obvious record of which nails received which treatment. Surface XRF analysis shows traces of zinc on some, but not all, of the nails whose visual appearances suggest previous conservation.

Such problems of surface zinc are unlikely to be confined to this assemblage. Electrochemical reduction with zinc was widely used in the twentieth century in both Europe and America (Drayman-Weisser 1994; Beale 1996; Beesley 2008; Paterakis 2003). Beesley (2008) tested different his-

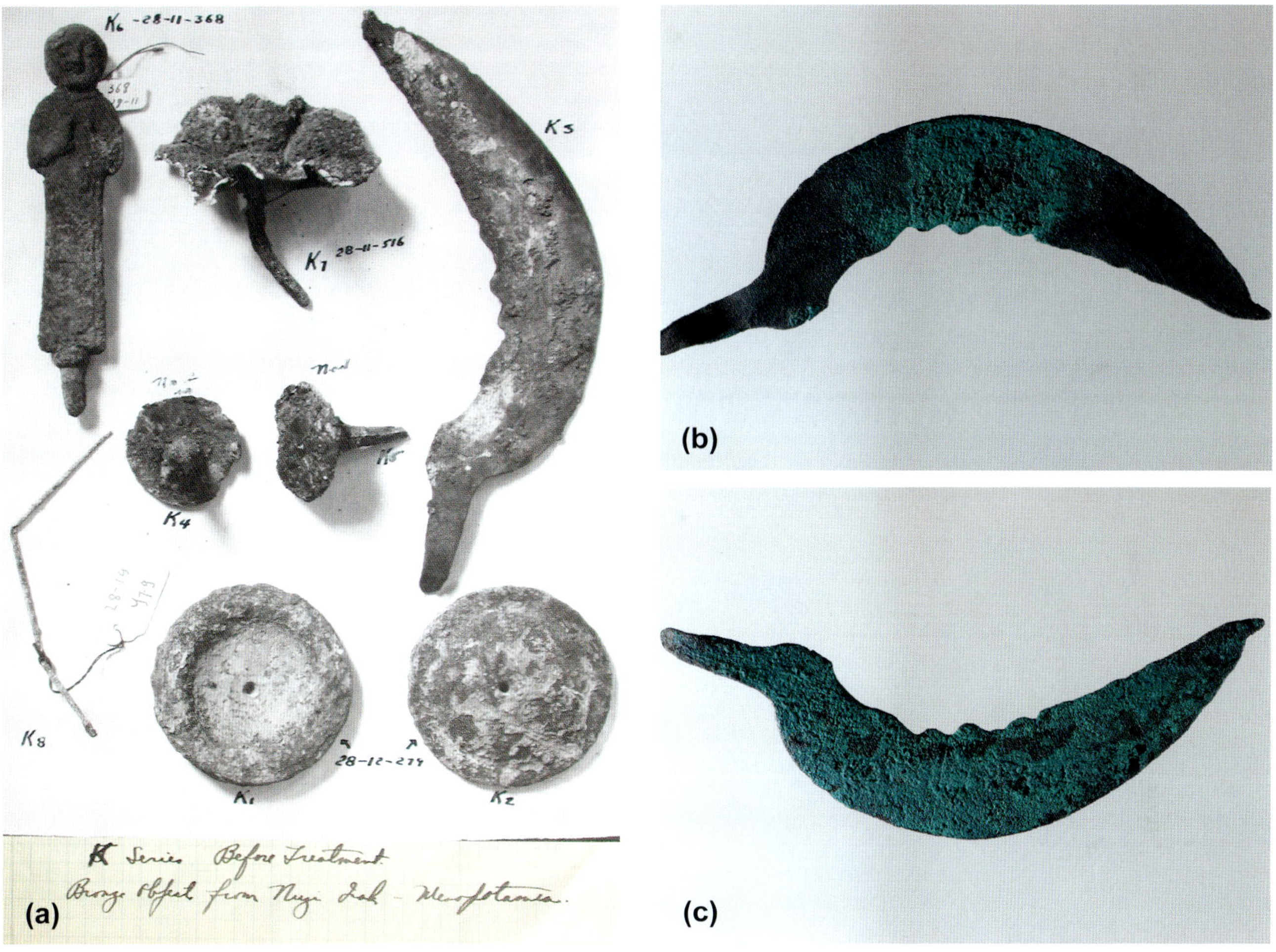

Figure 6. a. *Gettens's notes showing objects from Nuzi to be treated.* ***b*** *and* ***c.*** *Views of both sides of Sickle 1930.71.355 showing the current bright green surface. Courtesy of the Semitic Museum, Harvard University.*

torical methods of electrochemical reduction on copper and leaded bronze and showed zinc was deposited on the surface. Drayman-Weisser (1994) mentions deposition of zinc, and Paterakis (2003) suggests traces of zinc on an artifact in the Athenian Agora may have resulted from electrochemical reduction with zinc and sodium hydroxide. Hence, care is needed in interpreting traces of zinc on the surface of copper alloys when relying on nondestructive XRF data. Without consideration of the effects of early conservation treatment, trace zinc could be attributed to surface dezincification of a copper-zinc alloy, thus confusing the interpretation of alloy patterns. Careful investigation of the condition and previous history of the objects, analysis of multiple areas, and/or sampling to reveal the interior alloy type is required for true understanding of original alloy composition.

Conclusion

Study of the metalwork assemblage from Nuzi suggests that some caution is required in accepting the status of unusual artifacts from excavations dated to the nineteenth or early twentieth century. While it is clear that many objects from the site can be confidently assigned to the LBA, it is apparent that others, confidently attributed by the excavators at the time to the LBA, are either intrusive or, at best, of questionable date. Many areas contain later artifacts, reducing confidence in conclusions relying on finds without diagnostic typologies. Unfortunately, it is in such small items of jewelry and other items of uncertain original shape or function that the copper-zinc alloys tend to be found, both at Nuzi and from other similarly excavated sites. This is not to say that all of these early copper-zinc alloys are late or even questionable, merely to highlight the need for care when interpreting material from such early excavations. Further confusion comes from the uncertain history of many artifacts, since conservation and object records from the early twentieth century often lack crucial details about the artifacts' past history. In the absence of recorded histories, careful examination and analysis can provide clues about previous treatments that may have altered surface chemistry and hence interpretations based on this. Although the proportion of LBA copper-zinc alloys is considerably lower than the 4% indicated simply by surface XRF analysis, a number of LBA copper-zinc alloys are definitely present in the assemblage from Nuzi, and the alloy does appear to have been used deliberately for small decorative items at this time. Further studies of related assemblages and sampling of selected artifacts may shed further light on the distribution, use, and ultimately the origin of copper-zinc alloys in the Near East in the LBA.

Acknowledgments

We would like to thank the staff at the Semitic Museum, Harvard University, for their enthusiastic assistance with this study, particularly Adam Aja, James Armstrong, and Joe Greene. We are also grateful to Susanne Ebbinghaus, Lisa Anderson, and Bahadir Yildirim for tracking down Roman parallels and to Francesca Bewer for information on Rutherford J. Gettens and his documents. Susanne Ebbinghaus, Francesca Bewer, and Ariel O'Connor are thanked for comments on and assistance with the manuscript and presentation. Richard Newman at the Museum of Fine Arts, Boston, is thanked for assistance with SEM-EDS and discussion of the data. Finally, we would like to thank Elizabeth Beesley for discussing electrochemical reduction and allowing reference to her unpublished MSc dissertation.

References

Beale, A. 1996. Understanding, restoring, and conserving ancient bronzes with the aid of science. In *The fire of Hephaistos: Large classical bronzes from North American collections,* ed. C. Mattusch, 65–80. Cambridge, MA: Harvard University Art Museums.

Bedore (Ehlers), C., and C. Dixon. 1998. New discoveries in an old collection. *Context* 13:3–4.

Beesley, E. S. J. 2008. The impact of electrochemical reduction treatment on copper alloys: A historical and analytical perspective. MSc diss., Institute of Archaeology, University College London.

Bewer, F. G. 2001. Technical research and the care of works of art at the Fogg Art Museum (1900 to 1950). In *Past practice—Future prospects,* ed. A. Oddy and S. Smith, 13–18. Conservation Conference September 2001. British Museum Occasional Paper 145. London: The British Museum.

———. 2010. *A laboratory for art: Harvard's Fogg Museum and the emergence of conservation in America, 1900–1950.* Cambridge, MA, and New Haven, CT: Harvard Art Museum and Yale University Press.

Bliquez, L. J. 1994. *Roman surgical instruments and other minor objects in the National Archaeological Museum of Naples.* Mainz: Verlag Philipp von Zabern.

Craddock, P. T. 1978a. The composition of the copper alloys used by the Greek, Etruscan and Roman civilizations: The origins and early use of brass. *Journal of Archaeological Science* 5:1–16.

———. 1978b. Europe's earliest brasses. *MASCA Journal* 1(1):4–5.

———. 1980. The first brass: Some early claims reconsidered. *MASCA Journal* 1(5):131–33.

———. 1988. Problems and possibilities for provenancing bronzes by chemical composition, with special reference to Western Asia and the Mediterranean in the Early Iron Age. In *Bronze-working centres of Western Asia c. 1000–539 BC,* ed. J. Curtis, 317–27. London: Kegan Paul International.

———. 1998. Zinc in classical antiquity. In *2000 years of zinc and brass,* ed. P. T. Craddock, 1–6. British Museum Occasional Paper 50. London: British Museum Press.

De Ryck, I., A. Adriaens, and F. Adams. 2005. An overview of Mesopotamian bronze metallurgy during the 3rd millennium BC. *Journal of Cultural Heritage* 6(3):261–68.

Drayman-Weisser, T. 1994. A perspective on the history of the conservation of archeological copper alloys in the United States. *Journal of the American Institute for Conservation* 33(2):141–52.

Fink, C. G., and C. H. Eldridge. 1925. *The restoration of ancient bronzes and other alloys.* New York: Metropolitan Museum of Art.

Frame, L. 2010. Metallurgical investigations at Godin Tepe, Iran, part I: The metal finds. *Journal of Archaeological Science* 37(7):1700–715.

Gettens, R. J. 1931. The restoration of bronzes from Iraq. *Notes (Fogg Art Museum)* 2(6):273–83.

———. 1933. Mineralization, electrolytic treatment, and radiographic examination of copper and bronze objects from Nuzi. *Technical Studies in the Field of the Fine Arts* 1(3):118–42.

Kendall, T. 1974. Warfare and military matters in the Nuzi Tablets. PhD diss., Department of Mediterranean Studies, Brandeis University.

Kirk, S. 2009. The vitreous materials from the 2nd millennium BC city of Nuzi: Their preservation, technology and distribution. PhD diss., Centre for Archaeological and Forensic Analysis, Cranfield University.

Milne, J. S. 1907. *Surgical instruments in Greek and Roman times.* Oxford: Clarendon Press.

Moorey, P. R. S. 1994. *Ancient Mesopotamian materials and industries: The archaeological evidence.* Winona Lake, IN: Eisenbrauns.

Paterakis, A. B. 2003. The influence of conservation treatments and environmental storage factors on corrosion of copper alloys in the ancient Athenian Agora. *Journal of the American Institute for Conservation* 42(2):313–39.

Pigott, V. C. 1999. The development of metal production on the Iranian plateau: An archaeometallurgical perspective. In *The archaeometallurgy of the Asian Old World,* ed. V. C. Pigott, 73–106. Philadelphia: University of Pennsylvania Museum of Archaeology and Anthropology.

Shortland, A., K. Eremin, S. Kirk, and J. Armstrong. 2008. Reassessing Bronze Age manufacturing technologies at Nuzi. In *Materials issues in art and archaeology VIII,* Vol. 1047, ed. P. B. Vandiver, B. McCarthy, R. Tykot, J. L. Ruvalcaba-Sil, and F. Casadio, 217–32. Materials Research Society Symposium Proceedings. Warrendale, PA: Materials Research Society.

Starr, R. F. S. 1937. *Nuzi: Report on the excavations at Yorgan Tepa near Kirkuk, Iraq: Conducted by Harvard University in conjunction with the American School of Oriental Research and the University Museum of Philadelphia, 1927–1931,* Vol. 2, *Plates and maps.* 1st ed. Cambridge, MA: Harvard University Press.

———. 1939. *Nuzi: Report on the excavations at Yorgan Tepa near Kirkuk, Iraq: Conducted by Harvard University in conjunction with the American School of Oriental Research and the University Museum of Philadelphia, 1927–1931,* Vol. 1, *Text.* 1st ed. Cambridge, MA: Harvard University Press.

Stech, T. 1999. Aspects of early metallurgy in Mesopotamia and Anatolia. In *The archaeometallurgy of the Asian Old World,* ed. V. C. Pigott, 59–71. Philadelphia: University of Pennsylvania Museum of Archaeology and Anthropology.

Thornton, C. P. 2007. Of brass and bronze in prehistoric Southwest Asia. In *Metals and mines: Studies in archaeometallurgy*, ed. S. La Niece, D. Hook, and P. Craddock, 123–35. London: Archetype Publications in association with the British Museum.

Thornton, C. P., and C. B. Ehlers. 2003. Early brass in the ancient Near East. *Institute for Archaeo-Metallurgical Studies* 23:3–8.

Waldbaum, J. C. 1983. *Metalwork from Sardis: The finds through 1974.* Cambridge, MA: Harvard University Press.

Reconstructing Ancient Technologies: Chalcolithic Crucible Smelting at Tal-i Iblis, Iran

Lesley D. Frame

ABSTRACT Recent examination of coarseware ceramics and metallurgical crucibles from the copper smelting industry at Tal-i Iblis, Iran, dating to the fifth millennium BCE, has shown that these technologies are closely related. Similar fabrication methods employing the same clay sources as well as mineral and vegetal tempers were used for the production of the local Lalehzar coarseware and the copper smelting crucibles. The crucible smelting industry at Tal-i Iblis shows careful control of temperature and atmosphere, while at the same time showing high variability in the selection of copper ores used in the smelting operations. This paper discusses the continuity in technological transitions on the Iranian plateau despite the new and experimental aspects of this early example of copper pyrotechnology.

Introduction

It has been forty years since Joseph Caldwell's excavation of Tal-i Iblis, and despite the identification of dozens of prehistoric metallurgical sites on the Iranian plateau by recent archaeological projects, Iblis remains one of the earliest examples of copper smelting technology on the plateau and in the world. Further, Iblis is regarded as one of the most important early metallurgical production sites in the region (Craddock 2001; Pigott 1999a, 1999b, 1999c; Thornton 2009a, 2009b). The prestige of Tal-i Iblis is due to its early date (fifth millennium BCE), the large number of crucible fragments unearthed there (Caldwell 1967a), and its connections to other plateau sites (e.g., Tepe Yahya).

Tal-i Iblis, "the devil's mound," is an oval tell measuring 118 by 100 m at its largest, once standing over 11 m high (Caldwell 1967a, 9; Caldwell and Shahmirzadi 1966, 1). It is located in the Bard Sir valley southwest of the Kerman mountains in southeastern Iran (fig. 1). The site also includes a small dumping area to the northeast of

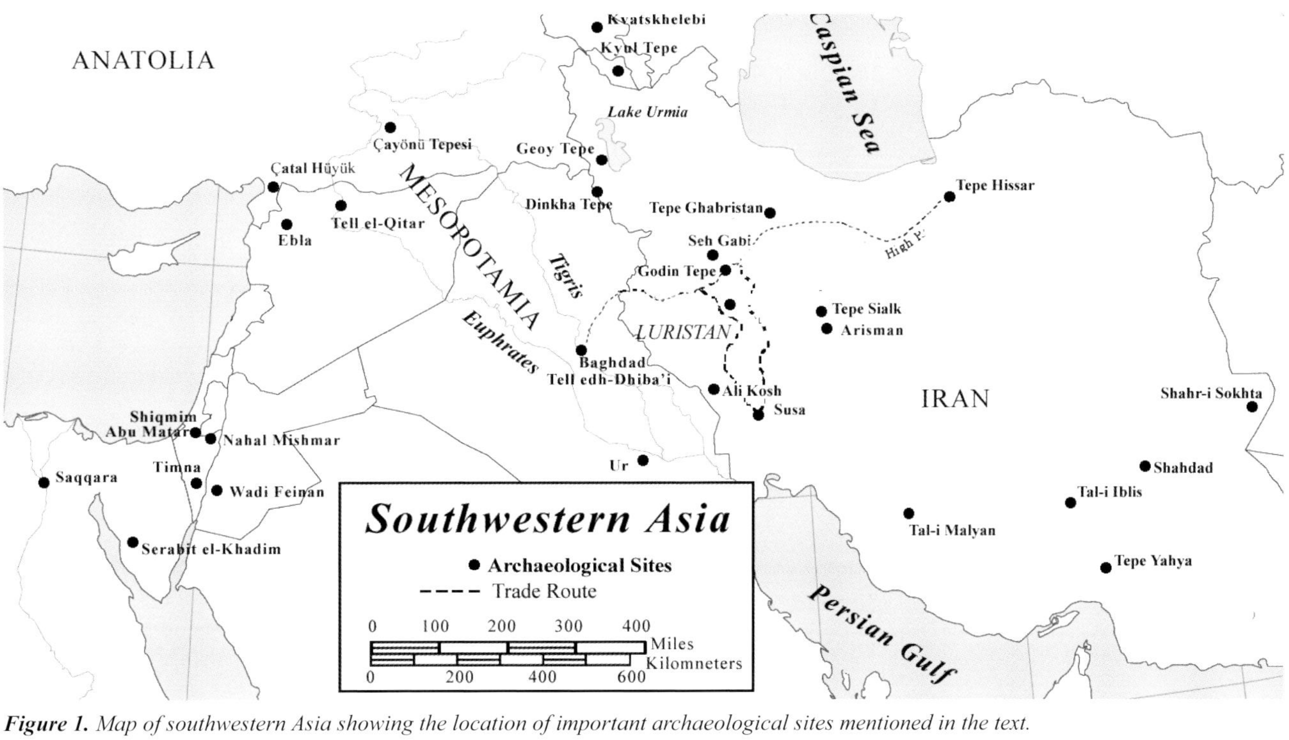

Figure 1. *Map of southwestern Asia showing the location of important archaeological sites mentioned in the text.*

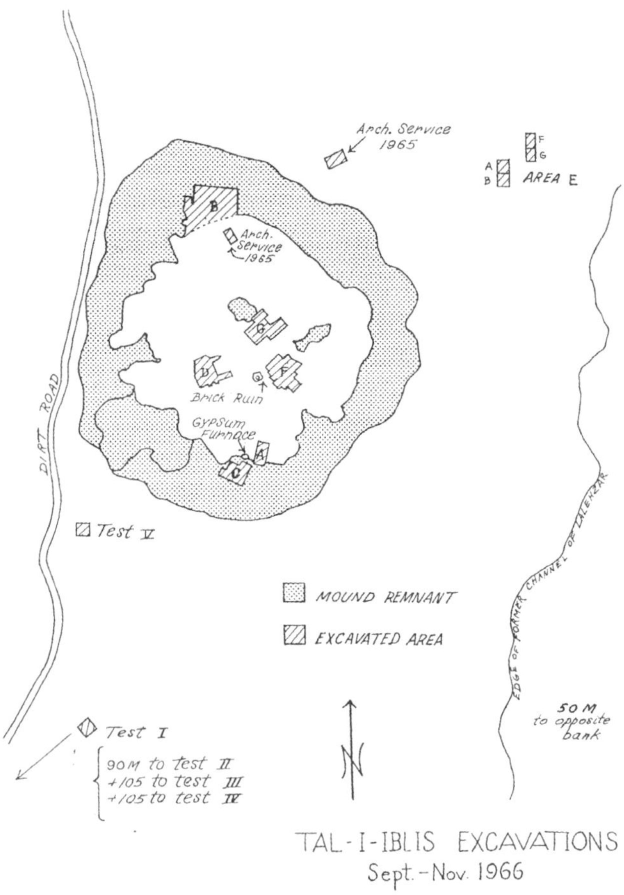

Figure 2. Map of Tal-i Iblis showing excavation areas and the remnant of the tell (Caldwell 1967, 31; reproduced here with permission from Illinois State Museum).

the mound (area E) and test pits south of the mound that revealed additional debris associated with the early levels of occupation (Caldwell 1967a) (fig. 2).

Tal-i Iblis was originally discovered and surveyed by explorer and archaeologist Sir Mark Aurel Stein in 1933 (Stein 1937), but it was not excavated until Joseph Caldwell rediscovered it during his search for archaeological sites in Iran in the 1960s. When he arrived at the site in 1964, Caldwell found "an archaeological tragedy" (Caldwell 1967a, 9). Much of the center of the mound had been

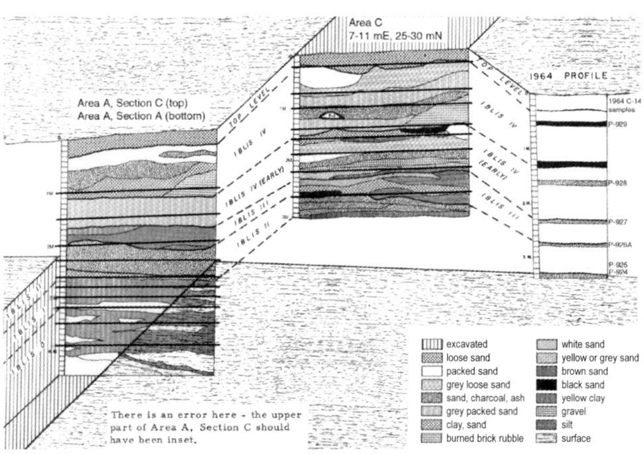

Figure 3. Profile map for areas A and C at Tal-i Iblis indicating excavation levels and locations for ^{14}C samples (adapted from Chase et al. 1967, 116; reproduced here with permission from Illinois State Museum).

removed by local farmers who were using the rich soil for fertilizer in their fields. During two seasons of excavation by Caldwell's team (1964 and 1966) under the auspices of the Illinois State Museum and one excavation by the Iranian Archaeological Service in 1965, much of the remaining tell and nearby dump area was excavated.

Caldwell's excavation revealed four houses and two dumps, and the cultural material from Iblis includes pottery styles distinct to the site as well as wares found elsewhere in Iran and Mesopotamia—Iblis period VI contained pottery similar to Sialk IV as well as beveled rim bowls that are typical of Mesopotamia (Caldwell 1967b, 25). In addition to the pottery and a prominent lithic industry, the site contained an unusually large amount of metallurgical material, especially given its early date. These artifacts include crucible fragments, copper ore fragments, and numerous small metal objects. It appears that most of these crucible and ore fragments were discarded, but approximately two hundred crucible fragments, over one hundred pieces of slag, seventy ore fragments, and eighteen copper implements found their way to the Natural History Museum at the University of Georgia, Athens, Georgia (UGA-NHM), and were available for examination in this study (for a detailed account of how the artifacts arrived in Athens, Georgia, see Frame 2009).

Archaeological Contexts

Due to the activity of the local farmers, much of the cultural materials from later levels at Tal-i Iblis was missing by the time Caldwell began his excavations in 1964. For this reason, the excavation returned a large amount of material from the earlier, undisturbed levels of the mound (Iblis I and II) and much less material from the later levels. The excavations revealed a continuous occupation sequence at the site beginning in the Chalcolithic (ca. 5500–3000 BCE) (Iblis O) and lasting into the Bronze Age (ca. 3000–1400 BCE) (Iblis VI). There is also evidence for later activity during the Old Elamite period (2700–1500 BCE)—a kiln dated to ca. 1100 BCE—and during the Achaemenid period (ca. 550–330 BCE) (Chase et al. 1967).

The metallurgical materials at Tal-i Iblis were spread throughout all of the periods present at the site, but higher concentrations of processing debris were recovered from the undisturbed period I and II contexts. Table 1 lists the metallurgical finds and corresponding periods according to excavation area and level. Depths in this table are given in centimeters below the surface of the excavation area and section. It should be noted that depth "0 cm" is not the same for all areas. Area C is located on top of the mound in an area not disturbed by the local farmers. However, area A is positioned at the edge of the farmers' handiwork. Area A, section C begins at the top of the mound, but sections A and B begin at the lowest level of section C (see fig. 2). Figure 3 is adapted from the profile map published in Chase et al. (1967, 116). This map has been relabeled with the excavation level intervals reported in the text. Period determinations were established by Caldwell through examination of stylistic variations in pottery at Tal-i Iblis.

Table 1.A: *Metallurgical finds from Areas A and C at Tal-I Iblis*

Area	Level	Depth	Period	Crucibles	Ore	Metal
Area A, Sect A	1	0-40	II	8	11	2
	2	-60	I	0	4	1
	3	-80	0[a]	None		
	4	-100	0[a]	None		
	5	-130	[b]	None		
	6	-170	[b]	None		
	--	-230	[c]	None		

Area	Level	Depth	Period	Crucibles	Ore	Metal
Area A, Sect B	1	0-20	II	2	1	11
	2	-50	I	None		
	3	-70	0[a]	None		
	4	-100	0[a]	None		

Area	Level	Depth	Period	Crucibles	Ore	Metal
Area A, Sect C[d]	1	0-100	IV	None		
	2	-150	Early IV	None		
	3	-200	Early IV to III	7	1	0
	4	-230	II	270	17	1
	5	-260	II	119	4	0

Area	Level	Depth	Period	Crucibles	Ore	Metal
Area C[f]	6	0-20	VII			1
	5c	-70	IV			1
	5g	-100	IV	1[e]		2
	5b	-140	IV			0
	5a	-160	Early IV			6
	4	-190	Early IV			3
	3	-220	Early IV			0
	2	-270	Early IV to III			5
	1	-310	II	300+[g]		0

[a] Only Lalehzar coarse ware is present, but this layer is drawn in the profile map as Iblis I.

[b] This layer is labeled in the profile as Iblis 0, but it contains no artifacts.

[c] This level is not mentioned in text, but it is drawn in the profile.

[d] "Slag" and crucible counts are added together for Area A, Section C.

[e] This was the large crucible fragment shown in Chase et al. 1967:186

[f] Area C was divided into at least four excavation areas, each dug with different intervals and slightly different level numbering. The depths indicated here are for 7-11 E, 25-30 N, the only square reaching Iblis II. All metal finds for this area are grouped by depth-intervals listed here. The numerous blanks are left because no sherd counts were done in the field for this area, so the values for each level are unknown.

[g] This is reported by Caldwell 1967b:34, not in the report by Chase et al. 1967.

Table 1.B: *Metallurgical finds from Area E at Tal-I Iblis*

Level	Depth	Area	Period	Crucibles	Ore	Metal	Area	Period	Crucibles	Ore	Metal	Area	Period	Crucibles	Ore	Metal	Area	Period	Crucibles	Ore	Metal
1	0-20	Section A	Early II[a]	3	2	0	Section B	II[b]	2	0	1	Section F	Early II	0	5	0	Section G	Late I	0	0	0
2	20-50		Early II[a]	0	1	0		Early II	2	3	0		Late I[c]	2	4	4		I[d]	1	9	2
3	50-70		Early II[a]	0	0	0		Early II	1	1	0		Late I[c]	0	1	2		I	0	10	1
4	70-100		Early II[a]	0	0	0		Early II	0	0	0		Late I[c]	2	4	4		I	0	6	0

All areas were excavated with same Level and Depth designations.

[a] This is called Late Iblis I by Chase et al. 1967:158), but I have called it Early Iblis II since the Iblis ware type is slighly more abundant than the Bard Sir pottery type.

[b] This section is also labeled Late Iblis I by Chase et al (1967:158), but there is a much higher ratio of Iblis to Bard Sir pottery (11.8) in this top level, so I believe it falls within the Iblis II period

[c] In these levels there is a greater number of Bard Sir sherds than the Iblis ware type, so I have accepted the Chase et al. (1967:158) designation of Late Iblis I.

[d] This section contain a much higher number of Bard Sir pottery sherds than Iblis ware sherds, suggesting that it could be placed slightly earlier than the other sections, therefore I would call this Iblis I period.

Table 1.C: *Metallurgical finds from Areas D, F, G, and B at Tal-I Iblis*

Area	Period (Floor)	Period (Fill)	Crucibles	Ore	Metal
D	Iblis 1	Late I or II	0	3	1
F	Iblis 1	Late I or II	0	5	2
G	Iblis 1	Late I or II	2	6	4
B[a]	Iblis 1	1	0	2	6[b]

[a] Crucible and ore from Area B is a count of the materials at UGA, total number of ore and crucible fragments was not reported for this area.

[b] Only metal from Iblis I and II contexts is included

***Table 2.** Radiocarbon dates for Tal-i Iblis based on Caldwell's original 1964 samples.*

Iblis Period	Caldwell (1967b, 24)	Voigt and Dyson (1992, 131)	OxCal 4 at 2s	Sample	Location
Iblis VII	1130 +/- 120	nd	nd	Gx 862	sand above area B structure
Iblis V	2869 +/- 57	3570–3365	3634–3358	P-929	1964 section
Iblis IV	3645 +/- 59	4415–4110	4440–4055	P-928	1964 section
Iblis III	3792 +/- 60	4460–4400	4532–4331	P-927	1964 section
Late Iblis II	4083 +/- 75	4925–4685	4906–4543	P-926	1964 section
Iblis II	4287 +/- 40	5205–4915	5196–4892	Gx 864	area A, section A, 0–40 cm
Late Iblis I	4110 +/- 150	5040–4555	5207–4405	Gx 865	area E, section F, 70–100 cm
Iblis I	3674 +/- 110	4430–4115	4520–4004	Gx 868	area B, room 5
Iblis I	3931 +/- 160	4735–4420	4954–4246	Gx 866	area D, floor
Iblis I	4003 +/- 160	4920–4440	5040–4333	Gx 863	area F, floor
Iblis I	4441 +/- 130	5290–4995	5468–4848	Gx 869	area G, floor
Iblis I	4410 +/- 165	5290–4945	5474–4771	Gx 867	area B, room 13

Dates are BCE. nd = no data available.

***Table 3.** Approximate date ranges for Tal-i Iblis periods.*

Date (BCE)	Tal-i Iblis
5500–4800	Iblis I
4800–4500	Iblis II
4500–4300	Iblis III
4300–3700	Iblis IV
3700–3300	Iblis V

Detailed discussion of these period determinations can be found elsewhere (Caldwell 1967a; Frame 2009).

The majority of metallurgical debris was found in the dump areas rather than the residential complexes. The earlier Iblis I and II dump, found in area E, returned just over one dozen crucibles (table 1), but the period II dump, located in area A section C and area C at the lowest level, contained well over six hundred crucible fragments and the majority of the metallurgical material at the site. Dumps for later periods were not identified during the 1964 and 1966 excavations, and without additional data, our focus is constrained to the earliest metallurgical activity at the site, represented by periods I and II.

Further archaeological investigation may reveal additional dump areas or workshops associated with the later periods, causing us to change the current interpretation of the site. As the data stand, it is impossible to determine whether metallurgical activity decreased after period II or whether the local farmers just carted off the period III and later evidence to fertilize their fields.

The dating for these periods is based on radiocarbon dates obtained from charcoal samples from many of the levels (Caldwell 1967b, 24). These dates, reported in table 2, have been recalibrated and adjusted by Voigt and Dyson (1992, 131), and they were recently run through the calibration program OxCal 4.0 (Frame 2009). From these dates it is clear that there is a large degree of overlap for the Iblis I and Iblis II dates. Based on differences in both stratigraphy and artifact assemblage, however, Caldwell argues that these are two distinct periods (Caldwell 1967b). An updated discussion of the chronology can be found elsewhere (Frame 2009), but it is summarized here in table 3.

Archaeological Materials

Every crucible fragment in the UGA-NHM collection was photographed, measured, and examined (Frame 2009). Forty-seven crucible fragments were taken back to the University of Arizona for sampling. In addition, twenty-two ore fragments, ten coarseware sherds, charcoal samples from three different contexts, and three corroded copper nuggets were also taken for sampling and analysis. A list of all samples analyzed in this project is reported in table 4. Sampling, mounting, and preparation of sample surfaces were carried out at the University of Arizona as outlined in Frame (2009). Analyses of the metals, ores, and charcoal are discussed in Frame (2009), and additional metal artifacts have been previously studied by Pigott and Lechtman (2003).

Opaque thick-sections as well as petrographic thin-sections were examined with both optical and electron beam techniques. Samples for electron microprobe analysis (EMPA) and scanning electron microscopy with energy-dispersive spectroscopy (SEM-EDS) were coated with approximately 250 Å of carbon using a DV-502A carbon evaporation coater. Microprobe analyses were performed on a Cameca Sx50 SEM-equipped electron microprobe. All analyses were conducted at 15 kV and 20 nA with a 1 mm beam size (Frame 2009). The ZAF (atomic number, absorption, and fluorescence) mathematical corrections applied to the analyses are based on single phase regions, but bulk analyses include multiple phases in the analyzed volume.

SEM and EDS analyses were performed using the Hitachi S-3400N in the University Spectroscopy and Imaging Facilities at the University of Arizona. This instrument is equipped with an UltraDry 30 Thermo Scientific EDS silicon drift detector (SDD) capable of detecting elements of atomic number 5 (Boron) and higher.

Table 4. *List of Tal-i Iblis artifacts analyzed in this study.*

Period	Sample No.	Artifact	Description
Iblis I (early)	Iblis-9-2008	sherd	coarseware ceramic (Lalehzar ware)
Iblis I (early)	Iblis-10-2008	sherd	coarseware ceramic (Lalehzar ware)
Iblis I (early)	Iblis-26-2008	slag	
Iblis I	Iblis-2-2008	orc	
Iblis I	Iblis-3-2008	crucible	rim
Iblis I	Iblis-4-2008	sherd	coarseware ceramic (Lalehzar ware)
Iblis I	Iblis-5-2008	ceramic	coarseware ceramic (Lalehzar ware)
Iblis I	Iblis-6-2008	ceramic	coarseware ceramic (Lalehzar ware)
Iblis I	Iblis-7-2008(a)	ore	
Iblis I	Iblis-7-2008(b)	ore	
Iblis I	Iblis-7-2008(c)	ore	
Iblis I	Iblis-7-2008(d)	crucible	large profile
Iblis I	Iblis-7-2008(e)	copper	prill
Iblis I	Iblis-7-2008(f)	slag	slag with ore attached
Iblis I	Iblis-7-2008(g)	crucible	rim
Iblis I	Iblis-7-2008(h)	crucible	rim
Iblis I	Iblis-7-2008(i)	crucible	base
Iblis I	Iblis-8-2008(a)	ore	
Iblis I	Iblis-8-2008(b)	crucible	base
Iblis I	Iblis-13-2008	crucible	rim
Iblis I	Iblis-15-2008	ore	
Iblis I	Iblis-18-2008	ore	
Iblis I	Iblis-20-2008	crucible	rim and slag
Iblis I	Iblis-28-2008	ore	
Iblis I	Iblis-30-2008	charcoal	
Iblis I	Iblis-32-2008	charcoal	
Iblis I (late)	Iblis-19-2008(a)	ore	
Iblis I (late)	Iblis-19-2008(b)	crucible	slag
Iblis I (late)	Iblis-29-2008	ore	
Iblis I or II	Iblis-33-2008(a)	crucible	base
Iblis I or II	Iblis-33-2008(b)	crucible	wall
Iblis II	Iblis-1-2008(a)	crucible	rim
Iblis II	Iblis-1-2008(b)	crucible	rim
Iblis II	Iblis-1-2008(c)	crucible	slag
Iblis II	Iblis-12-2008(a)	ore	dark
Iblis II	Iblis-12-2008(b)	ore	with native copper
Iblis II	Iblis-12-2008(c)	ore	
Iblis II	Iblis-12-2008(d)	sherd	coarseware ceramic (Lalehzar ware)
Iblis II	Iblis-12-2008(e)	copper	prill
Iblis II	Iblis-12-2008(f)	crucible	slumped rim
Iblis II	Iblis-12-2008(g)	crucible	rim
Iblis II	Iblis-12-2008(h)	crucible	rim
Iblis II	Iblis-12-2008(i)	crucible	base
Iblis II	Iblis-12-2008(j)	crucible	profile
Iblis II	Iblis-14-2008(a)	slag	slag with charcoal attached
Iblis II	Iblis-14-2008(b)	crucible	rim
Iblis II	Iblis-14-2008(c)	sherd	coarseware ceramic (Lalehzar ware)
Iblis II	Iblis-14-2008(d)	copper	object
Iblis II	Iblis-14-2008(f)	ore	

Period	Sample No.	Artifact	Description
Iblis II	Iblis-14-2008(g)	ore	
Iblis II	Iblis-14-2008(h)	crucible	profile
Iblis II	Iblis-14-2008(i)	crucible	rim with green stain
Iblis II	Iblis-14-2008(j)	crucible	rim
Iblis II	Iblis-14-2008(k)	crucible	rim
Iblis II	Iblis-14-2008(L)	crucible	base
Iblis II	Iblis-14-2008(m)	slag	slag with charcoal attached
Iblis II	Iblis-16-2008(a)	crucible	slag
Iblis II	Iblis-16-2008(b)	crucible	rim
Iblis II	Iblis-16-2008(c)	ore	
Iblis II	Iblis-16-2008(d)	ore	
Iblis II	Iblis-16-2008(e)	crucible	profile
Iblis II	Iblis-17-2008(a)	copper	prill
Iblis II	Iblis-17-2008(b)	crucible	unfired
Iblis II	Iblis-17-2008(c)	crucible	rim
Iblis II	Iblis-21-2008(a)	sherd	coarseware ceramic (Lalehzar ware)
Iblis II	Iblis-21-2008(b)	crucible	wall
Iblis II	Iblis-24-2008(a)	ore	
Iblis II	Iblis-24-2008(b)	sherd	low-fired ceramic chunk
Iblis II	Iblis-24-2008(c)	slag	with ore attached
Iblis II	Iblis-24-2008(d)	crucible	rim
Iblis II	Iblis-24-2008(e)	crucible	base
Iblis II	Iblis-24-2008(f)	crucible	rim
Iblis II	Iblis-24-2008(g)	crucible	base
Iblis II	Iblis-25-2008	ore	
Iblis III	Iblis-23-2008(a)	ore	
Iblis III	Iblis-23-2008(b)	crucible	rim
Iblis IV	Iblis-35-2008	copper	pin
Iblis V	Iblis-27-2008	ore	
date unknown	Iblis-11-2008(a)	slag	
date unknown	Iblis-11-2008(b)	slag	
date unknown	Iblis-34-2008	slag	

Bulk analyses were performed using one of three methods: (1) rastering over a 100 × 100 mm square region with a 1 mm electron beam using EMPA, (2) rastering over a 100 × 100 mm square region using the SEM-EDS, or (3) calculating the bulk composition using a weighted average of the compositions of the individual phases present in a 100 × 75 mm area. For this final method, the individual phases were analyzed using SEM-EDS, and each was weighted based on the area fraction of the phase. For all three methods, only areas that were free of large voids, ore chunks, prills, or other features that were not deemed to be part of the "background" slag material were considered. The totals for the bulk analyses were often less than 100% due to false assumptions in the ZAF correction statistics and the large number of voids in the ceramics and slags.

The elements examined along with the corresponding diffraction crystals, standards used for calibration, and detection limits for the microprobe analyses are reported in table 5. The microprobe was calibrated on the first morning of each session, and for sessions lasting more than one day, the calibrations were checked the following day to ensure accurate results. EDS measurements were calibrated with standards and samples previously analyzed with other methods, including EMPA.

Technological Reconstruction

The crucible fragments at UGA-NHM included rims, body/wall fragments, base fragments, entire profiles, and detached pieces of slag, and, in general, they are similar in form, fabric, and function. The variations among crucibles in the Iblis collection as well as the individual components of the crucibles have been examined in order to understand the uniformity of this early metallurgical industry and also to reconstruct the technology in place during the earliest levels of occupation at the site.

Table 5. *Microprobe standards and detection limits.*

	Element	Peak	Standard	Diffraction Crystal	Detection Limits (point)	Detection Limits (bulk)
Analysis of Metals	P	Ka	InP	TAP	0.01	0.02
	S	Ka	chalcopyrite	PET	0.03	0.03
	Pb	Ma	galena	PET	0.07	0.12
	Sb	La	stibnite	PET	0.04	0.04
	Sn	La	Sn	PET	0.05	0.07
	Ag	La	Ag	PET	0.06	nm
	Cu	Ka	Cu	LIF	0.14	0.14
	Fe	Ka	chalcopyrite	LIF	0.05	0.08
	As	Kb	As	TAP	0.05	0.11
	Zn	Ka	Zn	LIF	0.06	0.1
	Ni	Ka	Ni	LIF	0.03	0.09
Analysis of Oxides	Na	Ka	albite	TAP	0.03	0.03
	K	Ka	K-feldspar	PET	0.03	0.02
	Si	Ka	diopside	TAP	0.03	0.03
	Mg	Ka	diopside	TAP	0.03	0.03
	Al	Ka	anorthite	TAP	0.03	0.03
	Ca	Ka	diopside	PET	0.03	0.03
	Mn	Ka	rhodonite	LIF	0.05	0.06
	Fe	Ka	fayalite	LIF	0.09	0.08
	Cu	Ka	chalcopyrite	LIF	0.11	0.07
	Cl	Ka	scapolite	PET	0.02	0.02
	Ba	La	benitoite	PET	0.07	0.06
	S	Ka	barite	PET	0.03	0.03

LIF = lithium fluoride; nm = not measured; PET = pentaerythritol; TAP = thallium acid phthalate.

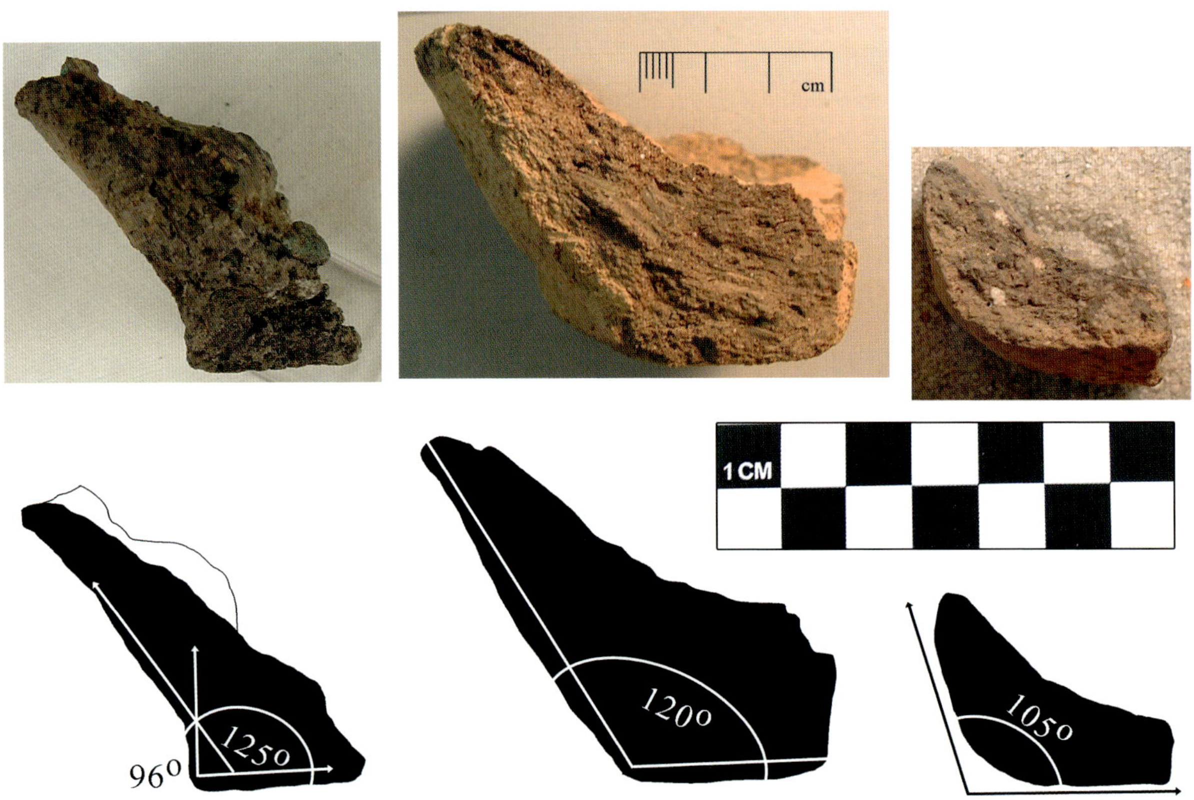

Figure 4. *Typical Iblis crucible profiles with a pinched foot (left), unpinched foot (center), and rounded foot (right). All profiles are shown at the same scale (indicated). The left and center profiles are representative of the majority of the Iblis crucibles, which have an average wall:base angle of 130°.*

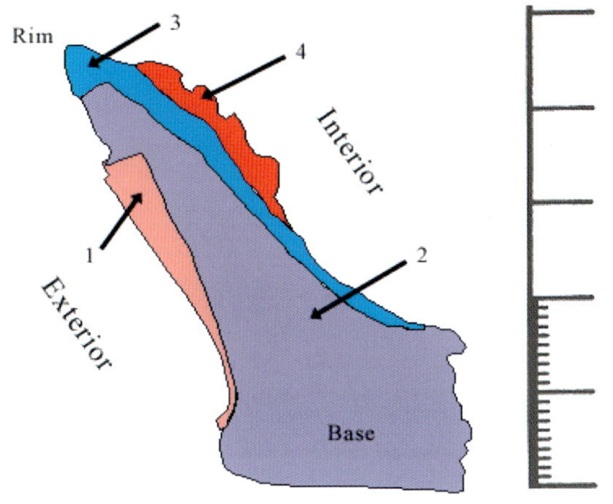

Figure 5. *Schematic drawing of the four zones identified for the crucible fabric (Frame 2004).*

Crucible Form

The Iblis crucibles are wide, shallow, bowl-like vessels ranging in height from 30 to 60 mm. Some of the crucibles exhibit a pinched foot, whereas others have a rounded base. The pinched or slightly curved foot is by far the most common. Most crucibles follow the schematic shown in figure 4 where the foot makes an 80°–100° angle with the base before curving outward to a wall that makes a 110°–150° angle with the base. Three basic crucible forms found in the Iblis collection are shown in figure 4. Such a difference in foot geometry would have little effect on the overall function of the crucible as a reaction vessel. Most pinched and nonpinched feet in the Iblis collection still contain flat bases, and even the few examples with curved bases (Iblis-12-2008(j); see fig. 4) have approximately the same wall-to-base angles and the same wall thicknesses. Because this difference in foot geometry does not greatly affect the overall function, it is possible that the variation represents different styles of production by different craftspeople.

Variations in the thickness of the walls and the geometry of the rims of these crucible fragments is minimal, but wall thickness at midway along the height of the crucible walls for 156 crucibles ranges from 8 to 20 mm (with 71% of samples falling within the 1σ range of 9.8 to 14.6 mm), and the rim styles include both thick-rounded and thin-pinched morphologies.

Crucible Fabric and Fabrication

For ease of discussion, the profile of the crucible sherds can be divided into four different zones. The characteristics of these zones have been described elsewhere (Frame 2004, 2009; Frame and Lechtman, in press), so they are only summarized here in table 6 and figure 5.

All crucible and coarseware ceramics were tempered with chaff—some variety of local grass—a technique that is common on the Iranian plateau in general (e.g., at Tepe Hissar, Tepe Ghabristan, Shar-i Sokhta). In oxidizing conditions, the organic matter burns off above 400°C, leaving behind voids in the shape of the temper. In reducing conditions, plant matter can remain in ceramics up to 800°C–1000°C (Johnson et al. 1988). Many crucibles still contain burnt remnants of the chaff, and both voids and remnant organic matter result in increased porosity, lower coefficient of thermal expansion, and better insulation (Rice 1987; Tylecote 1982). Such properties are highly desired for ceramics used in metallurgy.

The voids left behind by the chaff temper varied in size between 1 and 5 mm long in the Iblis crucibles. Most were aligned in the direction of the wall (pointing from base to rim), and these voids provide an indication of how the crucibles were fabricated. These crucibles were constructed by pinching to form the walls and occasionally adding additional clay to increase the height of the walls (Frame 2009). In addition, the flat base on these crucibles could be formed by constructing the crucible on a flat surface. The pinched foot, which appears on many crucible sherds, can be made by pressing inward at the base of the crucible, and this is consistent with the material evidence, since finger marks are visible at the lower exterior walls of the crucibles but not on the flat bases. The resulting "pinch-and-pull" method of construction would allow these crucibles to be constructed quickly.

The chaff temper, volume fraction porosity, mineral inclusions, and bulk chemistry of the crucible sherds from Tal-i Iblis were compared to the characteristics of the locally made Lalehzar coarseware, as shown in table 7 (Frame 2009). These similarities support the conclusion that the crucible fabrication methods were closely related to the existing ceramic technology. The same clay sources were used, and the same organic and mineral temper was added. This evidence ties metallurgy at Tal-i Iblis to another pyro-

Table 6. *Description of thermal regions of Iblis crucibles as shown in figure 5.*

Zone	Location	Fabric	Voids and Inclusions
1	exterior of the used crucibles	soft friable oxidized reddish/pink ceramic	chaff temper and elongated voids, full range of mineral inclusions present
2	core	friable reduced dark gray	chaff-shaped voids, occasional chaff temper, carbonate minerals not present
3	interior bottom surface, crucible rim, and between zone 2 and zone 4 on the upper crucible wall	slightly vitrified showing evidence of sintering in the ceramic fabric, and the fabric is even glassy (clear in plane-polarized transmitted light) in some of the sherds	pores range from slightly rounded to completely spherical in shape, and slight to moderate bloating of the ceramic material has caused increased volume fraction of pores for the higher temperature examples
4	interior of the crucibles along upper wall, not present in every crucible sherd	"crucible slag" has a denser appearance than zone 3; often black with green stains or ore fragments adhering to it; often contains prills of copper-base metal	few voids, inclusions represent crucible charge during metallurgical operation rather than the original mineral components of the ceramic

Table 7. *Chemical and petrographic characteristics of the ceramic fabrics of the Iblis crucibles and coarseware sherds.*

Artifact	Type	Volume Fraction Porosity	Mineral Assemblage		Bulk Composition								
			Major	Minor	Na_2O	BaO	K_2O	SiO_2	MgO	Al_2O_3	CaO	FeO	TiO_2
Iblis-1-2008(a)	crucible	0.29	weathered basalt (0.5 mm) quartz and plagioclase (<0.25 mm)	biotite tourmaline (0.4 x 0.1 mm)	2.08	0.61	2.12	57.70	3.76	18.49	7.62	7.62	0.00
Iblis-1-2008(b)	crucible	0.22	quartz and plagioclase (rounded and sub-angular), large basalt with 1 mm zoned feldspar and pyroxene augite	biotite	2.29	0.64	2.56	58.16	3.68	17.98	7.34	7.35	0.00
Iblis-3-2008	crucible	0.22	quartz and plagioclase (0.1 mm and around 25 μm), weathered basalt (up to 250 μm, mean: 100 μm)	biotite pyroxene	1.87	0.60	2.60	56.47	3.87	18.94	7.57	8.08	0.00
Iblis-6-2008	coarseware	0.22	plagioclase with few quartz (25–100 μm) weathered basalt (100–150 μm)	biotite (50–100 μm) pyroxene (augite)	1.78	0.64	2.91	56.22	4.08	19.39	7.11	7.88	0.00
Iblis-7-2008(d)	crucible	0.15	plagioclase and quartz (250+ μm or <100 μm), weathered basalt (0.3–1.5 mm)	biotite hornblende augite	2.00	0.60	2.66	62.14	2.96	16.73	6.68	6.22	0.00
Iblis-7-2008(g)	crucible	0.17	plagioclase (0.5 mm or <50 μm)	quartz biotite (0.5 mm)	2.24	0.67	2.38	55.27	3.60	19.52	8.25	8.07	0.00
Iblis-7-2008(i)	crucible	0.25	plagioclase (100–200 μm or <25 μm) basalt (100–500 μm) quartz	biotite hornblende	1.68	0.58	2.79	57.06	4.21	19.08	6.22	8.38	0.00
Iblis-8-2008(b)	crucible	0.09	plagioclase and quartz (75–150 μm or <50 μm) basalt (200–500 μm)	biotite (1 wt.%) hornblende calcite/ankerite at corner of base (in zone 1)	1.71	0.58	2.11	57.95	3.96	17.28	8.95	7.46	0.00
Iblis-12-2008(J)	crucible	0.22	plagioclase and few quartz (200–500 μm or <100 μm) basalt (200 μm)	biotite hornblende	1.89	0.00	3.56	57.65	2.94	17.97	6.15	8.94	0.90
Iblis-14-2008(J)	crucible	0.20	plagioclase and few quartz (100–200 μm or 50 μm) basalt (300–500 μm)	biotite hornblende	1.91	0.00	2.96	58.07	2.72	17.08	8.51	8.03	0.72
Iblis-16-2008(a)	crucible	NA	plagioclase (200 μm or <100 μm) basalt (400 μm)	hornblende (50 μm) (no biotite)	1.72	0.16	2.35	56.07	3.77	18.10	8.82	8.01	1.01
Iblis-16-2008(b)	crucible	0.24	plagioclase (<75 μm or 200–500 μm) weathered basalt (400 μm)	hornblende (50 μm) (no biotite)	1.63	0.11	2.17	56.22	3.28	17.66	9.18	8.87	0.89
Iblis-16-2008(e)	crucible	0.15	plagioclase and few quartz (<75 μm or 200–500 μm) basalt (200–500 μm)	biotite hornblende	1.89	0.00	1.98	55.55	2.89	16.61	12.76	7.54	0.77
Iblis-17-2008(b)	crucible	0.07	plagioclase (mostly 50 μm) basalt	biotite (common and up to 100 μm) hornblende (up to 200 μm)	2.02	0.16	2.87	58.64	2.81	17.22	6.41	9.03	0.82
Iblis-21-2008(a)	coarseware	0.21	plagioclase (200–500 μm) basalt (500 μm)	biotite and hornblende (<50 μm)	1.63	0.09	2.80	59.16	3.37	17.71	6.68	7.84	0.73
Iblis-21-2008(b)	crucible	0.22	plagioclase and few quartz (<75 μm or 200–700 μm) basalt (up to 1 mm)	biotite hornblende	2.13	0.02	2.49	60.46	3.20	15.73	8.16	7.22	0.59
Iblis-33-2008(a)	crucible	0.19	plagioclase and few quartz (<50 μm or 100–200 μm or 500 μm) basalt (100 μm up to 2 mm)	biotite hornblende calcite/ ankerite at corner of base (in zone 2)	1.67	0.28	3.22	58.67	3.15	17.52	6.71	8.18	0.61

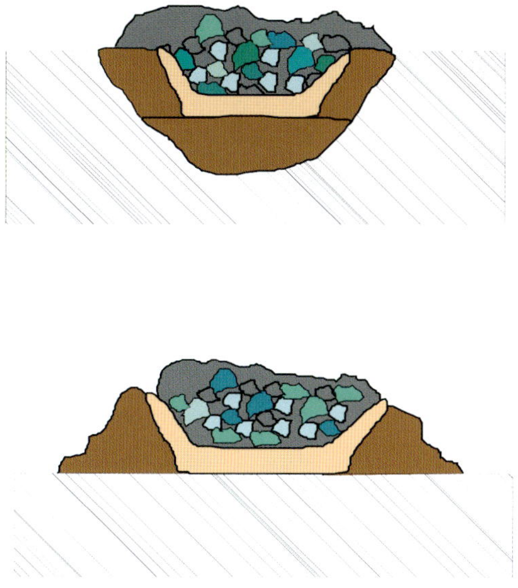

Figure 6. *Schematic drawing for the smelting setup showing **(a)** pit smelting with the crucible and **(b)** the crucible used on a flat surface.*

technology known to these people at the time: their ceramic technology.

Crucible Function

In addition to aiding the thermal properties, both voids and mineral inclusions can act as stress concentration centers to cause fracture in the crucible walls, especially at high temperatures. Lack of strength was most likely compensated by structurally supporting the crucibles as they were heated.

Such support could be provided by placing the crucible in a shallow, flat-bottomed pit or by surrounding the crucible with a small mound of earth (fig. 6). Using the crucibles on a flat surface with a small mound of dirt for support (see fig. 6b) would require less preparation, but there is limited evidence for either method at Tal-i Iblis. One "fire pit" is mentioned by Evett (1967) in the area G house, and this pit contained a few crucible fragments. However, this pit was located in the middle of a fill and not directly associated with any floor, prepared surface, or work area.

Regardless of which of these two setups was used, all of the crucibles were necessarily heated from above/inside. The profile of every single crucible sherd from the Iblis collection at UGA-NHM shows that the interior reached a higher temperature than the exterior. In addition, the rims and upper walls are commonly more vitrified and more bloated than the lower wall and interior base. Often, the interior base contained no slag and only a very thin layer of the zone 3 fabric. This suggests that the highest temperatures within the crucible occurred at the mid- to upper wall. An examination of the adhering crucible slags has allowed estimates of charge, temperature, and atmosphere for these operations.

Crucible Slags

The Iblis slags were grouped into four types based on their compositions and microstructures. The characteristics of these groups are summarized in table 8, and the chemical and microstructural features of each crucible slag are discussed at length in Frame (2009). The crucibles include calcium alumino-silicate glassy slag, barium-rich slag, aegerine-augite containing slag, and a slag high in iron and magnesium.

The compositions of the crucible slags are highly variable even within a single crucible. Bulk compositions are reported in table 9, and it is clear that sample location can severely impact the analytical results obtained. For example, Iblis-1-2008(a) shows a copper content of 10.39 wt.% CuO at the mid-upper wall 500 mm from a large copper prill, but less than 2 mm away, the concentration is only 2.2 wt.% CuO. Despite these intra-crucible variabilities, a few inter-crucible trends are also apparent. Some of the crucibles are significantly higher in lead, barium, or iron. This likely reflects the choice of ores used for the smelting operations. The alkali and alkaline earth contents are uniformly variable.

In addition to the differences in slag composition, different morphologies of crystalline phases commonly occur within and between crucible slags. The main factors influencing the microstructure of the slags are composition, temperature of the smelt, and the cooling rate. Within Iblis-7-2008(e), for example, near the interior surface of the slag the aegirine-augite crystals form as long acicular needles branching out into the glass from particles acting as nucleation centers (fig. 7a). The high temperatures reached in this region allowed easy growth of the needles. At a location only 2 mm away, the temperature was cooler, and the growth of these crystals was stunted as the glassy phase froze around them. The result is the short curly crystals seen in figure 7b. Similarly, in the crucible slag for Iblis-14-2008(m), aegirine-augite appears as both acicular and dendritic crystals (fig.

Table 8. *Descriptions of Iblis slag categories.*

Classification	Key Components	Description
glassy slag	copper prills, copper chloride veins, and very little else	glassy slag with low iron and magnesium
barium-rich slag	either a high-Ba glassy phase or a Ba-rich crystalline phase	also often contain copper silicate crystals (frequently dioptase) and magnetite crystals
aegerine-augite slags	aegerine-augite crystals surrounded by a calcium alumino-silicate glassy phase	largest group of slags in the collection; rich in Mg or Fe and Ca or Na; contain numerous copper prills ranging in composition from pure copper to copper-iron alloys and copper-arsenic alloys; glassy phase is commonly enriched in Cu and Fe
high Fe and Mg slags	fayalite crystals, hypersthene	contain various Fe-pyroxenes, but crystal morphology largely similar to the aegerine-augite slags

Table 9. *Bulk compositions of Iblis crucible slags.*

Crucible No.	Period	Analysis Locations	Location Details	Na_2O	BaO	K_2O	SiO_2	MgO	Al_2O_3	CaO	SO_4	CuO	FeO	MnO	PbO	Total
Iblis-7-2008(d)	Iblis I	upper wall	upper extent of slagged region	0.24	0.17	0.06	39.17	0.08	21.68	17.33	0.08	2.15	10.92	0.24	0.00	92.12
	Iblis I	mid-lower wall	middle of slagged region, adjacent to charcoal chunk (1 mm away) and undecomposed ore (250 µm away)	0.44	0.57	0.37	36.61	0.50	22.03	16.21	0.24	3.91	9.83	0.00	0.00	90.72
Iblis-7-2008(e)	Iblis I	detached slag chunk	at edge, heterogeneous slag with large crystals	0.79	0.13	1.44	45.50	0.87	2.00	11.10	0.07	2.49	17.75	0.09	0.00	82.23
	Iblis I	detached slag chunk	at edge, heterogeneous slag with small crystals	0.87	0.18	1.17	44.64	0.94	1.73	11.85	0.16	4.90	18.19	0.00	0.00	84.63
	Iblis I	detached slag chunk	interior, heterogeneous slag with small crystals	0.69	0.20	1.30	45.52	0.95	1.84	11.45	0.06	3.05	19.19	0.00	0.00	84.25
Iblis-7-2008(g)	Iblis I	mid-wall	vitrified interior wall surface	2.91	1.32	2.63	51.88	3.22	18.67	7.88	0.05	0.00	6.11	0.22	0.00	94.88
	Iblis I	mid-wall	vitrified interior wall surface	2.03	1.20	2.18	51.45	3.14	17.81	8.12	0.08	0.00	6.12	0.20	0.00	92.33
Iblis-7-2008(i)	Iblis I	mid-lower wall	vitrified interior wall surface just past slagged region	3.12	0.90	6.35	59.59	0.35	15.48	3.69	0.32	0.63	2.25	0.00	0.00	92.66
	Iblis I	mid-lower wall	under green crust	1.53	0.51	0.93	33.48	2.97	11.37	6.61	2.44	23.36	3.25	0.13	0.00	86.56
Iblis-8-2008(b)	Iblis I	mid-lower wall	lower extent of slagged region	0.78	1.44	0.54	38.19	6.62	12.91	19.03	0.11	1.22	11.87	0.22	0.00	92.92
	Iblis I	mid-lower wall	middle of slagged region	0.76	1.33	0.37	39.75	4.77	18.46	18.21	0.26	2.05	8.42	0.16	0.00	94.54
	Iblis I	upper wall	upper extent of slagged region	1.36	1.57	0.90	38.71	6.71	14.79	17.91	0.29	2.45	8.61	0.18	0.00	93.47
Iblis-20-2008	Iblis I	rim	slag near vitrified rim	1.29	0.28	2.78	51.34	1.53	13.88	11.65	0.00	8.14	7.80	0.12	0.02	98.83
Iblis-3-2008	Iblis I or II	rim tip	green crust/slag at rim tip	0.96	0.76	2.75	41.04	2.64	12.64	7.03	0.91	9.88	4.56	0.00	0.00	83.17
	Iblis I or II	rim	slag/vitrified ceramic at 500 µm into wall profile	2.58	0.63	4.85	58.06	1.33	16.45	4.57	0.33	1.21	2.84	0.00	0.00	92.84
Iblis-33-2008(a)	Iblis I or II	mid wall	surface of thin slag crust	1.23	1.02	0.69	47.06	2.15	4.67	16.99	0.04	8.68	14.48	0.21	0.00	97.21
Iblis-33-2008(b)	Iblis I or II	mid-lower wall	slag adjacent to ore gangue and vitrified ceramic	1.73	2.19	1.44	48.70	2.79	10.76	19.83	0.00	0.79	10.51	0.18	0.00	98.92
Iblis-1-2008(a)	Iblis II	rim tip	vitrified rim (no slag)	2.15	1.10	2.25	57.44	3.63	15.83	6.33	0.24	0.00	6.15	0.18	0.00	95.30
	Iblis II	mid-upper wall	upper extent of slagged region	0.82	0.59	1.44	42.19	1.53	8.78	9.37	0.46	2.20	4.35	0.13	0.00	71.86
	Iblis II	mid-upper wall	slag, 500 µm from large prill	0.39	0.56	0.73	42.26	2.44	6.02	6.88	1.29	10.39	4.03	0.11	0.00	75.09
	Iblis II	mid-lower wall	slag, near ore gangue	0.33	0.74	0.98	27.42	1.23	4.77	1.47	5.81	3.12	10.24	0.00	0.00	56.11
	Iblis II	mid-lower wall	slag, near undecomposed ore	1.60	0.77	1.29	47.09	2.26	12.52	14.88	0.06	4.22	4.82	0.10	0.00	89.61
Iblis-1-2008(b)	Iblis II	mid-wall	vitrified interior wall (no slag)	3.36	0.97	3.49	52.16	2.80	17.35	6.99	0.45	0.35	4.62	0.00	0.00	92.53
	Iblis II	mid-wall	vitrified interior wall (no slag)	2.86	5.92	2.99	46.86	5.53	16.54	4.80	0.35	0.36	9.47	0.35	0.00	96.03
	Iblis II	mid-wall	vitrified interior wall (no slag)	3.28	1.03	0.26	50.01	3.68	16.18	7.43	0.36	0.36	5.74	0.14	0.00	88.47

Iblis-12-2008(f)	Iblis II	rim	slag flowing over rim, surrounded by ore gangue	2.24	5.39	2.08	33.85	1.77	4.89	13.26	0.70	9.45	15.86	0.14	2.73	92.36
	Iblis II	mid-upper wall	slag at upper wall, middle of slagged region	0.74	7.14	2.91	37.04	0.88	7.34	1.35	0.01	14.27	12.69	0.07	0.37	84.81
Iblis-12-2008(g)	Iblis II	mid-wall	vitrified crucible at interior surface, on protrusion	3.55	0.10	1.02	49.20	5.45	17.47	7.60	0.17	0.22	6.88	0.17	0.46	92.31
	Iblis II	mid-wall	vitrified crucible at interior surface	3.63	0.10	3.07	47.54	4.55	13.46	13.30	0.30	0.57	6.71	0.16	0.12	93.51
Iblis-14-2008(a)	Iblis II	detached slag chunk	homogeneous slag surrounded by ore and prills	0.32	0.56	0.99	33.91	2.10	3.84	11.29	0.00	3.06	35.15	0.10	2.79	94.11
Iblis-14-2008(b)	Iblis II	rim tip	upper extent of slagged region (250 µm from ceramic interface and 250 µm from large copper sulfide prill)	1.77	0.16	1.17	40.98	2.59	7.05	23.86	0.00	0.38	18.23	0.25	0.00	96.44
	Iblis II	upper wall	slag area near surface	0.98	0.18	0.35	41.51	2.18	4.87	23.04	0.17	1.87	22.16	0.42	0.00	97.72
Iblis-14-2008(J)	Iblis II	rim	vitrified rim 500 µm below surface	2.00	0.00	2.59	57.87	2.84	17.04	8.43	0.00	0.00	7.74	0.22	0.00	98.73
	Iblis II	mid-lower wall	lower extent of slag crust (200 µm thick)	4.06	0.00	4.35	48.87	3.41	14.57	16.33	0.00	0.29	7.04	0.14	0.00	99.06
Iblis-14-2008(m)	Iblis II	detached slag chunk	homogeneous slag surrounded by ore and prills	1.10	2.81	1.20	40.20	2.37	4.08	20.04	0.00	5.29	19.50	0.10	1.02	97.71
Iblis-16-2008(a)	Iblis II	rim	slag at rim on interior surface	0.66	0.36	0.99	41.34	2.29	5.41	17.65	0.00	2.03	19.62	0.20	3.11	93.67
	Iblis II	mid-wall	slag 750 µm from ceramic interface (4 mm from interior surface) adjacent to large (3 × 4 mm) ore gangue chunk	0.04	0.06	0.12	39.26	2.12	1.40	24.80	0.00	1.31	19.92	0.23	0.00	89.25
Iblis-16-2008(b)	Iblis II	mid-upper wall	slag near ceramic interface	2.25	0.00	1.96	44.98	3.35	15.59	7.28	0.00	3.70	8.97	0.00	11.73	99.81
	Iblis II	mid-lower wall	vitrified region below lower extent of slagged region	1.85	0.00	2.26	56.16	3.31	17.34	8.61	0.09	0.00	9.03	0.00	0.00	98.65
Iblis-16-2008(e)	Iblis II	mid-upper wall	thin slag layer at interior surface	2.76	0.23	1.55	46.28	1.65	11.38	21.60	0.00	2.22	10.36	0.20	0.52	98.75
	Iblis II	mid-lower wall	thin slag layer at interior surface	1.64	0.49	0.94	41.78	1.71	10.30	21.30	0.18	1.55	13.77	0.23	2.17	96.05
Iblis-21-2008(b)	Iblis II	mid-wall	shin slag layer beneath green crust	4.21	0.00	0.56	48.17	1.67	20.74	7.71	0.00	0.11	6.51	0.15	0.92	89.84
Iblis-24-2008(d)	Iblis II	mid-lower wall	center of 6 ×1.5 mm slag region on interior surface	1.58	0.17	2.45	31.78	2.17	10.14	26.24	0.02	3.54	14.09	0.26	0.11	92.45

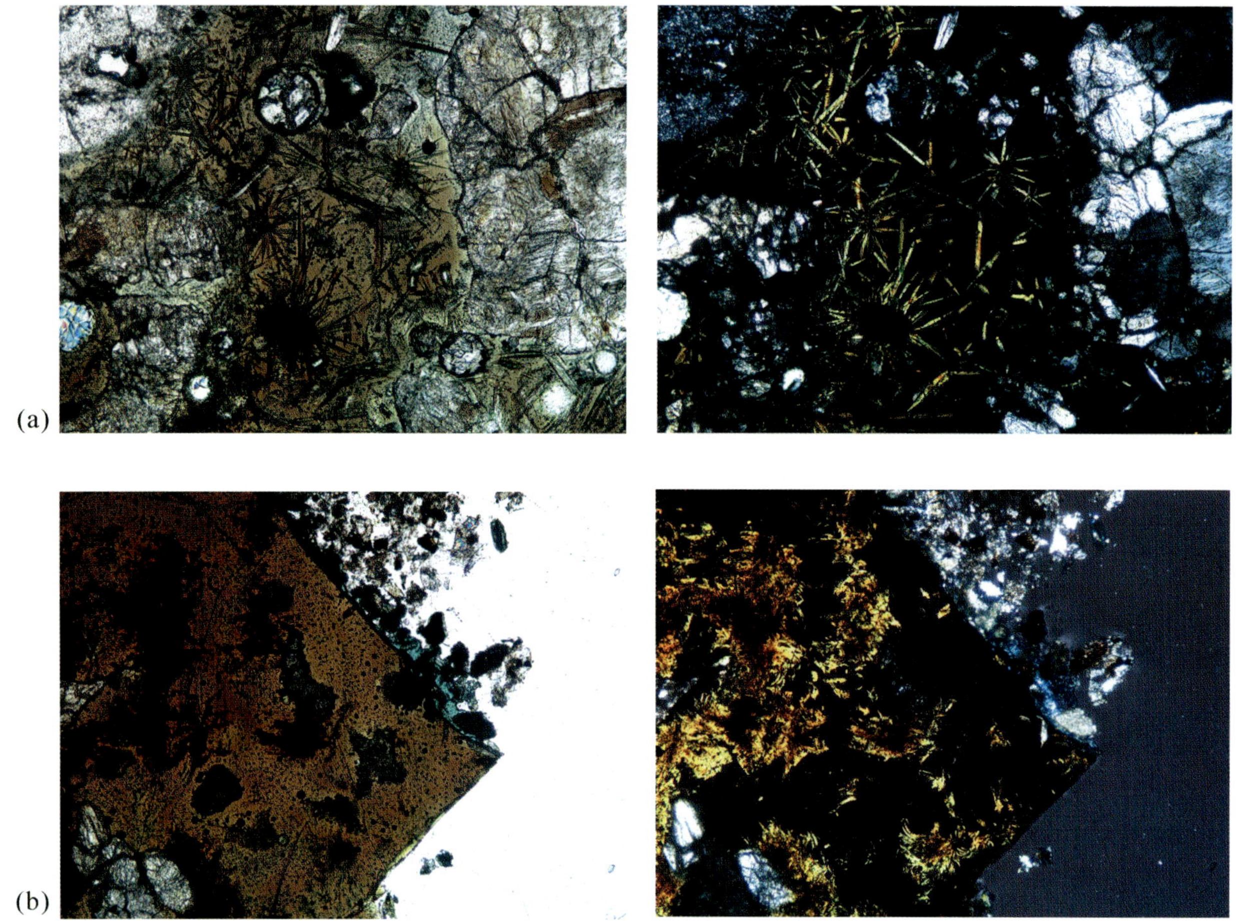

Figure 7. *Iblis-7-2008(e) crucible slag in transmitted plane-polarized light (left images) and cross-polarized light (right images) at two locations: **(a)** shows the slag microstructure at 2 mm from the interior surface of the crucible and **(b)** shows the microstructure at the interior surface of the crucible. The slag appears brown in the plane-polarized light (left image). Each image width: 1.2 mm.*

8). The feathery-looking dendrites are due to undercooling. As the liquid phase cooled below the temperature at which aegirine-augite can form, local instabilities formed near the nucleation centers. As long as the glassy phase remained liquid, aegirine-augite was able to solidify from the melt to form fine dendrites. In addition to showing the temperature gradients present within the crucibles, these microstructures suggest that the melting temperature for the glassy phase of the slag adhered to Iblis-7-2008(e) is slightly higher than that of the glassy phase of the Iblis-14-2008(m) slag. The compositions of each of these phases appear to support this hypothesis given the slightly higher number of network modifiers present in the Iblis-14-2008(m) glassy phase (1.6 wt.% Na, 1.6 wt.% Mg, 1.5 wt.% K, 13.9 wt.% Ca, 10.9 wt.% Fe, 9.2 wt.% Cu, and 4.1 wt.% Ba) compared to the glassy phase of Iblis-7-2008(e) (0.8 wt.% Na, 0.3 wt.% Mg, 1.2 wt.% K, 7.0 wt.% Ca, 12.3 wt.% Fe, 3.6 wt.% Cu, and 0.2 wt.% Ba). Generally speaking, more network modifiers result in a lower melting temperature.

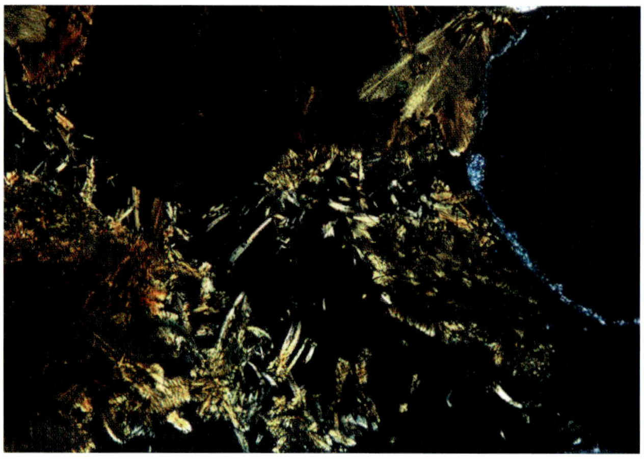

Figure 8. *Iblis-14-2008(m) crucible slag in transmitted cross-polarized light showing aegirine-augite as acicular crystals and as feathery dendrites. Image width: 1.2 mm.*

Another common feature in these crucible slags was partially decomposed ore chunks. These often appeared as bundles of iron oxide particles in areas replete with copper prills. A few crucible slags also include chunks of basalt and plagioclase decomposing into the surrounding glassy matrix, as is the case with Iblis-14-2008(b). It is unclear whether these chunks of basalt and plagioclase originated from the crucible wall or were added with the ore—either as gangue or as a flux—but the result of their addition to the slag was a lower melting temperature due to the presence of additional network modifiers (Na, K, Ca, Fe, and Mg).

The chunks of adhered gangue and the ore fragments associated with the crucible finds indicate raw material selection and sources. The charcoal held in cavities of the slag and also found scattered about the sites indicates the choice of fuel for these operations, and the compositions of prills embedded within the slaggy matrix provide an idea of the range of metal compositions produced in these crucibles. Ore and fuel selection are discussed at greater length in Frame (2009).

The prill compositions (table 10) are important in that they provide an idea of the kinds of alloys being smelted. In general, arsenic is uncommon in Iblis period I crucible prills. It does occur as high as almost 2 wt.% in Iblis-7-2008(f), but the only other prills containing arsenic from this period are found in Iblis-7-2008(e). This changes for period II, for which four of the ten crucibles analyzed contained significant (over 1 wt.%) amounts of arsenic. In Iblis-16-2008(b), this amount approaches 5 wt.% As. These results reveal a shift to a new alloy type. This shift to arsenic-rich sources is corroborated by the arsenic-rich ores from period II contexts (Frame 2009).

Several of the Iblis crucibles exhibited vitrified surfaces with no slag present. A comparison of the bulk compositions of the ceramic and vitrified regions (see tables 7 and 9) shows that the interior surfaces are merely melted (cf. Iblis-1-2008(a), Iblis-7-2008(g), and Iblis-14-2008(J)). The microstructures for these vitrified layers also contain homogeneous glassy regions speckled with a few alumino silicate grains, fine dispersions of iron oxide, and occasional copper prills or veins of copper chloride. Copper prills are commonly incorporated into the surfaces of melting crucibles when the melting temperature of the ceramic is near the melting temperature of the metal (Rehren 2009).

Atmosphere

The atmosphere within the crucible during operation is a key factor in the ability to smelt copper ore. Smelting copper between 1000°C and 1200°C requires an atmosphere in the range of $Po_2=10^{-8}$ to 10^{-6} (Frame 2009). We can estimate the atmosphere in the crucibles during operation by examining the mineral phases preserved in the crucible slags. These indicate both temperature and redox conditions. Two highly useful mineral phases include cuprite and delafossite. These two phases in particular are good indicators of weak reducing atmospheres since both can only form in the presence of oxygen at the expense of the iron silicate minerals. Cuprite will form at 1100°C in Po_2 of 10^{-5} or greater (Neumann et al. 1990). The absence of delafossite and cuprite in the majority of the Iblis crucibles indicates that the atmosphere for these smelts was highly reducing and well controlled.

The presence of iron silicates is another useful indicator of a reducing environment. With increasing oxidizing conditions, these silicates will decompose to form copper-iron oxides (Hauptmann 2007, 170). Most of the Iblis crucibles contain the iron-magnesium-silicate aegirine-augite rather than copper-iron oxides like delafossite. Furthermore, the presence of sulfide minerals is a good indication of a reducing atmosphere. In a reducing atmosphere, copper sulfide ores like covellite and chalcocite can melt, but they will not decompose. Such minerals require an oxidizing atmosphere to drive off sulfur as SO_2. Once the sulfur has combined with oxygen, it is possible to reduce the copper oxides. It is possible to smelt copper sulfide ores in a reducing atmosphere if there is sufficient oxygen supplied by the oxide and carbonate ores. Such a process is a co-smelting operation and has been discussed by Rostoker et al. (1989) and also by Doonan (1994). The presence of iron silicates and copper sulfides and the absence of copper oxides indicate that these crucibles saw well-controlled reducing conditions.

Temperature

The temperatures reached inside the crucible during operation can be determined based on the structure and compositions of both the crucible fabric and the crucible slag. A series of refiring experiments (Frame 2004) on a crucible replica constructed by Hildegard Wulff at Tal-i Iblis during the 1967 season of the Caldwell excavations at the site (Caldwell 1967b, 35) established that the fabric of the crucible begins to vitrify at 1150°C and the voids become spherical at 1175°C (fig. 9). The fabric is completely glassy and has slumped at 1200°C. The spherical voids in zone 3 for most Iblis crucibles suggest that this region reached at least 1150°C and the upper wall and rim region was closer to 1175°C for the crucibles.

In addition to this proxy study, samples from three of the Iblis crucibles of approximately 3 × 3 × 3 mm were taken for refiring experiments (Iblis-1-2008(c), Iblis-12-2008(f), and Iblis-14-2008(m)). These samples were heated in air in an electric furnace to 800°C to 1200°C for 15 minutes at each temperature. There was no change in the macrostructure of these samples until 1050°C. At this temperature, all of the slag samples showed signs of rounding at the edges (fig. 10) and both Iblis-1-2008(c) and Iblis-14-2008(m) were fused to the refractory block on which they were heated. Table 11 summarizes the morphological changes that took place during firing. These refiring experiments correlate well with the predictions made using other methods (described below), and they demonstrate the uniformity in the melting temperatures for at least three of the Iblis crucible slags.

The minerals in the body and slag regions of the crucibles were used as temperature indicators. The presence of organic matter is a poor indicator of temperature because, although it will oxidize at 500°C (Rice 1987, 103), when fired in a reducing environment, this carbon matter cannot oxidize and it will remain as charred material at much higher temperatures. The presence of remnant organic matter in the Iblis crucibles—even close to zone 3—does not mean that the temperature at the edge of zone 3 was lower than 500°C. It means that, if these crucibles were prefired in oxidizing conditions before their use as reaction vessels, they were not fired above 500°C. It is most likely that they were only heated at low temperatures to completely dry the clay before using them for smelting.

Carbonate minerals can be used to determine maximum firing temperatures in ceramics, and such minerals were identified in Iblis-8-2008(b), Iblis-33-2008(a), and Iblis-24-2008(e). Unfortunately calcite ($CaCO_3$), dolomite ($CaMg(CO_3)_2$), and ankerite ($CaFe(CO_3)_2$) are all uniaxial

Table 10. *Composition of prills in Iblis crucible slags.*

Sample	Period	Method	Analysis Locations	Location Details	Prill Diameter (μm)	S	Fe	Cu	As	Sb	Ag	Total
Iblis-7-2008(d)	Iblis I	EMPA	upper wall	upper extent of slagged region	100	22.32	0.40	76.79	bdl	bdl	bdl	99.51
Iblis-7-2008(d)	Iblis I	EMPA	upper wall	upper extent of slagged region	5	4.79	2.18	86.52	bdl	bdl	bdl	93.49
Iblis-7-2008(e)	Iblis I	EMPA	detached slag chunk	at edge, heterogeneous slag with large crystals	10	1.34	1.52	94.07	0.07	0.21	bdl	97.22
Iblis-7-2008(e)	Iblis I	EMPA	detached slag chunk	at edge, heterogeneous slag with large crystals	20	3.55	0.81	92.54	0.17	0.14	bdl	97.20
Iblis-7-2008(e)	Iblis I	EDS	detached slag chunk	at edge, heterogeneous slag with large crystals	2	0.44	1.86	96.13	bdl	bdl	bdl	98.43
Iblis-7-2008(e)	Iblis I	EMPA	detached slag chunk	at edge, heterogeneous slag with small crystals	20	1.11	0.84	96.65	0.45	0.53	bdl	99.59
Iblis-7-2008(e)	Iblis I	EMPA	detached slag chunk	at edge, heterogeneous slag with small crystals	15	3.26	0.92	92.07	0.09	0.10	bdl	96.43
Iblis-7-2008(e)	Iblis I	EDS	detached slag chunk	at edge, heterogeneous slag with small crystals	5	0.38	1.66	97.74	bdl	bdl	bdl	99.78
Iblis-7-2008(e)	Iblis I	EMPA	detached slag chunk	interior, heterogeneous slag with small crystals	10	0.36	1.26	97.73	bdl	bdl	bdl	99.35
Iblis-7-2008(e)	Iblis I	EMPA	detached slag chunk	interior, heterogeneous slag with small crystals	10	0.62	1.22	97.23	bdl	bdl	bdl	99.07
Iblis-7-2008(f)	Iblis I	EMPA	detached slag chunk	large prill near mineral inclusion		13.31	0.05	82.84	0.20	bdl	0.19	96.57
Iblis-7-2008(f)	Iblis I	EMPA	detached slag chunk	prill 1	5	0.00	1.18	96.83	1.89	0.03	0.04	99.96
Iblis-7-2008(f)	Iblis I	EMPA	detached slag chunk	prill 2	10	0.00	1.45	98.11	0.00	0.06	0.05	99.67
Iblis-7-2008(f)	Iblis I	EMPA	detached slag chunk	prill 3	7	0.01	1.60	98.11	0.00	bdl	0.05	99.77
Iblis-7-2008(f)	Iblis I	EMPA	detached slag chunk	prill 4	40	0.02	0.41	100.58	0.86	bdl	0.24	102.11
Iblis-7-2008(f)	Iblis I	EMPA	detached slag chunk	prill 5	5	0.01	1.93	96.75	0.00	bdl	bdl	98.70
Iblis-7-2008(g)	Iblis I	EMPA	upper wall	halo around void in green crust (not in slag)		25.58	3.63	67.24	0.00	bdl	0.02	96.47
Iblis-7-2008(i)	Iblis I	EMPA	mid-lower wall	vitrified interior wall surface just past slagged region		18.87	3.21	57.59	bdl	bdl	0.16	79.83
Iblis-7-2008(i)	Iblis I	EMPA	mid-lower wall	vitrified interior wall surface just past slagged region		17.44	4.55	57.06	bdl	bdl	0.28	79.33
Iblis-8-2008(b)	Iblis I	EMPA	upper wall	prill within large slag region	20	0.10	2.07	97.42	bdl	bdl	0.61	100.21
Iblis-8-2008(b)	Iblis I	EMPA	upper wall	prill within large slag region	40	0.24	1.44	97.98	bdl	0.14	0.62	100.41
Iblis-8-2008(b)	Iblis I	EMPA	upper wall	prill within large slag region	20	0.39	1.09	97.88	bdl	bdl	0.50	99.85
Iblis-8-2008(b)	Iblis I	EDS	rim	upper extent of slagged region	2	bdl	1.25	94.45	1.01	bdl	bdl	96.71
Iblis-20-2008	Iblis I	EDS	mid-upper wall	in large slag region	10	bdl	1.97	96.62	bdl	bdl	bdl	98.59
Iblis-20-2008	Iblis I	EDS	mid-upper wall	in large slag region	15	0.13	1.58	97.49	bdl	bdl	bdl	99.20
Iblis-20-2008	Iblis I	EDS	mid-upper wall	in large slag region	>100	0.39	0.41	97.82	bdl	bdl	bdl	98.61
Iblis-20-2008	Iblis I	EDS	mid-upper wall	in large slag region (sulfide inclusion of 100 μm prill)		16.63	0.33	76.21	bdl	bdl	bdl	93.17
Iblis-20-2008	Iblis I	EDS	mid-upper wall	in large slag region (sulfide inclusion of 100 μm prill)		16.53	bdl	79.65	bdl	bdl	bdl	96.18
Iblis-33-2008(a)	Iblis I or Iblis II	EDS	mid-wall	surface of thin slag crust	10	bdl	1.39	96.18	bdl	bdl	bdl	97.57
Iblis-33-2008(a)	Iblis I or Iblis II	EDS	mid-wall	surface of thin slag crust	15	0.34	1.65	97.80	bdl	bdl	bdl	99.79
Iblis-1-2008(a)	Iblis II	EMPA	mid-upper wall		40	0.23	bdl	97.29	bdl	bdl	bdl	97.52

Iblis-1-2008(b)	Iblis II	EMPA	mid-wall	vitrified interior wall (no slag)		27.31	16.80	44.91	1.30	0.01	0.01	90.35
Iblis-1-2008(c)	Iblis II	EMPA	detached slag chunk	prill 1		0.01	0.20	85.35	0.00	0.01	0.02	85.59
Iblis-1-2008(c)	Iblis II	EMPA	detached slag chunk	prill 2	2	0.02	1.04	97.40	2.33	bdl	0.06	100.86
Iblis-1-2008(c)	Iblis II	EMPA	detached slag chunk	prill 3	5	0.00	1.08	97.45	3.22	bdl	0.02	101.77
Iblis-1-2008(c)	Iblis II	EMPA	detached slag chunk	prill 4	5	0.04	1.08	97.37	0.29	0.04	0.03	98.84
Iblis-12-2008(g)	Iblis II	EDS	mid-wall	vitrified crucible at interior surface	5	22.40	2.73	74.68	bdl	bdl	bdl	99.81
Iblis-14-2008(J)	Iblis II	EDS	mid-lower wall	lower extent of slag crust (200 μm thick)	10	0.56	0.42	97.29	bdl	bdl	bdl	98.27
Iblis-14-2008(J)	Iblis II	EDS	mid-wall	upper extent of slagged region	5	0.09	0.67	97.48	bdl	bdl	bdl	98.24
Iblis-14-2008(m)	Iblis II	EDS	detached slag chunk	homogeneous slag	10	0.24	1.30	97.15	1.16	bdl	bdl	99.85
Iblis-14-2008(m)	Iblis II	EDS	detached slag chunk	homogeneous slag surrounded by ore and prills	10	bdl	1.71	97.74	bdl	bdl	bdl	99.45
Iblis-14-2008(m)	Iblis II	EDS	detached slag chunk	homogeneous slag surrounded by ore and prills	10	bdl	1.72	97.98	bdl	bdl	bdl	99.70
Iblis-14-2008(m)	Iblis II	EDS	detached slag chunk	homogeneous slag surrounded by ore and prills	>100	bdl	0.27	94.45	bdl	bdl	0.09	94.81
Iblis-16-2008(a)	Iblis II	EDS	rim	slag at rim on interior surface	10	bdl	1.84	97.22	bdl	bdl	0.12	99.18
Iblis-16-2008(a)	Iblis II	EDS	mid-wall	slag 750 μm from ceramic interface (4 mm from interior surface) adjacent to large (3 × 4 mm) ore gangue chunk	5	bdl	2.02	96.06	bdl	bdl	bdl	98.08
Iblis-16-2008(a)	Iblis II	EDS	mid-lower wall	slag at interior surface adjacent to large ore gangue	10	bdl	2.10	96.23	bdl	bdl	bdl	98.33
Iblis-16-2008(b)	Iblis II	EDS	mid-upper wall	slag near ceramic interface	150	bdl	0.62	92.91	bdl	bdl	bdl	93.52
Iblis-16-2008(b)	Iblis II	EDS	mid-upper wall	slag near ceramic interface	100	0.59	bdl	93.81	4.54	bdl	bdl	98.94
Iblis-16-2008(b)	Iblis II	EDS	mid-upper wall	slag near ceramic interface, inclusion in 100 μm prill		17.30	0.21	79.19	bdl	1.16	bdl	97.86
Iblis-16-2008(e)	Iblis II	EDS	mid-lower wall	thin slag layer at interior surface	1	0.52	1.88	96.80	bdl	bdl	bdl	99.20
Iblis-24-2008(d)	Iblis II	EDS	mid-lower wall	center of 6 × 1.5 mm slag region on interior surface	10	0.19	1.52	92.92	bdl	bdl	bdl	94.63
Iblis-24-2008(d)	Iblis II	EDS	mid-lower wall	center of 6 × 1.5 mm slag region on interior surface	10	0.31	1.71	92.01	bdl	bdl	bdl	94.03

bdl = below detection limit; EDS = energy-dispersive spectroscopy; EMPA = electron microprobe analysis.

(a)

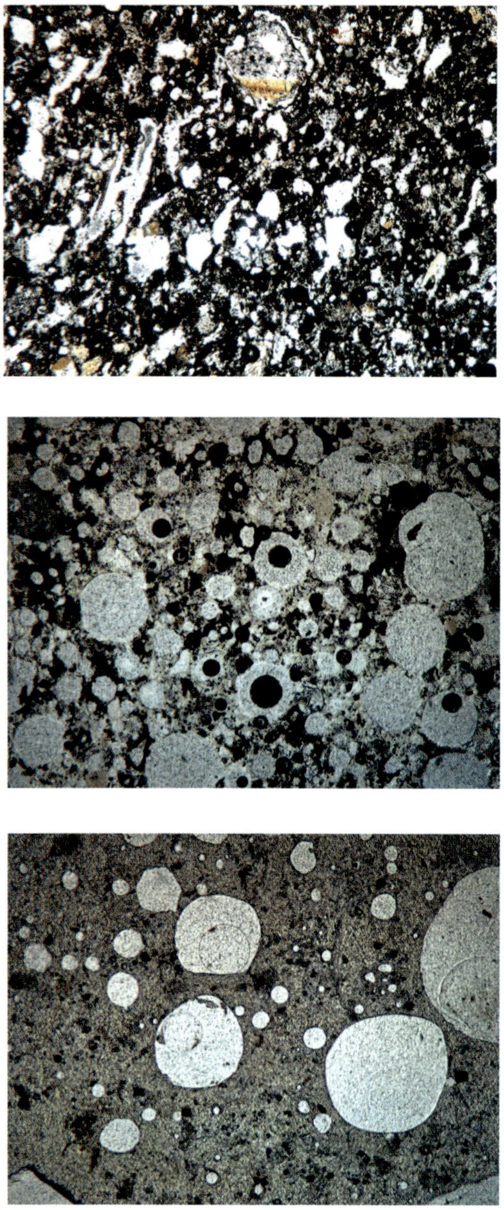

(b)

(c)

Figure 9. Transmitted plane-polarized light images of samples refired to (a) 1150°C, (b) 1175°C, and (c) 1200°C. Significant restructuring of the voids is apparent at 1175°C, while densification of the ceramic and bloating of the pores occurs by 1200°C. Image widths: 2.4 mm.

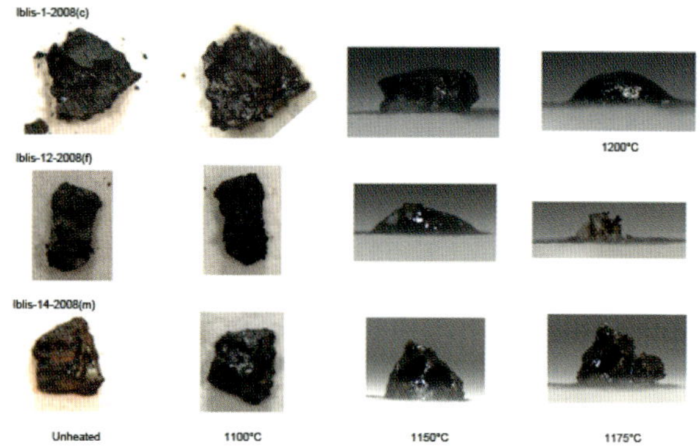

Figure 10. Refiring experiments for Iblis-1-2008(c) (top row), Iblis-12-2008(f) (middle row), and Iblis-14-2008(m) (bottom row). Each sample is shown before heating (left-most images), after heating to 1000°C (second from left) and 1150°C (third from left), and at 1200°C (for Iblis-1-2008(c)) or 1175°C (for the other two crucible slags). These refiring experiments revealed that slumping occurred at 1150°C for these regions of the Iblis-12-2008(f) and Iblis-14-2008(m) slags and at 1200°C for this region of the Iblis-1-2008(c) slag.

negative with high birefringence (Gribble and Hall 1993, 158), making them virtually indistinguishable in thin-section. Hauptmann (2007, 176) and Rice (1987) have both cited the decomposition of calcite at 800°C in reducing conditions. However, the decomposition of calcium carbonate was shown by Webb and Krüger (1970, 317) to vary between 860°C and 1010°C depending on grain size, the presence of salts, the presence of defects and other elements, and the equipment used for the measurement. Most analyses of calcite show its decomposition around 900°C. Dolomite begins to decompose at a slightly lower temperature, but it goes through two stages. First the magnesium and carbonate ions disassociate at around 780°C, then the calcium and carbonate ions disassociate at around 830°C (Gunasekaran and Anbalagan 2007). In addition, ankerite—the iron-rich end member of the dolomite-ankerite solid solution series—undergoes three distinct compositional and structural changes when heated. The first (between 675°C and 725°C) is the decomposition of carbonate ions associated with magnesium, the second is due to the formation of $Fe_2O_3{\cdot}CaCO_3$ at around 825°C, and the third is due to the decomposition of carbonate ions associated with calcium at just over 900°C (Webb and Krüger 1970, 324). If we take a conservative approach and assume that these minerals are ankerite, the full decomposition of the mineral is complete by 950°C. Therefore the presence of this carbonate mineral in the corners of these crucibles shows that 950°C was not reached in these locations—and it is possible that the temperatures reached in these parts of the crucibles were much lower than 950°C. Furthermore, the absence of carbonate minerals closer to the interior wall in these sherds proves that the reduced core (zone 2) and zone 3 reached higher temperatures during the use of the crucible. Therefore it is possible that the other crucible sherds contained these carbonate mineral inclusions at one point but they decomposed during firing or during use of the crucible so that they are no longer present.

Biotite, likewise, will decompose at elevated temperatures (Rice 1987, 431). The presence of this mineral was greatly reduced in the refired Iblis replica crucible at 1125°C, and it disappeared completely by 1150°C (Frame 2004, 72). This indicates that the regions containing biotite did not reach 1150°C. A few of the crucibles contained biotite in the zone 3 region, but most did not. This matches the conclusion based on the morphological changes occurring with the voids in the ceramics above 1100°C. In general, zones 1 and 2 saw temperatures around 950°C or less at the exterior of the crucible, and zone 3 reached 1125°C–1150°C. These temperature determinations result in a temperature gradient of 20°C/mm in the crucible wall.

Zone 4 and the crucible rims saw the highest temperatures during operation. Temperature estimates in zone 4 were made based on aegirine-augite, copper sulfides, and refiring experiments. Quartz can also be used as a temperature indicator mineral since it undergoes multiple crystalline

Table 11. *Macroscopic changes in Iblis slag samples according to firing temperature.*

Sample	Below 1050°C	1050°C	1100°C	1150°C	1175°C	1200°C
Iblis-1-2008(c)	no change	sample fused to brick, vitrified surface (shiny)	corners began to round	began to bloat, edges more rounded	more rounded and bloated, "wetted" the brick	slumped (flattened) against the refractory brick
Iblis-12-2008(f)	no change	sample fused to brick, otherwise no change	corners began to round, sample looked shiny	"wetted" the brick, slumped	completely melted, only quartz gangue remained	did not change from 1175°C
Iblis-14-2008(m)	slightly more shiny	corners and edges began to round, sample fused to brick	much more shiny, more rounded, slight bloating	glassy appearance, began to "wet" the brick	glassy phase was melting away from the gangue	did not change from 1175°C

transitions as it is heated—α to β transition at 573°C, tridymite formation at 870°C, and cristobalite formation at 1470°C (Deer et al. 1992, 457). The characteristic wedge-shaped tridymite features were identified in many cases of quartz gangue associated with the Iblis crucible slags, but the temperature range in which tridymite is stable is far too wide to be informative in this case.

According to the refiring experiments, the temperatures reached within the crucibles were at least 1150°C–1175°C in the hottest regions (at the rim and mid-wall). Temperature estimates based on the composition of the slags agree with these experiments. The glassy matrix phases for all of the crucible slags contain many network modifiers, including Na, K, Ca, and Mg, which all serve to reduce the melting temperature of the glass. The presence of so many elements in this system makes it difficult to calculate the melting temperature of the glass without the use of sophisticated computer software.[1] In addition, the highly heterogeneous nature of these crucible slags makes theoretical temperature estimates based on bulk chemistry meaningless. Therefore, a more informative estimate of temperature was made based on the mineral phases present in the slag at specific locations along the crucible profile. These mineral phases include pyroxenes ranging in compositions, including Ca, Mg, and Fe. Fortunately, these mineral phases have been heavily studied in geology. According to a melting study of basalts conducted at 1atm and with a furnace atmosphere controlled to the fayalite-magnetite-quartz oxygen buffer (reducing conditions), low Ca-pyroxene melts at 1113°C to 1153°C and augite melts in the range of 1167°C to 1182°C (Thy et al. 1999, 99). This study also shows that for conditions approaching equilibrium (their experimental samples were heated for 70 to 90 hours), iron content increases and magnesium decreases with decreasing temperature (Thy et al. 1999, 101, table 6). This characteristic feature is also apparent in the Iblis-33-2008(a) slag (fig. 11), which shows coring of a Mg-rich center and Fe-rich edge in the aegirine-augite euhedral crystals—the Mg-rich end member of augite crystallizes first. This suggests that increased iron over magnesium will lower the melting temperature of these phases. In order for these minerals to crystallize from the glassy slag matrix, the interior of the crucible must have reached at least 1180°C.

Evidence for Melting

Several crucibles did not contain any crucible slags on their interior surfaces, but they did show evidence of high heat—vitrified ceramic—and contact with copper—green staining—suggesting that they may have been used for a purpose other than smelting copper ore to obtain metal. Some of these crucible fragments may, however, just be nonslagged sections from smelting crucibles. The rim and mid-wall of the smelting crucibles typically contain larger amounts of slag and adhered ore than the interior bases. However, eight out of nine of the analyzed nonslagged crucible fragments also contain sections of the rim and wall, so it is most likely that these fragments come from crucibles that did not produce any slag. There are three possibilities for this difference in composition and microstructure at the interior surface of these crucibles.

1. These crucibles may represent smelting operations with very high purity carbonate ores containing little or no gangue material. Smelting these carbonate ores can be done at relatively low temperatures (ca. 900°C) provided the correct reducing atmosphere is achieved. Such a smelting operation would produce very little slag (Bourgarit 2007, 5). In this case, the presence of these unslagged crucibles represents either a more efficient process—recovering a higher percentage of copper from the ore while using up less fuel—or a wise (lucky?) selection of high grade ores.
2. These crucibles may represent a failed smelting process in which temperatures or reducing conditions were simply not high enough and no slag or copper could be produced.

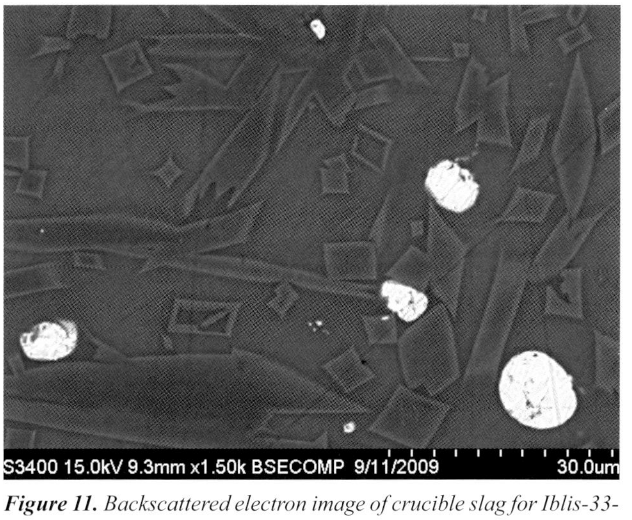

Figure 11. *Backscattered electron image of crucible slag for Iblis-33-2008(a) showing zoning in the aegirine-augite crystals with Fe-rich edges and Mg-rich cores.*

3. These crucibles could have been used for a different purpose—not smelting.

Options 1 and 2 are unlikely. These unslagged crucibles still exhibit heavy vitrification along the interior as seen by the morphological changes in the ceramic fabric. Comparison to the refired replica crucible puts their interiors close to 1150°C, which is certainly high enough to smelt even low-grade ore. Further, even in an oxidizing environment, ore would melt or adhere to the crucible surface, as seen with the many pieces of copper sulfide attached to the Iblis crucibles, and a slag formed in a weak reducing, or oxidizing atmosphere would contain large amounts of delafossite, cuprite, and magnetite. These crucible interiors contain very little copper or iron. Option 3 is most likely. I believe that the separate purpose for which these crucibles may have been used is melting the smelted copper (Frame 2009).

Conclusion

The smelting technology in place at Tal-i Iblis during period I and period II was a well-controlled operation that saw only minor variability. The crucibles were constructed using clay local to Tal-i Iblis, to which a basalt-based temper was added along with large portions of chaff. The raw material selection for the crucibles mirrors the ceramic technology already in place at the site prior to the advent of metallurgical activity. The Iblis crucibles show slight differences in wall thickness and foot geometry, which may be due to multiple craftspeople using the techniques with which they were familiar when constructing similar-style crucibles.

Technological reconstruction of the activities shows that two processes were taking place: melting and smelting. A likely scenario for the use of these crucibles as smelting vessels is that they were placed on a flat surface with dirt or sand piled around to support the sides. They were charged with charcoal and copper carbonate, copper chloride, and copper oxide ores that were often associated with copper sulfides. Another pile of charcoal above the mixture would ensure a reducing environment. The smelting operation proceeded at elevated temperatures (the interior of these crucibles saw temperatures in the range of 1175°C to 1200°C). Such temperatures can be achieved with the use of blow-pipes, tuyeres, and bellows. Tuyeres are rare on the Iranian plateau, and none were recovered from Tal-i Iblis. The lack of archaeological evidence for the use of tuyeres does not mean that no such devices were used. It is possible to use ephemeral material as blow tubes to direct the flow of oxygen to the interior of the crucible, and the tips for these blow tubes may have been constructed of a more friable ceramic that did not survive as well as the slagged crucibles.

The temperature estimate reported here for the interior of the crucibles directly contradicts the evidence published by Dougherty and Caldwell (1966) as well as the on-site experiment organized by Cyril Smith (Pleiner 1967, 375; Smith et al. 1967; Wertime 1967, 1968). Dougherty and Caldwell estimated that smelting "was probably below 1000°C" and that the crucible fabric melted at 1000°C (1966). However, these measurements were performed using an optical pyrometer, and the reported "melting points are uncorrected" (Dougherty and Caldwell 1966), making these estimates unreliable. The experiment by Smith, Wertime, and Pleiner achieved 1100°C at the tuyere tip, and it resulted in a melted crucible and a small amount of copper (Pleiner 1967, 375; Smith et al. 1967). However, their experimental setup placed the tuyere tip outside of the crucible rim (Pleiner 1967, 369), which would have directed the heat to the outer edge of the crucible rather than the middle. Furthermore, they used an automatic bellows, which surely provided a different flux of air than would be available with lung-powered blow-pipes. These factors may have created an inaccurate set of conditions and caused the ceramic to reach higher temperatures than during use in prehistory.

The smelting operations at Iblis likely only resulted in small centimeter-sized nuggets of copper, such as that produced by the Wertime, Smith, Pleiner experiment (Pleiner 1967, 375). The experimental evidence shows that it would be impossible to produce a bun of copper in such a small reaction vessel without adding additional ore and charcoal charge. Furthermore, the archaeological evidence did not include large buns of copper or any metal objects larger than a pin. The small nuggets could be recovered from the crucible by winnowing through the bits of loose unburnt charcoal and undecomposed ore. Such a process would not necessarily result in the destruction of the crucible itself, though these vessels were more likely to break after being heated to such high temperatures. The crucibles at Iblis do not show evidence for reuse, and most are in small pieces, so it is likely that they were destroyed during the copper-recovery process.

In general, the temperature and atmosphere for smelting was well controlled and uniform from crucible to crucible, and there is greater variability among the crucible compositions than is seen for temperature and atmosphere estimates. The ceramic compositions (both in chemistry and the suite of mineral inclusions present) of these crucibles are remarkably consistent, so the compositional variability seen in the slagged regions (zone 4) is more likely due to differences in raw material selection. Barium-rich ores resulted in the Ba-rich slags, arsenical copper ores resulted in increased arsenic in the prills of some crucibles, and variable gangue material resulted in the formation of Fe- and Mg-rich mineral phases within the crucible slags. Such aspects of the metallurgical process at Tal-i Iblis suggest that the craftspeople were familiar with the pyrotechnological details of this technology, but ore selection was somewhat experimental, especially during period I. Period II saw an increased preference for As-bearing ores, as indicated by the presence of some Talmessi algodonite (Cu_6As) (Frame 2009). Furthermore, the Iblis crucibles saw steep thermal gradients from the interior to the exterior of the crucibles and very heterogeneous slag compositions.

The metallurgical industry at Tal-i Iblis drew from existing pyrotechnical knowledge held by the craftspeople, specifically knowledge of clay sources, a history of creating refractory coarseware ceramics, and an understanding of controlling temperature and atmosphere when firing ceramic. The experimental aspect of this early copper smelting technology lies with the selection of the copper ores. Ores that are rich in Ba, Pb, and As lie in the mountain regions immediately surrounding the site of Tal-i Iblis, and these could have easily been collected as animals grazed in the hills. The bright green and blue minerals that are typical of arsenic, lead, and barium ores are easily spotted and mistaken for the bright green and blue ores of copper.

As with most studies of archaeological material, it is clear from these results that additional excavation at Tal-i Iblis may provide further insight into this very early metallurgical process. As the evidence stands, Iblis continues to be one of the most important archaeological sites for understanding early metallurgical activity on the Iranian plateau, but perhaps more importantly, this study reinforces the importance of cross-craft investigations of prehistoric technological processes. New technologies are very rarely entirely novel, and comparison to existing technological practices may shed light on important technological transitions seen in the archaeological record.

Acknowledgments

This research would not have been possible without the advice, suggestions, and support of many colleagues and friends. I would like to thank the National Science Foundation for funding much of this research through a Dissertation Improvement Grant (SBE 0923714). Thanks also go to Drs. Pamela Vandiver, David Killick, Michael Schiffer, Nancy Odegaard, Vincent Pigott, and Christopher Thornton for their advice and mentorship throughout this project, and thank you to Heather Lechtman for introducing me to Tal-i Iblis many years ago! This project would not have been possible without the support of the Natural History Museum at the University of Georgia in Athens, Georgia—a huge thank you to Drs. Mark Williams and Jared Wood for granting access to the Iblis collection. Finally, thank you to Blythe McCarthy and the editors and organizers of the Forbes Symposium for inviting me to participate.

Note

1. Programs like FACT Sage are available for determining liquidus temperatures in these kinds of complex systems, but this software was not available for this investigation.

References

Bourgarit, D. 2007. Chalcolithic copper smelting. In *Metals and mines: Studies in archaeometallurgy,* ed. S. La Niece, D. R. Hook, and P. Craddock, 3–14. London: Archetype Publications in association with the British Museum.

Caldwell, J. R., ed. 1967a. *Investigations at Tal-i Iblis*. Illinois State Museum Preliminary Reports No. 9. Springfield: Illinois State Museum.

———. 1967b. The setting and results of the Kerman project. In *Investigations at Tal-i Iblis,* ed. J. R. Caldwell, 21–40. Illinois State Museum Preliminary Reports No. 9. Springfield: Illinois State Museum.

Caldwell, J. R., and S. M. Shahmirzadi. 1966. *The Kerman range and the beginnings of smelting*. Illinois State Museum Preliminary Reports No. 7. Springfield: Illinois State Museum.

Chase, D. W., J. R. Caldwell, and I. Fehérvári. 1967. The Iblis sequence and the exploration of excavation areas A, C, and E. In *Investigations at Tal-i Iblis,* ed. J. R. Caldwell, 111–201. Illinois State Museum Preliminary Reports No. 9. Springfield: Illinois State Museum.

Craddock, P. 2001. From hearth to furnace: Evidences for the earliest metal smelting technologies in the eastern Mediterranean. *Paléorient* 26(2):151–65.

Deer, W. A., R. A. Howie, and J. Zussman. 1992. *An introduction to the rock forming minerals*. 2nd ed. Essex: Longman Scientific & Technical.

Doonan, R. 1994. Sweat, fire and brimstone: Pre-treatment of copper ore and the effects on smelting techniques. *The Journal of the Historical Metallurgy Society* 28(2):84–97.

Dougherty, R., and J. R. Caldwell. 1966. Evidence of early pyrometallurgy in the Kerman range in Iran. *Science* 153(3739):984–85.

Evett, D. 1967. Artifacts and architecture of the Iblis I period: Areas D, F, and G. In *Investigations at Tal-i Iblis,* ed. J. R. Caldwell, 202–55. Illinois State Museum Preliminary Reports No. 9. Springfield: Illinois State Museum.

Frame, L. D. 2004. Investigations at Tal-i Iblis: Evidence for copper smelting during the Chalcolithic period. BSc thesis, Department of Materials Science and Engineering, Massachusetts Institute of Technology.

———. 2009. Technological change in southwestern Asia: Metallurgical production styles and social values during the Chalcolithic and Early Bronze Ages. PhD diss., Materials Science and Engineering, University of Arizona.

Frame, L. D., and H. Lechtman. In press. Early Chalcolithic crucible smelting of copper ores at Tal-i Iblis, Iran. *Journal of Field Archaeology*.

Gribble, C. D., and A. J. Hall. 1993. *Optical mineralogy: Principles and practice*. London: CRC Press.

Gunasekaran, S., and G. Anbalagan. 2007. Thermal decomposition of natural dolomite. *Bulletin of Materials Science* 30(4):339–44.

Hauptmann, A. 2007. *The archaeometallurgy of copper: Evidence from Faynan, Jordan*. Heidelberg and New York: Springer-Verlag.

Johnson, J. S., J. Clark, S. Miller, V. D. Robins, M. B. Schiffer, and J. M. Skibo. 1988. Effects of firing temperature on the fate of naturally occurring organic matter in clays. *Journal of Archaeological Science* 15(4):403–14.

Neumann, J. P., T. Zhong, and Y. A. Chang. 1990. Cu-O (copper-oxygen). In *Binary alloy phase diagrams,* Vol. 2, ed. T.B. Massalski and H. Okamoto, 1446–48. Materials Park: ASM International.

Pigott, V. C. 1999a. The archaeometallurgy of the Asian Old World: Introductory comments. In *The archaeometallurgy of the Asian Old World,* ed. V. C. Pigott, 1–14. Philadelphia: University of Pennsylvania Museum Press.

———. 1999b. The development of metal production on the Iranian plateau. In *The archaeometallurgy of the Asian Old World,* ed. V. C. Pigott, 73–106. Philadelphia: University of Pennsylvania Museum Press.

———. 1999c. A heartland of metallurgy: Neolithic/Chalcolithic metallurgical origins on the Iranian plateau. In *The beginnings of metallurgy,* ed. A. Hauptmann, E. Pernicka, T. Rehren, and Ü. Yalçin, 107–20. Der Anschnitt, Beiheft 9. Bochum: Deutsches Bergbau-Museum.

Pigott, V. C., and H. Lechtman. 2003. Chalcolithic copper-base metallurgy on the Iranian plateau: A new look at old evidence from Tal-i Iblis. In *Culture through objects: Ancient Near Eastern studies in honour of P. R. S. Moorey,* ed. T. Potts, M. Roaf, and D. Stein, 291–312. Oxford: Griffith Institute.

Pleiner, R. 1967. Preliminary evaluation of the 1966 metallurgical investigations in Iran. In *Investigations at Tal-i Iblis,* ed. J. R. Caldwell, 340–405. Illinois State Museum Preliminary Reports No. 9. Springfield: Illinois State Museum.

Rehren, T. 2009. From mine to microbe: The Neolithic copper melting crucibles from Switzerland. In *From mine to microscope,* ed. A. Shortland, I. Freestone, and T. Rehren, 155–62. Oxford: Oxbow Books.

Rice, P. M. 1987. *Pottery analysis: A sourcebook*. Chicago: The University of Chicago Press.

Rostoker, W., V. C. Pigott, and J. R. Dvorjak. 1989. Direct reduction to copper metal by oxide-sulfide and mineral interaction. *Archaeomaterials* 3(1):69–87.

Smith, C. S., T. A. Wertime, and R. Pleiner. 1967. Preliminary reports of the metallurgical project. In *Investigations at Tal-i Iblis,* ed. J. R. Caldwell, 318–26. Illinois State Museum Preliminary Reports No. 9. Springfield: Illinois State Museum.

Stein, Sir Aurel. 1937. *Archaeological reconnaissances in north-western India and south-eastern Iran*. London: Macmillan.

Thornton, C. P. 2009a. The emergence of complex metallurgy on the Iranian plateau: Escaping the Levantine paradigm. *Journal of World Prehistory* 22(3):301–27.

———. 2009b. The Chalcolithic and Early Bronze Age metallurgy of Tepe Hissar, northeast Iran: A challenge to the "Levantine Paradigm." PhD diss., University of Pennsylvania.

Thornton, C. P., and C. C. Lamberg-Karlovsky. 2004a. A new look at the prehistoric metallurgy of southeastern Iran. *Iran* 42:47–59.

———. 2004b. Tappeh Yahya and the prehistoric metallurgy of south-eastern Iran. In *Persiens antike pracht,* ed. T. Stöllner, R. Slotta, and A. Vatandoost, 265–73. Bochum: Deutsches Bergbau-Museum.

Thornton, C. P., C. C. Lamberg-Karlovsky, M. Liezers, and S. M. Young. 2002. On pins and needles: Tracing the evolution of copper-base alloying at Tepe Yahya, Iran, via ICP-MS analysis of common-place items. *Journal of Archaeological Science* 29(12):1451–60.

Thy, P., C. E. Lesher, and J. D. Mayfield. 1999. Low pressure melting studies of basalt and basaltic andesite from the southeast Greenland continental margin and the origin of dacites at site 917. In *Proceedings of the Ocean Drilling Program, scientific results*, Vol. 163, ed. H. C. Larsen, R. A. Duncan, J. F. Allan, and K. Brooks, 95–112. College Station: Texas A&M University Press.

Tylecote, R. F. 1982. Metallurgical crucibles and crucible slags. In *Archaeological ceramics,* ed. A. Franklin and J. Olin, 231–43. Washington, DC: Smithsonian Institution Press.

Voigt, M. M., and R. H. Dyson. 1992. The chronology of Iran, c. 8000–2000 BC. In *Chronologies in Old World archaeology,* Vol. 2, ed. R. W. Ehrich, 125–53. Chicago: University of Chicago Press.

Webb, T. L., and J. E. Krüger. 1970. Carbonates. In *Differential thermal analysis,* Vol. 1, ed. R. C. MacKenzie, 303–41. London and New York: Academic Press.

Wertime, T. A. 1967. A metallurgical expedition through the Persian Desert. In *Investigations at Tal-i Iblis,* ed. J. R. Caldwell, 327–39. Illinois State Museum Preliminary Reports No. 9. Springfield: Illinois State Museum.

———. 1968. A metallurgical expedition through the Persian desert. *Science* 159(3818):927–35.

Metallurgy during the Middle Chalcolithic Period in the Southern Caucasus: Insight through Recent Discoveries at Mentesh-Tepe, Azerbaijan

Antoine Courcier, Bertille Lyonnet, and Farhad Guliyev

ABSTRACT The way in which metallurgy started and developed in the Caucasus is still poorly known, but a few occurrences of metal objects in settlements situated in the Southern Caucasus date it from the middle sixth millennium BCE at the time of the Shomu-Shulaveri culture. During the fifth millennium BCE, metallurgy developed progressively in this same area, while the first evidence of metal production appeared in the northern Caucasus (Meshoko culture). Since 2008, excavations at Mentesh-Tepe in the valley of the Kura River in western Azerbaijan (Tovuz district, 45°49'58"/40°56'30.8") have revealed several diverse metallurgical artifacts. They illustrate the level of metallurgy reached in the Caucasus at the time of the Sioni culture, during the second half of the fifth millennium BCE. The exceptional character of these discoveries led us to conduct archaeometallurgical research on these artifacts. Our aim is to characterize the metallurgical activities at Mentesh-Tepe and, further, to better understand the origins and development of ancient metallurgy in the Caucasus since it seems to have played a major role in the relationships between the cultures of this area and those of neighboring zones, including northern Mesopotamia.

Introduction

The settlement of Mentesh-Tepe is located in western Azerbaijan at the foot of the Smaller Caucasus in the Kura River valley (figs. 1–3). It was first visited in the 1960s by I. Narimanov, who dated it to the Chalcolithic period (ca. 5000–3000 BCE), mentioning the presence of combed pottery. At that time, the site was a small and already very damaged mound 40–45 m in diameter and 1 m high (Narimanov 1987, 32–33). During the 1970s, the mound was totally leveled in order to make way for a vineyard. Thanks to inhabitants' indications, its location was rediscovered in 2007 near the village of Ashagi Ayibli in the Tovuz district (45°49'58"/40°56'30.8") (fig. 4). A small sounding revealed Chalcolithic structures related to the Sioni culture below the actual surface (table 1). With this culture being still very poorly known, excavations have been programmed since 2008 and are

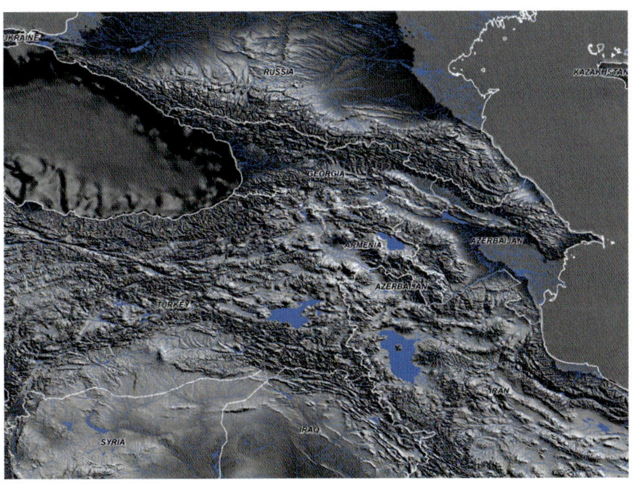

Figure 1. *General map of the Caucasus including political boundaries and hydrography. CNRS-BRGM, Caucasus-GIS Project.*

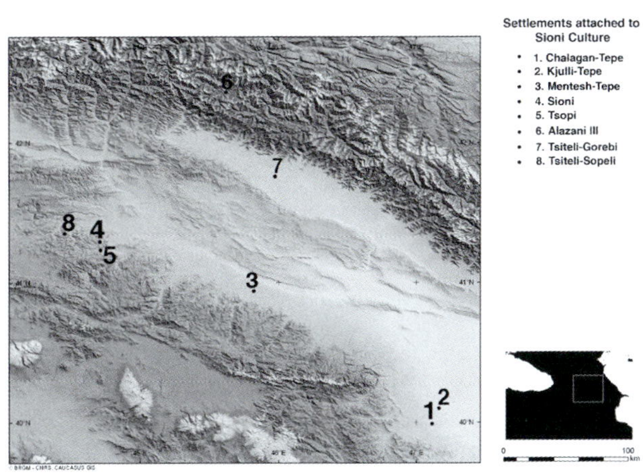

Figure 2. *Map of the settlements attributed to the Sioni culture (second half of the fifth millennium BCE). CNRS-BRGM, Caucasus-GIS Project.*

Table 1. *Chronological scheme of the cultures in the Caucasus from the Neolithic to the Early Bronze Age.*

Period		Absolute Date	Caucasus						
			North		Northeast		North/ Northwest	South (Transcaucasia)	
								Georgia / Azerbaijan	Armenia
Early Bronze Age		2000	North Caucasus Culture		Ginchi EBA	Velikent IV		Bedemi Martkopi	
		2100							
		2200							
		2300				Velikent II			
		2400							
		2500							
		2600	Majkop	Majkop_phase Novosvobodnaja	?	Velikent II		Kuro-Araxe	
		2700							
		2800							
		2900							
Late Chalcolithic	LC 5	3000				Velikent I			
		3100							
	LC4	3200							
		3300			Ginchi Chalcolithic				
		3400							
	LC3	3500							
		3600		Majkop_ phase Majkop				Berikldeebi Leila-Tepe	
	LC 2	3700							
		3800							
	LC 1	3900	Svobodnoe-Meshoko-Zamok		hiatus		Darkveti	Sioni	
		4000							Aratshen 0
		4100							
Ancient and Middle Chalcolithic		4200							
		4300			Chokh				
		4400							
		4500							
		4600	?					Alikemektepe Kul- Tepe I ?	Aratashen I
		4700							
		4800							
		4900							
Neolithic		5000						Shulaveri-Shomu	Aratashen III-II-I (Artsh. III-II-I)
		5100							
		5200							
		5300							
		5400							
		5500							
		5600							
		5700							
		5800							
		5900							
		6000			Chokh D				
		6100							
		6200							
		6300							
		6400							
		6500							

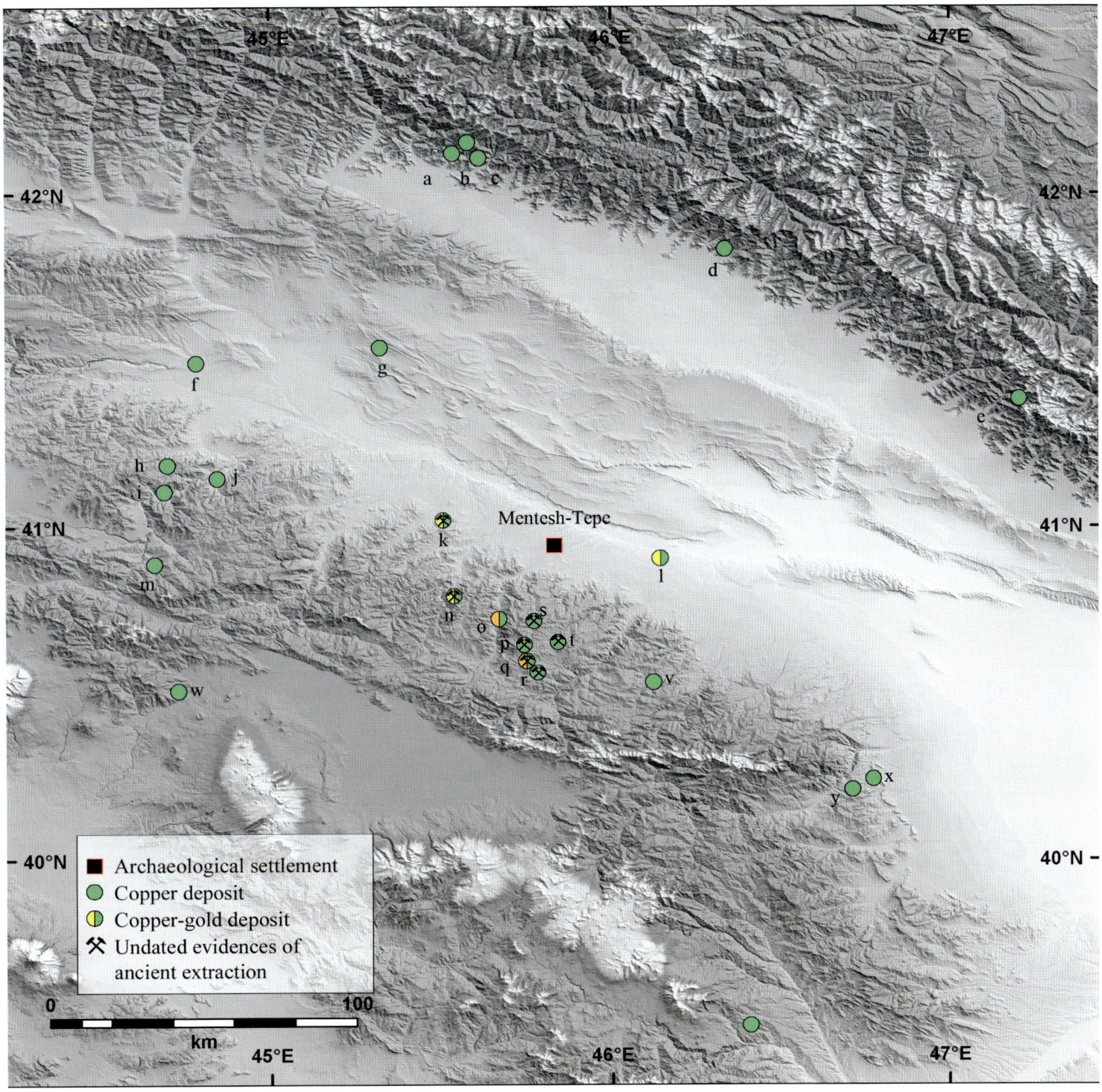

Figure 3. *Map centered on the Mentesh-Tepe settlement showing the main nearby copper ore deposits and undated evidence of ancient extraction (mining), Sioni culture (second half of the fifth millennium* BCE*). a = Akhotiskali; b = Artana; c = Kokhta; d = Mazimskoe; e = Salvat; f = Madneuli; g = David-Garedji; h = Shamlug; i = Alaverdi; j = Tekhut; k = Dagkesaman; l = Dzhagirchaj; m = Ankadzor; n = Gosha; o = Bitty-Bulaq; p = Siny-Yar; q = Kedabek; r = Misdag-Maghara; s = Karabakh; t = Sojudlu; u = Dashkesan; v = Mekhami; w = Agdara; x = Akht'ala; y = Ajgedzor. CNRS-BRGM, Caucasus-GIS Project.*

still in process. Sponsored originally by the Ministry of Foreign Affairs in France, the excavations are now part of a joint Agence Nationale de la Recherche and Deutsche Forschungsgemeinschaft (ANR-DFG) project, "Ancient Kura,"[1] directed by Bertille Lyonnet and Barbara Helwing.[2]

Excavations at Mentesh-Tepe have unveiled a completely new and previously unknown cultural phase in this area of the middle Kura River valley. They also brought to light evidence of metallurgy during the second half of the fifth millennium BCE.[3]

The aim of this article is to present the first results of our research on the metallurgical artifacts.

The Settlement of Mentesh-Tepe

In 2007, the remains of Mentesh-Tepe could only be guessed from the few visible sherds on the actual surface of gardens at the northern periphery of the village of Ashagi Ayabli (see fig. 4). Three campaigns of excavations have, however, established with certainty the presence of a settlement dating to the Middle Chalcolithic period (second half of the fifth millennium BCE) into which large pits and graves were dug during the Early Bronze Age (ca. 3000–2000 BCE) (fig. 5). These intrusions, combined with the sinking of cement poles to great depths during the recent establishment of a vineyard, have badly damaged the site.

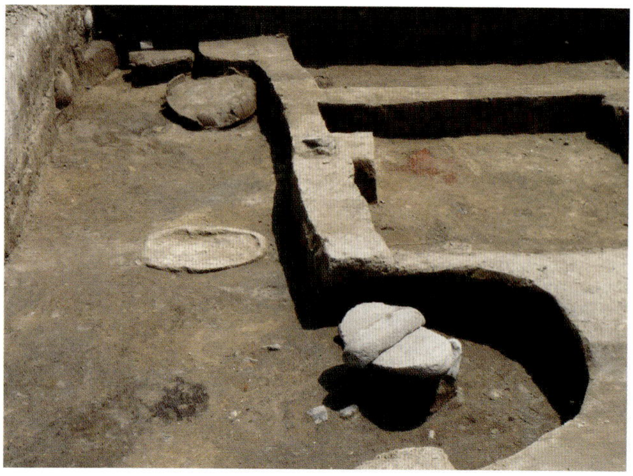

***Figure 4**. View of a Chalcolithic level at the Mentesh-Tepe settlement, Sioni culture (second half of the fifth millennium* BCE*). Foreground: intrusive Kuro-Araxe pit with a grinding stone; middle ground: Chalcolithic silo (left) and Chalcolithic floor with red ochre and, below, crushed azurite ore (right); background: Chalcolithic jar (left) (Lyonnet and Gulyev 2010).*

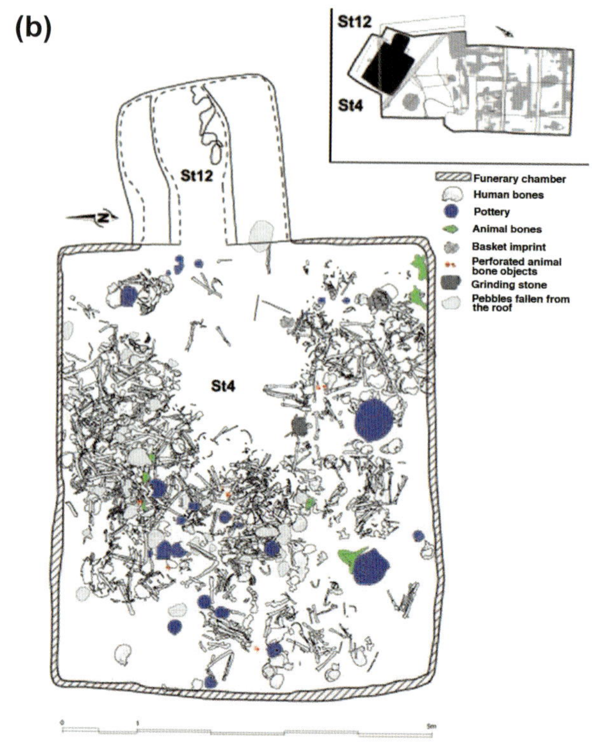

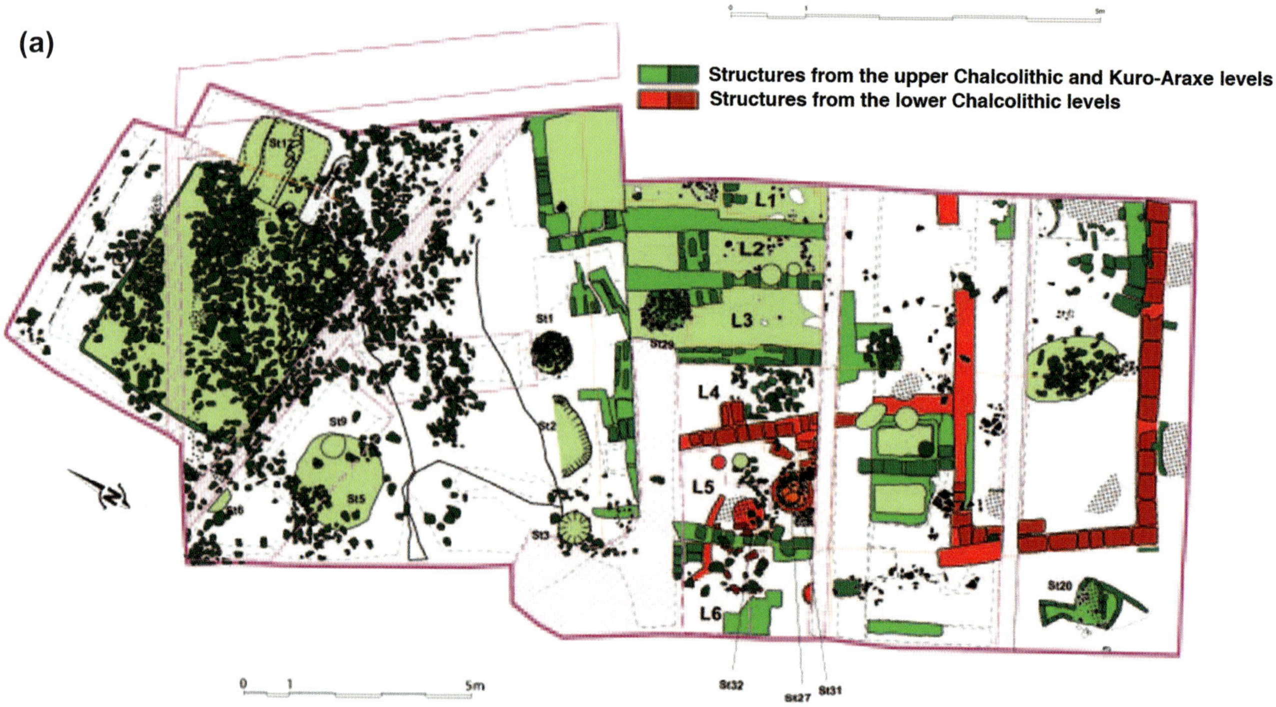

***Figure 5**. Plans of the 2009 excavations at Mentesh-Tepe, Sioni culture (second half of the fifth millennium* BCE*). **a.** General plan. **b.** Detailed plan of the Kuro-Araxe collective tomb under the kurgan (Lyonnet and Gulyev 2010).*

The Chalcolithic settlement rests on virgin soil and reaches a maximum of 2 m of stratigraphy. It apparently did not last very long, and ^{14}C dates put it around 4300 BCE. Evidence shows that after a long period of abandonment at the site, the site was again occupied. A large collective tomb containing about fifty bodies, located under a kurgan (i.e., tumulus), was excavated (see fig. 5). It is dated to the very end of the fourth millennium/beginning of the third millennium BCE and related to the Kura-Araxes culture. In addition to this tomb, ritual places and other individual graves dating to various later phases of this culture, as well as large pits whose function is yet unclear, were dug into the site. Very few architectural remains and little material can confidently be related to this last period of occupation. The pottery consists mostly of vessels found complete in the graves or broken in large pieces in the pits, but a few grinding stones and animal figurines attest to living areas. It seems that during the last period the site was in seasonal use. It is important to note that no metal implements have been found in the graves dated to the Early Bronze Age.

***Table 2**. List of metallurgical artifacts discovered at Mentesh-Tepe (Azerbaijan) during the 2008–10 excavations.*

Inv. No.	Analysis No.	Type of Analyses	Figure Reference	Type of Artifact	Archaeological Context
MET 01_2008	not analyzed	—	fig. 6o	slag	zone 1, above pebbles, upper structure of kurgan
MET 02_2008	not analyzed	—	—	ore fragment	zone 1, above pebbles, upper structure of kurgan
MET 03_2008	4180/09	ICP-MS; LIA	fig. 6r	awl	sounding 1
MET 04_2008	4181/09	ICP-MS; LIA	fig. 7e	ring	zone 2/3 West, structure 2
MET 05_2008	4182/09	OM; EDS; ICP-MS; LIA	fig. 6a	fragment of unfinished awl	zone 1, West
MET 06_2008	4183/09	ICP-MS; LIA	fig. 6g	awl fragment	zone 1, West
MET 07_2008	4184/09	ICP-MS; LIA	fig. 8e	agglomerate of metal prills	zone 1, West
MET 08_2008	4185/09	XRD; ICP-MS; LIA	fig. 8c	semismelted ore	zone 1, West
MET 09_2008	not analyzed	—	fig. 9c	ingot mold	zone 1, West
MET 10_2008	4186/09	OM; EDS; ICP-MS; LIA	fig. 6t	awl	zone 1, West
MET 11_2008	4187/09	ICP-MS; LIA	fig. 8f	agglomerate of metal prills	zone 2 West, sounding 6
MET 01_2009	not analyzed	—	fig. 8c	semismelted ore	zone 1, East level 4
MET 02_2009	not analyzed	—	—	metal ware	out of context
MET 03_2009	not analyzed	—	fig. 7h	undetermined object	zone 8, under pebbles, upper structure of kurgan
MET 04_2009	not analyzed	—	fig. 6l	awl	zone 6
MET 05_2009	not analyzed	—	fig. 7b	ring	zone 5, level 12
MET 06_2009	not analyzed	—	fig. 6m	awl fragment	zone 6, locus 2
MET 07_2009	not analyzed	—	fig. 6q	awl fragment	zone 7, locus 4
MET 08_2009	not analyzed	—	fig. 6f	awl fragment	zone 8, locus 3, level 29
MET 09_2009	not analyzed	—	—	undetermined object	out of context
MET 10_2009	not analyzed	—	fig. 9d	mold (?)	zone 6, locus 3, level 29
MET 11_2009	not analyzed	—	fig. 9b	crucible with entrapped prills	zone 14, level 20
MET 01_2010	4613/10	ICP-MS	fig. 8h	thin metal plate with prills unmelted	zone 17
MET 02_2010	4614/10	ICP-MS	fig. 6p	awl fragment	zone 14, structure 20
MET 03_2010	4615/10	ICP-MS	fig. 7i	blade fragment	zone 13
MET 04_2010	4616/10	ICP-MS	fig. 6d	awl fragment	zone 15
MET 05_2010	not analyzed	—	fig. 8i	thin metal plate with unmelted prills	zone 12, locus 7
MET 06_2010	4617/10	ICP-MS	fig. 6h	awl fragment	zone 15
MET 07_2010	4618/10	ICP-MS	fig. 8k	thin metal plate with prills unmelted	zone 14, structure 44
MET 08_2010	4619/10	ICP-MS	fig. 6n	fragment of awl	zone 19, structure 58
MET 09_2010	not analyzed	—	fig. 8l	thin metal plate with prills unmelted	zone 14, structure 44
MET 10_2010	not analyzed	—	fig. 6i	awl fragment	zone 24
MET 11_2010	4620/10	ICP-MS	fig. 8d	agglomerate of metal prills	zone 13
MET 12_2010	4627/10	ICP-MS	fig. 8m	thin metal plate	zone 13
MET 13_2010	not analyzed	—	—	soil with metal fragments	zone 24, level 74
MET 14_2010	4621/10	ICP-MS	fig. 6s	awl	zone 19, hearth 58
MET 15_2010	not analyzed	—	—	soil with metal fragments	zone 24, level 74
MET 16_2010	4622/10	ICP-MS	fig. 7g	undetermined object	zone 24, level 74
MET 17_2010	not analyzed	—	—	soil with metal fragments	zone 20
MET 18_2010	not analyzed	—	fig. 6c	awl fragment	zone 13, level 50
MET19_2010	not analyzed	—	fig. 8j	thin metal plate	zone 17
MET 20_2010	4623/10	ICP-MS	fig. 6b	awl fragment	zone 24, level 74
MET 21_2010	not analyzed	—	fig. 9a	crucible with slag encrustation	zone 21

cont'd

***Table 2**. List of metallurgical artifacts discovered at Mentesh-Tepe (Azerbaijan) during the 2008–10 excavations* cont'd.

Inv. No.	Analysis No.	Type of Analyses	Figure Reference	Type of Artifact	Archaeological Context
MET 22_2010	4624/10	ICP-MS	fig. 6o	awl	zone 15, level 13
MET 23_2010	4625/10	ICP-MS	fig. 7c	ring	zone 22, locus 97
MET 24_2010	not analyzed	—	—	soil with metal fragments	zone 21, level 102
MET 25_2010	not analyzed	—	—	soil with metal fragments	zone 24, level 111
MET 26_2010	not analyzed	—	—	soil with metal fragments	zone 24, level 111
MET 27_2010	not analyzed	—	fig. 6k	awl fragment	zone 24, level 111
MET 28_2010	4628/10	ICP-MS	fig. 8b	semismelted ore	zone 23, locus 103
MET 29_2010	4626/10	ICP-MS	fig. 8g	thin metal plate with prills unmelted	zone 14, structure 45
MET 30_2010	4629/10	ICP-MS	fig. 8n	thin metal plate	zone 20
MET 31_2010	4630/10	ICP-MS	fig. 7a	ring	zone 20
MET 32_2010	4631/10	ICP-MS	fig. 7f	ring or curved awl (?)	zone 19, locus 117
MET 33_2010	not analyzed	—	fig. 6j	awl fragment	zone 21, level 102-107
MET 34_2010	not analyzed	—	fig. 7d	ring, heavily corroded	zone 24, locus 11
MET 35_2010	not analyzed	—	fig. 6e	awl fragment	zone 24, locus 11
MET 36_2010	4632/10	XRD; ICP-MS	fig. 8a	crushed ore	zone 21, locus 97

EDS = energy-dispersive x-ray spectrometry; ICP-MS = inductively coupled plasma–mass spectrometry; LIA = lead isotope analysis; OM = optical microscopy; XRD = x-ray diffraction.

By far, most of the material—over 97%—found during the excavations dates to the Chalcolithic period. It mainly consists of pottery, lithic tools (obsidian and flint), and grinding stones. A great part of it has been found disturbed or reused by the later inhabitants for ritual places around the kurgan or as oven floors, but some has also been found *in situ* on the floors of rectangular rooms with walls built of unbaked bricks. A large rectangular building is now starting to appear, with round hearths, large pithoi, and silos *in situ*. We still need to excavate a larger area to better understand the architecture.

Contrary to the Neolithic period (ca. 9000–5500 BCE; Shomu-Shulaveri culture), bone artifacts are extremely rare at Mentesh-Tepe and seem to have been replaced by metal implements. Some, including tools, molds, or azurite ore, indisputably come from Chalcolithic floors, but many of the metal artifacts discovered in the course of the excavations come from disturbed levels or areas. Nevertheless, since there is so little pottery related to the Early Bronze Age, since no metal artifacts have been discovered in the graves of this period, and since the evidence of living at that time is also so scanty, we are inclined to think that most of these metal implements (or objects linked to metal production) date to the Chalcolithic period.

Evidence of Metallurgy at Mentesh-Tepe

Forty-nine artifacts related to metalworking activities were discovered on the site. As mentioned previously, we believe that most, if not all, are related to the Chalcolithic levels. The 2010 excavations have revealed several structures and artifacts related to different stages of the metallurgical process (*chaîne opératoire*) within the undisturbed Chalcolithic levels of the site (table 2).

Altogether, three campaigns of excavation have revealed not only metal artifacts, including awls, rings, and other undetermined objects (figs. 6 and 7), but also the remains of metal production, including metallurgical waste (slags, ore, corroded agglomerates), objects used in melting metal or smelting ore (plates, ingot mold, crucibles, and perhaps a mold), and even crushed ore (azurite) (figs. 8 and 9).

All the metal objects are strongly corroded, and therefore, in some cases, their type and/or function are difficult to determine. However, the variety of shapes is minimal.

The Metal Objects

To date, only two main artifact types have been recognized—awls and rings—while others are not yet understood completely.

Awls

Awls with a square section represent the majority of the metal objects (see fig. 6), and most are represented only by fragments. The state of corrosion of most of them is rather severe and prevents, in some instances, complete certainty of their shape. Complete awls are rare (see fig. 6 r–t). One awl, characterized by a square section but thicker than the others, could be a semimanufactured and unfinished one (see fig. 6a).

Awls were already known in the previous cultures of the sixth/early fifth millennium BCE like those of the Shomu-Shulaveri and Alikemek-Tepesi/Kul'tepe I (Courcier, in press), and they are frequently found at sites contemporaneous to Mentesh and also related to the Sioni culture: several examples have been found in Georgia at Alazani III (Kiguradze 2000, 323), Tsiteli-Gorebi (Varazashvili 1992, 32), and Kviriastskali (Varazashvili 1992, 68–69) and in Azerbaijan at Chalagan-Tepe (Narimanov 1985, 490).

Rings

Rings of different sizes are also well represented at Mentesh-Tepe (see fig. 7 a–f). Their advanced state of corrosion induces some doubts in the classification of some of them: one could also be a curved awl (see fig. 7 f). Similarly, some fragments with a semicircular section could be fragments of rings (see fig. 6 f–h). In the current state of research, no ring has ever been discovered in other settlements related to the Sioni culture. A ring was, however, discovered in level I of Kül'tepe I, which is related to the Alikemektepe/Kul'tepe I culture and predates the Sioni culture (Abibullaev 1965, 161–63; Abibullaev 1982, 78–79) (see table 1). Furthermore, rings are known from north of the Caucasus from the Meshoko culture (Kruglov et al. 1941, 76; Formozov 1965, 70–71), which was contemporaneous with the Sioni culture (see table 1).

Unidentified artifact types

Other undetermined metal artifacts were found (see fig. 7 g and h) that could be unfinished objects. For example, figure 7i could be either a fragment of blade or an ingot.

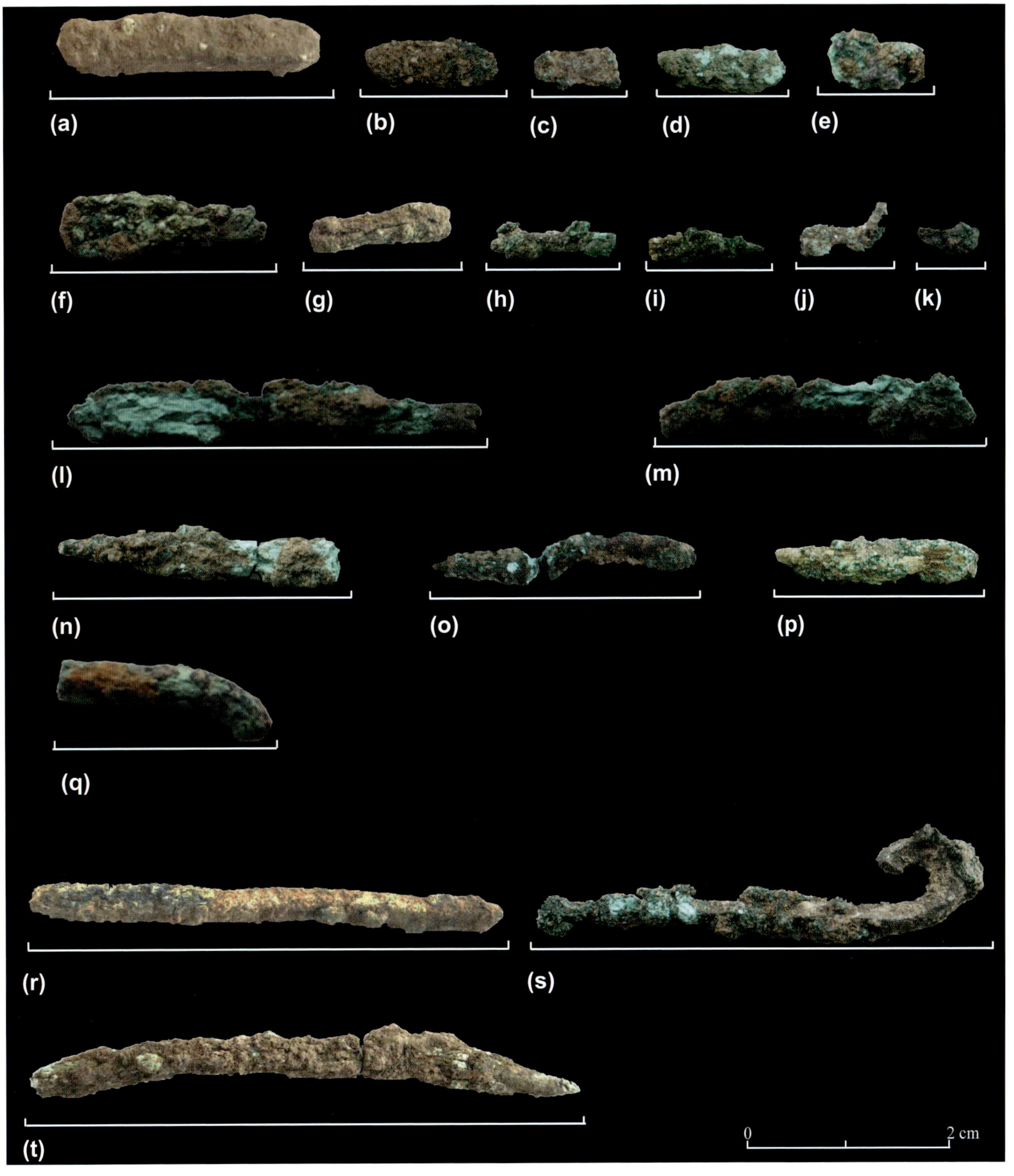

*Figure 6. Metal objects, fragments, and complete awls, Sioni culture (second half of the fifth millennium BCE), Mentesh-Tepe. **a.** Semimanufactured awl. **b–q.** Awl fragments. **r–t.** Complete awls. Institute of Ethnology and Archaeology of Baku.*

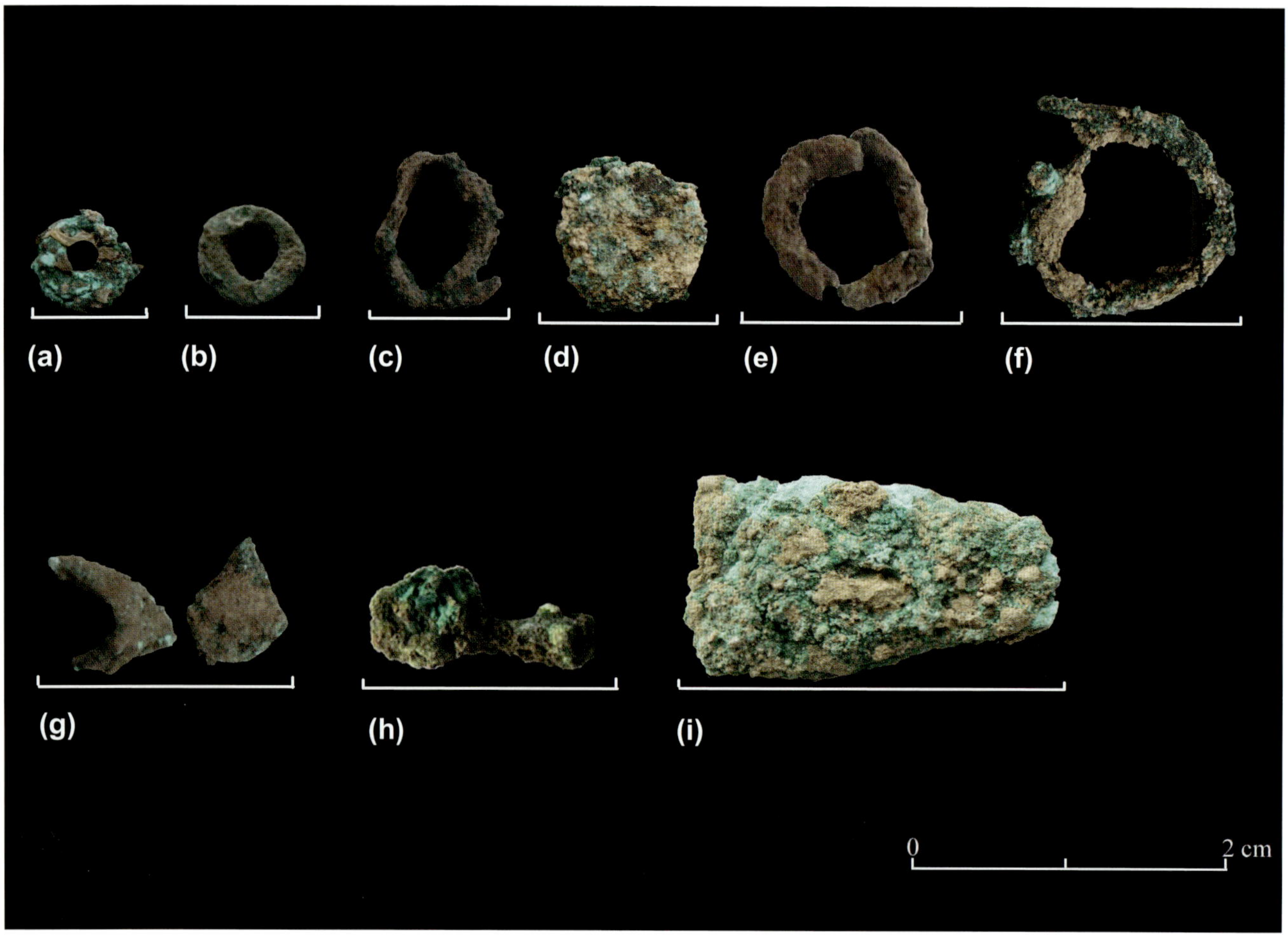

Figure** 7. Metal objects, rings, and other objects, Sioni culture (second half of the fifth millennium* BCE*), Mentesh-Tepe.* ***a–f. *Rings.* ***g*** *and* ***h.*** *Undetermined objects.* ***i.*** *Blade fragment. Institute of Ethnology and Archaeology of Baku.*

Remains of Metal Production

Several pieces of metallurgical waste and items for melting or smelting have been discovered, including ore, slags, prills, thin metal plates, one fragment of a mold, and two fragments of crucibles (see figs. 8 and 9).

Metallurgical waste

In 2008, on the surface of the tepe, a small and light slag was collected (see fig. 8o). This discovery suggested the possibility of metal activities at the site, and this was confirmed by further excavations during which fragments of copper ore were found. The ore can be separated into two types. One corresponds to small fragments of crushed ore (see fig. 8a), the height of which is mainly between 1 and 3 mm and does not exceed 5 mm. The second type of ore is composed of aggregates that have an oxidized surface (see fig. 8 b and c). They correspond to a mixture of oxide products and copper minerals associated with quartz.

Heavily corroded undetermined aggregates were also discovered (see fig. 8 d–i). They consist of small metallic spherical or oval particles (prills) agglomerated by oxide elements.

Another proof of metal production on the site is represented by thin metal plates (see fig. 8 j–n). Some are characterized by small protuberances (see fig. 8 j–l) that could be metallic prills that were only partially melted during formation of the metal plate. Two other plates exhibit flat edges and a rectangular shape (see fig. 8 m and n) and could indicate the use of additional steps in a more elaborate metallurgical process if they are indeed semimanufactured artifacts.

Metallurgical waste is rarely mentioned among the discoveries of ancient settlements of the Caucasus area, and it was not the subject of specific research until recently. Slag and a copper lens have been found at Tsiteli-Sopeli, a site that is linked to the Sioni culture (Lordkipanidze 1989, 104, n. 11), and a few pieces of metallurgical waste (slags, prills, and metal fragments lost during melting processes) are known from Leilatepe (Akhundov 2007, 100–108), where they date to the first half of the fourth millennium BCE. Recently, however, fragments of copper ore have been discovered in Armenia at Aratashen in a level dated to the second half of the sixth millennium BCE (Badalyan et al. 2007, 52–53), and, in 2007, a slag "cake" made of a gangue of ore and some metal prills was found in Azerbaijan at Göy-Tepe (Courcier, in press), a settlement dated to the middle of the sixth millennium BCE and related to the Shomu-Shulaveri culture (see table 1).

Molds and crucibles

In addition to the fragments of metallurgical waste discussed above, a few other items linked to metallurgical activities were found.

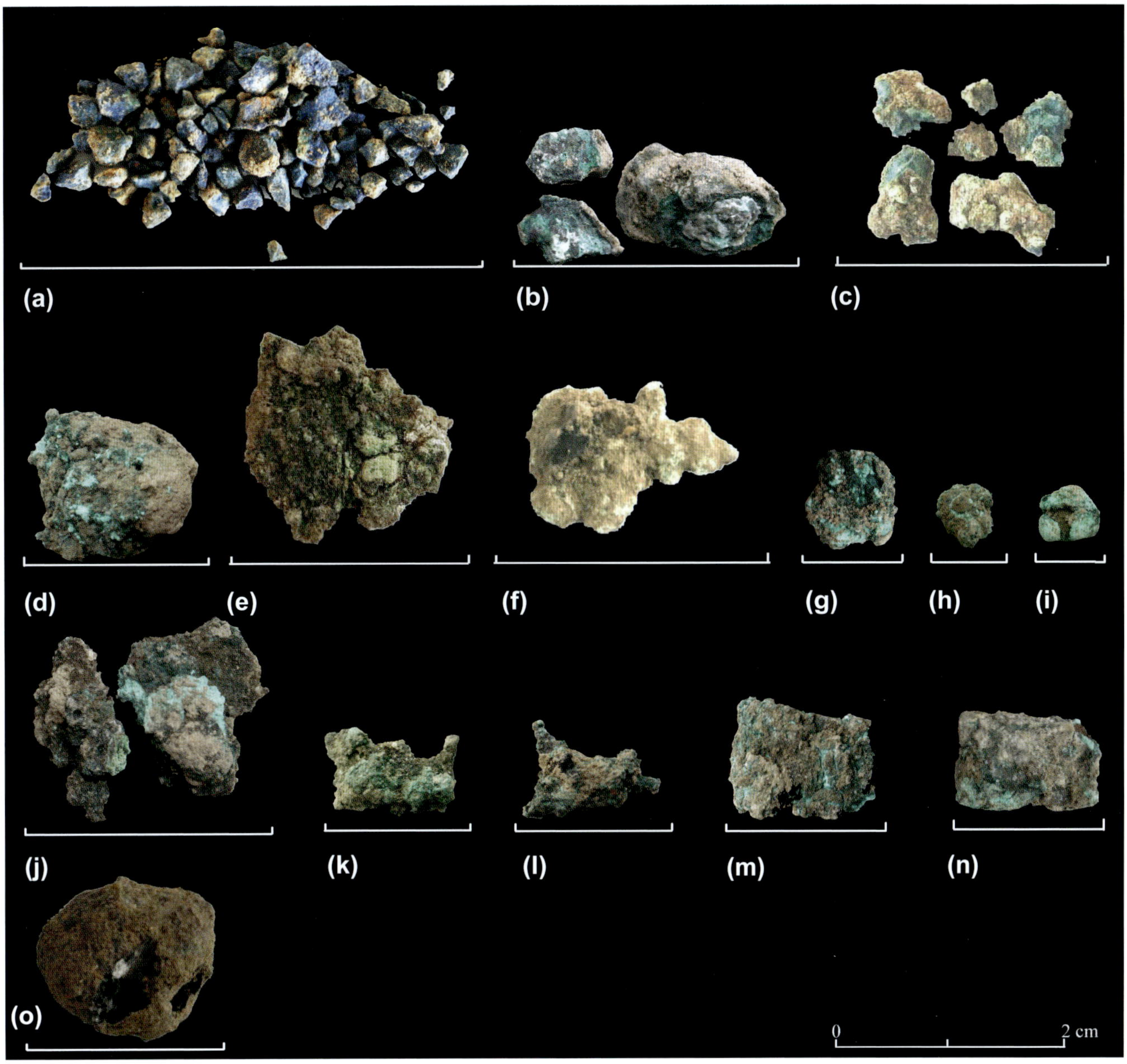

Figure 8**. Remains of metal production, metallurgical waste (ores, prills, slag, plates), Sioni culture (second half of the fifth millennium* BCE*), Mentesh-Tepe.* ***a. *Crushed azurite ore.* ***b*** *and* ***c.*** *Agglomerate of semismelted ore.* ***d–i.*** *Corroded agglomerate of metal prills.* ***j–l.*** *Thin metal plates with semismelted prills.* ***m–n.*** *Rectangular thin metal plates.* ***o.*** *Slag. Institute of Ethnology and Archaeology of Baku.*

A fragment of an ingot mold was discovered during the first campaign (see fig. 9c). No traces of metallurgical use are visible to the naked eye. Only two other ingot molds are known at present in the Caucasus. They were discovered at Dikha-Guzube (Kuftin 1950, 241, 243) and at Kul'tepe II (Abibullaev 1965, 68), settlements both related to the Kura-Araxe culture, which dates to the third millennium BCE (see table 1).

In 2009, a stone fragment thought to be a mold (see fig. 9d) was also discovered at Mentesh, but there is disagreement with this interpretation and doubts remain regarding its use.

Two fragments of crucibles were discovered during the two last campaigns (see fig. 9 a and b). The first one (see fig. 9a) was discovered among many other Chalcolithic potsherds and consists of a round bottom of a Chalcolithic pot with vegetal temper. This sherd, like all the others, had been reused to make a floor for a Kura-Araxe furnace. This first fragment of crucible is characterized by metal prills trapped in the porosity of the sherd (see fig. 9a). The second fragment, which is made of clay, was discovered in zone 21 (see fig. 9b). Its rectilinear edges, which form a right angle, suggest a general rectangular shape. It is encrusted with slag and has small copper prills on one of its rims (see fig. 9b). It seems to have endured higher temperatures than the first crucible because of cracks on its external edge. The slag penetrates into the rim 1 cm in depth. Red color in the crucible suggests oxidized firing.

Very few crucibles are known in the Caucasus. One example was discovered in Veselogo in the northern Caucasus, a settlement related to the Meshoko culture (Formozov 1965, 116), but, unfortunately, the context of this discovery is uncertain. Another was found near the Afips River, also in the northern Caucasus, but it is undated (Iessen and Degen-Kovalevskij 1935, 36). Finally, a few other crucibles have been found in Kura-Araxe sites in the southern Caucasus and are all dated to the second half of the fourth millennium BCE (Courcier, in press).

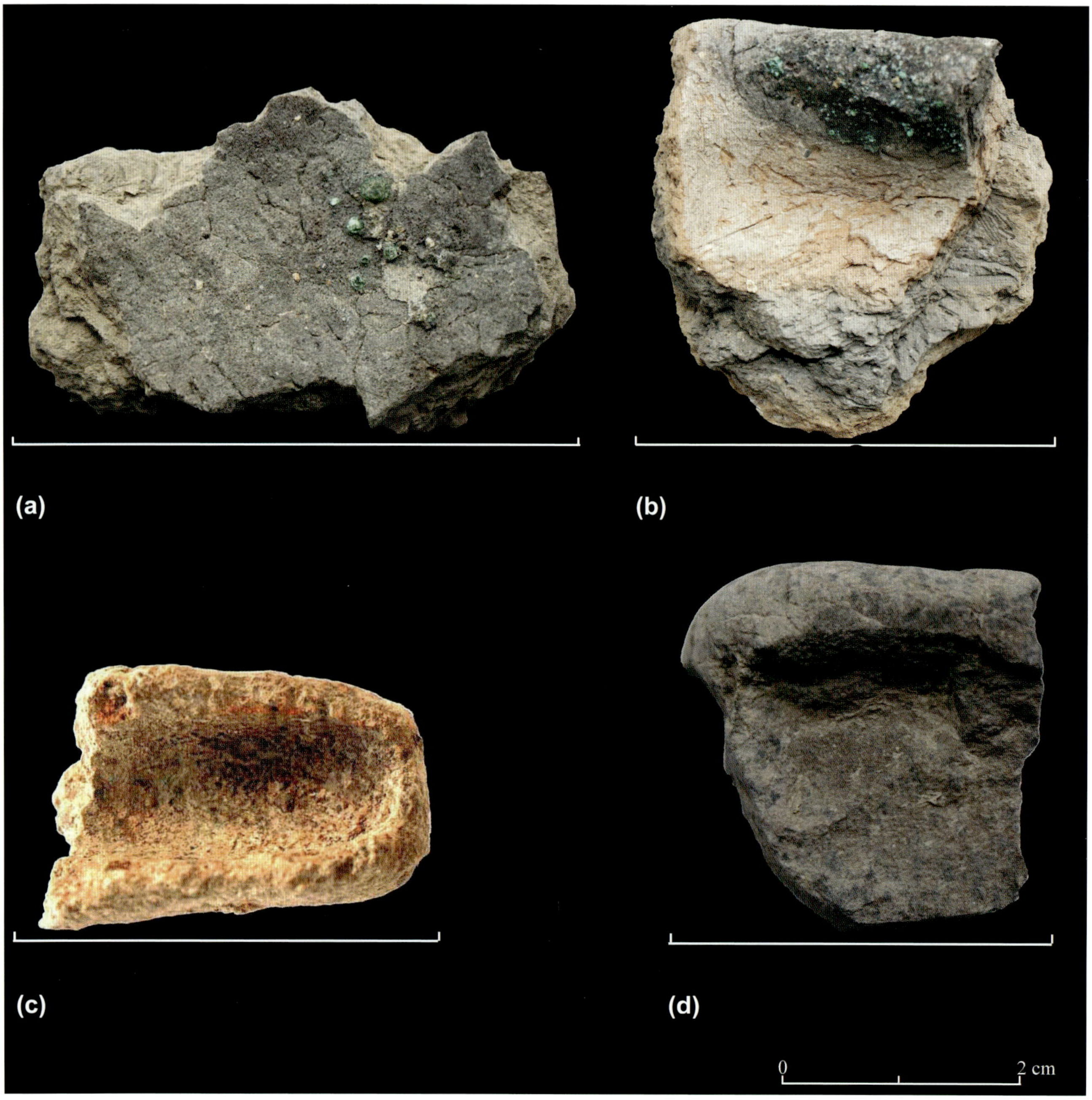

Figure 9. *Remains of metal production, crucibles, ingot mold, mold, Sioni culture (second half of the fifth millennium* BCE*), Mentesh-Tepe.* ***a.*** *Crucible with entrapped pills.* ***b.*** *Crucible with a slag encrustation.* ***c.*** *Ingot mold.* ***d.*** *Mold (?). Institute of Ethnology and Archaeology of Bakou.*

The discoveries at Mentesh-Tepe are exceptional. Indeed, it is the first time that so many metallurgical artifacts—not only metal objects but also production waste and other elements testifying to metal production—have been discovered in a settlement of such ancient date in the Caucasus. In order to better understand and characterize the metallurgy at Mentesh-Tepe and, more generally, the metallurgy of the Sioni culture, we decided to engage in archaeometallurgical studies, the major results of which are presented here.

Archaeometallurgical Studies on the Artifacts of Mentesh-Tepe

These studies have been made by A. Courcier under the direction of A. Hauptmann[4] and M. Prange[5] in the laboratory of the Deutsches Bergbau-Museum at Bochum in Germany. A first series of analyses was done in spring 2009. A second series was completed in September 2010. A new series was planned for March–April 2011.

The Analytical Protocol

Metallography

Metallographic analyses have been done on the few objects that contain uncorroded metal. Unfortunately, the great majority of the artifacts are totally oxidized, and therefore the number of cross-sections is limited. Furthermore, some internal technical problems in the laboratory in September 2010 did not allow completion of the second series of analyses; they will be done later. We used an optical microscope

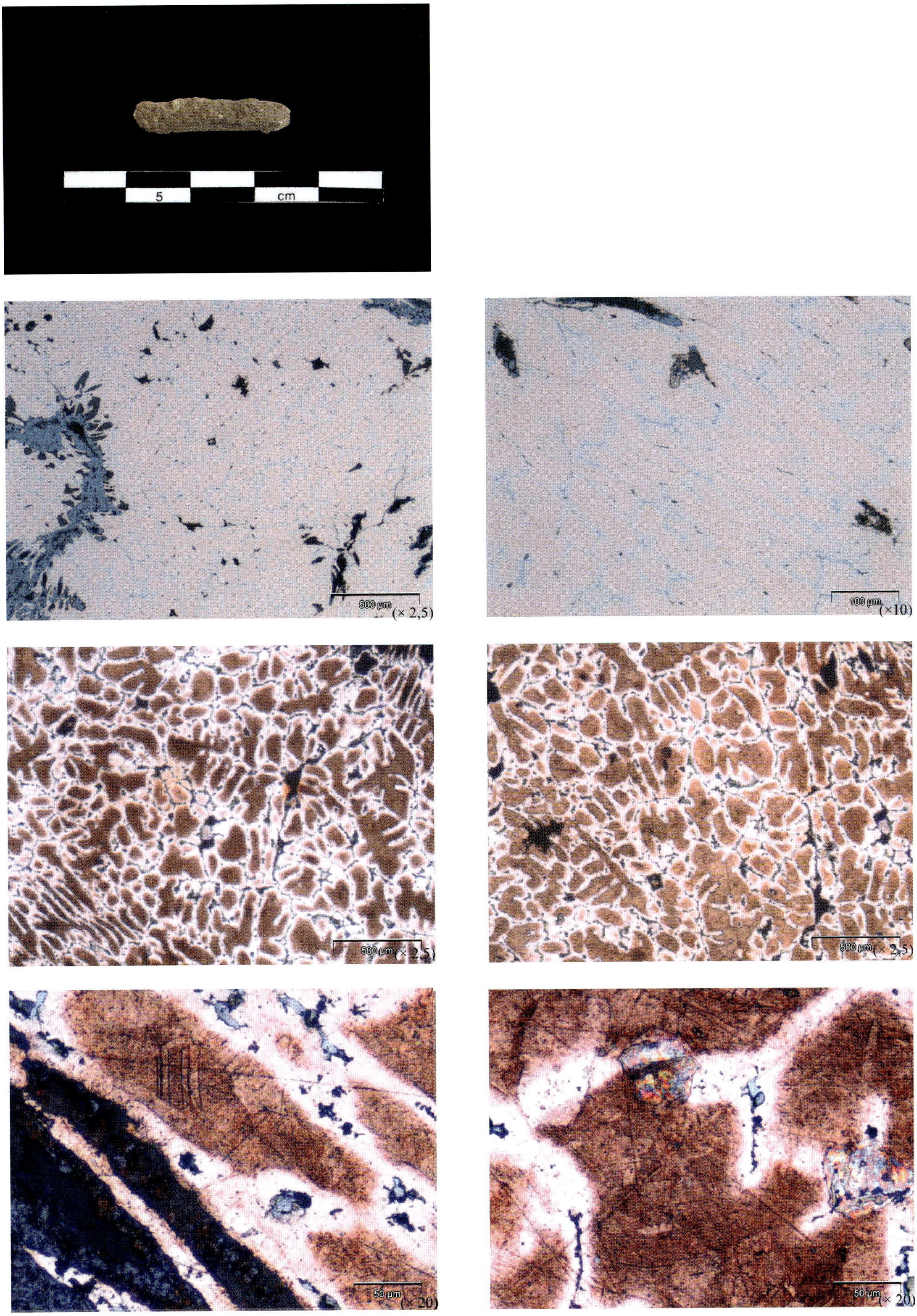

Figure 10*. Metallographic analysis of the undetermined object showing a dendritic, eutectoidal structure and annealing twins, Sioni culture (second half of the fifth millennium* BCE*), Mentesh-Tepe. Deutsches Bergbau-Museum.*

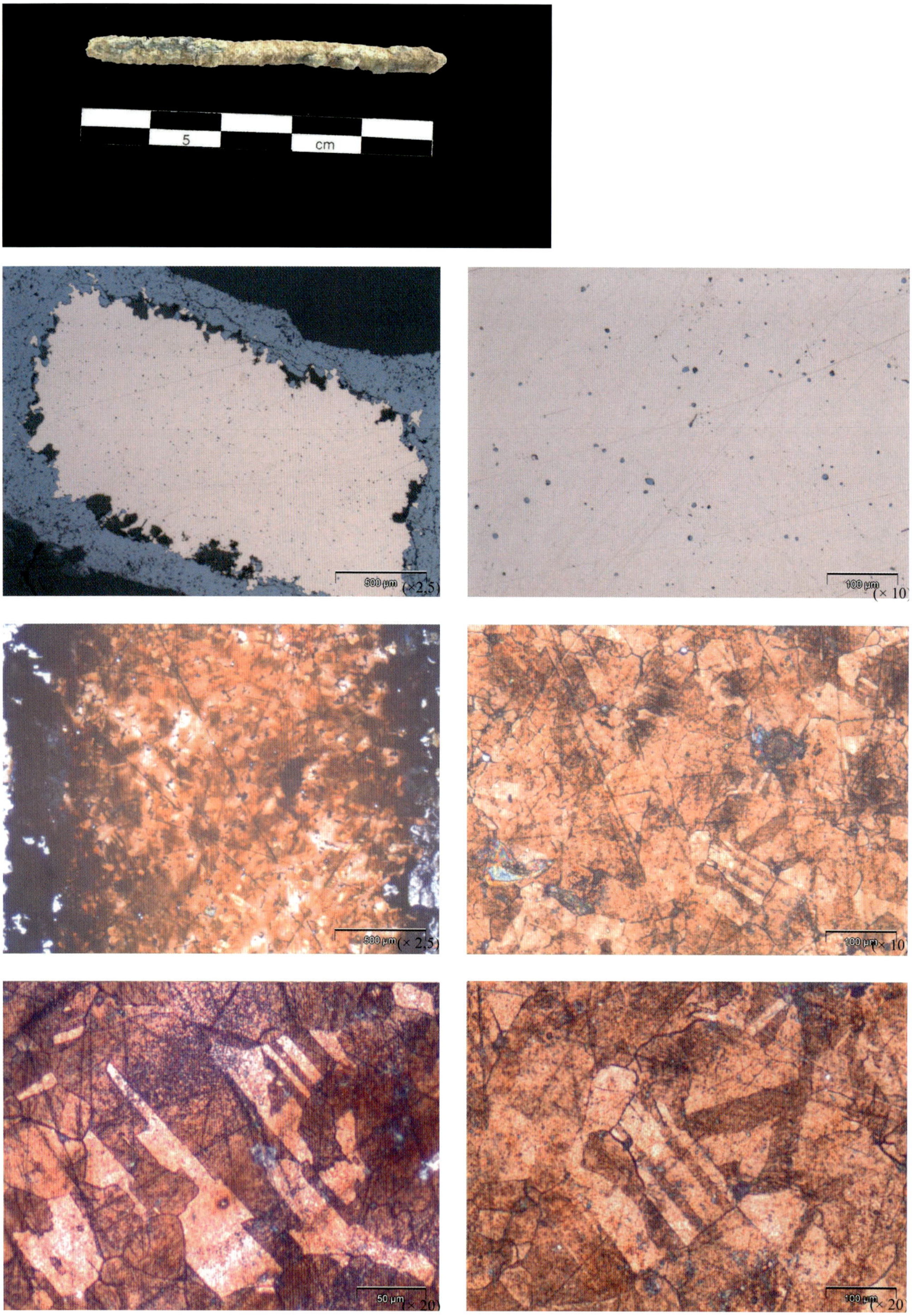

***Figure 11**. Metallographic analysis of an awl showing distorted dendrites, annealing twins, and thin strain lines, Sioni culture (second half of the fifth millennium* BCE*), Mentesh-Tepe. Deutsches Bergbau-Museum.*

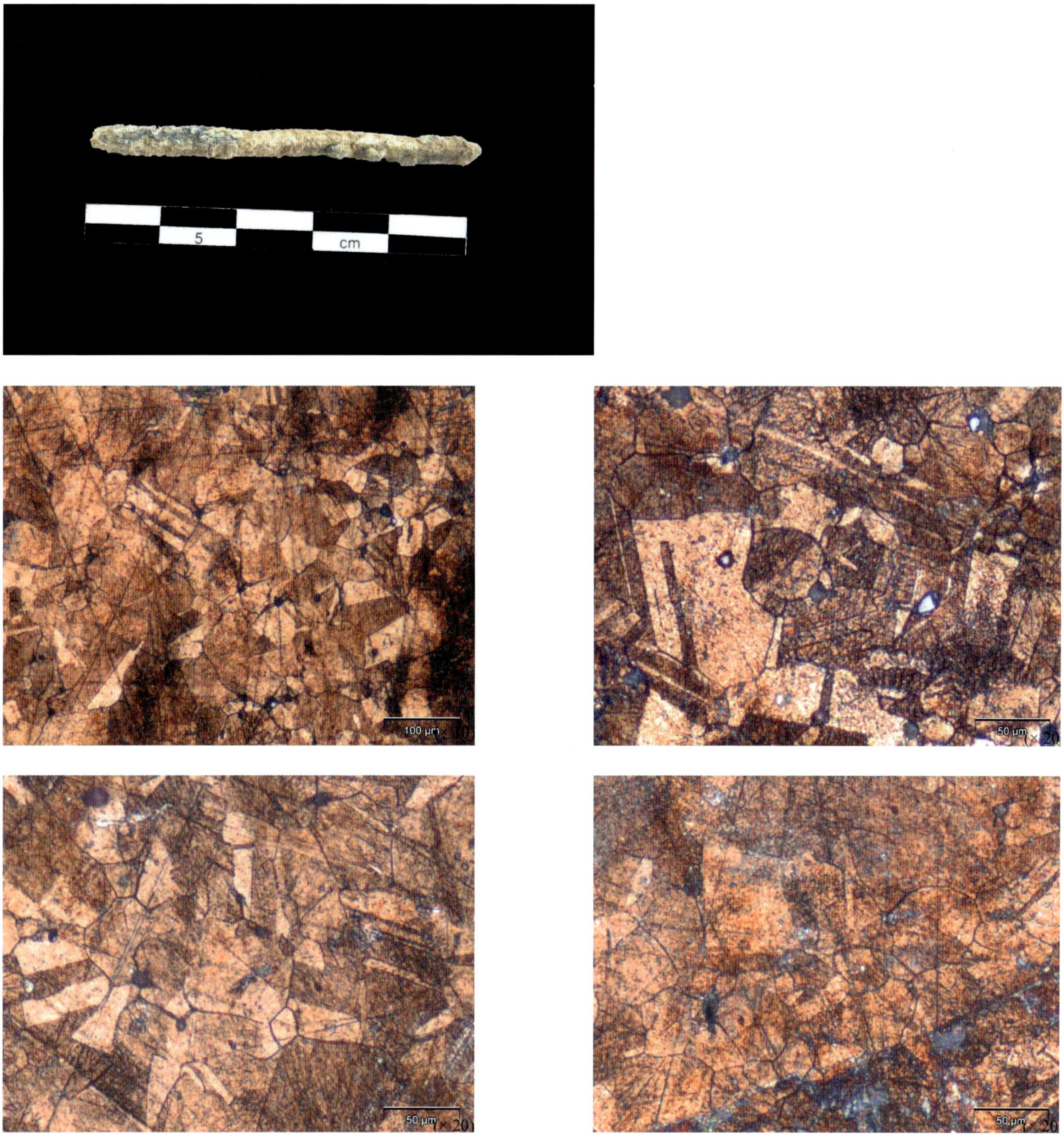

Figure 12. *Metallographic analysis of an awl showing details of the annealing twins and thin strain lines, Sioni culture (second half of the fifth millennium* BCE*), Mentesh-Tepe. Deutsches Bergbau-Museum.*

(OM) with Zeiss Axiophot capable of several magnifications (×12.5, ×25, ×50, ×100, ×200, and ×500). The etching agent selected for copper objects was hydrated ammonium peroxide.[6]

X-ray diffraction

X-ray diffraction (XRD) was carried out on two artifacts—small fragments of ore and an undetermined corroded agglomerate (fig. 8 a and f)—using an XpertPro Panalytical instrument with Cu Kα (1.5406 λ) radiation. We wanted to know if these artifacts contained a metallic or mineral matrix, or, in others words, we wanted to determine their nature: metal or ore.

Energy-dispersive x-ray spectrometry

In order to rapidly determine the composition of some objects, we used standardless energy-dispersive x-ray spectrometry (EDS) on a scanning electronic microscope (SEM). The instrument was a Noran System Seven Thermo UltraDry Silicon Drift X-ray Detector (SDD) on a Zeiss Supra 40VP SEM. We analyzed two objects: an object of unknown type and an awl (see fig. 6 a and r).

Inductively coupled plasma–mass spectrometry

In order to precisely characterize the chemical composition of the metal objects, the samples were quantitatively analyzed with inductively coupled plasma–mass spectrometry (ICP-MS) using an Element XR Thermo spectrometer. These quantitative analyses measure the trace elements. Samples taken for this analysis method by micro-drilling or by cutting exhibited significant or even total corrosion for the majority of the metal artifacts.

For analysis, we selected eight objects from the first campaign at Mentesh-Tepe and twenty other artifacts from among the thirty-six discovered during the last campaign of excavation. Our choice was determined by the quality of these artifacts and the possibility of measuring copper metal rather than copper oxides. We therefore rejected all the artifacts that were completely corroded or were, unfortunately, damaged in storage prior to analysis.

Lead isotope analysis by high resolution–single collector–inductively coupled plasma–mass spectrometry

We also began lead isotope analyses (LIA) with an Element XR Thermo single-collector spectrometer. The lead isotopes (^{206}Pb, ^{207}Pb, and ^{208}Pb) were analyzed separately. The internal precision is between 0.09% and 0.5%. The selected artifacts are the same as those that were studied by ICP-MS, and we began with the eight objects discovered during the first campaign.

Results

Metallographic analysis

Metallographic analysis of the unfinished awl (see fig. 6a) revealed a dendritic and eutectoidal structure ($Cu{+}CuO_2$) (fig. 10), which is a characteristic of a cast object. Etching permitted the identification of a few hexagonal grains with annealing twins in the alpha dendrites. They prove that a soft hammering and annealing step had been executed. Metallographic analysis done on the awl (see fig. 6r) showed some distorted dendrites (fig. 11), suggesting cold-working. Etching the section confirmed this hypothesis by revealing the presence of several hexagonal grains with annealing twins as well as grains with thin strain lines (see figs. 11 and 12). In sum, this object was cast, cold-hammered, annealed, and finally cold-worked again.

XRD analysis

XRD analysis of the small fragments (see fig. 8a) confirmed that this is a copper ore (fig. 13). This ore is characterized by a mineral complex of azurite ($Cu_3(OH)_2(CO_3)_2$), muscovite ($KAl_2AlSi_3O_{10}(OH)_2$), clinochlore ($(Mg,Fe^{2+})_5Al[(OH)_8AlSi_30_{10}]$), and quartz ($SiO_2$).

XRD analysis of the undetermined corroded agglomerate (see fig. 8f) gave evidence of metal oxides (fig. 14): cuprite (Cu_2O), paratacamite ($Cu_3(Cu,Zn)[(OH)_3Cl]_2$), atacamite ($Cu_2[(OH)_3Cl]$), and malachite ($Cu_2[(OH)_2CO_3]$). This result proves that these are metal particles that were agglomerated by oxidation. Our initial visual observation had led us to suppose that the spherical or oval particles were metal prills.

EDS analysis

According to the EDS analysis, the unfinished awl (see fig. 6a) is made of relatively "pure" copper (90.0–96.4 wt.% Cu) with minor amounts of arsenic (0.8–3.8 wt.% As) (fig. 15).

A similar study on an awl (see fig. 6r) revealed that it also was made of relatively "pure" copper (91.5–97.9 wt.% Cu) with traces of arsenic (0.7–1.4 wt.% As) and sulfur (0.5–0.9 wt.% S). EDS mapping shows homogeneous distribution of sulfur, which may correspond to small inclusions seen within the grains (fig. 16). Arsenic is concentrated at the edge of the object, a consequence of the natural segregation of this element.

ICP-MS analysis

The ICP-MS analysis performed on the objects discovered during the 2008 campaign confirm the EDS results (table 3). The unfinished awl (see fig. 6a) and one of the awls (see fig. 6r) are "pure" copper (96.9 and 100 wt.% Cu, respectively) with minor amounts of arsenic (3.5 and 0.7 wt.% As, respectively). The unfinished awl also contains traces of antimony (0.2 wt.% Sb). Except for another awl (see fig. 6t) that also

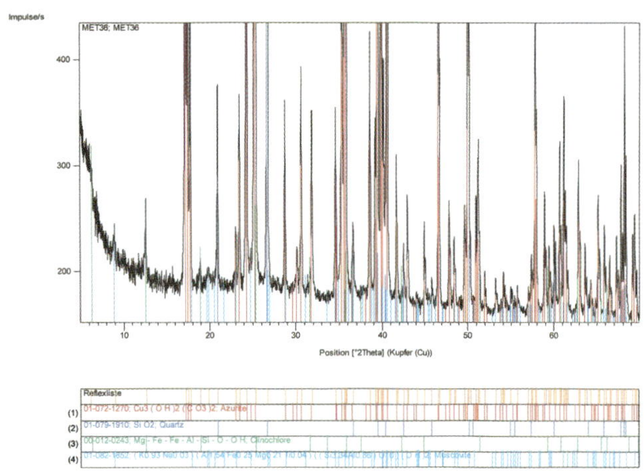

Figure 13. *Phases found by XRD analysis of small fragments of azurite ore included (1) Azurite, reference pattern 01-072-1270 in red, (2) Quartz, reference pattern 01-079-1910 in dark blue, (3) Clinichlore, reference pattern 00-012-0243 in green, and (4) Muscovite, reference pattern 01-082-1862 in blue. Sioni culture (second half of the fifth millennium bce), Mentesh-Tepe. Deutsches Bergbau-Museum.*

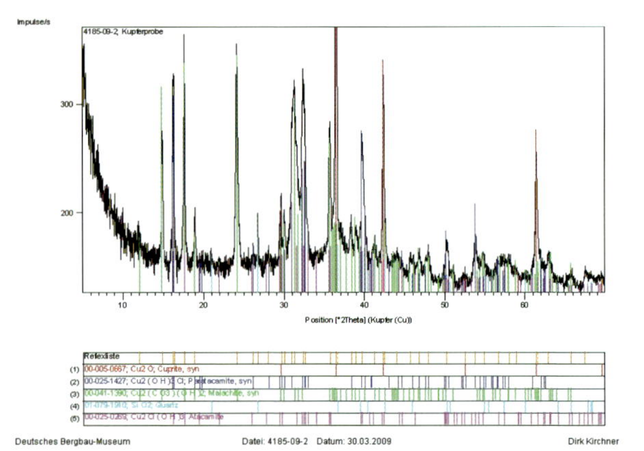

Figure 14. *Phases found by XRD analysis on undetermined corroded amalgam of oxidized prills included (1) Cuprite, reference pattern 00-005-0667in red, (2) Paratacamite, reference pattern 00-025-1427 in purple, (3) Malachite, reference pattern 00-041-1390 in green, (4) Quartz, reference pattern 01-079-1910 in blue, and (5) Atacamite, reference pattern 00-025-0269 in pink. Sioni culture (second half of the fifth millennium bce), Mentesh-Tepe. Deutsches Bergbau-Museum.*

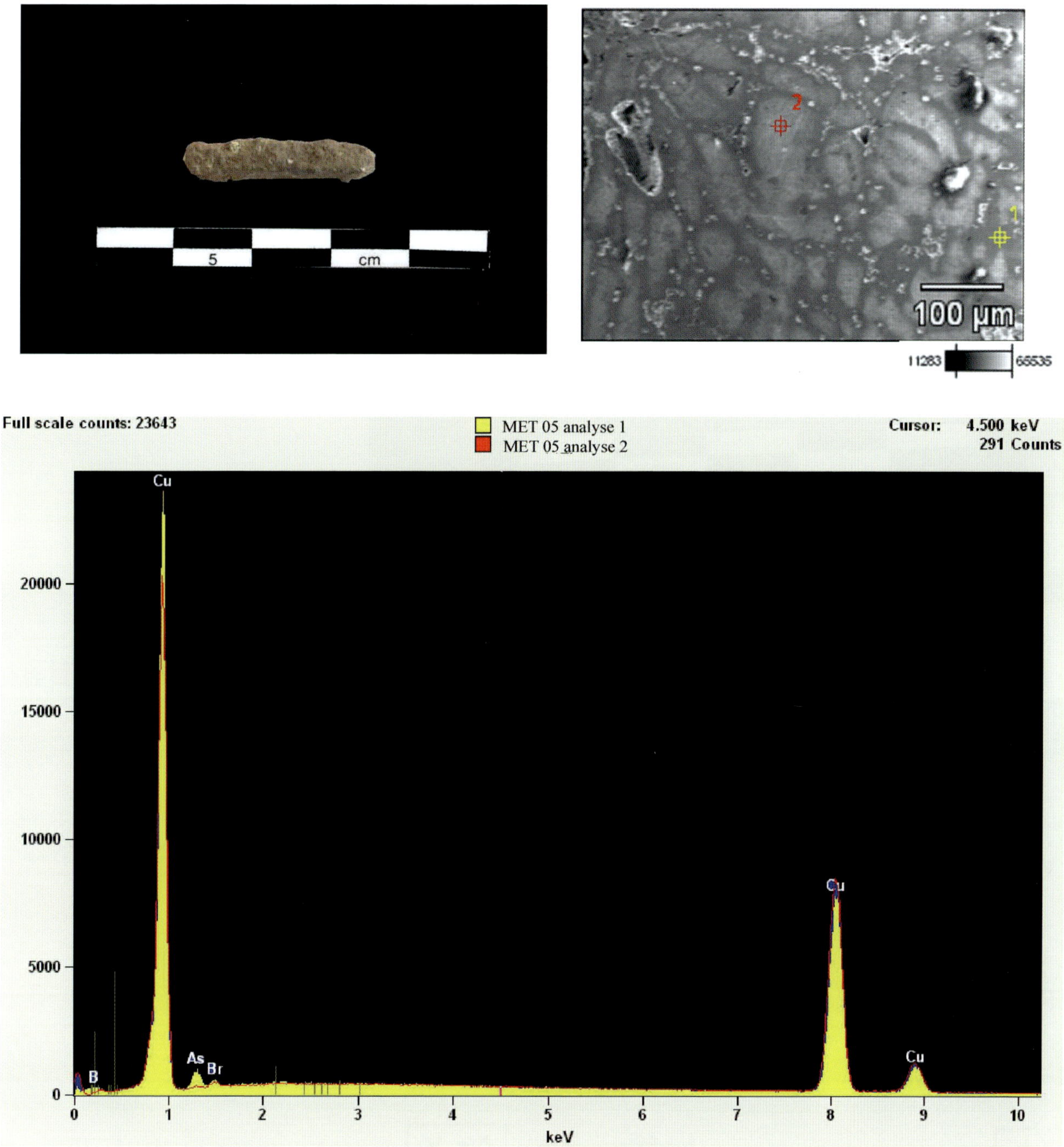

Weight %

	B-K	C-K	Al-K	Cu-K	As-K	Br-K
MET 05 analyse 1	6.22			89.96	3.82	0.00
MET 05 analyse 2		2.23	0.57	96.38	0.82	

***Figure 15**. Results of standardless EDS analysis on an undetermined object, spectra and table of results, Sioni culture (second half of the fifth millennium BCE), Mentesh-Tepe. Deutsches Bergbau-Museum.*

contains minor amounts of arsenic (1.2 wt.% As), all the other objects are made of relatively "pure" copper (see table 3). In two cases (see fig. 6 c and e), the amount of copper is less than the others. Their degree of corrosion explains this difference. The metal objects are probably not alloyed; the amounts of arsenic detected in some can be explained by the type of ore used.

The amount of sulfur previously detected by EDS in one of the awls is probably too high given the ICP-MS results (see table 3). The ICP-MS analyses only found trace amounts of sulfur, ranging between 0.0110 and 0.0180 wt.% S, for the artifacts analyzed.

The ICP-MS data from the objects from the 2010 campaign agree with these results. The artifacts are "pure" copper (table 4). The amount of copper is, however, less than that of the previous series of objects (see table 3). Indeed, some artifacts discovered during the last excavations are more corroded than the earlier finds. A minor amount of arsenic (1.45 wt.%) was detected in the fragment of blade. It is worth noticing the higher amount of sulfur (8.10 wt.%), that was measured in a thin metal plate with attached prills (see fig. 8g; table 4). This characteristic may be explained by the ore that was used.

Lead isotope analysis

Lead isotope results are presented in table 5. Their number is, however, insufficient to allow any hypotheses. In order to establish eventual correlations, we need LIA of ores from deposits close to Mentesh-Tepe. Furthermore, the data have a relatively large error range (ratio RSD: 0.08% to 0.5%) for the analyses done with high resolution–single collector–inductively coupled plasma–mass spectrometry (HR-SC-ICP-MS), making it impossible to compare it with data taken in other analysis sessions on ancient metal artifacts from Georgia and Armenia.

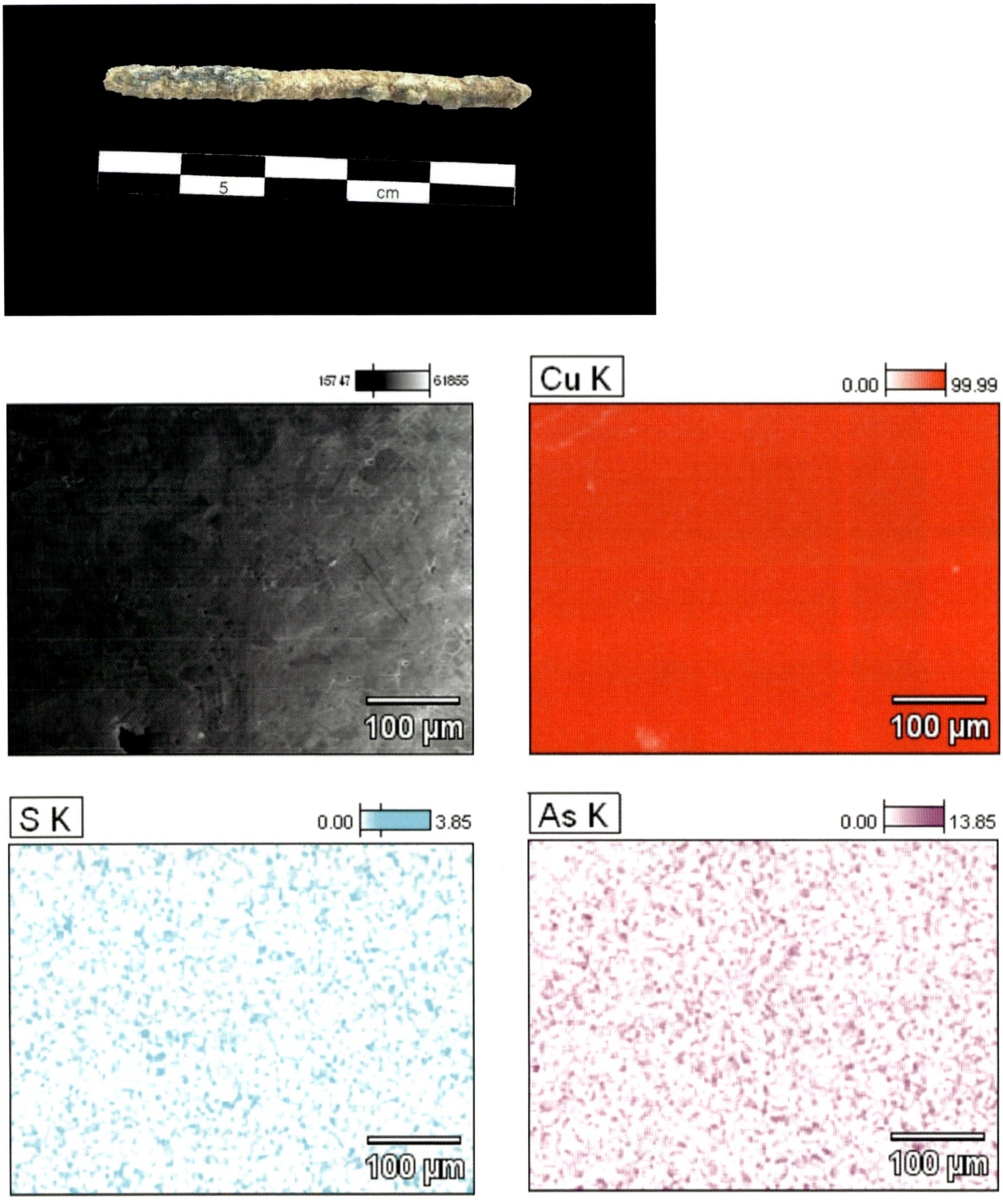

***Figure 16**. EDS mapping of a transverse section of an awl showing a homogeneous distribution of sulfur and arsenic throughout the core, Sioni culture (second half of the fifth millennium BCE), Mentesh-Tepe. Deutsches Bergbau-Museum.*

Table 3. *ICP-MS chemical compositions of selected metallurgical artifacts discovered at Mentesh-Tepe during the 2008 campaign.*

General Information			Composition (wt.%)																	
Inv. No.	Type of Artifacts	Analysis No.	Cu	Sn	Pb	Ag	As	Au	Bi	Co	Fe	Hg	Ni	P	S	Sb	Se	Te	Sn	Total
MET 03_2008	awl	4180/09	96.9	0.098	0.001	0.042	1.220	< LOD	< LOD	< LOD	0.001	0.001	0.003	< LOD	0.013	< LOD	0.001	< LOD	< LOD	98.24
MET 04_2008	ring	4181/09	99.0	0.091	0.007	0.072	0.009	< LOD	< LOD	< LOD	0.003	0.001	0.003	0.001	< LOD	< LOD	0.001	< LOD	< LOD	99.22
MET 05_2008	fragment of unfinished awl	4182/09	96.3	0.093	0.002	0.051	3.572	< LOD	0.007	< LOD	0.009	0.001	0.019	< LOD	< LOD	0.268	0.055	0.003	< LOD	100.37
MET 06_2008	fragment of unfinished awl	4183/09	98.3	0.093	0.010	0.017	0.020	< LOD	< LOD	< LOD	0.237	0.001	0.001	0.019	0.015	< LOD	0.006	< LOD	0.001	98.72
MET 07_2008	agglomerate of metal prills	4184/09	59.2	0.092	< LOD	< LOD	0.002	< LOD	< LOD	< LOD	< LOD	0.001	< LOD	< LOD	< LOD	< LOD	< LOD	< LOD	< LOD	59.34
MET 08_2008	semismelted ore	4185/09	66.0	0.098	0.029	0.032	0.433	< LOD	0.006	< LOD	0.143	0.001	0.007	0.021	0.018	0.002	0.003	< LOD	0.001	66.84
MET 10_2008	awl	4186/09	100.7	0.093	0.002	0.019	0.764	< LOD	< LOD	< LOD	0.051	0.001	0.003	0.004	0.003	< LOD	0.001	< LOD	0.001	101.68
MET 11_2008	agglomerate of metal prills	4187/09	98.5	0.094	0.008	0.035	0.111	0.001	0.001	< LOD	0.186	0.001	0.001	0.026	0.011	< LOD	0.007	< LOD	0.002	99.01

LOD = limit of detection.

Table 4: *ICP-MS chemical compositions of selected metallurgical artifacts discovered at Mentesh-Tepe during the 2010 campaign.*

General information			Composition																
Inventory number	Artifact type	Analysis number	Cu (wt%)	Fe (wt%)	As (wt%)	S (wt%)	Pb (wt%)	Ag (wt%)	P (wt%)	Sb (ppm)	Te (ppm)	Au (ppm)	Bi (ppm)	Hg (ppm)	Co (ppm)	Ni (ppm)	Zn (ppm)	Sn (ppm)	Se (ppm)
MET 01_2010	Thin metal plate with unmelted prills	4613-10	49.8	0.19	0.66	0.03	0.06	0.004	0.05	87	1.0	2.9	0.7	<1	3.1	7.9	11	2.1	85
MET 02_2010	Awl fragment	4614-10	58.6	0.04	0.01	0.08	0.01	0.01	0.01	6.5	4.0	36	3.8	<1	0.3	1.8	12	1.4	23
MET 03_2010	Blade fragment	4615-10	58.6	0.50	1.45	0.05	0.21	0.002	0.04	170	<1	0.8	16	<1	1.5	145	29	<1	<1
MET 04_2010	Awl fragment	4616-10	61.1	0.10	0.09	0.03	0.09	0.02	0.03	42	<1	1.9	4.7	<1	0.7	38	7.6	<1	54
MET 06_2010	Awl fragment	4617-10	88.3	0.04	0.02	<0.001	0.03	0.02	0.01	6.8	<1	0.7	0.4	<1	0.2	63	7.8	<1	2.8
MET 07_2010	Thin metal plate with unmelted prills	4618-10	57.9	0.29	0.46	0.15	0.11	0.01	0.12	67	3.2	3.6	3.5	<1	0.7	14	9.8	<1	18
MET 08_2010	Awl fragment	4619-10	55.3	0.03	0.01	0.06	0.01	0.002	0.02	1.6	<1	0.4	0.1	<1	0.1	1.2	2.0	<1	16
MET 11_2010	Agglomerate of metal prills	4620-10	89.1	0.002	0.40	<0.001	0.02	0.02	0.000	130	<1	1.7	10	<1	0.5	450	2.7	<1	1.9
MET 12_2010	Thin metal plate	4627-10	96.3	0.01	0.01	<0.001	0.34	0.03	0.002	185	<1	90	28	<1	<0.01	3.3	20	13	11
MET 14_2010	Awl	4621-10	50.9	0.15	0.02	0.005	0.004	0.01	0.06	4.0	<1	2.2	0.4	<1	2.5	5.6	5.2	2.2	38
MET 16_2010	Undetermined object	4622-10	91.2	0.01	0.002	0.01	0.03	0.03	0.002	51	<1	21	0.7	<1	<0.01	23	54	<1	19
MET 20_2010	Awl fragment	4623-10	71.3	0.04	0.02	0.05	0.06	0.02	0.02	14	<1	1.1	0.9	<1	0.3	56	<1	<1	44
MET 22_2010	Awl	4624-10	85.7	0.07	0.01	0.01	0.29	0.02	0.03	5.0	<1	0.5	0.1	<1	0.1	3.1	35	<1	56
MET 23_2010	Ring	4625-10	57.0	0.07	0.03	<0.001	0.03	0.01	0.03	12	<1	0.3	0.4	<1	0.1	12	<1	<1	<1
MET 28_2010	Semi-smelted ore	4628-10	34.4	15.3	0.002	8.10	0.06	0.02	0.02	<1	<1	1.5	0.1	<1	115	11	7.1	<1	33
MET 29_2010	Thin metal plate with prills unmelted	4626-10	57.0	0.07	0.02	0.001	0.02	0.00345	0.027	10	<1	0.1	0.3	<1	0.2	0.4	4.0	<1	4.3
MET 30_2010	Thin metal plate	4629-10	80.4	0.06	0.45	0.003	0.02	0.02	0.01	165	<1	0.3	3.4	<1	29	2150	3.9	<1	5.1
MET 31_2010	Ring	4630-10	63.6	0.45	0.004	<0.001	0.01	0.004	0.03	2.8	<1	0.2	0.1	<1	1.4	6.6	12	50	64
MET 32_2010	Ring or curved awl (?)	4631-10	90.3	0.03	0.000	<0.001	0.03	0.01	0.004	<1	<1	0.1	<0.05	<1	0.50	25	1.9	<1	9.6

***Table 5**. HR-SC-ICP-MS lead isotope results of selected metal objects discovered at Mentesh-Tepe.*

General Information			Results					
Inv. No.	Artifact Type	Analysis No.	Pb208/Pb207	Pb208/Pb206	Pb206/Pb204	Pb207/Pb204	Pb208/Pb204	Pb207/Pb206
MET 03_2008	awl	4180/09	2.445	2.024	19.342	16.015	39.149	0.828
MET 04_2008	ring	4181/09	2.477	2.056	18.881	15.668	38.813	0.830
MET 05_2008	fragment of unfinished awl	4182/09	2.467	2.064	18.805	15.729	38.806	0.836
MET 06_2008	fragment of unfinished awl	4183/09	2.473	2.070	18.808	15.738	38.928	0.837
MET 07_2008	agglomerate of metal prills	4184/09	2.446	2.004	19.218	15.743	38.505	0.819
MET 08_2008	semismelted ore	4185/09	2.471	2.070	18.780	15.728	38.870	0.837
MET 10_2008	awl	4186/09	2.466	2.048	19.063	15.834	39.049	0.831
MET 11_2008	agglomerate of metal prills	4187/09	2.472	2.075	18.702	15.699	38.802	0.839

Mentesh-Tepe: A Settlement Specialized in Metallurgical Production?

Mentesh-Tepe seems to have been a site with intensive metallurgical activities. We believe that local copper deposits were exploited (see fig. 3).

The discovery of fragments of crushed azurite (see fig. 6a) and heavy stone tools (mortars, pestles, and stone with cupula; see Lyonnet and Gulyev 2009, 39) suggest that crushing activities were probably practiced at the site. It is probable that the copper ore used was not limited to carbonates (azurite) but that ores containing arsenic (i.e., enargite, luzonite, domeykite, or sinnerite) were also used as they are present in local copper deposits (Bitty-Bulaq, Kedabek; see Cassard et al. 2009).

The agglomerates of ore and oxide products (see fig. 8 b and c) could be the result of the beginnings of a smelting process that separated the matrix from the metal, which concentrated into prills. According to their relative densities, the slag remains on the surface whereas the prills form on the bottom. The discovery of two fragments of crucibles seems to support this (see fig. 9 a and b). Additionally, all the prills are collected together. This next step is illustrated by the strongly corroded agglomerates of prills (see fig. 8 d–i).

The fragment of ingot mold (see fig. 9c) suggests the practice of melting, in spite of the fact that the mold did not show any trace of use. Indeed, some of the plates (see fig. 8 j–n) seem to be an intermediate product between prills and ingot.

The examined metallographic sections of the undetermined object and the awl (figs. 10–12) demonstrate the use of casting and the alternation of cold hammering and annealing. They also suggest a manufacturing scheme for the awls in which the undetermined object could be an unfinished awl. After casting, cold hammering alternating with annealing would have given a square section to the object. The practice of cold hammering and annealing formed a thinner and elongated shape that in the end resulted in an awl.

The objects are probably of unalloyed copper. The minor amount of arsenic identified in some of the objects can be explained by the ores used. Indeed, some of the local copper deposits are copper ores naturally associated with arsenic. This question of whether the alloys are intentional or accidental will be re-examined in the future when we have additional results.

Conclusion: Future Research

Our future research will follow two main directions correlated with each other. First, we would like to complete our archaeometallurgical investigations. Second, we would like to intensify our fieldwork at Mentesh-Tepe and in the surrounding areas.

The first steps of a metallographic analysis have been done on a fragment of awl (see fig. 6 n, p, and s), a ring (see fig. 7c), and a thin plate (see fig. 8l). We have already detected that they still contain uncorroded metal. Completion of these studies should permit us to understand metal manufacturing processes during the Chalcolithic period.

We also wish to conduct further research on the microstructures of semismelted ores and corroded agglomerates with an optical microscope (see fig. 8 b and d–i) to improve our understanding of the smelting process.

We will make thin-sections on the two fragments of crucibles (see fig. 9 a and b) and determine the composition of the prills trapped in the first fragment of crucible by EDS analysis. The aim is twofold for the second crucible: first, we would like to identify the different constitutive phases of the slag encrustations using observations made with an optical microscope, and second, we want to analyze these phases and the prills trapped inside by EDS. In doing so, we will be able to compare the composition of the prills in these two crucibles. We also hope to be able to estimate if a correlation exists between the prills trapped in the crucibles and those found in the agglomerates. The EDS on the slaggy encrusting remains should give us some information about the gangue of the ore used.

We will continue the LIA of the other metal artifacts. In order to reduce the error mentioned above, we will use an isotope ratio multicollector ICP-MS (HR-MP-ICP-MS) (Thermo Scientific NEPTUNE equipped with multicollector platform TRITON Thermo). The aim is to obtain a sufficient number of results in order to establish an isotopic signature of the metal objects at Mentesh-Tepe.

The second direction of research will concern fieldwork at Mentesh-Tepe. During the next campaign, we will increase our efforts to understand the links between the different zones of the site where metallurgical activities have been identified in order to eventually determine a *chaîne opératoire* between the different practices. We also hope to discover more metallurgical evidence and, in particular, furnaces.

We will also focus efforts on the identification of ancient traces of exploitation in the ore deposits close to the site (i.e., Dagkesaman, Gosha). In a previous archaeometallurgical survey in 2007, we observed evidence suggesting the possibility of ancient activities in these deposits. We need to verify this by detailed investigations using methods like soundings. We will collect ore in order to make lead isotope analyses and compare the results with those of the objects and of the ore previously discovered at Mentesh-Tepe.

Acknowledgments

I present my warmest thanks to Bertille Lyonnet and Farhad Guliyev for their authorization to study the metal artifacts of Mentesh-Tepe. I also address my deepest gratitude to Professor Andreas Hauptmann and Dr. Michael Prange for their generosity and their scientific support during these analyses at Bochum.

Notes

1. The general aim of the ANR Project—Ancient Culture of the Southern Part of the Caucasus, "Ancient Kura"—is to understand the paleo-environment, way of life, and exploitation of resources in the middle valley of the Kura River, throughout the period spanning the sixth millennium BCE to the third millennium BCE.
2. Dr., Deutsches Archäologisches Institut, Eurasien-Abteilung, Berlin, Germany.
3. ^{14}C results of the ancient levels of Mentesh-Tepe: **Mentesh07-07** (sounding 2008, hearth close to the wall, Z: -1,60m): 5330 ± 50 BP, [2 sigma] 4320–4290 cal BCE (6270–6240 cal BP), 4270–4030 cal BCE (6220–5980 cal BP), 4020–3990 cal BCE (5970–5940 cal BP); **Mentesh08-22** (under roof collapse): 5390 ± 40 BP, [2 sigma] 4340–4220 cal BCE (6290–6170 cal BP), 4200–4160 cal BCE (6150–6110 cal BP), 4120–4110 cal BCE (6070– 6060 cal BP), 4100–4070 cal BCE (6050–6020 cal BP); **Mentesh08-44** (under level of bricks, zone 3): 4590–4450 cal BCE.
4. Prof. Dr., Deutsches Bergbau-Museum, Archaeometallurgy, Bochum, Germany.
5. Dr., Deutsches Bergbau-Museum, Archaeometallurgy, Bochum, Germany.
6. 25 ml H_2O, 25 ml NH_4OH, 5–25 ml H_2O_2.

References

Abibullaev, O. A. 1965. K voprosu o drevnej metallurgii Azerbajdzhana [About the ancient metallurgy in Azerbaijan]. *Materialy i isselovanija po arkheologii SSR* [Materials for the Study of Archaeology in the SSR]125:65–73.

———. 1982. *Eneolit i bronza na territorii Nakhichevanskoj ASSR* [Chalcolithic and Early Bronze Age in the region of Nakhichevan in Soviet Azerbaijan]. Bakou: Nauka.

Akhundov, T. 2007. Sites de migrants venus du Proche-Orient en Transcaucasie. In *Les cultures du Caucase (VIe—IIIe millénaires avant notre ère). Leurs relations avec le Proche-Orient,* ed. B. Lyonnet, 95–121. Paris: CNRS Editions/Editions Recherche sur les Civilisations.

Badalyan, R., P. Lombard, P. Avetisjan, C. Chataigner, J. Chabot, E. Vila, R. Hovsepyan, G. Willcox, and H. Pessin. 2007. New data on the late prehistory on the southern Caucasus. The excavations at Aratashen (Armenia): Preliminary report. In *Les cultures du Caucase (VIe–IIIe millénaires avant notre ère). Leurs relations avec le Proche-Orient,* ed. B. Lyonnet, 37–62. Paris: CNRS Editions/Editions Recherche sur les Civilisations.

Cassard, D., A. Courcier, M. Billa, C. Gateau, A. Lafitte, B. Lyonnet, F. Maldan, and J.-F. Mallon. 2009. *SIG*: *Les cultures du Caucase et la métallurgie, 6ᵉ–3ᵉ millénaires (Sud Russie, Républiques du Nord Caucase, Géorgie, Arménie, Azerbaïdjan, Iran du NO, Turquie orientale).* Projet CNRS-BRGM.

Courcier, A. In press. Ancient metallurgy in the Caucasus from the middle of the Vth/beginnings of the IVth to the IIIrd millennium BC. In *A global perspective in early metallurgy*, ed. B. W. Roberts and C. P. Thornton. New York.

Formozov, A. A. 1965. *Kamennyj vek i eneolit Prikuban'ja* [Neolithic and Chalcolithic in the Kuban region]. Moscow: Nauka.

Iessen, A. A., and B. E. Degen-Kovalevskij. 1935. *Iz istorii drevnej metallurgii Kavkaza* [About the history of the ancient metallurgy in the Caucasus]. Moscow: Gosudarstvennoe Sotsial'no Ekonomicheskoe Izdatel'stvo "Ogiz."

Kiguradze, T. V. 2000. The Chalcolithic–Early Bronze Age transition in the eastern Caucasus. In *Chronologie des pays du Caucase et de l'Euphrate au IVᵉ–IIIᵉ millénaires*, ed. C. Marro and A. Hauptmann, 321–28. Istanbul and Paris: IFEA, de Boccard.

Kruglov, A. P., J. J. Piotrovskij, and G. V. Podgaetskij. 1941. Mogil'nik v g. Nal'chike [The cemetery close to the town of Nal'chik]. *Materialy i isselovanija po arkheologii SSR* [Materials for the Study of Archaeology in the SSR] 3:67–135 and Pl. I–XI.

Kuftin, B. A. 1950. *Arkheologicheskie izyskanija v Rionskoj nizmennosti i na chernomorskom poberezh'i 1935 i 1936 godov* [Archaeological surveys in the inferior valley of the Rioni River in the soils of the Black Sea revealed during the years 1935 and 1936]. In *Materialy k Arkheologii Kolkhidy* [Colchian Materials and Archaeology], Vol. 2. Tbilisi: Tekhnika da Shroma.

Lordkipanidze, O. 1989. *Naselenie drevnej Gruzii* [The ancient population of Georgia]. Tbilisi: Metsniereba.

Lyonnet, B., and F. Gulyev. 2009. Mission "Boyuk Kesik", Azerbaidjan (campagnes 2006–2008). Report to the Ministère des Affaires Etrangères, Paris.

———. 2010. La 2ème campagne de fouilles (2009) sur Mentesh-Tepe. Report to the Ministère des Affaires Etrangères.

Narimanov, I. G. 1985. Raskopki poselenija Chalagan Tepe [The excavations in the settlement of Chalagan Tepe]. *Arkheologicheskie otkrytija 1983 goda* [Archaeological discoveries in 1983]1985: 489–90.

———. 1987. *Kul'tura drevnejshego zemledel'chesko-skotovodcheskogo naselenija Azerbaidzhana* [The ancient cultures of the farming and cattle-breeding inhabitants of Azerbaidjian]. Baku: Elm.

Varazashvili, V. L. 1992. *Rannezemledel'cheskaja kul'tura jugo-alazonskogo bassejna* [The ancient cultures in southern basin of the Alazani River]. Tbilisi: Metsniereba.

The Casting Techniques of Antique South Arabian Large Bronze Statues

Benoît Mille

ABSTRACT South Arabian large bronze statues have garnered little scientific interest despite their extraordinary features. The bronze statues presented here cover the whole period of the South Arabian kingdoms, from the seventh century BCE to the third century CE. These bronzes were studied in the frame of a cooperative agreement between the General Organization of Antiquities and Museums in Yemen and the Louvre Museum. The most prominent statue is Hawtar'athat, a 1.4 m male standing figure discovered in Nashqum in the Jawf region, dated by its inscription to the time from the late seventh century to the first half of the sixth century BCE. In 2008, a French mission to the Yemeni museums was organized in order to put Hawtar'athat into a better techno-historical context by studying a comprehensive corpus of South Arabian statues and statuettes. The results achieved so far show that the manufacturing methods used for South Arabian large bronzes share two main technical principles with those of Greek and Roman large bronze statues—namely, an indirect lost-wax process is used for hollow casting, and primary castings are assembled by means of flow fusion welding. Nevertheless, while South Arabian large bronze statues were constructed using the same technical methods as the Greek and Roman ones, almost every detail in the way these methods were carried out is different, thus defining a proper South Arabian technology. The Mediterranean influence is therefore noticeable only in style and iconography, not in the technology. As a correlated consequence, this study raises the delicate problem of the origin of the techniques used to make the antique large bronze statues. Archaic South Arabian and Greek bronze statues share two main principles of manufacture, suggesting a common technological inheritance. The Middle East and Egypt, two areas that had a long history of manufacturing large bronze statues, are appropriate candidates to explore in future investigations.

Introduction

Ancient South Arabia, the region known in antiquity as *Arabia Felix* (present-day Yemen), is most renowned for the illustrious queen of Saba, the long-distance camel caravan trade for frankincense, and fine alabaster sculptures (Robin and Vogt 1997). In contrast, South Arabian large bronze statues have received little scientific interest despite their extraordinary features. Our knowledge of South Arabian cultural heritage is still very limited, since the first large-scale excavations only date back sixty years. Nonetheless, the bronze statues uncovered so far are already numerous enough to predict that metal statuary was of special importance in the South Arabian civilization, particularly in comparison to its role in other contemporaneous cultural areas, such as Egypt or Greece. The bronze statues that will be presented in this study cover the entire time of the South Arabian kingdoms, from the seventh century BCE to the third century CE. Saba (capital, Marib) was the most prominent kingdom of South Arabia; other kingdoms to be mentioned are Qataban (capital, Tamna) and Hadramawt (capital, Shabwa). Also to be noted is the Jawf region, composed of a few small independent kingdoms which were defeated at the end of the eighth century BCE by the Sabaean king Karib'il Watar, son of Dhamar'ali. The Jawf was then colonized by the Sabaeans (fig. 1). Recently, intensive illegal excavation activities have occurred in the Jawf leading to the discovery and subsequent sale on the art market of many bronzes.

The Louvre–General Organization of Antiquities and Museums in Yemen Cooperation

In 2006, thanks to the initiative of the French embassy in Yemen, the Louvre Museum signed a cooperative agreement with the General Organization of Antiquities and Museums in Yemen (GOAM). The aim of the cooperation was both to promote the Yemeni cultural heritage in France through the organization of several exhibitions in the Louvre Museum and to bring some assistance and training to Yemeni museums. A decision was made to focus on South Arabian bronzes, and the first scheduled operation was the restoration and study of Hawtar'athat, a 1.4 m male standing figure discovered in Nashqum in the Jawf and acquired by the National Museum of Sanaa in 2004 (Mille et al. 2010) (fig. 2). This statue is the tallest known bronze figure of an Archaic style. The statue arrived in 2006 in France and was restored at the Centre de Recherche et de Restauration des Musées de France (C2RMF) in collabora-

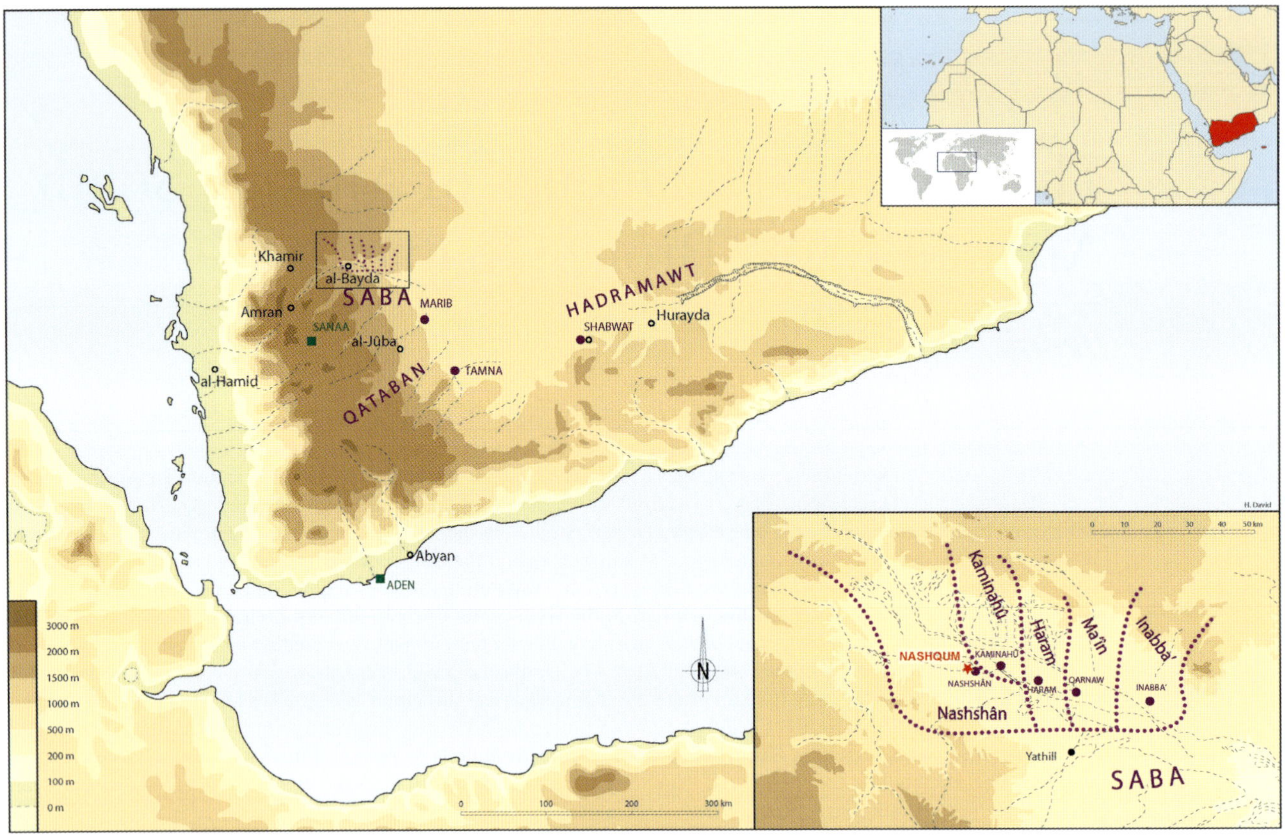

Figure 1. Arabia Felix *(present-day Yemen), between the eighth and the fifth century* BCE. *South Arabian kingdoms and their capitals are named in purple. A detailed view of the Jawf region is shown. Taken from a map by H. David in Robin and Vogt (1997, 88).*

tion with Yemeni restorers from the National Museum of Sanaa. It was exhibited to the public in the Department of Oriental Antiquities of the Louvre Museum (May–October 2007). Concurrently, technological, stylistic, and epigraphic studies were conducted. A Sabaean inscription carved on the stomach was discovered during the cleaning of the statue. Its decipherment provided key information for putting this extraordinary bronze statue into context. From the inscription, it was learned that a settler from Marib living in the Sabaean colony of Nashqum located in the Jawf decided to offer a bronze image to Almaqah, the main god of the powerful kingdom of Saba. The dedicant was Hawtar'athat, son of Radawil, from the lineage of Shalalum, and he lived during the reign of the king Yada'il Bayan. Further study established that the corresponding king was very probably in power during the late seventh century or the first half of the sixth century BCE, thus providing a date for the statue.

Because the study of the Hawtar'athat statue was successful in producing results about both its history and the technology used to create it, a three-week mission to the Yemeni museums was organized in 2008 under the direction of the Louvre Museum for the purpose of additional research. Fifty-one bronze statues and statuettes (statues less than 30 cm in height) were studied in the National Museum of Sanaa, the Military Museum of Sanaa, the National Museum of Ibb, and the Regional Museum of Baynun. Ten of them were selected for more in-depth analytical examinations. Thirty-two samples of the metal were collected for elemental analysis, and four samples of the clay cores were collected for petrographic analysis. The samples were then officially exported to France and characterized at the C2RMF.

At the end of 2008, two more objects were studied and restored at the C2RMF and then exhibited between June and October 2009 in the Louvre Museum (fig. 3). The first object is a lion statue from the National Museum of Ibb that was discovered in the mountain settlement of Jabal al-awd. The second object is a lion appliqué of an Oriental style, probably from Tamna and dated to 50–25 BCE (Arbach 2005), which was recently acquired by the National Museum of Sanaa. The last action of the Louvre–GOAM cooperation took place in March 2009 at the National Museum of Aden, where a course in the conservation of archaeological artifacts was held. Technical staff of museums from all over Yemen took part in this training. The cooperation has since been in hiatus due to the extreme political and military tensions in Yemen.

South Arabian Bronze Statuary

The excavation of the Awam temple near Marib by Wendell Phillips and William Foxwell Albright under the auspices of the American Foundation for the Study of Man (AFSM) in 1951 and 1952 was the first to reveal the extraordinary richness of the South Arabian bronze statuary (Phillips Hodgson 2005; Maraqten 2005). In addition to the outstanding Madikarib statue, two male standing figures as well as a large number of inscribed stone statuary bases with

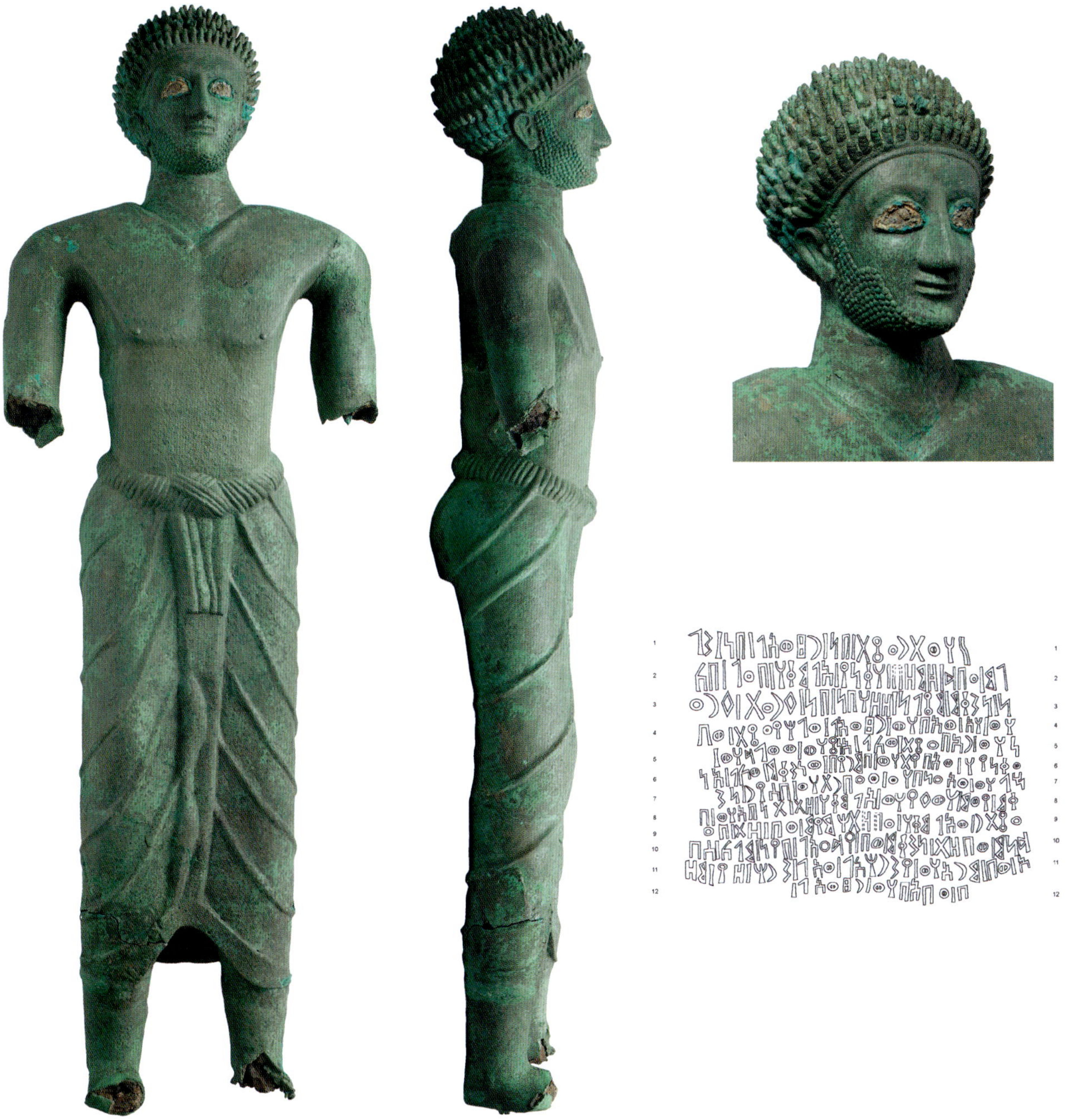

Figure 2. *The Hawtar'athat statue discovered in Nashqum (present-day al-Bayda), end of the seventh–sixth century* BCE, *h. 140 cm. National Museum of Sanaa, YM 23206. © Musée du Louvre, A. Chauvet. Drawing of the inscription carved on the stomach: © H. David.*

foot-shaped indentations on top were discovered, testifying that as early as the Archaic period (seventh–sixth centuries BCE) numerous large bronze statues stood in the Awam temple. The persons portrayed in these Archaic statues have facial features typical of local populations, defining a South Arabian indigenous style. Furthermore, the inscription on the Madikarib statue points to a possible relationship with that of Hawtar'athat: the same Sabaean king is mentioned, and the two dedicants both possessed lands in Marib and Nashqum (Mille et al. 2010, 19). Unfortunately, the Awam bronze statues never underwent any technological study. Therefore, we were only able to include two large bronze statues of the Archaic period in our study: the aforementioned Hawtar'athat statue and a headless male standing figure from the National Museum of Sanaa (fig. 4a). The context and place of discovery of this second statue still needs to be clarified, and its inscription has not yet been deciphered.

The importance of the long-distance contacts with the Mediterranean area was rapidly reflected in South Arabian artistic styles, probably as soon as the beginning of the third century BCE, perhaps even earlier. The unquestionable Greek influence led a number of scholars to define three main styles for the antique bronzes discovered in Yemen (see, for example, Ali Aqil and Antonini 2007, 94): (1) bronze of a South Arabian type and iconography locally produced, (2) Greek or Roman bronze import, and (3) bronze of an Oriental, Hellenistic, or Roman style, locally produced. The

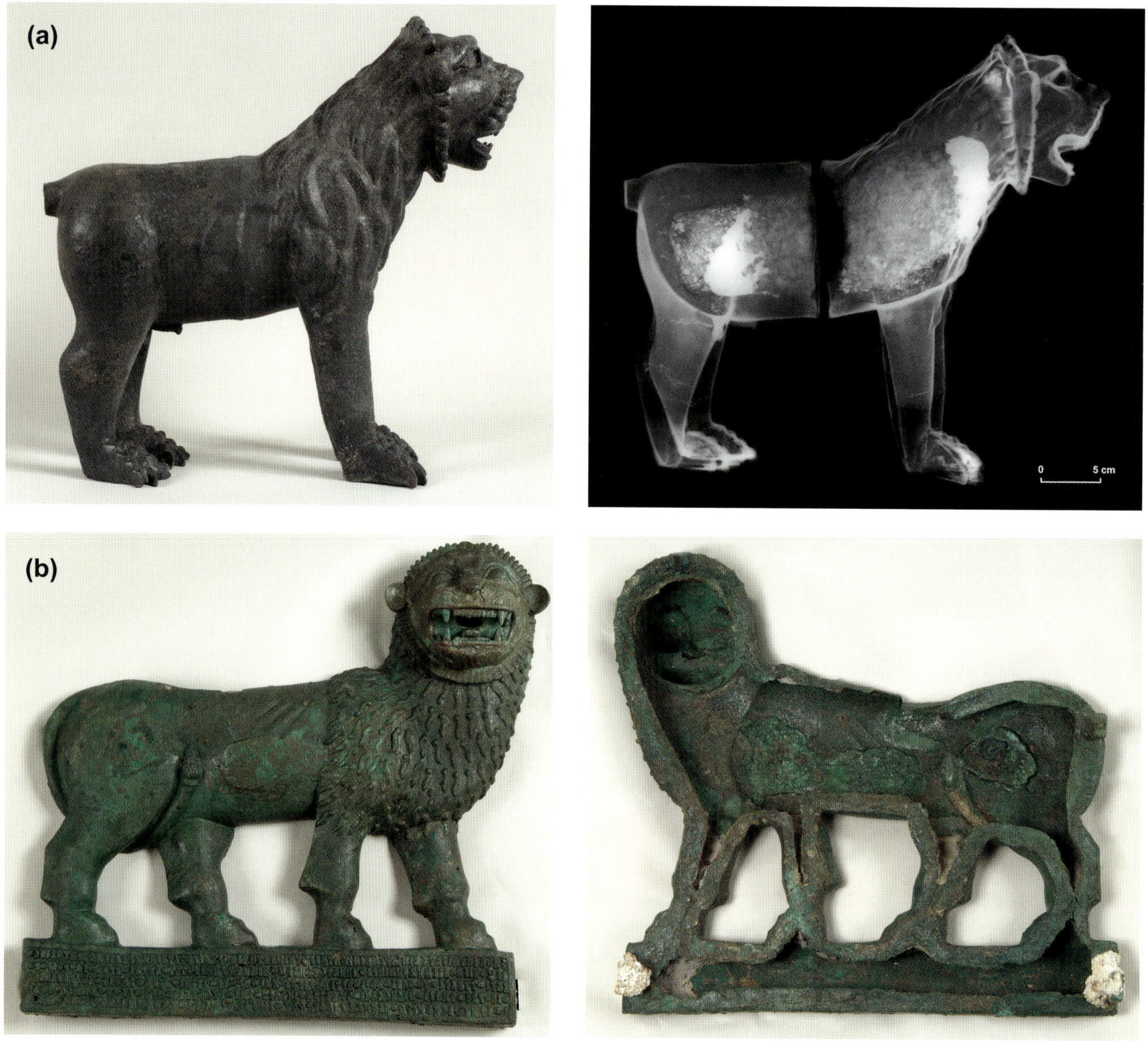

Figure 3. a. *Lion statuette discovered in Jabal al-awd, third century* BCE*–third century* CE*, h. 37 cm. National Museum of Ibb, IM24. © C2RMF, A. Chauvet (photography) and T. Borel (x-ray radiography).* ***b.*** *Lion appliqué probably discovered in Tamna, third quarter of the first century* BCE*, h. 67 cm. National Museum of Sanaa, YM36526. © C2RMF, A. Chauvet.*

latter case seems to be the most common from the third century BCE to the third century CE, leading to a bronze production of a syncretic Hellenistic/South Arabian style. As a consequence, it is often impossible to stylistically place such bronzes more narrowly within this large time span, and the presence of an inscription is often the only way to provide an accurate date.

The case of the Jabal al-awd mountain settlement illustrates the situation well. More than one hundred bronzes have been discovered since 1996, with the early discoveries from looting and the later discoveries from excavations conducted by the German Archaeological Institute (DAI). It was established that the bronzes mostly originated from a prestigious building occupying a central position in the settlement (Hitgen 2005). They probably result from a very long process of artifact accumulation in ancient times. A large amount of the Jabal al-awd bronze statuary is marked by a Hellenistic influence, but it is extremely difficult to determine if individual sculptures are imported or to measure the percentage of Mediterranean imports among the sculptures. For that reason, most of the pieces have broad stylistic dates—third century BCE–third century CE (Antonini 2007).

In addition to the aforementioned lion statuette (see fig. 3a), we had the opportunity to analyze three more bronzes from Jabal al-awd, currently on display at the National Museum of Ibb (fig. 4 g–i),[1] including two female statuettes. The first female statuette is considered from a stylistic point of view to be an example of the syncretic Hellenistic/South Arabian style (fig. 4i), whereas the second one is considered imported (fig. 4h). We also studied a fragmentary life-size bronze statue from the National Museum of Sanaa that exemplifies the Hellenistic influence on South Arabian large bronze production (fig. 4b). Coming back to the Oriental lion appliqué from Tamna (see fig. 3b), despite a very different style, it shares both a close similarity in date and an identical provenance with the exquisite Hellenistic pair of lion riders discovered by the AFSM during the earliest 1950–51 excavations in Tamna (fig. 5). This raises some doubts about the ability to use style to determine the chronology of late South Arabian bronzes.

The South Arabian Lost-Wax Casting Process

The technological study carried out in 2007 on the Hawtar'athat statue (late seventh or early sixth century BCE) established that this large bronze was hollow cast from an indirect lost-wax process and that the statue was initially manufactured in six separate pieces (the body, the head, the arms, and the legs), then joined by a flow fusion welding process (Demange et al. 2007; Mille et al. 2010, 29) (fig. 6).[2] This discovery is of fundamental importance because we now have ample evidence that these two techniques formed the basis of the manufacture of most, if not all, of the Greek and Roman large bronze statues from the middle of the sixth century BCE to the end of the fifth century CE (see Steinberg 1973; Mattusch 1988; Lie and Mattusch 1996; Bouquillon et al. 2006; Mille 2007). It has sometimes been assumed that South Arabian craftsmen adopted not only Hellenistic style and iconography but also Greek techniques. If so, how does one interpret the fact that the same technical features are found on a South Arabian statue slightly older than its earliest Greek analogues? Were these lost-wax casting and joining processes really identical? A closer examination of the technology of South Arabian bronze statuary is necessary before any comparison with Greek and Roman statues.

Thin Hollow Casts

The first distinctive feature of the South Arabian bronzes is that every bronze figure is a hollow casting, even the smallest statuettes (fig. 4j). Whereas statuettes less than 20 to 30 cm in height from the cultural areas of classical antiquity are typically solid cast, even during Roman times (Mille and Borel 2007), in Yemen, figurines as small as 5 cm high are hollow.

From a more general point of view, and as early as the Archaic period, the South Arabian bronzes are mostly free of casting defects, and the metal walls are extremely thin and even (from less than 1 to 3 mm) (table 1). For example, the walls of the Hawtar'athat large bronze statue range between 2 and 2.5 mm in thickness, including details in marked relief like the nose, the ears, and the belt (see fig. 6). The x-ray radiography of the Jabal al-awd lion statuette is still more impressive (see fig. 3a): cast without any defects, the metal walls are extraordinarily thin and even (only 0.5 mm on areas such as the mane and ears). The fragmentary life-size Hellenistic-style statue (see fig. 4b) and the Oriental-style lion appliqué (see fig. 3b) also have these characteristically thin walls.

Working the Wax

The large bronze statues studied provided us with evidence of the use of the indirect lost-wax process. The first feature is related to the thinness and the evenness of the metal walls, as mentioned before. Indeed, it would be extremely difficult to obtain such a result by directly carving the wax, and the result was more likely achieved by lining a layer of wax inside pieces of a mold taken from a model. The indirect lost-wax process requires that the wax be worked from the inside of the statue. Three techniques are conventionally used to apply the wax in the mold, and they are often combined with each other: (1) some wax sheets are placed in the hollow parts of the mold, or (2) the wax is applied in the liquid state with a brush, or (3) some liquid wax is poured into the mold to deposit a layer of uniform thickness, the excess liquid wax being subsequently removed by inverting the mold (for this reason this method is called *au renversé* in French). Considering the general case of the South Arabian statuary, none of these methods can be ruled out. Nevertheless, concordant observations do favor the frequent use of wax sheets. For example, the impressions from superimposed wax sheets and the resulting wax-to-wax joins can be seen on the flanks of the two Archaic male standing figures examined (Mille et al. 2010, 37). From the back side of the Tamna lion appliqué, it is also possible to see the geometric profiles of the wax sheets that were folded in on themselves (see fig. 3b).

The wax hairdressing of Hawtar'athat is an interesting case to discuss. Each curl is undercut, and it was obviously not possible to piece-mold them individually. The most probable scenario is that a naked skull was initially formed by the indirect process, as with all other parts of the statue. Concentric rows of curls, each one carved directly into the wax, were then applied to the skull. Absolute proof for this is found through the examination of the hair in the nape area (fig. 7). The last two rows of curls are not part of the mold piece corresponding to the skull but are from the mold of the lower face. The curls here are not represented with the same relief. The reason is simple, since here the inner profile of the metal wall perfectly follows the outer one (see fig. 6). The contrast of these curls, either very flat and lacking detail or modeled in strong relief, validates all of our proposals: this contrasted manufacture proves, on the one hand, that the curls in high relief were individually modeled in wax by the direct process, but on the other hand—and this is the most important point—the flat curls demonstrate that the indirect lost-wax method was the main process for the manufacture of the different wax pieces of the statue. We uncovered other details directly manufactured in wax: the teeth of the Tamna lion appliqué (see fig. 3b), the relief letters of the inscription, and some decorative parts of the cloth of the headless Archaic male standing figure (see fig. 4a), or the braids decorating the cloth of the fragmentary life-size Hellenistic statue (see fig. 4b). In summary, the South Arabian craftsmen systematically used the indirect lost-wax process for their large bronzes, but the statues studied so far were then often adorned with details directly carved in wax. The lost-wax process seems to have remained unchanged during the whole period. Thin, even metal walls are present throughout.

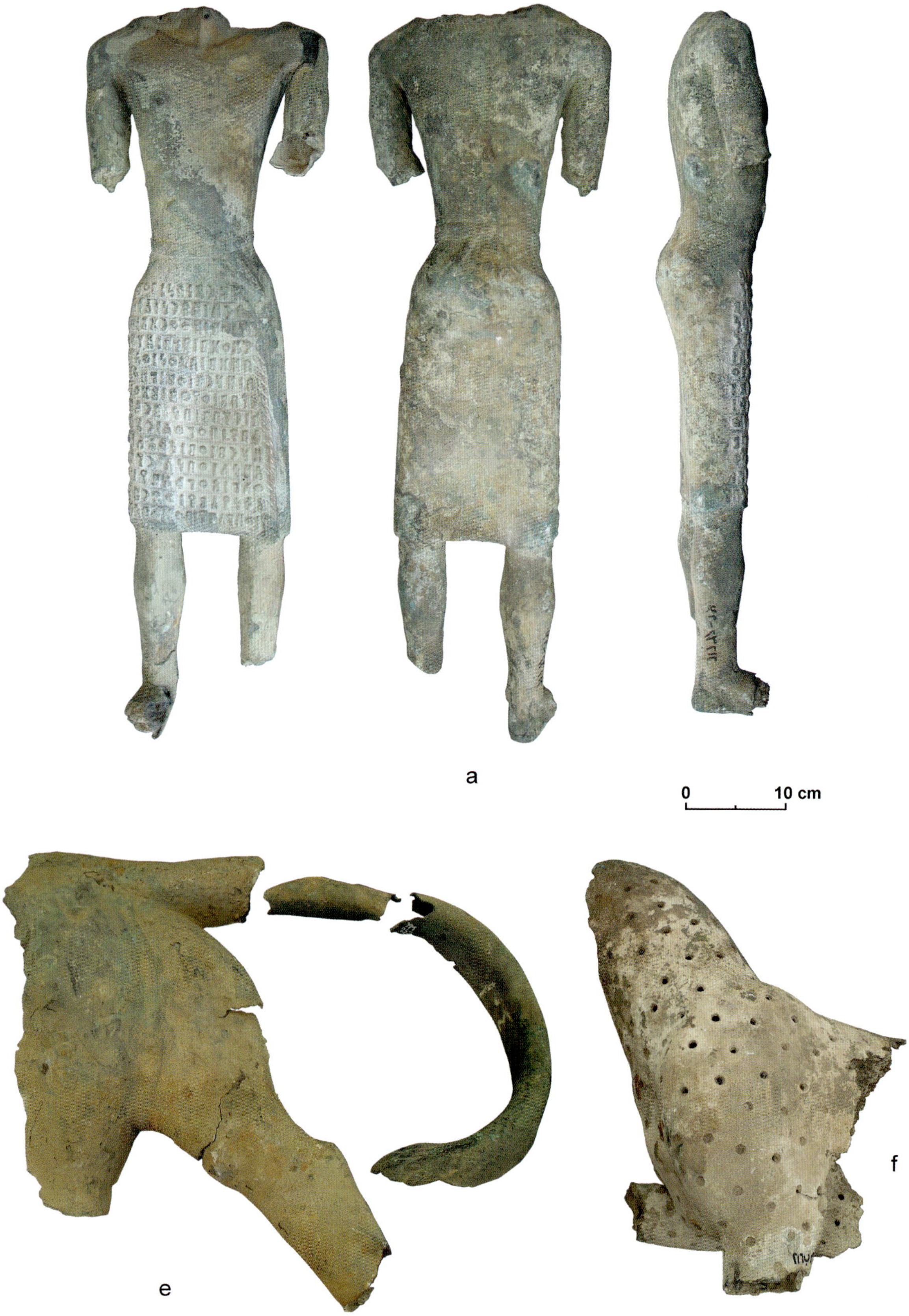

Figure 4. *South Arabian statues and statuettes selected during the 2008 mission to Yemen for in-depth analytical study, full references given in table 1. © C2RMF, B. Mille.* ***a.*** *Headless standing male statue, National Museum of Sanaa, YM23212.* ***b.*** *Fragmentary life-size standing statue (National Museum of Sanaa):* ***b1.*** *Cloth decorated with braids (YM238);* ***b2.*** *Knee, vertical folds, cloth (YM241);* ***b3.*** *Cloth with long vertical folds (YM239).* ***c.*** *Life-size left arm fragment, National Museum of Sanaa, YM218.* ***d.*** *Life-size leg fragment, National Museum of Sanaa, YM220.* ***e.*** *Fragmentary lion appliqué, National Museum of Sanaa, YM261–YM252.* ***f.*** *Forepart of a leopard statue, Regional Museum of Baynun, MB216.* ***g.*** *Ibex protome, National Museum of Ibb, IM25.* ***h.*** *Draped female statuette, National Museum of Ibb, IM8 and IM19.* ***i.*** *Draped female statuette (IM9).* ***j.*** *Female statuette, Military Museum of Sanaa, 9570.*

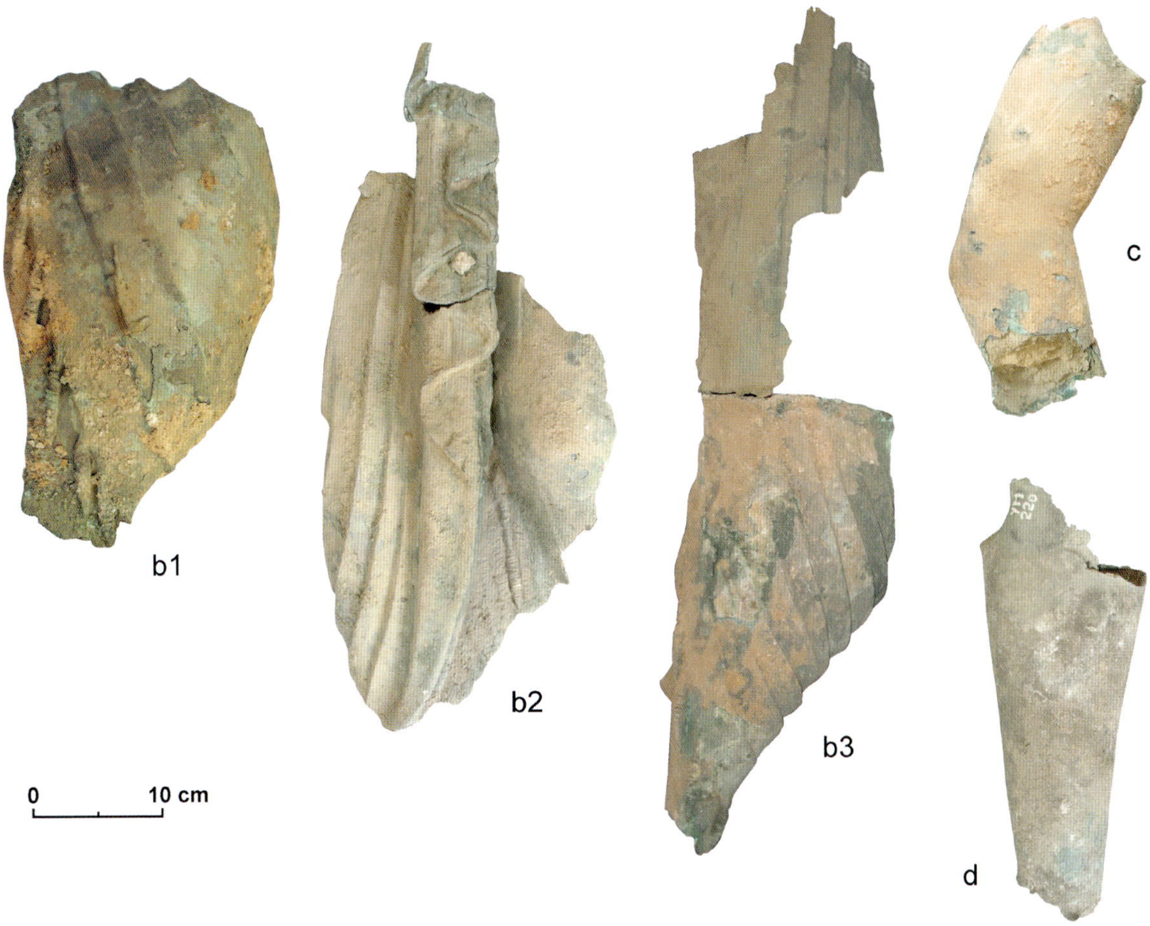
b1
b2
b3
c
d
0
10 cm

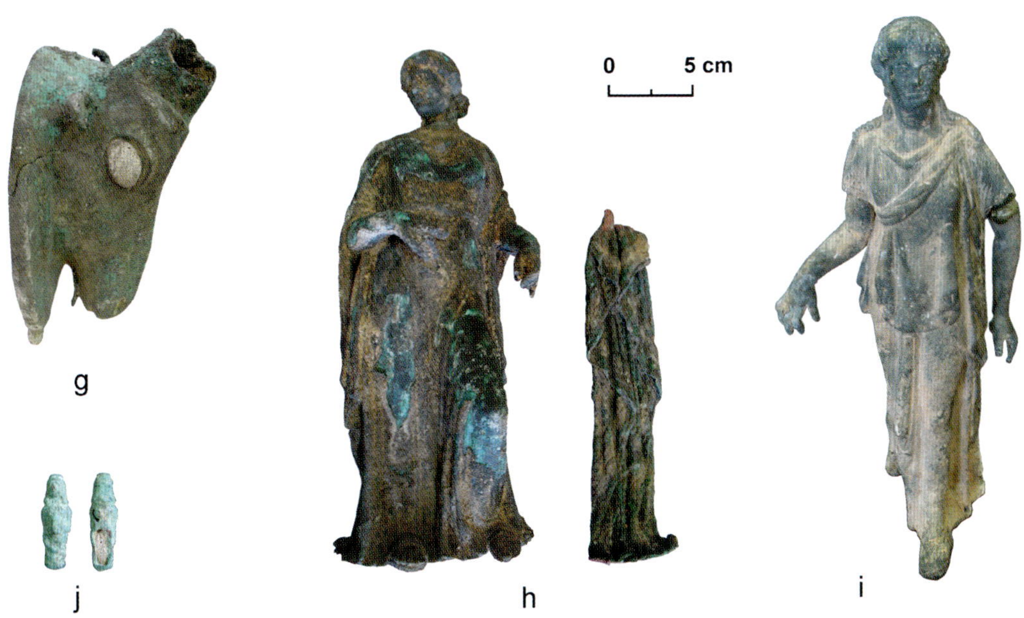
g
j
h
i
0
5 cm

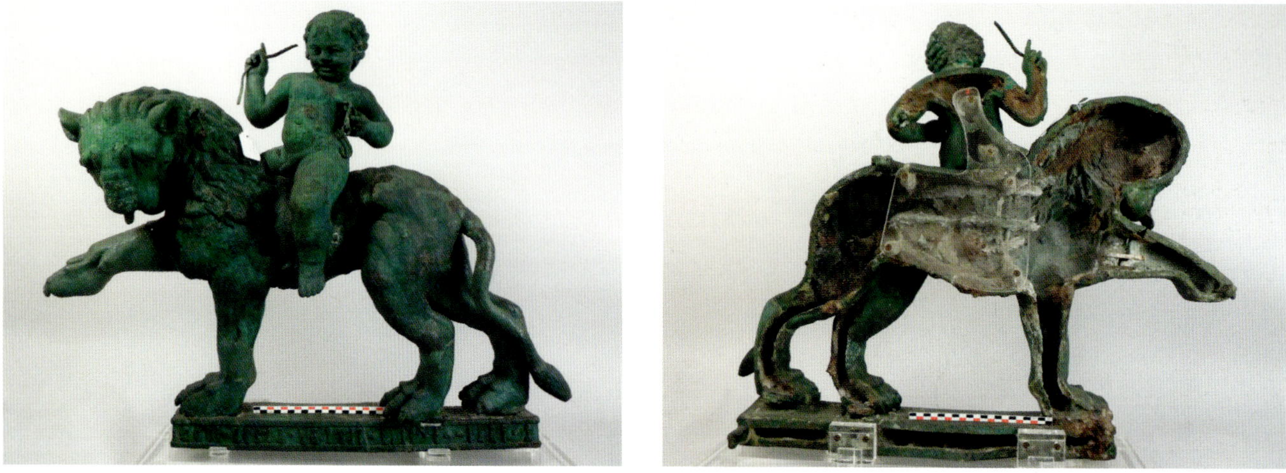

Figure 5. Lion rider B from the pair discovered in Tamna, early first century BCE–mid-first century CE, h. 61 cm. American Foundation for the Study of Man, TS 152B. © UMR7055, B. Mille.

Hollow Clay Cores

The clay core inside the body of Hawtar'athat is hollow (thickness: 30 to 35 mm on the dorsal side, 35 to 60 mm on the ventral side) (see fig. 6). To our knowledge, such a technique had, until this study, never been seen in a bronze statue, whatever its origin. After examining other South Arabian statues, we now know that Hawtar'athat was not an isolated case. It seems that this technique of using a hollow core was used whenever the dimensions and internal volumes of the statues allowed (see table 1). A horse statuette from the National Museum of Sanaa made possible the direct observation of this unusual method (fig. 8): similar to Hawtar'athat, a 35 to 45 mm-thick core was formed by pressing clay on the inner side of the wax walls. The Tamna Oriental lion appliqué and the numerous ibex and bovid protomes from Jabal al-awd (see fig. 4g) testify to a very similar technique: their backsides were not entirely filled with clay, but only lined by a "core" layer.

Clay cores are usually removed after casting in order to avoid possible crack formation in the bronze (caused by later expansion of the hydrophilic clay core) and to lighten the object. In South Arabian statues, the core is still present. Why did the Sabaean foundrymen keep the core within the statue? The answer may lie with the extremely fine metal walls, for which the core may play a structural role. For example, it is impossible to remove the core of the Jabal al-awd lion without revealing many holes, the metal thickness being often less than a millimeter. The clay core was therefore most likely left to act as structural reinforcement. In this case, the use of a hollow core is fully justified; the statues would have been far too heavy with a solid core.

The petrographic study of the clay cores is still in progress. The results obtained so far are related to that seen for the core of Hawtar'athat (Mille et al. 2010, 24) and show that a marl was used, tempered with both vegetal fibers (5 to 20% in volume) and sand with a 250–300 μm grain size. The sand is composed of quartz, calcite, feldspars, pyroxenes, and a few fragments of volcanic rock. Sourcing happens to be straightforward given the discriminating composition of the sand: either it comes from the Sanaa basin and its extension in the direction of the Jawf near Nashqum, or from the area around Marib.

Alloy Composition

The bulk metal composition was obtained from forty-eight samples taken from the different parts of the twelve South Arabian statues selected for in-depth study. Analyses were performed by inductively coupled plasma–atomic emission spectrometry (ICP-AES) at the C2RMF. An approximate 20 mg metal sample was removed by drilling (1 mm diameter, 10 mm deep) after eliminating most surface corrosion products; the drillings were carefully examined under the stereomicroscope to avoid the inclusion of any corrosion product or dust. About 10 mg of the drillings was precisely weighed and digested in 5 ml aqua regia solution (hydrochloric and nitric acids). The solution was then nebulized in the argon plasma, and twenty-nine chemical elements were quantified. The operating conditions are described in Bourgarit and Mille (2003). The detailed results of the ICP-AES analyses can be found in the appendix. The comments here will first be limited to the primary castings, that is to say the different parts of a statue that were separately lost-wax cast (thirty analyses). The eighteen remaining analyses are mainly of the metal used to weld together these different parts (see next section). The published results from four more statues (six analyses) have been added to the corpus analyzed in this work in order to enlarge the total number of compositions available for statistical analysis (fig. 9).[3]

Almost all of the analyzed statues and statuettes are made of leaded bronze (see fig. 9). The standard South Arabian composition corresponds to a low-tin bronze containing a noticeable amount of lead. The average alloy composition seems to remain very stable over time, despite the paucity of analyses from the Archaic period: neither the type of alloy (leaded bronze) nor the average amount of tin and lead show any significant evolution from the Archaic to Hellenistic (late fourth century–second century BCE) and Roman (first century

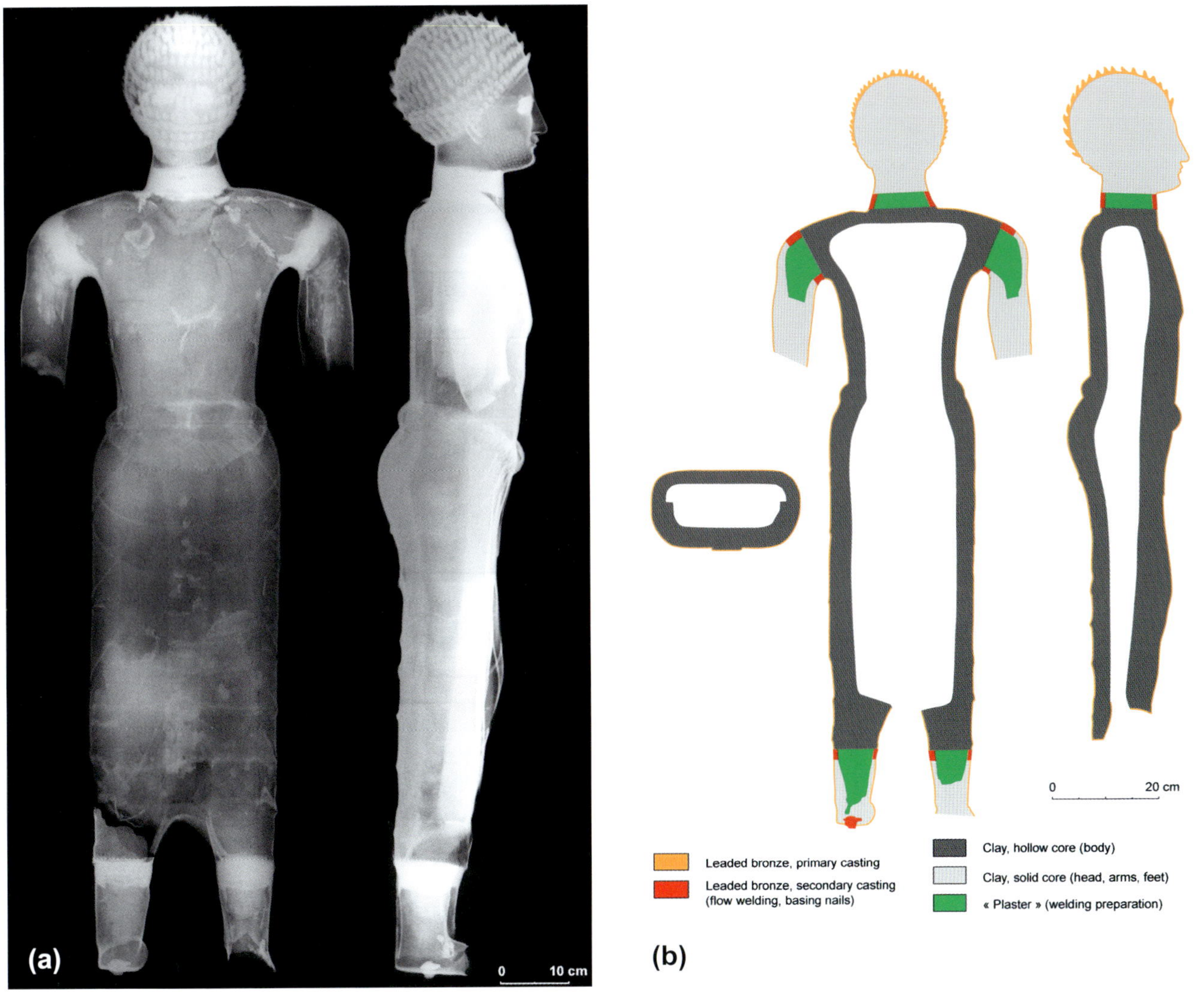

Figure 6. a. *Hawtar'athat, x-ray radiography. © C2RMF, T. Borel.* ***b.*** *Hawtar'athat, sagittal, frontal and transversal sections obtained from the x-ray radiography. © UMR7055, B. Mille.*

BCE–third century CE) times. There is only one real exception—the female statuette from Jabal al-awd (see fig. 4h). Stylistically considered a Mediterranean import, this statuette is a simple binary bronze with a relatively high tin level (around 10 wt.%). The analogous statuette of a mixed South Arabian/Hellenistic style that was analyzed is, on the contrary, a bronze highly alloyed with lead (see fig. 4i).

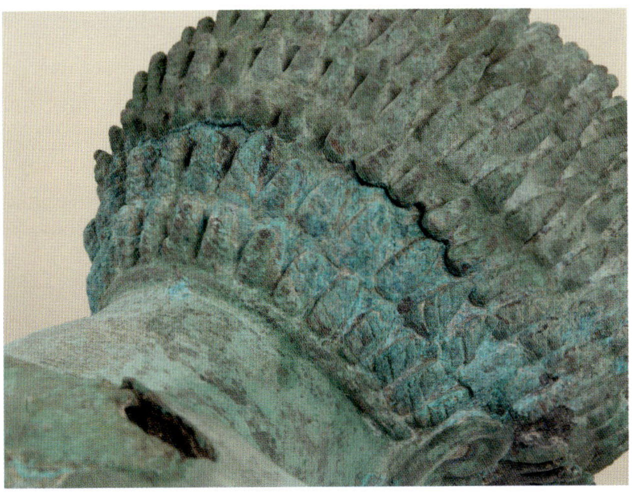

Figure 7. *Hawtar'athat, detail of the hair loops. Note the difference between the two first rows of curls manufactured by the indirect lost-wax process and those in relief formed by direct carving in the wax. © C2RMF, A. Chauvet.*

Particularly noteworthy is the variation of the alloy composition within the same statue. Generally, it affects only one of the two alloying elements, while more rarely both tin and lead are concerned: for example, the Hawtar'athat body reaches 12 wt.% tin, whereas the legs, arms, and head only contain 6 to 8 wt.%; the lead level of the Oriental Tamna lion appliqué ranges from 4.9 to 12 wt.%; and the fragmentary lion appliqué from the National Museum of Sanaa (see fig. 4e) shows varying amounts of lead (4.2 to 24 wt.%) and tin (2.9 to 7.9 wt.%). In the next section about welding, we will come back to this particularly important feature of the South Arabian bronze statuary.

It is beyond the scope of this inquiry to discuss *in extenso* the metal provenance; we will only point out here the very high cobalt levels reported in a large number of statues, often strongly correlated with nickel. This impurity pattern is compatible with ophiolitic Omani copper sources (Hauptman 1985; Weeks 2003).

Table 1. *South Arabian statues and statuettes selected for an in-depth analytical study.*

Lab. No.	Inv. No.	Designation	Museum	Discovery Place	Dating	Lost-Wax Process	Average Wall Thickness (mm)	Core Type/Thickness	Weld Geometry
C2RMF 64057	YM23206	Hawtar'athat statue	National Museum of Sanaa	Nashqum	late 7th–6th cent. BCE	indirect (+ direct: upper curls)	2–2.5	hollow (body: 30–60 mm) solid (other parts)	ring shape, casting on (nails)
C2RMF 67747	YM36526	lion appliqué	National Museum of Sanaa	Tamna	3rd quarter 1st cent. BCE	indirect	1.5–2.5	layer (body: 20 mm) solid (head)	linear (head), casting on (tongue)
C2RMF 67748	IM24	lion statuette	National Museum of Ibb	Jabal al-awd	3rd cent. BCE–3rd cent. CE	indirect	0.5 (body) 1–1.5 (legs)	hollow (20 mm)	spot welded
Y9	YM23212	headless man standing statue	National Museum of Sanaa	unknown	6th cent. BCE	indirect (+ direct: inscription, cloth)	2–3	solid (body: hollow?)	ring-shape
Y10	YM238	fragment. life-size standing statue, cloth decorated with braids	National Museum of Sanaa	unknown	3rd cent. BCE–3rd cent. CE	indirect (+ direct: braids)	3	layer (35 mm)	—
	YM239	*ibid.*, cloth with long vertical folds				indirect	2.5	removed (modern?)	—
	YM241	*ibid.*, large fragment (knee, vertical folds, cloth)				indirect	2.5	partly removed (modern), layer	linear
Y13	YM220	life-size leg fragment	National Museum of Sanaa	unknown	3rd cent. BCE–3rd cent. CE	indirect	2.5–3.5	removed (modern?)	linear
Y14	YM218	life-size left arm fragment	National Museum of Sanaa	unknown	4th cent. BCE–3rd cent. CE	indirect	2.5–3.5	solid	linear
Y19	YM261-YM252	fragmentary lion appliqué	National Museum of Sanaa	unknown	5th cent. BCE–3rd cent. CE	indirect	3.5–4 (body) 2.5 (tail)	removed (modern?)	half-ring shape
Y27	MB216	forepart of a leopard statue	Regional Museum of Baynun	unknown	6th cent. BCE–3rd cent. CE	indirect	4.5–5	partly removed (modern)	linear
Y30	IM25	ibex protome	National Museum of Ibb	Jabal al-awd	7th cent. BCE–3rd cent. CE	indirect	2	layer (20–35 mm)	—
Y32	IM9	draped female statuette	National Museum of Ibb	Jabal al-awd	8th cent. BCE–3rd cent. CE	direct	2.5–3 solid (folds)	solid	—
Y33	IM8 & IM19	draped female statuette	National Museum of Ibb	Jabal al-awd	9th cent. BCE–3rd cent. CE	indirect? (statuette) direct (mantle)	2–3.5 (stat.) 7 (mantle)	removed	soldering Pb-Sn (mantle)

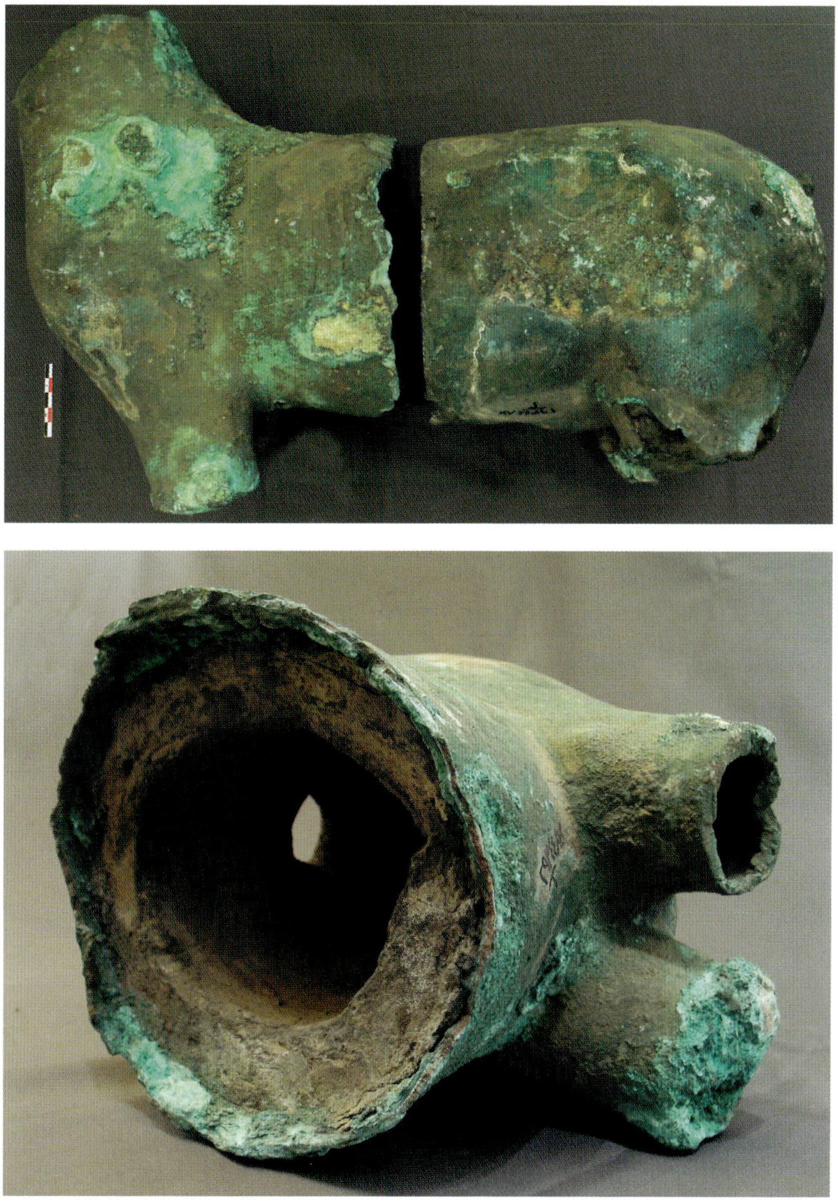

Figure 8. *Horse statuette, unknown provenance, h. 37 cm. Note the thick hollow clay core. National Museum of Sanaa, YM29344. © C2RMF, B. Mille.*

Flow Fusion Welding

All South Arabian large bronze statues studied so far have systematically shown the use of a flow fusion welding process in order to join the various bronze parts previously made in the primary castings. Even some statuettes, at least the largest ones such as the Jabal al-awd lion, were found to apply this technique. Flow fusion welding is a specific kind of secondary casting,[4] in which liquid bronze is poured between the two pieces to be joined, with the bond formed by local fusion of the edges of the two primary castings (Azema and Mille 2010, 2011). The corresponding joins on the Archaic statue of Hawtar'athat appear at the surface as rings of a 40–50 mm width, and x-ray radiography established that the rings are considerably thicker than the metal walls (15 to 25 mm) (see fig. 6). The remains of a join formerly existing between the head and the body of the second Archaic statue studied (see fig. 4a) strongly suggest a similar ring-shape welding. The fragmentary lion appliqué from the National Museum of Sanaa (see fig. 4e) and the pair of Tamna lion riders (see fig. 5) illustrate a variant of this technique, in which numerous weld joints form a half-ring shape. The other South Arabian statues examined from the Hellenistic and Roman periods do not show evidence of ring-shape welding. Most of them display linear joins (see, for example, the Tamna Oriental lion appliqué and the fragmentary life-size statue). A noticeable exception is the welding in numerous tiny spots to join the two parts of the lion statuette from Jabal al-awd. The use of spot welding in antiquity was totally unknown prior to this study. Welding in basins (or *a vaschette*), the emblematic technique used by Greek and Roman foundrymen at least from the fifth century BCE to the fourth century CE (Formigli 1984), is underrepresented in Yemen. Among the fifty-three South Arabian statues and statuettes examined, only three displayed this kind of join: a leopard hind leg from Jabal al-awd (National Museum of Ibb, MI 70), and the sculpture of King Dhamar alay Yuhabirr and that of his son Tharan, both 2.4 m high sculptures from

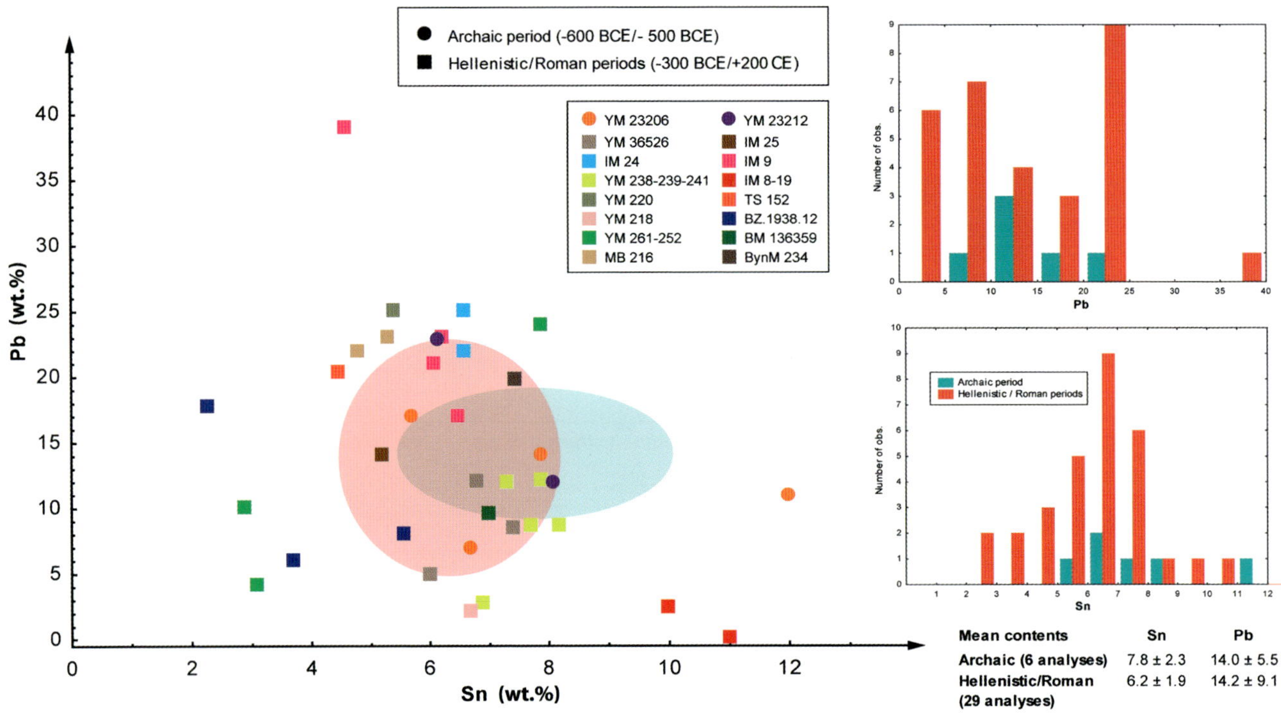

Figure 9. *Alloy composition of the South Arabian statuary; only the primary castings have been reported. Left: Lead-tin binary plot. Pink ellipse: Archaic period mean composition. Blue ellipse: Hellenistic/Roman periods mean composition. Right: tin and lead distribution histograms. © UMR7055, B. Mille.*

the third or fourth century CE that were discovered in Naklat al-Hamra (National Museum of Sanaa, YM 196 and 306) (Weidemann 1983). We know from inscriptions on the two latter statues that their manufacture resulted from the cooperation of two foundrymen: a Greek native named Phokas and a South Arabian named Lahay'Amm. It is therefore not surprising to find some technological features typical of the Mediterranean area on these two statues.

The South Arabian welding techniques are not only characterized by distinctive weld join geometry, such as the ring shapes found on the Archaic statues, but also by complex preparation of the pieces before joining. By closely examining the x-ray radiography of the joined areas on Hawtar'athat, we discovered that the solid cores of the five pieces assembled to the body have been hollowed out after casting to a depth of 60 to 90 mm in the arms and legs and 10 mm maximum in the head (fig. 10, radiography, and fig. 10 a and b). The cavities thus formed were next filled with a ball of a "plaster type" material (fig. 10c). Each piece and its plaster were then pressed on the core of the body by using the openings provided for this purpose (fig. 10d). Such a plaster would be used for both temporary bonding, providing a support to maintain the parts to be welded in position, and as a core for the secondary casting to come (i.e., the ring-shape join). The next steps were to build a casting mold to enclose the assembly (fig. 10e), and then to pour the filler metal into the vacant space (fig. 10f). After solidification, the mold was removed and the effectiveness of the welding process was checked (fig. 10g). The Archaic headless male standing figure (see fig. 4a) also shows material removal in the core at the neck opening, and some remnants of a plaster type material were found in the corresponding hole. The later lion statuette from Jabal al-awd provides another example of the use of a plaster piece for the preparatory phases of welding (see radiography on fig. 3a). In this statue, the "plaster" was directly observed and analyzed. It corresponds to a lime mortar containing a large amount of ferruginous inclusions. Such a complex preparation has never been observed for linear weld joins in South Arabian statues of the late periods (Hellenistic and Roman times).

Concerning flow fusion welding, the last point to be discussed is the elemental composition of the added metal in the weld. A plot showing the tin and lead levels of welding metal as compared to the levels in the primary castings indicates that alloy compositions were more controlled than previously thought when looking only at the primary castings (fig. 11). Very often, the welding metal has an intermediate composition between the two primary castings that were joined. Nevertheless, the pattern, if any, is not straightforward: neither the increase nor the decrease of lead and tin contents followed any discernible trend. Further investigation (elemental analyses and metallographic examinations) is needed to proceed on this topic. A possible explanation could rely on the sequential order of welding, as alloy formulation changes have an obvious effect on the melting temperature; the first piece to be joined would therefore correspond to the less alloyed primary casting.

South-Arabian versus Greek and Roman Large Bronze Statues

South Arabian large bronzes clearly share two important manufacturing techniques that are found in the overwhelming majority of Greek and Roman large bronze

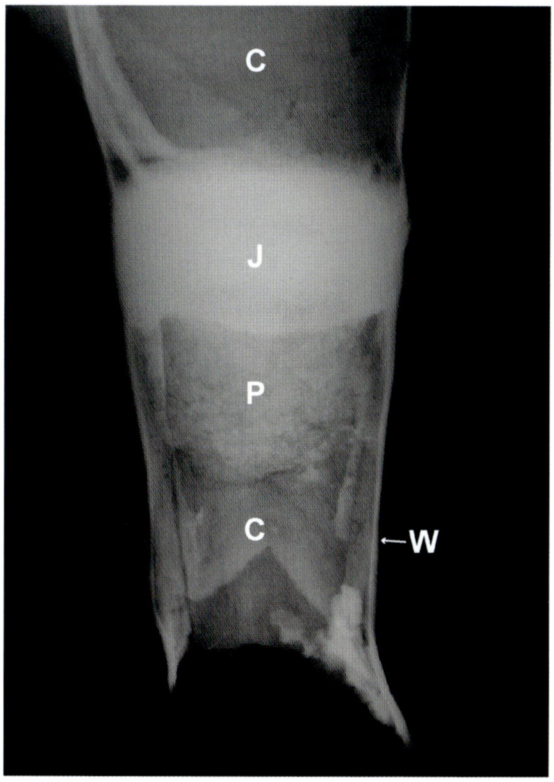

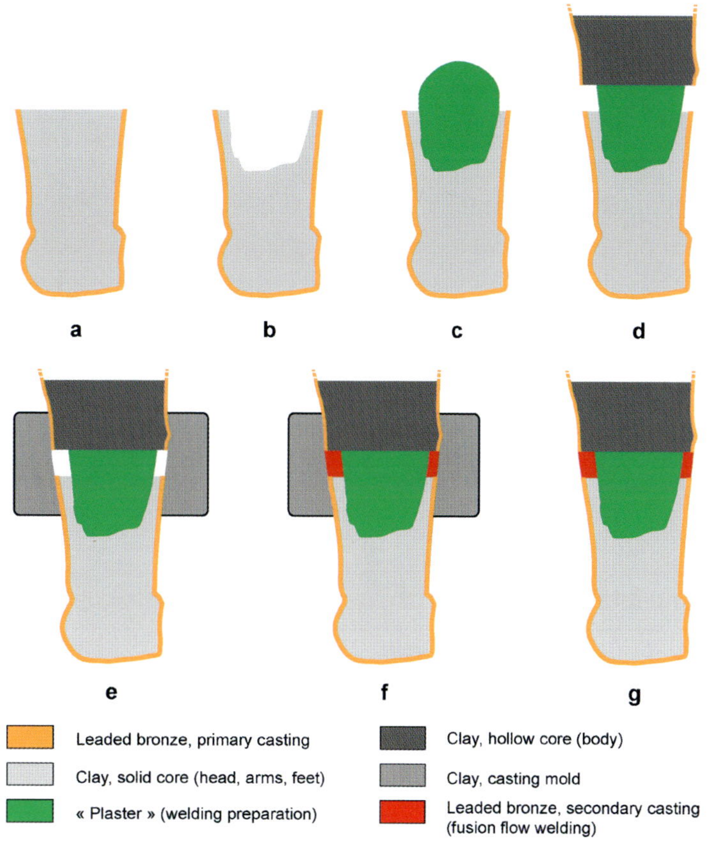

Figure 10. *Left: Hawtar'athat, detail of the left foot, x-ray radiography. W = metal walls; C = clay core; J = ring-shape weld join; P = "plaster type" material.* © C2RMF, T. *Borel. Right: Proposed reconstruction of the flow fusion welding of the left foot to the body.* © *UMR7055, B. Mille.*

statues—namely, the use of an indirect lost-wax process for hollow casting and the assembly of the primary castings by means of flow fusion welding. The question remains whether the South Arabian techniques are derived from the Greek ones. This question is all the more legitimate, since at the end of the fourth century BCE, Greek influence is obvious in the statues in Yemen, at least with regard to style and iconography. The fundamental issue raised here concerns the dating of the South Arabian Archaic statues (Hawtar'athat, the headless figure from the National Museum of Sanaa, and the Awam temple statues). If the proposed date, from the late seventh century to the first half of the sixth century BCE, is accepted, these South Arabian large bronze statues would be earlier than their Greek counterparts. By expanding the date range to the whole of the sixth century BCE, South Arabian statues would be contemporaneous with the emergence of this phenomenon in Greece.[5]

Yet, while the South Arabian statues share the two main techniques that form the basis of the construction of Greek and Roman statues, almost every detail of their manufacture is different (table 2).

- The metal walls of the large Greek bronzes of the late sixth century and first half of the fifth century BCE were approximately three times thicker than the metal walls of the Archaic South Arabian statues. In the Mediterranean world, it was not until the end of the Hellenistic period, but more commonly during the Roman period, that metal walls became 3 mm or thinner.
- Core systems were much more complex in the case of South Arabian bronzes. This is well illustrated by the hollow cores that were noticed in the largest internal cavities of the statues. Such a technique has never been documented elsewhere. In addition, the cores are always left in the statue after manufacture, probably to reinforce the structure given the extreme thinness of the bronze walls. This is the opposite of the more usual method, which was used in Greek and Roman large bronzes, in which the clay cores are removed after casting.
- The alloy used during the South Arabian periods under study was a leaded bronze, and its average composition remained constant over time. However, there are large differences in the compositions of the primary castings of a given statue (see figs. 9 and 11). In contrast, the large metal statues from the Greek Archaic (sixth century BCE) and Classical (fifth century–mid-fourth century BCE) periods were made of a binary bronze (copper-tin). The use of ternary alloys in which lead was added was not seen until the Hellenistic period (Mille et al. 2009).
- Greek and Roman large bronze statues were assembled after careful cold-work preparation of the join in order to achieve a butt joining of the two primary castings (filing to adjust the pieces to each other, engraving to prepare the basins) (Azema and Mille 2010, 2011).

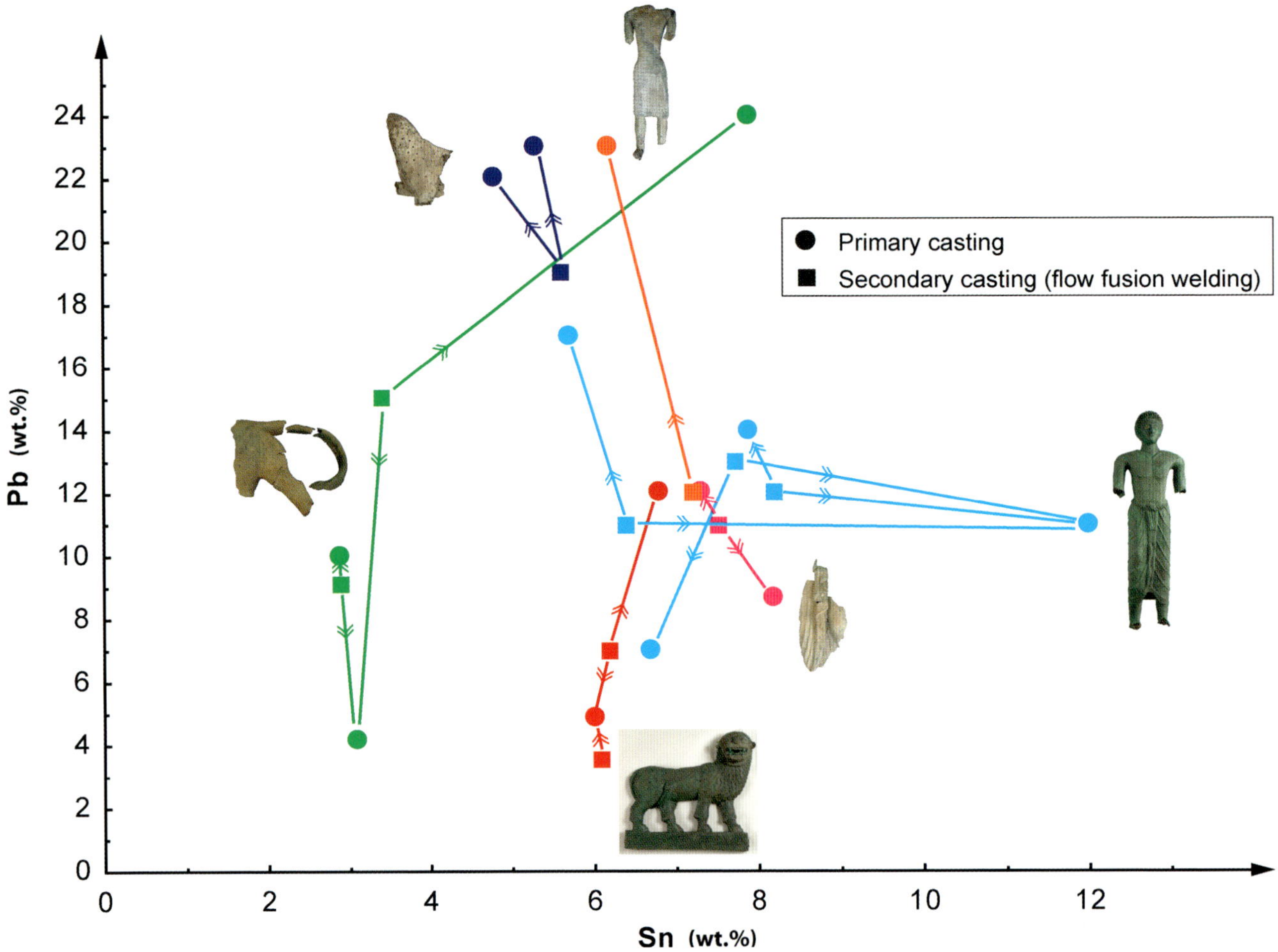

Figure 11. Alloy composition of the South Arabian filler metal used for flow fusion welding compared to the alloy composition of the corresponding primary castings. Note that the composition of the welding metal is generally an intermediate of the two pieces to be joined. © UMR7055, B. Mille.

In South Arabia, flow fusion welding was carried out in a manner completely unknown to the Greek world, at least during the earliest periods of bronze statuary production. In the case of ring-shape joins, the preparation was made as soon as the wax was worked since the rings were not part of the primary castings. The constraints were therefore totally different: on the one hand, butt joins were not possible given the empty space separating the two parts to be welded; on the other hand, clay cores were to be completely preserved. Some plaster or lime mortar was then used, both to maintain in position the two pieces to be joined and to create the intermediate core piece that was missing. Such complex preparation techniques were probably useless when linear welding became widespread in South Arabia. Nevertheless, the elemental composition of the filler metal definitely distinguishes the South Arabian flow fusion welding technique from the Mediterranean one (see figs. 11 and 12). The South Arabian filler metal usually has an intermediate composition between the two primary castings that were joined, whereas the Greek and Roman rule was to use the same alloy both for primary castings and welding. In the South Arabian case, the alloy variations could possibly be connected to a search to use the subtle variations of the melting temperature of the different alloys.

In the Greek and Roman case, the homogeneous composition of the whole statue was probably related to the goal of achieving a homogeneous color and appearance, the success of which was most dependent on the care used in the fabrication and casting of the statue.

Conclusion

This study is the first extensive technological investigation conducted on a comprehensive set of South Arabian bronze statues. Despite a mutual use of both indirect lost-wax casting and flow fusion welding, the technology used in South Arabia is totally different from the technology used in Greece and Rome at the same time. As a consequence, it defines a proper South Arabian technology for the manufacture of large bronze statues. The Mediterranean influence during Hellenistic times is therefore noticeable only in the style and iconography, not in the technology. Moreover, the South Arabian technology remained nearly unchanged for a very long time, thus suggesting a very stable and conservative bronze industry, at least for the production of large statues. Ultimately, this study raises the delicate problem of the origin of the techniques used to produce South Arabian and Greek antique large bronze statues. Archaic South

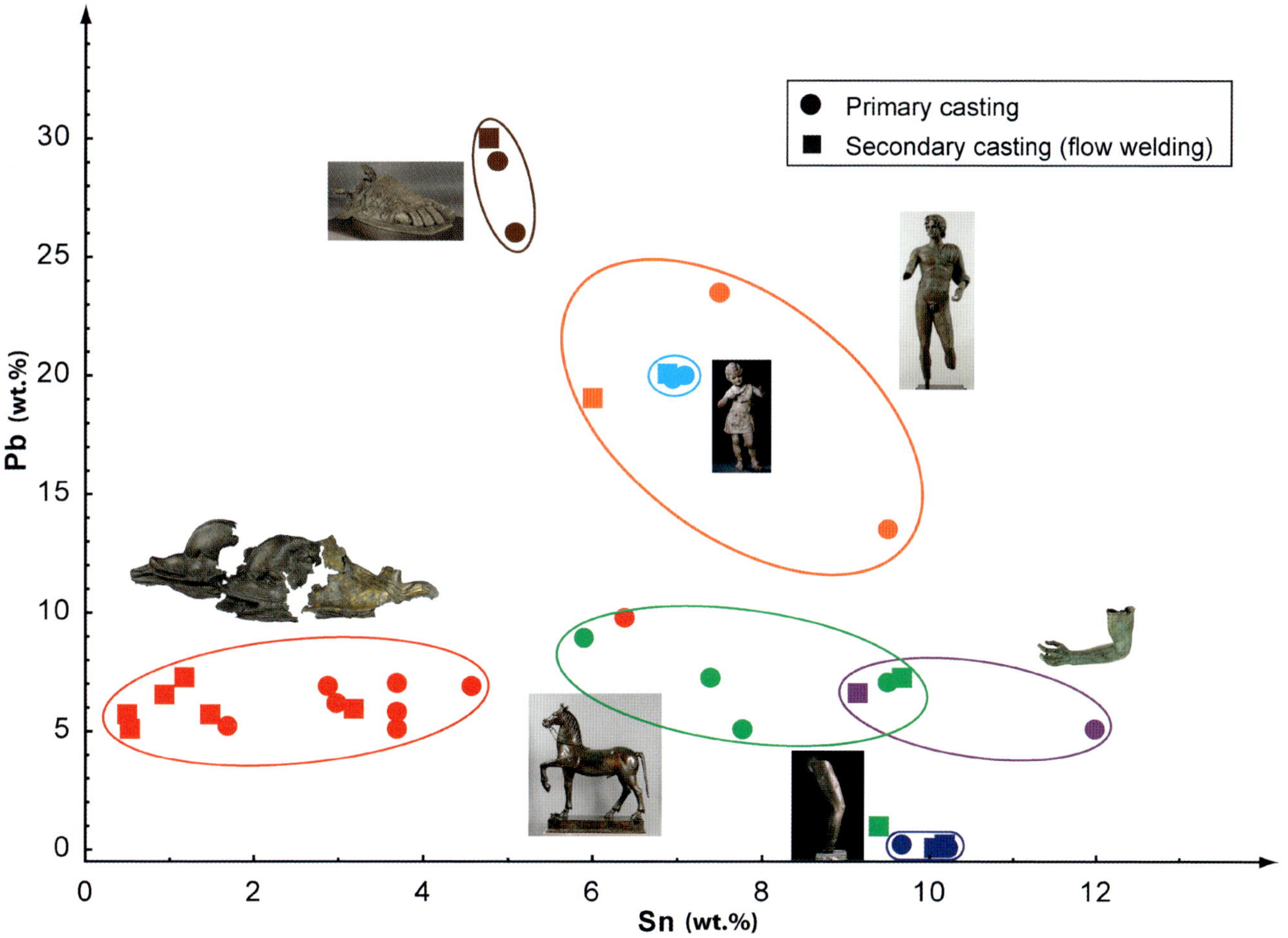

Figure 12. *Alloy composition of the Greek and Roman filler metal used for flow fusion welding compared to the alloy composition of the corresponding primary castings. Data from recent studies conducted at the C2RMF. A trend in favor of identical composition is visible (Chattsworth Apollo leg, Agde Royal statue of a boy, Clermont-Ferrand colossal foot, Vienne dolphins). A lack of control possibly explains the relative discrepancy noticed for the Neuvy-en-Sullias horse and the Essegney arm. The Agde statue of a youth may relate to a different rule (study in progress). © UMR7055, B. Mille.*

Arabian and Greek bronze statues share two main principles of manufacture (indirect lost-wax casting and flow fusion welding), suggesting a common technological inheritance. In addition, the flourishing and complex bronze industry revealed by Hawthar'athat and the statues from the Awam temple, dated most likely as early as the end of the seventh century BCE, presupposes the acquisition of a high level of metallurgical knowledge and skill. Cultures of the Middle East and Egypt, with their long experience of manufacturing large metal statues, are appropriate candidates to explore in a continued search for the sources of innovation and methods of transmission of the metallurgical techniques seen in South Arabian and Greek large bronze sculpture.

Acknowledgments

This study is a part of a wider project not only concerning the work presented here but also many other aspects regarding archaeology, cultural heritage, and conservation, among others. The archaeometallurgical study would not have been possible outside the frame of this cooperation. The missions to Yemen have been funded by the Louvre Museum. The French team wishes to thank Mrs. M. Barret, formerly of the French Embassy in Yemen, and Remy Audouin, formerly of UNESCO, who initiated the cooperation; Mr. A. Bawazir, director of the GOAM; Mr. Al Gendary, director of the National Museum of Sanaa; Mr. H. Loyrette, director of the Louvre Museum; and Mrs. B. André-Salvini, head of the Department of Oriental Antiquities in the Louvre Museum.

I am indebted to the team members of the mission to Yemen, without whose support this study would not have been possible: Mrs. F. Demange, curator at the Department of Oriental Antiquities in the Louvre Museum and head of the Louvre–GOAM cooperation; Mrs. A. Ali Aqil, South Arabian art historian; and Mrs. C. Pariselle, restorer. I am also very grateful to Mrs. I. Gajda, epigraphist at CNRS; Mr. Y. Coquinot, geologist at C2RMF; and Mrs. E. Porto and Mr. A. Zink, geochronologists at C2RMF (thermoluminescent dating analyses), for their participation in the Hawtar'athat study. Lastly I wish to thank Paul Jett, head of the Department of Conservation and Scientific Research of the Freer Gallery of Art, who kindly invited me to this conference and secured permission from Merilyn Phillips Hodgson, president of the American Foundation for the Study of Man, to examine the lion riders from Tamna.

***Table 2.** Main technological features for the manufacture of a large bronze statue, South Arabia compared to the Mediterranean world.*

South Arabian	Greek and Roman
first large bronze statues: ca. 600 BCE	first large bronze statues: ca. 550 BCE
indirect lost-wax process	indirect lost-wax process
flow fusion welding	flow fusion welding
details in relief with direct lost-wax process (common)	details in relief with direct lost-wax process (rare)
few casting defects	numerous casting defects
alloy composition: leaded bronze during the whole period	alloy composition: bronze, moving to leaded bronze (Hellenistic period)
thin metal walls seen from 600 BCE to the latest periods (1 to 3 mm)	thick metal walls in early sculptures (6 to 10 mm), progressively thinning (2 to 4 mm)
core was hollow whenever the internal volumes of the statue allowed, core was not removed	core was generally removed at least partially to allow for joining
complex setup in order to prepare for flow welding, involving a "plaster type" material	mechanical alignment of the metal pieces to be welded
ring-shape joins/linear joins	linear joins often punctuated by small basins (vaschettes)
welding filler metal ≠ metal of primary castings	welding filler metal = metal of primary castings

Notes

1. A comprehensive achaeometallurgical study of the Jabal al-awd bronzes is currently in progress (Hitgen, pers. comm.).
2. We have proposed and use the term "flow fusion welds" to describe these joins. This is a combination of the terminology classically used in archaeometallurgy (flow welds) and that used by modern metallurgists (fusion welds).
3. Few South Arabian bronzes have been analyzed so far, and the available results are listed below:

- Horse from the Dumbarton Oaks Research Library and Collection in Washington (BZ.1938.12), discovered in Ghayman and dated from the second century CE. Analyses performed at the laboratory of the Freer Gallery by W. T. Chase in 1971 (emission spectroscopy), and results kindly communicated by P. Jett:
    - a. Inscription B: 83.9 wt.% Cu, 5.96 wt.% Pb, 3.69 wt.% Sn, <0.18 wt.% Zn.
    - b. Saddle blanket: 80.3 wt.% Cu, 7.97 wt.% Pb, 5.57 wt.% Sn, <0.14 wt.% Zn.
    - c. Nostril: 74.05 wt.% Cu, 17.71 wt.% Pb, 2.29 wt.% Sn, <0.15 wt.% Zn.

- Pair of lion riders of the American Foundation for the Study of Man (TS 152 A–B) discovered in Tamna and dated from the end of the first century BCE to the beginning of the first century CE. Analysis reported in Ali Aqil and Antonini (2007):
    Sample taken from a lion: 53.88 wt.% Cu, 20.36 wt.% Pb, 4.35 wt.% Sn.

- Male head in the British Museum collection (BM 136359), discovered in Ghayman and dated from the first or second century CE. Analysis reported in Ali Aqil and Antonini (2007):
    wt.% Cu, 9.6 wt.% Pb, 7.0 wt.% Sn, 0.02 wt.% Zn.

- Fragment of a male statue of the Baynun museum (Byn M234), dated from the end of the first century CE. Analysis performed at the Florence University by scanning electron microscopy–energy-dispersive x-ray spectrometry (Chiarantini 2009):
    67.51 wt.% Cu, 19.82 wt.% Pb, 7.43 wt.% Sn.
    NB: Two more statues from the Baynun museum were analyzed by Chiarantini, but the results will not be taken into account here as one, a lion statuette, was a heavily corroded sample, and the other, a leopard statue, was also analyzed in the present work.

4. Other types of secondary casting visible in the South Arabian bronzes include the repair of large casting defects (foremost on the Oriental Tamna appliqué but also on Hawtar'athat), and casting on, that is the direct casting of a part on a pre-existent one (for example, nails to attach Hawtar'athat to a base were cast on the lower side of the feet).
5. The earliest Greek hollow-cast large bronze sculptures would date back to the middle of the sixth century BCE according to ancient texts. The oldest archaeological evidence discovered so far are the mold fragments of a Kouros from the Athenian Agora, indeed corresponding to this period (Mattusch 1988, 54), and two bronze statues roughly dated 520 BCE (Zeus head from Olympia and the Ugento Zeus; Rolley 1994, 66).

References

Ali Aqil, A., and S. Antonini. 2007. *Bronzi Sudarabici di periodo pre-islamico*. Vol. 3. Repertorio Iconografico Sudarabico. Paris: Académie des Inscriptions et Belles-Lettres.

Antonini, S. 2007. *Le opere del Jabal al-'awd*. In *Bronzi Sudarabici di periodo pre-islamico,* Vol. 3, Appendix. Repertorio Iconografico Sudarabico, ed. A. Ali Aqil and S. Antonini, 93. Paris: Académie des Inscriptions et Belles-Lettres.

Arbach, M. 2005. Un lion en bronze avec un nouveau synchronisme mineo-qatabanite. In *Sabaean studies. Studies in honour of Yusuf Abdallah, Alessandro de Maigret and Christian J. Robin in the occasion of their 60th birthdays,* ed. A. Sholan, S. Antonini, and M. Arbach, 21–33. Naples: Sanaa.

Azema, A., and B. Mille. 2010. La grande sculpture antique en bronze assemblée par soudage par fusion. *Soudage et techniques connexes* March–April 2010: 20–22.

———. 2011. The flow fusion welding process: A study of the joining techniques used on the Greek and Roman large bronze statues. Paper presented at the Third International Conference "Archaeometallurgy in Europe," 29 June–1 July, Deutsches Bergbau-Museum, Bochum, Germany.

Bouquillon, A., S. Descamps, A. Hermary, and B. Mille. 2006. Une nouvelle étude de l'Apollon Chatsworth. *Revue Archéologique* 42 (2006/2) :227–261.

Bourgarit, D., and B. Mille. 2003. The elemental analysis of ancient copper-based artifacts by inductively-coupled-plasma atomic-emission-spectrometry (ICP-AES): An optimized methodology reveals some secrets of the Vix crater. *Measurement Science and Technology* 14(9):1538–55.

Chiarantini, L. 2009. Microstructural and chemical analyses of the bronze statuettes. In *Art and technique in Yemen: The bronzes from the Museum of Baynun,* ed. A. Avanzini, 107–9. Pisa: Bandecchi & Vivaldi.

Demange, F., I. Gajda, B. Mille, C. Pariselle, and O. Tavoso. 2007. *L'Homme de bronze. Hawtar 'athat, fils de Radaw'il, du lignage de Shalalum*. Paris: Musée du Louvre.

Formigli, E. 1984. La tecnica di construzione delle statue di Riace. *Due bronzi da Riace. Rinvenimento, restauro, analisi ed ipotesi di interpretazione I,* Bolletino d'Arte, Serie speciale 3:107–142.

Hauptmann, A. 1985. *5000 jahre kupfer in Oman: Die entwicklung der kupfermetallurgie vom 3. jahrtausend bis zur neuzeit*. Der Anschnitt, Beiheft 4. Bochum: Deutsches Bergbau-Museum.

Hitgen, H. 2005. An early Himyarite mountain settlement on Jabal al-Awd. In *Caravan kingdoms: Yemen and the ancient incense trade,* ed. A. C. Gunter, 42–51. Washington, DC: Arthur M. Sackler Gallery.

Lie, H., and C. C. Mattusch. 1996. Introduction to the catalogue entries and technical observations. In *The fire of Hephaistos: Large classical bronzes from North American collections,* ed. C. C. Mattusch, 162–79. Cambridge, MA: Harvard University Press.

Maraqten, M. 2005. The Awam temple: Recent epigraphic discoveries at Mahram Bilqis. In *Caravan kingdoms, Yemen and the ancient incense trade,* ed. A. C. Gunter, 68–73. Washington, DC: Arthur M. Sackler Gallery.

Mattusch, C. C. 1988. *Greek bronze statuary: From the beginnings through the fifth century* B.C., Ithaca, NY: Cornell University Press.

Mille, B. 2007. Etude technique du cheval de bronze de Neuvy-en-Sullias. *Le cheval et la danseuse, à la redécouverte du trésor de Neuvy-en-Sullias,* ed. C. Gorget and J.-P. Guillaumet, 88–99 and 264–65. Catalogue de l'exposition présentée au musée des Beaux-Arts d'Orléans du 13 mars au 26 août 2007. Paris: Somogy.

Mille, B., and T. Borel. 2007. Composition élémentaire et radiographie des statuettes de Neuvy-en-Sullias. In *Le cheval et la danseuse, à la redécouverte du trésor de Neuvy-en-Sullias,* ed. C. Gorget and J. P. Guillaumet, 194–99 and 268–69. Catalogue de l'exposition présentée au musée des Beaux-Arts d'Orléans du 13 mars au 26 août 2007. Paris: Somogy.

Mille, B., I. Gajda, F. Demange, C. Pariselle, Y. Coquinot, E. Porto, O. Tavoso, and A. Zink. 2010. Hawtar'athat, fils de Radaw'il du lignage de Shalalum, une grande statue de bronze du royaume de Saba'. In *Monuments et mémoires de la fondation Eugène Piot,* Vol. 89, 5–68. Paris: Académie des Inscriptions et Belles-Lettres.

Mille, B., L. Rossetti, C. Rolley, D. Bourgarit, E. Formigli, and M. Pernot. 2009. Les deux statues d'enfant en bronze (Cap d'Adge): Étude iconographique et technique. Paper presented at the conference "Bronzes grecs et romains, recherches récentes, hommage à Claude Rolley," 16–17 June, Paris.

Phillips Hodgson, M. 2005. The Awam temple: Excavations at the Mahram Bilqis near Marib. In *Caravan kingdoms: Yemen and the ancient incense trade,* ed. A. C. Gunter, 60–66. Washington, DC: Arthur M. Sackler Gallery.

Robin, C. J., and B. Vogt, ed. 1997. *Yémen, au pays de la reine de Saba.* Paris: Flammarion.

Rolley, C. 1994. *La sculpture grecque,* Vol. 1, *Des origines au milieu du Ve siècle*. Paris: Picard.

Steinberg, A. 1973. Joining methods on large bronze statues: Some experiments in ancient technology. In *Application of science in examination of works of art,* ed. W. J. Young, 103–38. Boston: Museum of Fine Arts.

Weeks, L. 2003. *Early metallurgy of the Persian Gulf: Technology, trade, and the Bronze Age world.* Boston: Brill Academic Publishers.

Weidemann, K. 1983. *Könige aus dem Yemen: Zwei spätantike bronzestatuen*. Mainz: Römisch-Germanischen Zentralmuseums.

Appendix. Results of the ICP-AES Analyses Made at the C2RMF for the South Arabian Statues Studied in the Frame of the Louvre–GOAM Cooperation Project

Designation	Analysis No.	Sample Location	Casting type	Cu	Pb	Sn	Ag	As	Fe	Ni	S
				wt.%	wt.%	wt.%	wt.%	wt.%	wt.%	wt.%	wt.%
Hawtar'athat statue; Inv. No. YM23206	C2RMF 64057-7	weld, left arm–body	SC (welding)	X	11 ±1	6.4 ±0.6	0.052 ±0.005	0.32 ±0.03	0.011 ±0.001	0.056 ±0.006	0.11 ±0.01
	C2RMF 64057-8	left arm	PC	X	17 ±2	5.7 ±0.6	0.039 ±0.004	0.26 ±0.03	0.029 ±0.003	0.043 ±0.004	0.11 ±0.01
	C2RMF 64057-10	head (curl)	PC	X	14 ±1	7.9 ±0.8	0.04 ±0.004	0.24 ±0.02	0.036 ±0.004	0.047 ±0.005	0.099 ±0.016
	C2RMF 64057-11*	body (right flank)	PC	X	11 ±1	12 ±1	0.042 ±0.004	0.28 ±0.03	0.053 ±0.005	0.046 ±0.005	0.1 ±0.01
	C2RMF 64057-12	welding, head–body	SC (welding)	X	12 ±1	8.2 ±0.8	0.053 ±0.005	0.34 ±0.03	0.03 ±0.003	0.055 ±0.005	0.12 ±0.02
	C2RMF 64057-13	welding, right foot–body	SC (welding)	X	13 ±1	7.7 ±0.8	0.051 ±0.005	0.33 ±0.03	0.03 ±0.003	0.055 ±0.005	0.11 ±0.01
	C2RMF 64057-14	right foot	PC	X	7 ±0.7	6.7 ±0.7	0.043 ±0.004	0.3 ±0.03	0.031 ±0.003	0.047 ±0.005	0.1 ±0.02
	C2RMF 64057-15	base attachment nail under the right foot	SC (basing)	X	12 ±1	3.7 ±0.4	0.03 ±0.003	0.26 ±0.03	0.11 ±0.01	0.046 ±0.005	0.097 ±0.015
Lion appliqué; Inv. No. YM36526	C2RMF 67747-a	attachment hook, top of the head	PC	X	8.5 ±0.9	7.4 ±0.7	0.13 ±0.01	0.43 ±0.04	0.056 ±0.006	0.12 ±0.01	0.038 ±0.004
	C2RMF 67747-b	weld, head–body	SC (welding)	X	7 ±0.7	6.2 ±0.6	0.072 ±0.007	0.5 ±0.05	0.054 ±0.005	0.095 ±0.009	0.084 ±0.008
	C2RMF 67747-c	repair, body	SC (repair)	X	4.1 ±0.4	7.3 ±0.7	0.083 ±0.008	0.55 ±0.05	0.051 ±0.005	0.1 ±0.01	0.086 ±0.009
	C2RMF 67747-d	tongue	SC (casting on)	X	3.5 ±0.4	6.1 ±0.6	0.064 ±0.006	0.48 ±0.05	0.054 ±0.005	0.098 ±0.010	0.043 ±0.004
	C2RMF 67747-e	repair of the repair (analysis d)	TC (repair)	X	13 ±1	6.2 ±0.6	0.057 ±0.006	0.4 ±0.04	0.33 ±0.03	0.07 ±0.007	0.16 ±0.02
	C2RMF 67747-f	head (tooth)	PC	X	4.9 ±0.5	6 ±0.6	0.067 ±0.007	0.47 ±0.05	0.052 ±0.005	0.094 ±0.009	0.1 ±0.01
	C2RMF 67747-g	circular rod (to fill the hole left by a spacer)	—	X	6.7 ±0.7	7 ±0.7	0.07 ±0.007	0.53 ±0.05	0.063 ±0.006	0.092 ±0.009	0.072 ±0.007
	C2RMF 67747-h	body, foreleg	PC	X	12 ±1	6.8 ±0.7	0.23 ±0.02	0.51 ±0.05	0.05 ±0.005	0.25 ±0.03	0.07 ±0.007
Lion statuette; Inv. No. IM24	C2RMF 67748-5	right hindleg	PC	X	22 ±2	6.6 ±0.7	0.032 ±0.003	0.14 ±0.01	0.033 ±0.003	0.19 ±0.02	0.067 ±0.007
	C2RMF 67748-6	left foreleg	PC	X	25 ±3	6.6 ±0.7	0.031 ±0.003	0.13 ±0.01	0.038 ±0.004	0.093 ±0.009	0.038 ±0.004

Sb wt.%	Zn wt.%	Au ppm	Ba ppm	Bi ppm	Co ppm	Hg ppm	In ppm	Mg ppm	Mn ppm	P ppm	Se ppm	Te ppm	Ti ppm	U ppm
0.079 ±0.008	nd <0.00085	33 ±3	nd <0.2	85 ±8	178 ±18	2.7 ±0.7	11 ±2	20 ±2	nd <0.3	nd <32	58 ±6	75 ±8	nd <0.8	nd <4.4
0.093 ±0.009	nd <0.0010	36 ±4	nd <0.3	104 ±10	429 ±43	1.7° <3.1 >0.9	15 ±1	29 ±3	nd <0.4	nd <38	16 ±3	54 ±5	nd <1.0	nd <5.2
0.12 ±0.01	nd <0.00088	34 ±3	nd <0.3	82 ±8	354 ±35	2.6° <2.7 >0.8	13 ±3	15 ±1	nd <0.3	nd <33	50 ±5	74 ±7	nd <0.8	nd <4.5
0.12 ±0.01	0.006 ±0.0006	38 ±4	nd <0.3	71 ±7	266 ±27	2.2° <3.7 >1.1	14 ±2	179 ±18	nd <0.5	nd <46	51 ±5	90 ±9	nd <1.2	nd <6.2
0.079 ±0.008	nd <0.00085	31 ±3	nd <0.2	83 ±8	216 ±22	2.4° <2.6 >0.8	14 ±3	15 ±1	nd <0.3	nd <32	64 ±6	85 ±9	nd <0.8	nd <4.4
0.074 ±0.007	nd <0.00087	30 ±3	nd <0.3	80 ±8	222 ±22	3.1 ±0.5	12 ±2	15 ±2	nd <0.3	nd <33	64 ±6	81 ±8	nd <0.8	nd <4.5
0.11 ±0.01	nd <0.0018	41 ±4	nd <0.5	103 ±10	506 ±51	2.2° <5.5 >1.7	19 ±2	161 ±16	nd <0.7	nd <68	12° <22 >7	59 ±6	nd <1.8	nd <9.4
0.067 ±0.007	0.0025° <0.0029 >0.0009	53 ±5	nd <0.3	56 ±6	277 ±28	2.6° <2.7 >0.8	15 ±2	15 ±1	nd <0.3	nd <33	4.9° <10.8 >3.2	39 ±4	nd <0.8	nd <4.5
0.15 ±0.02	0.019 ±0.002	48 ±5	nd <0.4	97 ±10	189 ±19	3.4 ±0.3	16 ±4	nd <0.5	nd <0.3	nd <11	50 ±5	92 ±9	nd <0.4	nd <11
0.27 ±0.03	0.047 ±0.005	52 ±5	nd <0.3	108 ±11	166 ±17	4.3 ±0.4	14 ±2	5 ±0.5	nd <0.2	15° <30 >9	91 ±9	143 ±14	nd <0.3	nd <8.8
0.3 ±0.03	0.049 ±0.005	58 ±6	14 ±1	115 ±11	179 ±18	3.8 ±0.4	17 ±7	14 ±1	nd <0.3	nd <13	90 ±9	155 ±15	nd <0.4	19° <41 >12
0.25 ±0.02	0.053 ±0.005	54 ±5	nd <0.4	98 ±10	177 ±18	4.3 ±0.5	15 ±5	nd <0.5	nd <0.3	nd <12	60 ±6	105 ±10	nd <0.4	nd <11
0.22 ±0.02	0.035 ±0.004	43 ±4	nd <0.5	159 ±16	152 ±15	2.7 ±0.3	15 ±3	54 ±5	nd <0.4	nd <14	28 ±5	66 ±7	nd <0.5	nd <14
0.25 ±0.03	0.045 ±0.005	50 ±5	nd <0.4	102 ±10	167 ±17	4.1 ±0.4	13 ±2	nd <0.5	nd <0.3	12° <36 >11	75 ±8	136 ±14	nd <0.4	nd <11
0.29 ±0.03	0.043 ±0.004	52 ±5	nd <1.0	114 ±11	164 ±16	3.3° <4.7 >1.4	21 ±8	nd <1.2	nd <0.7	29° <92 >28	85 ±9	158 ±16	nd <0.9	40° <91 >27
0.19 ±0.02	0.054 ±0.005	57 ±6	nd <0.4	112 ±11	166 ±17	3.8 ±0.4	15 ±4	nd <0.5	nd <0.3	10° <35 >10	76 ±8	110 ±11	nd <0.3	nd <10
0.06 ±0.006	0.028 ±0.003	19 ±2	nd <0.4	46 ±5	244 ±24	3.2 ±0.5	9.7 ±3.1	0.8° <1.5 >0.4	nd <0.3	20° <33 >10	nd <2.1	19 ±11	nd <0.3	nd <9.8
0.067 ±0.007	0.0038° <0.0045 >0.0013	20 ±2	nd <0.4	39 ±5	147 ±15	3.4 ±0.3	4.7° <6.4 >1.9	nd <0.4	nd <0.3	nd <9.7	2.2° <6.7 >2.0	14 ±8	nd <0.3	nd <9.5

cont'd

Appendix. Results of the ICP-AES Analyses Made at the C2RMF for the South Arabian Statues Studied in the Frame of the Louvre–GOAM Cooperation Project *cont'd*

Designation	Analysis No.	Sample Location	Casting type	Cu	Pb	Sn	Ag	As	Fe	Ni	S
				wt.%	wt.%	wt.%	wt.%	wt.%	wt.%	wt.%	wt.%
Headless man standing statue; Inv. No. YM23212	Y9-1	body, base of the neck	PC	X	23 ±2	6.2 ±0.6	0.024 ±0.002	0.28 ±0.03	0.056 ±0.006	0.05 ±0.005	0.16 ±0.02
	Y9-2	weld, head–body	SC (welding)	X	12 ±1	7.2 ±0.7	0.1 ±0.01	0.26 ±0.03	0.035 ±0.004	0.03 ±0.003	0.15 ±0.02
	Y9-3	right foot	PC	X	12 ±1	8.1 ±0.8	0.028 ±0.003	0.38 ±0.04	0.06 ±0.006	0.061 ±0.006	0.14 ±0.01
Fragmentary life-size standing statue, cloth decorated with braids; Inv. No. YM238	Y10-238-1	cloth	PC	X	12 ±1	7.9 ±0.8	0.27 ±0.03	0.24 ±0.02	0.25 ±0.03	0.2 ±0.02	0.082 ±0.008
	Y10-238-2	fill from the hole left by a spacer	SC (repair)	X	9.2 ±0.9	8.7 ±0.9	0.31 ±0.03	0.28 ±0.03	0.28 ±0.03	0.2 ±0.02	0.11 ±0.01
Fragmentary life-size standing statue, cloth with long vertical folds; Inv. No. YM239	Y10-239-1	cloth	PC	X	8.7 ±0.9	7.7 ±0.8	0.053 ±0.005	0.31 ±0.03	0.1 ±0.01	0.1 ±0.01	0.072 ±0.007
	Y10-239-2	spacer	—	X	1.2 ±0.1	6.9 ±0.7	0.016 ±0.002	0.13 ±0.01	0.1 ±0.01	0.031 ±0.003	0.051 ±0.005
Fragmentary life-size standing statue, large fragment (knee, vertical folds, cloth); Inv. No. YM241	Y10-241-1	vertical folds	PC	X	2.7 ±0.3	6.9 ±0.7	0.039 ±0.004	0.078 ±0.008	0.046 ±0.005	0.11 ±0.01	0.11 ±0.01
	Y10-241-2	knee	PC	X	12 ±1	7.3 ±0.7	0.24 ±0.02	0.22 ±0.02	0.24 ±0.02	0.2 ±0.02	0.071 ±0.007
	Y10-241-3	cloth	PC	X	8.7 ±0.9	8.2 ±0.8	0.25 ±0.02	0.24 ±0.02	0.23 ±0.02	0.2 ±0.02	0.076 ±0.008
	Y10-241-4	weld, cloth–knee folds	SC (welding)	X	11 ±1	7.5 ±0.7	0.24 ±0.02	0.23 ±0.02	0.32 ±0.03	0.2 ±0.02	0.068 ±0.007
Life-size leg fragment; Inv. No. YM220	Y13-1	leg	PC	X	25 ±3	5.4 ±0.5	0.031 ±0.003	0.046 ±0.005	0.008 ±0.0008	0.025 ±0.002	0.07 ±0.007
Life-size left arm fragment; Inv. No. YM218	Y14-1	arm	PC	X	2.2 ±0.2	6.7 ±0.7	0.07 ±0.007	0.31 ±0.03	0.012 ±0.001	0.13 ±0.01	0.12 ±0.01
Fragmentary lion appliqué; Inv. No. YM261-YM252	Y19-1	right hind leg	PC	X	10 ±1	2.9 ±0.3	0.007 ±0.0007	0.17 ±0.02	1.2 ±0.1	1.2 ±0.1	0.14 ±0.01
	Y19-2	body	PC	X	4.2 ±0.4	3.1 ±0.3	0.016 ±0.002	0.17 ±0.02	0.24 ±0.02	0.91 ±0.09	0.28 ±0.03
	Y19-3	weld, leg–body	SC (welding)	X	9.1 ±0.9	2.9 ±0.3	0.009 ±0.0009	0.19 ±0.02	0.049 ±0.005	1.1 ±0.1	0.15 ±0.01
	Y19-4	weld, tail–body	SC (welding)	X	15 ±1	3.4 ±0.3	0.016 ±0.002	0.2 ±0.02	0.026 ±0.003	0.95 ±0.10	0.17 ±0.02
	Y19-5	tail	PC	X	24 ±2	7.9 ±0.8	0.043 ±0.004	0.31 ±0.03	0.071 ±0.007	0.59 ±0.06	0.16 ±0.02

Sb	Zn	Au	Ba	Bi	Co	Hg	In	Mg	Mn	P	Se	Te	Ti	U
wt.%	wt.%	ppm	ppm	ppm	ppm	ppm	ppm	ppm	ppm	ppm	ppm	ppm	ppm	ppm
0.046 ±0.005	0.011 ±0.001	27 ±3	nd <0.3	51 ±5	738 ±74	3.3 ±0.6	9.5 ±1.0	nd <0.4	nd <0.2	nd <8.6	16 ±2	44 ±4	nd <0.3	nd <8.5
0.04 ±0.004	0.006 ±0.0006	20 ±2	nd <0.4	55 ±10	697 ±70	3.7 ±0.4	12 ±5	nd <0.5	nd <0.3	nd <11	20 ±3	46 ±5	nd <0.4	nd <11
0.066 ±0.007	0.012 ±0.001	30 ±3	nd <0.4	44 ±7	932 ±93	3.6 ±0.6	13 ±2	nd <0.5	nd <0.3	27° <39 >12	8.1° <8.1 >2.4	40 ±9	nd <0.4	nd <12
0.093 ±0.009	0.012 ±0.001	68 ±7	nd <0.4	98 ±10	272 ±27	3.8 ±0.8	10 ±1	nd <0.5	nd <0.3	14° <37 >11	35 ±4	53 ±5	nd <0.4	nd <11
0.11 ±0.01	0.01 ±0.001	75 ±7	nd <0.5	107 ±11	241 ±24	4 ±0.4	9.9 ±3.1	152 ±15	4.2 ±0.4	37° <42 >13	37 ±4	55 ±6	37 ±4	nd <12
0.1 ±0.01	0.033 ±0.003	43 ±4	nd <0.3	67 ±7	241 ±24	3.9 ±0.4	13 ±1	nd <0.4	nd <0.2	11° <28 >8	47 ±5	69 ±7	nd <0.3	nd <8.3
0.034 ±0.003	0.0032° <0.0094 >0.0028	19 ±2	nd <0.8	27° <33 >10	81 ±8	4.3 ±0.5	5.0° <13.6 >4.1	94 ±9	nd <0.5	nd <20	nd <4.2	20° <21 >6	11 ±1	nd <20
0.055 ±0.005	0.006 ±0.0006	19 ±2	nd <0.3	40 ±5	150 ±15	4.6 ±0.5	5.4 ±0.9	2.3 ±0.2	nd <0.2	9.8° <22.6 >6.8	19 ±2	17 ±2	nd <0.2	nd <6.7
0.087 ±0.009	0.014 ±0.001	66 ±7	nd <0.3	87 ±9	265 ±27	3.6 ±0.4	9.4 ±4.4	nd <0.4	nd <0.2	9.4° <29.7 >8.9	37 ±4	44 ±6	nd <0.3	nd <8.8
0.089 ±0.009	0.013 ±0.001	72 ±7	nd <0.4	90 ±9	261 ±26	4 ±0.5	6.6° <7.1 >2.1	nd <0.5	nd <0.3	27° <36 >11	38 ±4	52 ±5	nd <0.4	nd <11
0.088 ±0.009	0.015 ±0.001	73 ±7	nd <0.4	85 ±9	272 ±27	4.1 ±0.4	7.0° <7.2 >2.2	nd <0.5	nd <0.3	17° <36 >11	33 ±3	50 ±5	nd <0.4	nd <11
0.075 ±0.008	0.013 ±0.001	9 ±0.9	nd <0.3	37 ±4	8.3 ±0.8	3.5 ±0.4	2.3° <5.3 >1.6	nd <0.4	nd <0.2	18° <27 >8	28 ±3	11 ±11	nd <0.3	nd <7.9
0.11 ±0.01	0.0038° <0.0050 >0.0015	49 ±5	nd <0.4	81 ±8	181 ±18	4.8 ±0.5	6.2° <7.2 >2.2	4.4 ±0.4	nd <0.3	20° <36 >11	67 ±7	77 ±8	3 ±0.3	nd <11
0.006 ±0.0006	0.009 ±0.0008	2.5° <3.2 >1.0	nd <0.4	18 ±5	2 466 ±247	3.8 ±0.4	18 ±5	nd <0.5	nd <0.3	nd <11	68 ±7	9.3° <11.0 >3.3	nd <0.4	nd <10
0.004 ±0.0006	0.007 ±0.0007	15 ±1	nd <0.4	20 ±9	1 419 ±142	4.5 ±0.5	14 ±1	3.9 ±0.4	nd <0.3	nd <12	60 ±6	31 ±8	nd <0.4	nd <12
0.006 ±0.0006	nd <0.0016	6.6 ±0.7	nd <0.4	15° <19 >6	1 143 ±114	3.9 ±0.6	10 ±1	nd <0.5	nd <0.3	nd <12	57 ±6	19 ±3	nd <0.4	nd <12
0.019 ±0.002	nd <0.0015	7.7 ±0.9	nd <0.4	27 ±3	691 ±69	3.7 ±0.4	12 ±3	11 ±1	nd <0.3	nd <11	45 ±4	18 ±2	nd <0.4	nd <11
0.05 ±0.005	nd <0.0016	37 ±4	nd <0.4	54 ±5	720 ±72	2.8 ±0.5	8.1 ±2.6	nd <0.5	nd <0.3	13° <38 >12	8.5 ±2.3	18 ±8	nd <0.4	nd <11

cont'd

Appendix. Results of the ICP-AES Analyses Made at the C2RMF for the South Arabian Statues Studied in the Frame of the Louvre–GOAM Cooperation Project *cont'd*

Designation	Analysis No.	Sample Location	Casting type	Cu wt.%	Pb wt.%	Sn wt.%	Ag wt.%	As wt.%	Fe wt.%	Ni wt.%	S wt.%
Forepart of a leopard statue; Inv. No. MB216	Y27-1	head	PC	X	22 ±2	4.8 ±0.5	0.009 ±0.0008	0.08 ±0.008	0.49 ±0.05	1.2 ±0.1	0.28 ±0.03
	Y27-2	weld, head–body	SC (welding)	X	19 ±2	5.6 ±0.6	0.013 ±0.001	0.081 ±0.008	0.071 ±0.007	1 ±0.1	0.24 ±0.02
	Y27-3	body	PC	X	23 ±2	5.3 ±0.5	0.023 ±0.002	0.073 ±0.007	0.77 ±0.08	1.2 ±0.1	0.25 ±0.03
	Y27-4	spacer	—	X	0.2 ±0.02	2.1 ±0.2	0.015 ±0.001	0.15 ±0.02	0.16 ±0.02	0.99 ±0.10	0.03 ±0.003
Ibex protome; Inv. No. IM25	Y30-1	ibex protome	PC	X	14 ±1	5.2 ±0.5	0.042 ±0.004	0.33 ±0.03	0.01 ±0.001	0.39 ±0.04	0.089 ±0.009
Draped female statuette; Inv. No. IM9	Y32-1	left forearm	PC	X	21 ±2	6.1 ±0.6	0.006 ±0.0006	0.036 ±0.004	0.066 ±0.007	0.011 ±0.001	0.007 ±0.0022
	Y32-2	armature in the left forearm	—	X	23 ±2	6 ±0.6	0.006 ±0.0006	0.035 ±0.003	0.067 ±0.007	0.01 ±0.0010	nd <0.0012
	Y32-3	right forearm	PC	X	39 ±4	4.6 ±0.5	0.04 ±0.004	0.13 ±0.01	0.023 ±0.002	0.17 ±0.02	0.068 ±0.007
	Y32-4	left foot	PC	X	17 ±2	6.5 ±0.7	0.007 ±0.0007	0.037 ±0.004	0.064 ±0.006	0.011 ±0.001	nd <0.0014
	Y32-5	head	PC	X	23 ±2	6.2 ±0.6	0.006 ±0.0006	0.035 ±0.004	0.055 ±0.006	0.010 ±0.0010	0.015 ±0.002
Draped female statuette; Inv. No. IM8 & IM19	Y33-1	cloth	PC	X	2.4 ±0.2	10 ±1	0.047 ±0.005	0.07 ±0.007	0.025 ±0.002	0.043 ±0.004	0.058 ±0.006
	Y33-2	mantle	PC	X	0.18 ±0.02	11 ±1	0.013 ±0.001	0.014 ±0.001	0.085 ±0.009	0.018 ±0.002	nd <0.0014

Eight samples were taken from Hawtar'athat in 2007. Thirty-two samples were made on ten more statues during the 2008 mission to Yemen (including the Jabal al-awd lion statuette). The Tamna lion appliqué was sampled during its 2009 stay at the C2RMF (eight samples). Results in wt.% or wt. ppm. The elements systematically below the limit of detection are not reported individually; the corresponding limits of detection are as follows (in ppm): Cd: 1.1, Cr: 2.4, Ge: 9.6, Mo: 3.1, V: 1.3, W: 7.5. nd = below the limit of detection; PC = primary casting; SC = secondary casting; TC = tertiary casting; X = base metal (not quantified); ° = below the limit of quantification.

Sb	Zn	Au	Ba	Bi	Co	Hg	In	Mg	Mn	P	Se	Te	Ti	U
wt.%	wt.%	ppm	ppm	ppm	ppm	ppm	ppm	ppm	ppm	ppm	ppm	ppm	ppm	ppm
0.006 ±0.0006	0.44 ±0.04	8.2 ±0.8	nd <0.3	12° <15 >4	1 882 ±188	2.9 ±0.3	13 ±3	nd <0.4	nd <0.2	nd <8.9	28 ±3	16 ±6	nd <0.3	nd <8.8
0.013 ±0.001	0.018 ±0.002	11 ±1	nd <0.4	20 ±3	1 170 ±117	2.8 ±0.3	14 ±3	14 ±1	nd <0.3	13° <32 >10	22 ±2	9.3° <10.1 >3.0	nd <0.3	nd <9.5
0.006 ±0.0006	0.22 ±0.02	9.1 ±1.3	nd <0.5	28 ±6	2 404 ±240	2.9 ±0.3	16 ±2	11 ±1	nd <0.3	nd <12	23 ±2	10° <13 >4	nd <0.4	nd <12
0.004 ±0.0004	6.3 ±0.6	12 ±1	nd <0.3	23 ±3	509 ±51	5.1 ±0.7	15 ±3	nd <0.4	nd <0.2	16° <28 >8	54 ±5	16 ±7	nd <0.3	nd <8.1
0.081 ±0.008	0.01 ±0.001	33 ±3	nd <0.4	66 ±7	264 ±26	3.8 ±0.8	10 ±2	13 ±1	nd <0.3	25° <36 >11	32 ±4	55 ±16	nd <0.4	nd <11
0.012 ±0.001	nd <0.0013	3 ±0.3	nd <0.3	34 ±3	149 ±15	3.5 ±0.6	6.1 ±0.6	nd <0.4	nd <0.2	9.3° <30.4 >9.1	2.7° <6.3 >1.9	13 ±4	nd <0.3	nd <9.0
0.011 ±0.001	nd <0.0014	2.0° <3.1 >0.9	nd <0.4	32 ±3	147 ±15	3.2 ±0.3	4.8° <6.7 >2.0	nd <0.5	nd <0.3	nd <10	nd <2.1	9.6° <10.5 >3.2	nd <0.3	nd <9.9
0.037 ±0.004	nd <0.0014	20 ±2	nd <0.4	63 ±6	138 ±14	1.8 ±0.3	3.0° <6.6 >2.0	nd <0.4	nd <0.3	17° <33 >10	25 ±3	35 ±5	nd <0.3	nd <9.7
0.012 ±0.001	nd <0.0016	1.8° <3.6 >1.1	nd <0.4	32 ±3	160 ±16	3.1 ±0.6	8.1 ±1.5	nd <0.5	nd <0.3	nd <12	nd <2.5	16 ±6	nd <0.4	13° <39 >12
0.011 ±0.001	nd <0.0018	1.3° <4.0 >1.2	nd <0.5	36 ±4	148 ±15	3.5 ±0.6	8.6° <8.7 >2.6	nd <0.6	nd <0.3	17° <44 >13	nd <2.7	22 ±2	nd <0.4	19° <43 >13
0.055 ±0.006	0.004 ±0.0004	28 ±3	nd <0.4	58 ±6	101 ±10	4 ±0.4	5.6° <6.2 >1.9	nd <0.4	nd <0.2	13° <31 >9	13 ±3	23 ±7	nd <0.3	nd <9.2
0.017 ±0.002	nd <0.0016	nd <1.1	nd <0.4	32 ±7	nd <0.4	4.1 ±0.6	4.8° <7.9 >2.4	nd <0.5	nd <0.3	15° <40 >12	5.2° <8.2 >2.5	15 ±9	nd <0.4	nd <12

Medieval Islamic Copper Alloys

Susan La Niece, Rachel Ward, Duncan Hook, and Paul Craddock

ABSTRACT This paper builds on the research into Islamic metal technology presented at the first Forbes symposium and reviews the published analyses of medieval Islamic copper alloys. The repertoire of copper alloy types used to make vessels between the seventh century and approximately the sixteenth century CE shows remarkable consistency across the regions under Islamic rule in terms of the major alloy compositions, though the trace element picture is much less clear cut. The most pronounced division is between the relatively pure brass vessels made by hammering (sinking or raising) and the more mixed alloys of the items made by casting. Additionally there is a small but distinctive, high-status group of high-tin bronze vessels with approximately 20 wt.% or more tin and a larger group of utilitarian vessels made of copper with only minor levels of other elements. Two groups of objects that are distinguished from the majority of cast copper alloys by their tin and zinc compositions are identified: a group of cast mirrors of the same form as the round mirrors from China but with Arabic inscriptions and a group of sand-cast candlesticks dated to the late thirteenth and fourteenth century.

Introduction

At its peak, Islamic rule extended from Spain in the west, through North Africa, and on to Afghanistan, Bengal, and beyond to the east and was centered on the regions of the Middle East. The copper-alloy objects discussed here are those made in the extensive area under Islamic influence; there is no implication of religious use for most of the metalwork. The objects range in date from the seventh century CE to approximately the sixteenth century. This paper builds on the research into metal technology presented at the first Forbes symposium (La Niece 2003) and reviews the published analyses of medieval Islamic copper alloys. The long-running study of the Islamic metalwork collections at the British Museum continues to produce new results; in particular a group of Islamic mirrors is discussed here.

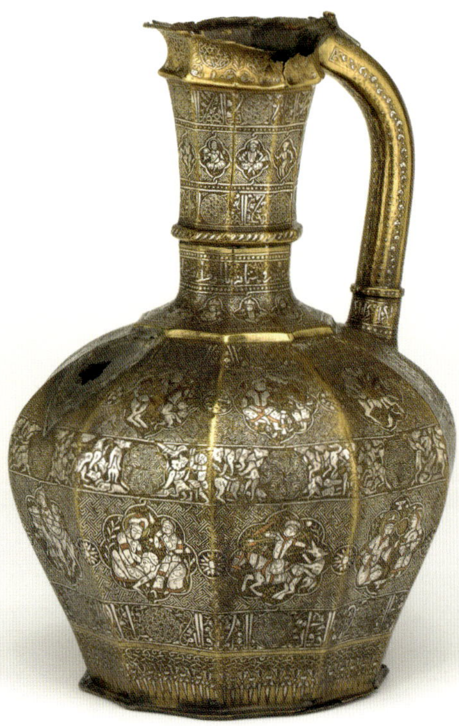

Figure 1. *Brass ewer inlaid with silver and copper. The inscription around its neck gives the date and place of the inlay work as the month of Rajab 629 AH (April 1232 CE) in Mosul, h. 30.4 cm. British Museum, 1866,1229.61.*

Major Element Alloy Compositions

It was established some years ago that the favored copper alloy of the Islamic world is brass, an alloy of copper and zinc (Roberts-Austin 1892; Allan 1979; Craddock 1979; Craddock and Eckstein 2003). The technology for making brass in large quantities was developed by the Romans, and there was continuity in this technology after the decline of the empire. Romans made both brass and bronze (copper alloyed with tin) as well as a copper alloy with both tin and zinc, sometimes termed gunmetal (Hook and Craddock 1996). A few items dating to the beginning of the Islamic era have been found to be bronze (Schweizer 1994), but, with the exception of the groups discussed below, bronze became almost unknown. Atil, Chase, and Jett (1985, 37) concluded from the results of x-ray fluorescence (XRF) analyses that the Islamic copper-alloy objects in the Freer Gallery of Art "can fairly be called 'brasses'; they contain more zinc than tin."

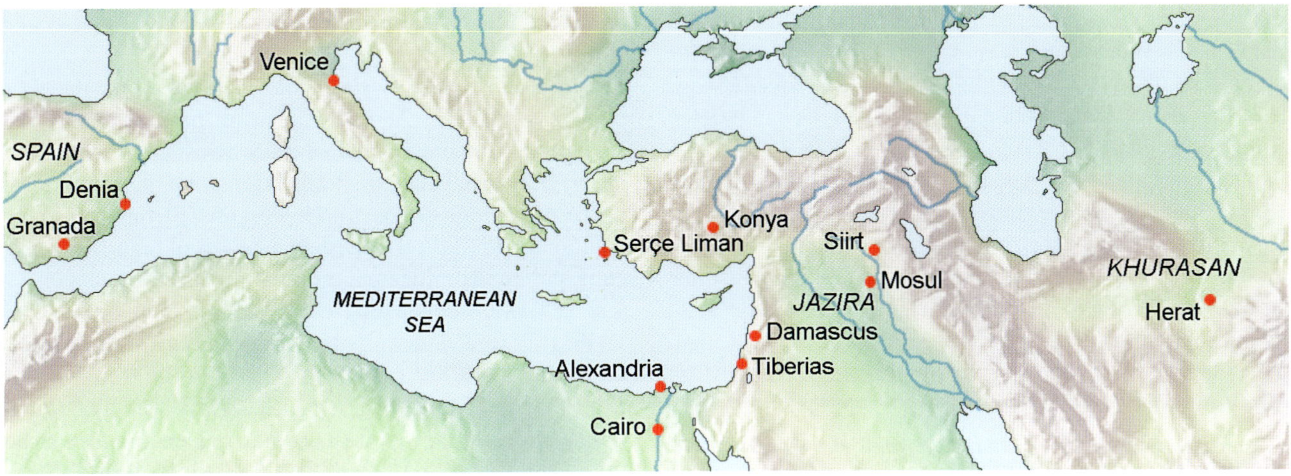

Figure 2. *Map locating places referred to in the text.*

Unalloyed copper continued to be used for simple hammered wares. Tinning of copper vessels to prevent contamination of food and drink by copper corrosion became common in the fourteenth century. Most of the fine vessels that were shaped by hammering are relatively pure brasses and inlaid brass is uniquely characteristic of Islamic metalwork from the twelfth century (fig. 1). Casting was also used to make items for inlaying as well as those decorated more simply. Cast vessels are of a mixed alloy with several percent of lead and often several percent tin in addition to zinc. This alloy is not strictly brass and is often anachronistically called leaded gunmetal to distinguish it from true brass (e.g. Hook 1988), but nevertheless zinc is an important alloying ingredient of these castings. This general picture is based on the published analyses by atomic absorption spectrometry (AAS) (Hughes et al. 1976) and inductively coupled plasma–atomic emission spectroscopy (ICP-AES) (Hook 1998) of more than three hundred items of Islamic metalwork from the collections of the British Museum (London), the Ashmolean Museum (Oxford), the Courtauld Institute (London), and the Archaeological Museum (Granada, Spain) (Craddock, La Niece, et al. 1998). The museum artifacts analyzed for these and other studies (Allan 1999, 123–24) are mainly prestige pieces, with few everyday items. However, this gap has been partly filled by the analysis of 153 items, including utilitarian pieces, all dated to the Fatimid period (eleventh to early twelfth century) from four important archaeological sites (Ponting 2003). The analyses, by ICP-AES and scanning electron microscopy–energy-dispersive x-ray spectrometry (SEM-EDX), were carried out on a representative selection from the approximately one thousand metal artifacts found in three ceramic storage jars buried beneath the floor of a workshop at Tiberias (fig. 2); from the 118 artifacts excavated at Caesarea (100 km south of Tiberias); from another hoard buried in a large earthenware jar in the coastal town of Denia, a center for the Arab administration of southeast Iberia during Islamic rule; and from a small group of metal objects found among the cargo of the Byzantine merchantman wrecked off the southwest tip of Turkey at Serçe Liman, apparently on the return trip to Constantinople from Fatimid Syria. Among the Fatimid metalwork, Ponting was able to identify groups based on major alloying element composition, which equate to the established groups discussed previously. In general terms, with some exceptions, utilitarian objects were cast in leaded gunmetal, and beaten items like pots and buckets were made of copper. Fine decorative wares were made in brass with just a few items of high-tin bronze.

It can be concluded from all the published analyses that the levels of lead seen in the beaten brass objects are almost always below 2 wt.% and probably occurred as an accidental impurity from the zinc and copper minerals or came into the mix with recycled metal. In the case of castings, however, it is clear from the analyses that lead was deliberately added to the alloy. Lead has very limited miscibility with copper, and it is precipitated late during freezing of the casting melt, segregating into the areas that freeze last, some of which might otherwise have become shrinkage porosity (Heine et al. 1967). The addition of lead to copper/zinc/tin alloys greatly extends their freezing range (Hanson and Pell-Walpole 1951, 213). Lead also lowers the viscosity of the casting melt, which must be an advantage for complex mold forms. An alloy composition with lead up to about 2 wt.% has been found to improve casting properties (Staniaszek and Northover 1982). At percentages above this, lead, which was and is cheaper than copper, was probably added for economic reasons: it is not uncommon to find 15 wt.% or more lead in these casting alloys.

One obvious reason for the prevalence of zinc over tin is the shortage of tin in the Middle East; in early medieval times, tin seems to have come from Southeast Asia, and later also from Spain and Britain. Deposits of sphalerite (ZnS), the major ore of zinc, were much more locally available than tin within the Islamic world (Allan 1979; Craddock, in this vol.). Metallic zinc production probably started around one thousand years ago in India with distillation furnaces designed to condense the vapor of this volatile metal in sealed containers (Craddock, Freestone, et al. 1998), but this technology seems not to have spread to the Middle East.[1] Large-scale brass production was, however, carried out by reacting reduced zinc mineral with copper metal rather than adding zinc to the melt (Craddock and Eckstein 2003). The accounts written in the Islamic world indicate that the process was carried out in a crucible that had a lid but was not sealed and that the copper was molten (Allan 1979, 42). This method resulted in a relatively low uptake of zinc, always less than 28 wt.% and frequently much less, as can be seen from the published analyses.

A copper alloy that has not yet been identified among Islamic metalwork, although it is referred to by Abu'l-qasim Kashani writing in Tabriz about 1300, is copper added to molten gold to make black gold (Allan 1979, 9). This must surely be referring to the alloy of copper with a little gold, which takes a deep black patina, equivalent to the Roman Corinthian bronze and Egyptian black copper—*hmty km* (Giumlia-Mair and Craddock 1993).

Tin in Islamic Copper Alloys

As stated above, few bronzes that could be classed as Islamic are later in date than the tenth century, and those pieces that do contain significant amounts of tin fall into very specific classes, discussed in the following sections.

High-Tin Bronze Cups and Bowls

There is a small class of vessels made of bronze with a very high tin content of around 20 wt.% or more (Melikian-Chirvani 1974; Lakpour 1997). The characteristics of such an alloy are well known. It can be extremely brittle unless it is quenched to limit the formation of the brittle delta phase. At red heat the ductile beta phase predominates, which allows it to be forged, and on rapid cooling an acicular martensitic phase develops. Artifacts thus fashioned produce a ringing tone when struck, hence the use of this alloy in many cultures worldwide for making bells and gongs (Craddock and Hook 2007). Another property of this alloy that marks it out for specialist purposes is its color. Remarkably, it can resemble silver or pale gold, depending on the proportion of tin and the manufacturing process (Fang and McDonnell, in press), and is more resistant to corrosion than other copper alloys—both reasons why high-tin bronze was used for high-status drinking vessels, particularly in the Islamic world (fig. 3). Published analyses of high-tin bronze cups and bowls indicate a range of tin content between 19 and 25 wt.% (Schweizer 1994; Craddock, La Niece, et. al 1998, 109). It is generally thought that many of these high-tin bronze cups are from Iran. The high-tin bronze items from Tiberias, Denia, and Serçe Liman analyzed by Ponting (2003) indicate that high-tin bronze objects are also to be found outside Iran, at least in the Fatimid period (909–1171), though he suggests that the metal from which they were made may have originated in Iran.

Mirrors

Cast round mirrors with relief decoration and a perforated knob in the center of the back through which a cord could be threaded are well known from China (Kerr 1990; Zhu and He 1993). Mirrors of similar form that bear Arabic inscriptions have also been found. One distinctive group, examples of which are found in several collections, share the decorative motif of a pair of sphinxes and Kufic inscriptions that run round the rim (fig. 4). Other decorative designs, astrological symbols or geometric designs for example, also occur, and some of the mirrors have handles (fig. 5). Few have come from dated excavation contexts, but it is believed they are from the eastern regions, perhaps Khorasan, and a broad date range from the eleventh to the early thirteenth century has been suggested (Melikian-Chirvani 1982, 130–31), but even this is uncertain.

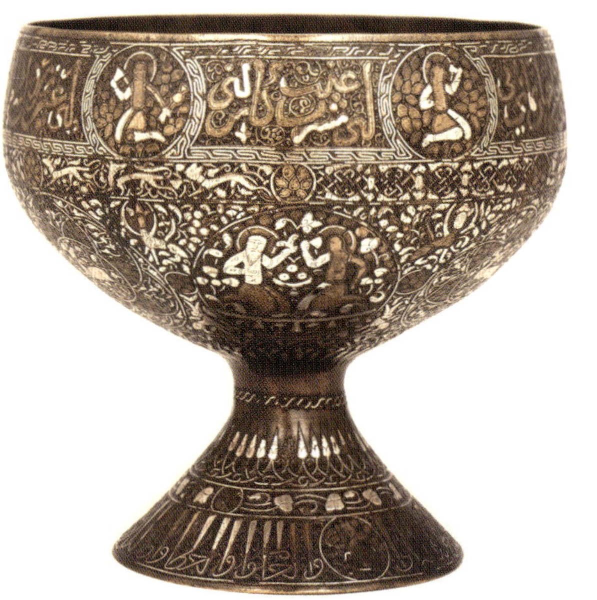

Figure 3. *Fourteenth-century high-tin bronze cup inlaid with silver and gold, h. 12.7 cm. British Museum, 1891,0623.4.*

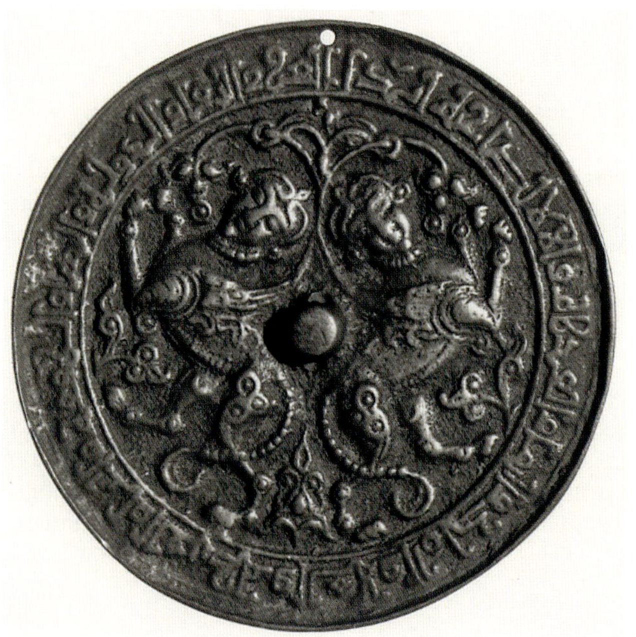

Figure 4. *Back of a cast mirror with a pair of sphinxes either side of the central perforated lug and Kufic script around the rim, diam. 10.5 cm. British Museum, 1866,1229.76.*

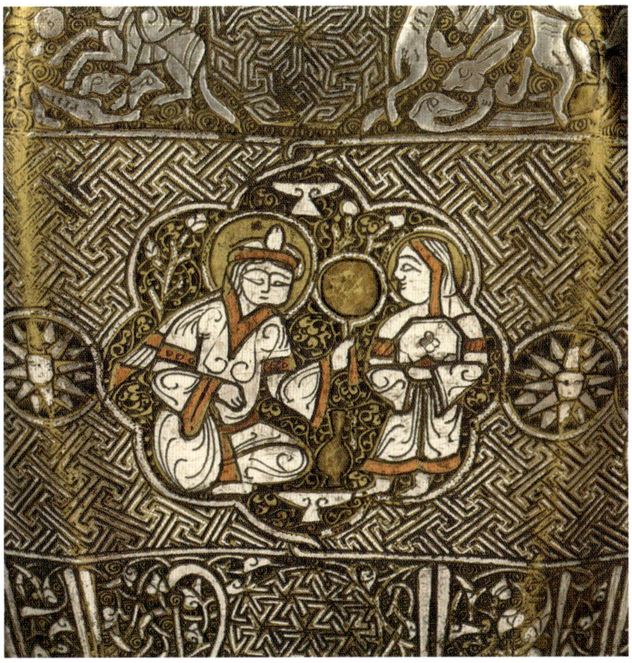

Figure 5. Detail from the thirteenth-century ewer in figure 1 showing a mirror with a handle.

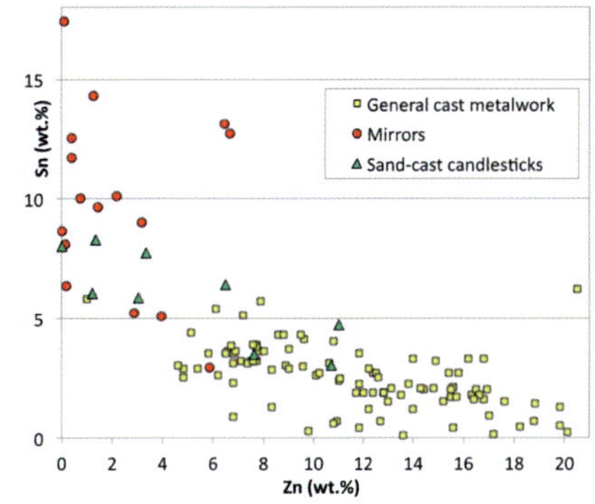

Figure 6. *Plot of zinc vs. tin content for sand-cast candlesticks, mirrors, and twelfth- to fourteenth-century cast items analyzed at the British Museum (Craddock, La Niece, et al. 1998).*

Many of the Islamic mirrors with the sphinx decoration share similar dimensions, as would be expected of items made with a standard model or mold for repeat casting. The relief decoration on some of this group is not sharply defined, indicating either the use of a worn model for making the mold or, as suggested by Melikian-Chirvani (1982) in discussing two similar mirrors in the Victoria and Albert Museum, wear on a reused mold. There is no direct evidence for how these Islamic mirrors were cast. In China, clay molds and some stone molds for mirrors survive from the Han (206 BCE–220 CE) and Warring States (475–221 BCE) periods (Zhu and He 1993, 52), well before the beginning of the Islamic era, but the description of mirror casting in the late Ming dynasty (1368–1644) in a technical handbook by Sung Ying-Hsing published in 1637 (E-tu Zen Sun and Shiou-Chan Sun 1966) indicates that later mirrors

Table 1. *Analyses of Islamic mirrors from the British Museum.*

Registration No.		Cu	Zn	Pb	Sn	Ag	Fe	Sb	Ni	Co	As	Bi	S	Method
Round Mirrors														
1904,1202.2	Sphinx group	73.9	3.96	15.8	5.06	0.080	0.154	0.46	0.061	0.023	0.55	0.036	0.22	ICP-AES
1889,0425.218	Sphinx group	80.8	0.21	10.9	6.36	0.141	0.011	0.12	0.051	0.004	0.25	0.025	0.04	ICP-AES
1866,1229.76	Sphinx group	81.0	0.40	5.50	12.5	0.06	0.90	0.15	0.06	0.050	0.50	0.01		AAS
1866,1229.77	Sphinx group	74.6	2.89	14.3	5.20	0.087	0.251	0.18	0.042	0.012	0.36	0.011	0.16	ICP-AES
1949,0217.6		78.1	0.39	10.8	11.7	0.090	0.022	0.38	0.050	0.007	0.26	0.022	0.09	ICP-AES
1963,0718.1		76.9	6.46	2.10	13.1	0.18	0.100	1.26	0.100	0.016	0.20	<0.01	0.09	ICP-AES
1866,1229.75		76.0	2.20	10.2	10.1	0.06	0.26	0.45	0.06	0.035	0.30	0.01		AAS
1959,1218.1		86.7	0.13	3.66	8.07	0.064	0.162	0.08	0.022	0.002	0.13	<0.01	0.08	ICP-AES
1964,0212.1		77.3	1.43	10.2	9.66	0.104	0.281	0.32	0.070	0.017	0.45	0.016	0.19	ICP-AES
1891,0418.41		73.2	3.18	13.7	8.98	0.083	0.096	0.24	0.074	0.010	0.53	0.014	0.27	ICP-AES
1914,0513.2		79.8	6.67	2.41	12.7	0.049	0.295	0.14	0.071	0.041	0.42	<0.01	0.10	ICP-AES
Mirrors with Handle														
1922,0812.122		82.9	0.77	6.19	10.0	0.086	0.331	0.17	0.043	0.010	0.27	<0.02	0.11	ICP-AES
1992,0102.1		91.4	<0.01	0.31	8.63	0.040	0.133	0.12	0.133	0.048	0.46	<0.02	0.20	ICP-AES
1866,1229.79		80.9	0.12	3.57	17.4	0.157	0.156	0.32	0.060	<0.007	0.25	<0.04	<0.04	ICP-AES
1866,1229.78		76.6	1.27	6.07	14.3	0.107	0.169	1.90	0.116	0.014	0.55	0.033	0.12	ICP-AES
1921,0220.1		68.7	5.87	20.4	2.91	0.079	0.133	0.35	0.087	0.014	0.70	0.013	0.18	ICP-AES

The analyses were carried out over a number of years at the British Museum, initially using atomic absorption spectrophotometry (AAS) but later using inductively coupled plasma–atomic emission spectrometry (ICP-AES).

The figures (expressed in wt.%) have a precision of ca. ±1–2 wt.% for copper, and ca. ±5–10 wt.% for zinc, tin and lead when present at levels greater than 1 wt.%. The remaining minor and trace elements have a precision of ca. ±10–30 wt.%, deteriorating to ±50 wt.% at the detection limit.

< denotes an element not present above the quoted detection limit. A blank (missing value) denotes an element not found or sought by AAS.

Sulfur cannot be measured directly using AAS; bismuth was analysed using AAS throughout. In addition, gold (<0.005 wt.%), cadmium, and manganese (<0.001 wt.%) were also sought but not found to be present at levels above their respective detection limits.

were made by sand casting, and Allan (1979, 62) suggests that the Islamic mirrors may also have been cast by this method. Sand casting was used in China during the Tang period (618–907) for coins (Professor Zhou Weirong, China Numismatic Museum, pers. comm.), and it was certainly known in the Middle East by the tenth century (La Niece 2003). Additionally, the shape of these mirrors is well suited to this method.

Early Chinese mirrors were cast in high-tin bronze, with at least 20 wt.% tin. This level of tin in the alloy forms the alpha + delta eutectoid, which has excellent reflective properties when polished (Zhu and He 1993). None of the round Islamic mirrors analyzed has this level of tin content (table 1), though one with a handle has 17 wt.% tin. From the Song dynasty (960–1279), the tin content of Chinese mirrors declined, the lead content increased, and brass mirrors began to be made (Zhu and He 1993, 52). These would have been less reflective, and thirteenth-century literary references indicate that at that period in China mirrors had to be regularly repolished and coated with a tin-mercury amalgam to produce a bright reflective finish: analytical evidence has been found for this treatment (Kerr 1990, 92). No evidence of mercury tinning has been found on Islamic mirrors, but it is a feature to be looked for as the form of these Islamic mirrors would seem to have originated in China and their technology may also have traveled into the Middle East. Certainly their alloy composition, with less zinc and more tin, differs from the run of cast Islamic metalwork (fig. 6).

Sand-Cast Candlesticks

A thirteenth-century inlaid candlestick in the Nuhad Es-Said collection has been reported as leaded bronze with no zinc detectable by XRF analysis (Allan 1999, 123–24). It belongs to a group attributed to Konya, in Anatolia (Melikian-Chirvani 1982, 356–68), and candlesticks of this distinctive form are thought to have been manufactured by sand casting (La Niece 2003, 93). The results of XRF analyses by Catherine Mortimer of three candlesticks from the Nuhad Es-Said collection (Allan 1999, cat. nos. 7, 8, and 9), one from the Freer Gallery of Art (Atil et al. 1985, cat. no. 19), four from the British Museum (fig. 7) (Craddock, La Niece, et al. 1998, 103), and one from Harvard Art Museums are listed in table 2. Except for one, the candlesticks do have zinc in the alloy, though only three have more zinc than tin (see fig. 6).

Interestingly, not only do the metal compositions of this apparently mass-produced form show fairly broad ranges, but the inlay work also encompasses different styles and attributed dates. It might be suggested that some were traded as undecorated items to be inlaid at other metalworking centers, and perhaps the inconsistencies in metal composition reflect a long production period during which the supply of raw materials suffered from fluctuations.

Trace Elements and Isotope Analysis

As discussed above, zinc, tin, and lead could be deliberate additions to the copper. All the other elements seen from analysis of the metal are likely to be accidental impurities that were introduced with the minerals from which the alloying metals were extracted and are described here as trace.

In theory, both the trace element content of the alloy and the lead isotope signature (Pollard and Heron 1996) might be expected to allow the metal to be matched to provenance. In practice, there are three main factors that

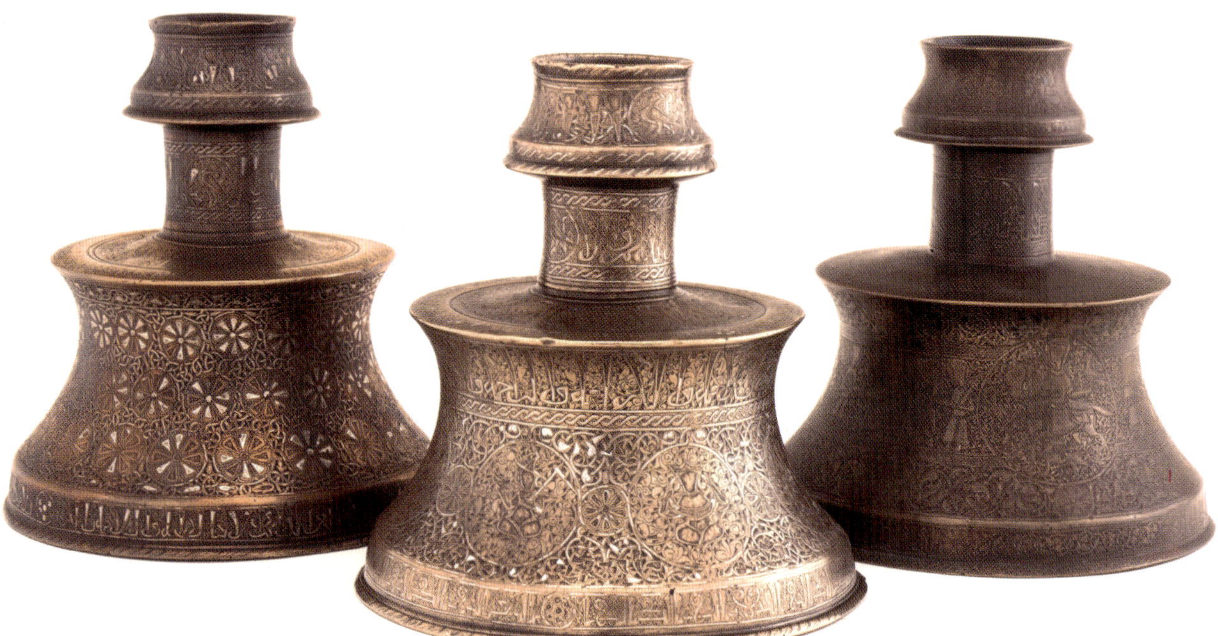

Figure 7. *Three sand-cast candlesticks from Konya, h. 20–22 cm. British Museum, 1885,1010.8; 1955,0214.1; and 1955,0214.2.*

Table 2. Analyses of thirteenth-century sand-cast candlesticks from several collections.

Collection	Registration No.	Cu	Zn	Pb	Sn	Ag	Fe	Sb	Ni	Co	As	Method
British Museum	1885,1010.8	71.7	7.61	16.5	3.47	0.09	0.41	0.31	0.08	0.043	0.49	ICP-AES
British Museum	1955,0214.1	73.1	3.03	17.1	5.86	0.09	0.22	0.32	0.06	0.015	0.35	ICP-AES
British Museum	1955,0214.2	73.0	3.33	13.1	7.71	0.08	0.37	0.22	0.08	0.025	0.54	ICP-AES
British Museum	1954,1016.1	72.0	1.23	18.4	6.01	0.15	0.32	0.22	0.06	0.009	0.33	ICP-AES
Freer Gallery	81.27	78.4	11.0	5.2	4.7							XRF
Nuhad Es-Said	cat. no. 7	73	10.7	12.0	3.0	0.6	0.2	0.2			0.3	XRF*
Nuhad Es-Said	cat. no. 8	76	nd	14.0	8.0	0.7	0.1	0.2	0.2		0.8	XRF*
Nuhad Es-Said	cat. no. 9	76.5	6.5	9.0	6.4	0.9	0.2	tr	nd		0.3	XRF*
Harvard	349.1983	73.25	1.34	13.17	8.27	0.12	0.06	3.12	0.04	0.50	0.52	ICP-MS**

A blank (missing value) denotes an element not found or sought by XRF. ICP-AES = inductively coupled plasma–atomic emission spectrometry; ICP-MS = inductively coupled plasma–mass spectroscopy; nd = not detected; tr = trace; XRF = x-ray fluorescence.
* Semiquantitative surface analysis by C. Mortimer. ** Unpublished analysis by J. Riederer.

make provenancing of the metal in these Islamic copper alloys problematic. The first is the unpredictable extent to which the processes of extracting metals from their minerals result in changes in composition. The second, and perhaps the greatest obstacle to provenancing, is the mixing of different metals to make an alloy; for example lead is often added deliberately to brass, but lead may also come as an impurity in both the zinc and copper minerals used to make the brass, so any signature will be confused. Further mixing would occur if the copper used to make the brass had been subjected to the liquation process by which lead is added to extract low levels of silver from copper. The extent to which it was practiced in the Islamic world is not documented, but liquation was carried out to extract silver from copper in late medieval Europe (Craddock 2002), which, as discussed in the following paragraph, was the source of much of the copper used in the Islamic world in the later period. The third and most obvious problem is the unknown extent of recycling. Metal has always been a valuable commodity, and the melting down, trade, and reuse of metal has been practiced throughout history, resulting in the inevitable mixing of metals from different sources. Therefore attempting to accurately assign the sources of the metals by either trace element or lead isotope analysis is unlikely to be fruitful. However, this is not to say that the trace metal content of the objects cannot offer other insights, for example, into authenticity (al-Saa'd 2000; Northover 2008) or into the smelting technology employed in extracting the metal: the low levels of iron in Islamic brass and more erratic levels in contemporary European brass have been attributed to the different treatment of the zinc ores in Europe from those in the Middle East (Craddock, La Niece, et al. 1998).

In the case of the finished artifacts, closely similar trace element signatures might be expected to indicate groupings on the basis of date or origin. The nickel content of brass was found to increase to a median value of 0.16 wt.% in vessels dated to the fifteenth and sixteenth centuries compared to a median value of 0.06 wt.% in those of the preceding centuries (Ward et al. 1995). Nickel is an impurity of copper ore, and this change in concentration can be interpreted as change to a different source of copper or processing technology. This does indeed seem to be corroborated by the documentary evidence from both European and Muslim sources that indicate the main supplier of copper to the Eastern Mediterranean switched from the Middle East to Europe in the fifteenth and sixteenth centuries (Braunstein 1977).

There is, however, surprisingly little variation in trace element chemistry in much of the analyzed material, a conclusion also reached by Ponting that "most of the objects from Denia and the Serçe Liman merchant ship came from the same metal supply pool as the objects from the Tiberias workshop" (Ponting 2003, 93).[2]

Conclusion

The repertoire of copper alloy types used to make vessels shows remarkable consistency across the regions under Islamic rule in terms of the major alloy compositions, though the trace element picture is much less clear cut.

There is a large group of utilitarian vessels made of copper, with only minor levels of other elements. Among the higher quality items, a clear division exists between the relatively pure brass vessels made by hammering (sinking or raising) and the more mixed alloys of the items made by casting. Brass was the generally preferred alloy, presumably because of the availability of zinc minerals and lack of tin. There was, however, a continuing presence of several percent of tin in the zinc-containing leaded gunmetal alloys because of the good casting properties of the mixed alloy. Three object types stand out from the general output of medieval Islamic wares because of their tin content: cast mirrors, hot-forged drinking vessels, and a distinctive group of sand-cast candlesticks. It is suggested that these three object types were made in specialist workshops.

Acknowledgments

The authors are grateful to Antony Simpson for the map and Katherine Eremin for information concerning the unpublished candlestick from Harvard Art Museums.

Notes

1. An astrolabe containing between 40 and 50 wt.% zinc from seventeenth-century Lahore has been reported (Newbury et al. 2003). This might indicate metallic zinc being used by Islamic metalworkers in India to make spelter brass.
2. Lead isotope analysis was carried out on just two leaded cast items from the metalsmith's workshop at Tiberias and four items from the Serçe Liman wreck. Ponting suggests that they indicate that the lead and copper could be sourced to either Iran or Turkey even though the objects were made in the Syrio-Palestinian area.

References

al-Saa'd, Z. 2000. Technology and provenance of a collection of Islamic copper-based objects as found by chemical and lead isotope analysis. *Archaeometry* 42(2):385–97.

Allan, J. 1979. *Persian metal technology 700–1300* AD. Oriental Institute Monographs 2. Oxford: Ithaca Press.

———. 1999. *Islamic metalwork: The Nuhad Es-Said Collection*. London: Philip Wilson Publishers.

Atil, E., W. T. Chase, and P. Jett. 1985. *Islamic metalwork in the Freer Gallery of Art*. Washington, DC: The Smithsonian Institution.

Braunstein, P. 1977. Le marché du cuivre à Venise à la fin du moyen-age. In *Schwerpunkte der kupferproduktion und des kupferhandels in Europa 1500–1650*, ed. H. Kellenbenz, 78–94. Cologne: Böhlau.

Craddock, P. T. 1979. The copper alloys of the medieval Islamic world: Inheritors of the classical tradition. *World Archaeology* 11(1):68–79.

———. 2002. De-silvered copper ingots from Central Europe. *Journal of the Antique Metalware Society* 10:43–46.

Craddock P. T., and K. Eckstein. 2003. Production of brass in antiquity by direct reduction. In *Mining and metal production through the ages*, ed. P. T. Craddock and J. Lang, 216–30. London: British Museum Press.

Craddock, P. T., I. C. Freestone, L. K. Gurjar, and L. Willies. 1998. Zinc in India. In *2000 years of zinc and brass*, ed. P. T. Craddock, 27–72. British Museum Occasional Paper 50. London: British Museum.

Craddock, P. T., and D. R. Hook. 2007. The bronzes of the south of India: A continuing tradition? In *Scientific research on the sculptural arts of Asia: Proceedings of the Third Forbes Symposium at the Freer Gallery of Art*, ed. J. G. Douglas, P. Jett, and J. Winter, 75–89. London: Archetype Publications.

Craddock, P. T., S. C. La Niece, and D. R. Hook. 1998. Brass in the medieval Islamic world. In *2000 years of zinc and brass*, ed. P. T. Craddock, 73–101. British Museum Occasional Paper 50. London: British Museum.

E-Tu Zen Sun and Shiou-Chan Sun, trans. 1966. *Chinese technology in the seventeenth century*, translation of Sung Ying-Hsing, *Tien-Kung K'ai-Wu*. University Park: Pennsylvania State University.

Fang, J. L., and G. McDonnell. In press. The colour of copper alloys. *Historical Metallurgy* 45(1).

Giumlia-Mair, A. R., and P. T. Craddock. 1993. *Das schwarze Gold der alchimisten: Corinthium aes*. Mainz am Rhein: Philipp von Zabern.

Hanson D., and W. T. Pell-Walpole. 1951. *Chill cast tin bronzes*. London: Edward Arnold.

Heine R. W., C. R. Loper, and P. C. Rosenthal. 1967. *Principles of metal casting*. New York: McGraw-Hill.

Hook, D. R. 1988. Scientific analysis of the copper based medals, Appendix. In *A catalogue of the French medals in the British Museum*, Vol. 2, *1600–1672*, ed. Mark Jones, 305–12. London: British Museum Publications.

———. 1998. Inductively coupled plasma atomic emission spectrometry and its role in numismatic studies. In *Metallurgy in numismatics*, Vol. 4, ed. W. A. Oddy and M. R. Cowell, 237–52. London: Royal Numismatic Society.

Hook, D. R., and P. T. Craddock. 1996. The scientific analysis of the copper alloy lamps: Aspects of classical alloying practices. In *A catalogue of the lamps in the British Museum*, Vol. 4, *Lamps of metal and stone, and lampstands*, ed. D. M. Bailey, 144–64. London: British Museum Press.

Hughes, M. J., M. R. Cowell, and P. T. Craddock. 1976. Atomic absorption techniques in archaeology. *Archaeometry* 18(1):19–36.

Kerr, R. 1990. *Later Chinese bronzes*. London: Bamboo Publishing in association with the Victoria and Albert Museum.

La Niece, S. 2003. Medieval Islamic metal technology. In *Scientific research in the field of Asian art: Proceedings of the First Forbes Symposium at the Freer Gallery of Art*, ed. P. Jett with J. G. Douglas, B. McCarthy and J. Winter, 90–96. London: Archetype Publications.

Lakpour, S. 1997. *Sefidruy*. Teheran: National Cultural Heritage Organization.

Melikian-Chirvani, A. S. 1974. The white bronzes of early Islamic Iran. *The Metropolitan Museum Journal* 9:123–51.

———. 1982. *Islamic metalwork from the Iranian World, 8th–18th centuries*. London: HMSO.

Newbury, B., B. Stephenson, J. Almer, M. Notis, D. Haeffner, B. Stephenson, and G. S. Cargill III. 2003. Synchrotron applications in archaeometallurgy: Analysis of high zinc brass astrolabes. In *Archaeometallurgy in Europe*, Vol. 2, 359–68. Milan: Associazione Italiana di Metallurgia.

Northover, J. P. 2008. Alloy choice in Chinese and Islamic forgeries. In *Original—Copy—Fake?*, ed. E. Pernicka, 70–73. Stiftung Situation Kunst/Ruhr-Universität, Bochum, 2007. Mainz: Philipp von Zabern.

Pollard, A. M., and C. Heron. 1996. *Archaeological chemistry*. London: Royal Society of Chemistry, 302–40.

Ponting, M. 2003. Damascus to Denia: The scientific analysis of three groups of Fatimid period metalwork (11th–early 12th century AD). *Historical Metallurgy* 37(2):85–105.

Roberts-Austin, W. C. 1892. *Report on the analysis of various examples of Oriental metalwork etc in the South Kensington Museum and other collections*. London: Eyre and Spottiswoode.

Schweizer, F. 1994. Composition et technologie des objet metalliques. In *L'oevre d'art sous le regard des sciences*, 193–209. Editions Slatkine: Geneva.

Staniaszek, B. E. P., and J. P. Northover. 1982. The properties of leaded bronze alloys. In *Proceedings of the 22nd International Symposium on Archaeometry*, ed. A. Aspinall and S. E. Warren, 262–72. Bradford: University of Bradford.

Ward, R., S. La Niece, D. R. Hook, and R. White. 1995. Veneto-Saracenic metalwork: An analysis of the bowls and incense burners in the British Museum. In *Trade and discover: The scientific study of artefacts from post-medieval Europe and beyond*, ed. D. R. Hook and D. M. Gaimster, 235–58. British Museum Occasional Paper 109. London: British Museum.

Zhu, S., and T. He. 1993. Studies in ancient Chinese mirrors and other bronze artefacts. In *Metal plating and patination*, ed. S. La Niece and P. Craddock, 50–62. Oxford: Butterworth.